Advances in Direct Simulation Monte Carlo: From Micro-Scale to Rarefied Flow Phenomena

Ehsan Roohi · Hassan Akhlaghi · Stefan Stefanov

Advances in Direct Simulation Monte Carlo: From Micro-Scale to Rarefied Flow Phenomena

Springer

Ehsan Roohi
Department of Mechanical and Industrial Engineering
University of Massachusetts Amherst
Amherst, MA, USA

Hassan Akhlaghi
Department of Aerospace Engineering
College of Interdisciplinary Science and Technologies
University of Tehran
Tehran, Iran

Stefan Stefanov
Institute of Mechanics
Bulgarian Academy of Sciences
Sofia, Bulgaria

ISBN 978-981-96-8199-0 ISBN 978-981-96-8200-3 (eBook)
https://doi.org/10.1007/978-981-96-8200-3

This Springer imprint is published by the registered company Springer Nature Singapore Pte Ltd.
The registered company address is: 152 Beach Road, #21-01/04 Gateway East, Singapore 189721, Singapore

Dedicated to our wives, Zahra, Sara, and Elena

Preface

This book presents a comprehensive account of recent developments in the Direct Simulation Monte Carlo (DSMC) method, focusing on advanced collision algorithms designed to ensure accuracy even with low particle-per-cell ratios. Alongside methodological advances, the book explores fluid flow and heat transfer phenomena at the micro- and nano-scales, domains where rarefied gas dynamics and non-equilibrium effects dominate.

The growing relevance of micro/nano-electromechanical systems (MEMS/NEMS) and renewed interest in space travel have significantly advanced our understanding of fluid mechanics and heat transfer at small scales and at rarefied gas conditions. Micro- and nano-systems, which integrate mechanical and electrical elements with features ranging from nanometers to hundreds of microns, have created new challenges and opportunities for research. Chapter 1 introduces the concept of micro- and nano-flows, detailing their key features—including slip and jump conditions—and their applications. It also defines critical terms such as Knudsen number and explores the unique behaviors observed in rarefied gases, including cold-to-hot heat transfer, the Knudsen minimum, nonlinear pressure distribution, deviations in the Poiseuille number, and thermal polarization.

Chapter 2 offers a foundational review of kinetic theory, establishing its relevance to non-continuum and rarefied gas flows. It contrasts classical continuum mechanics with molecular-level approaches and traces the historical development of atomic theory. The chapter introduces different flow regimes based on the Knudsen number. It explains how molecular behavior, collisions, and interaction potentials (e.g., Lennard-Jones, hard sphere, and variable hard sphere models) determine macroscopic properties such as pressure and viscosity. It also introduces statistical models such as the Poisson and Gaussian distributions to describe molecular fluctuations, laying the groundwork for understanding advanced kinetic methods like DSMC.

Chapter 3 introduces the DSMC method itself, which uses a finite number of model particles to simulate gas dynamics stochastically. The method discretizes space and time and alternates between particle motion and collisions within computational cells. Stefanov's novel N-particle equation is presented, along with the core concepts behind DSMC. Traditional collision algorithms—including Majorant Collision Frequency, Bernoulli Trials, and No Time Counter—are reviewed, along with discussions on boundary conditions and the simulation of low-speed flows. The chapter concludes by presenting multiscale DSMC schemes and recent algorithms developed for simulating complex 2D and 3D domains using low particle-per-cell counts.

Chapter 4 delves into advanced DSMC collision models, particularly those based on the Boltzmann equation and Kac master equation. The focus lies on the BT family of collision algorithms (SBT, ISBT, GBT, NTC-GBT-SBT-TAS hybrid, SSBT, and SGBT), derived from Stefanov's framework. Each scheme is discussed regarding algorithmic structure, performance, and comparison with traditional DSMC results. Additionally, emerging techniques aimed at optimizing DSMC performance in near-continuum regimes are presented, including DSMC-Fokker-Planck (DSMC-FP), Low Diffusion DSMC (LD-DSMC), Low Variance Deviational Simulation Monte Carlo (LVDSMC), and the use of collision limiters.

Chapter 5 focuses on pressure-driven rarefied flows, specifically Poiseuille flow in micro- and nano-channels. It examines analytical solutions across a range of Knudsen numbers, evaluates the performance of different velocity slip boundary conditions, and discusses heat transfer, wall behavior, and experimental findings. Additional topics include step flow analysis and porous microchannel modeling.

Chapter 6 addresses thermally driven flows at the micro/nano scale, where temperature and gradients cause flow in rarefied gases. It details various thermal flows, including thermal creep, reverse thermal creep, thermal stress slip flow, nonlinear thermal stress flow, thermal edge flow, and radiometric flow.

Chapter 7 covers rarefied Couette and cavity flows, examining velocity profiles and thermal flows like Fourier flow. Analytical solutions in different rarefaction regimes are presented, highlighting the unique characteristics of these flows.

Chapter 8 discusses airfoil aerodynamics under rarefied conditions, introducing a semi-analytical approach for analyzing lift and drag in free-molecular flow. Parametric studies evaluate how rarefaction affects aerodynamic performance as a function of Mach number, Reynolds number, angle of attack, camber, and airfoil thickness.

Chapter 9 introduces non-equilibrium flows characterized by instabilities and transitions between different attractors. It centers on the Rayleigh–Bénard convection problem, a canonical system used to study self-organization, chaos, and turbulence. In rarefied regimes, where conventional Rayleigh and Prandtl numbers lose applicability, alternative dimensionless groups must be used. This chapter highlights kinetic-based analysis of the transition from conduction to convection in a 2D system, focusing on hard-sphere gases and molecular-level modeling.

Chapter 10 provides a concise study of rarefied supersonic flow over wedges and cylinders, key geometries in aerospace engineering. The shock wave detection (SWD) method is discussed, and results for various Mach and Knudsen numbers are presented to illustrate the effects of rarefaction.

Ferdowsi University of Mashhad Press previously published two shorter Persian-language versions of this work: "Fluid Mechanics at Micro- and Nano-Scales" and "Rarefied Gas Dynamics and Direct Simulation Monte Carlo." With permission from this publisher, selected materials from those works were adapted and widely expanded in this book.

A repository containing DSMC codes in various scientific programming languages—used to generate the results presented in this book as well as for general-purpose applications—will soon be available at the following link: https://github.com/sn-code-inside/dsmc-python-codes. The content of the repository will be updated gradually.

Some other chapters and sections will be added in the second edition of this book, which includes, but not limited to: detailed discussion on slip and jump boundary conditions, practical DSMC implementations and Python version of standard DSMC codes for typical problems, more details on Fokker-Plank, discrete unified gas kinetic scheme (DUGKS), Information Preservation (IP), and other alternative methods, as well as discussions on the machine learning approaches for rarefied gas dynamics applications and DSMC method.

The first author, Ehsan Roohi, would like to express his deep appreciation to his Ph.D. advisor, Dr. Masoud Darbandi (Sharif University of Technology), for introducing him to rarefied gas dynamics, micro- and nano-flows, and the DSMC method. He would also like to dedicate this book to his wife, Zahra, whose unwavering support has been with him throughout their marriage. Roohi also acknowledges Dr. Tobias Baier from the Technical University of Darmstadt for reviewing and editing Chap. 6 on thermally driven flows. He also thanks Dr. Fahimeh Ebrahimzadeh and Dr. Ahmad Shojasani for their help in preparing some parts of this book. Finally, he would like to thank all his students in Ferdowsi University of Mashhad who contributed in preparation and proofreading of the Persian version of his two books mentioned previously.

The third author, Stefan Stefanov, wishes to express his heartfelt appreciation for the comprehensive support provided by the Institute of Mechanics at the Bulgarian Academy of Sciences during the time of work on this book. He also extends deep gratitude to his beloved wife, Elena, for her understanding and unwavering support throughout their life together.

Amherst, USA — Ehsan Roohi
Tehran, Iran — Hassan Akhlaghi
Sofia, Bulgaria — Stefan Stefanov

Competing Interests The authors have no competing interests to declare that are relevant to the content of this manuscript.

Contents

Chapter 1
Introduction to Micro- and Nano-Flows

Abstract Due to the emerging applications in micro/nano-electromechanical systems (MEMS/NEMS), knowledge of fluid mechanics and heat transfer behavior at nanoscale dimensions has advanced significantly over the last few decades. MEMS/NEMS refers to small devices that combine mechanical and electrical components with feature sizes ranging from a few nanometers to hundreds of microns. This chapter introduces the micro- and nano-flows, and their applications will be described. Basic flow concepts related to micro- and nano-flows will be reviewed and discussed. The flow features related to such flows will be presented. The main features are slip and jump conditions. The discussion will focus on rarefied gas flows, which occur in gas flows at micro- and nano-scales. Different concepts related to dilute gases, rarefied flow, and the Knudsen number will be discussed. Following that, the gas flow classification based on the Knudsen number is introduced. In the final section of this chapter, unique flow phenomena of rarefied gas flows will be discussed and reviewed. They are cold-to-hot heat transfer, Knudsen minimum, nonlinear pressure distribution, Poiseuille number deviation, and thermal polarization.

1.1 Introduction

Nowadays, with the increasing demand for creating various devices and equipment in small dimensions (micro- and nano-), investigating fluid flow in micro- and nano-channels is crucial. On the other hand, the flow behavior in these dimensions has different characteristics compared to the flow behavior in the macro scale. Therefore, the study of fluid mechanics in micro- and nano-dimensions corresponds to investigating other transport mechanisms compared to flow on macro scales. For the first time, American researcher Richard Feynman stated in an article published in 1960, "There's Plenty of Room at the Bottom" [1]. In his scientific speech, he revealed new perspectives and new creative applications in engineering and micro/nanosciences to scientists. Nevertheless, documented research in the field of fluid mechanics in microchannels began in the 1980s.

E. Roohi et al., *Advances in Direct Simulation Monte Carlo: From Micro-Scale to Rarefied Flow Phenomena*, https://doi.org/10.1007/978-981-96-8200-3_1

A key purpose of this book is to introduce and analyze flow features at micro- and nano-scales. The current chapter provides an overview of micro- and nano-technologies, the basic concepts of fluid dynamics from a molecular viewpoint, the idea of rarefaction, and the specific characteristics of rarefied gas flows. The initial topics of the chapter begin with a review of introductory concepts.

1.1.1 Conventional Length Scale Units

Firstly, familiarity with the dimensions related to the meter is necessary for a more accurate understanding of the concept of micro-scale flows. For this purpose, Table 1.1 presents common length measurement units.

For a more intuitive understanding of some of the length units introduced in Table 1.1, examples of daily or scientific applications of these units are shown in Fig. 1.1. The distance between cities or long distances in kilometers, the height of children in meters, the dimensions of a sugar cube in centimeters, the dimensions of gravel in millimeters, the width of human hair or dust particles in micrometers, and the dimensions of DNA and viruses (such as Coronavirus) are in the nanometer range.

1.2 Micro- and Nano-Technologies

With advances in microfabrication technology, it has become possible to produce micro- and nano-scale systems, known as micro- and nano-electromechanical systems (MEMS-NEMS). Employing small subsystems enables the production of smaller final products, which in turn creates a better market for the final product. A

Table 1.1 Conventional length scale units

Prefix	Magnitude (m)	Scientific index
Kilo	1000	1×10^{3}
Hecta	100	1×10^{2}
Deca	10	1×10^{1}
Meter	1	1×10^{0}
Deci	0.1	1×10^{-1}
Centi	0.01	1×10^{-2}
Milli	0.001	1×10^{-3}
Micro	0.000001	1×10^{-6}
Nano	0.000000001	1×10^{-9}
Pico	0.000000000001	1×10^{-12}
Femto	0.000000000000001	1×10^{-15}

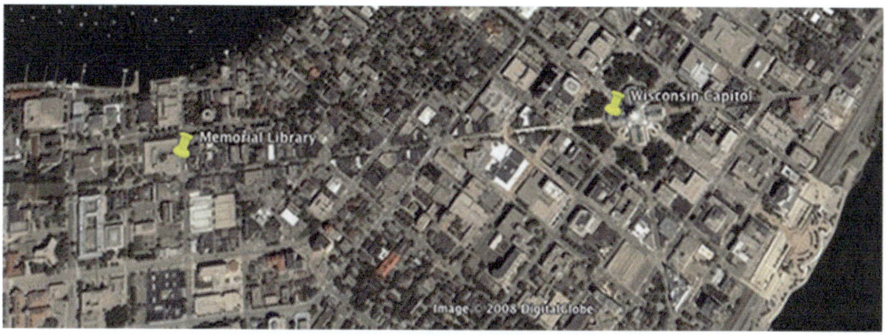

(a)

(b)

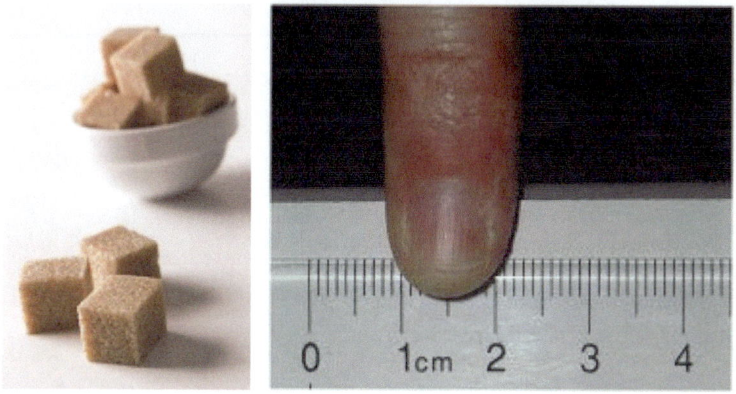

(c)

Fig. 1.1 Real examples of different length scale units **a** the distance between two cities as indicated by kilometers **b** the height of children as indicated by meters **c** dimensions of sugar cubes in the range of centimeters **d** the width of the coin and dimensions of gravel indicated by millimeters **e** the width of a human hair (40–50 microns) and the size of mite (400 microns) indicated by micrometers **f** DNA (2.5 nm) and Coronavirus (diameter: 40–160 nm, tentacle length: 9–12 nm) dimensions indicated by nanometers

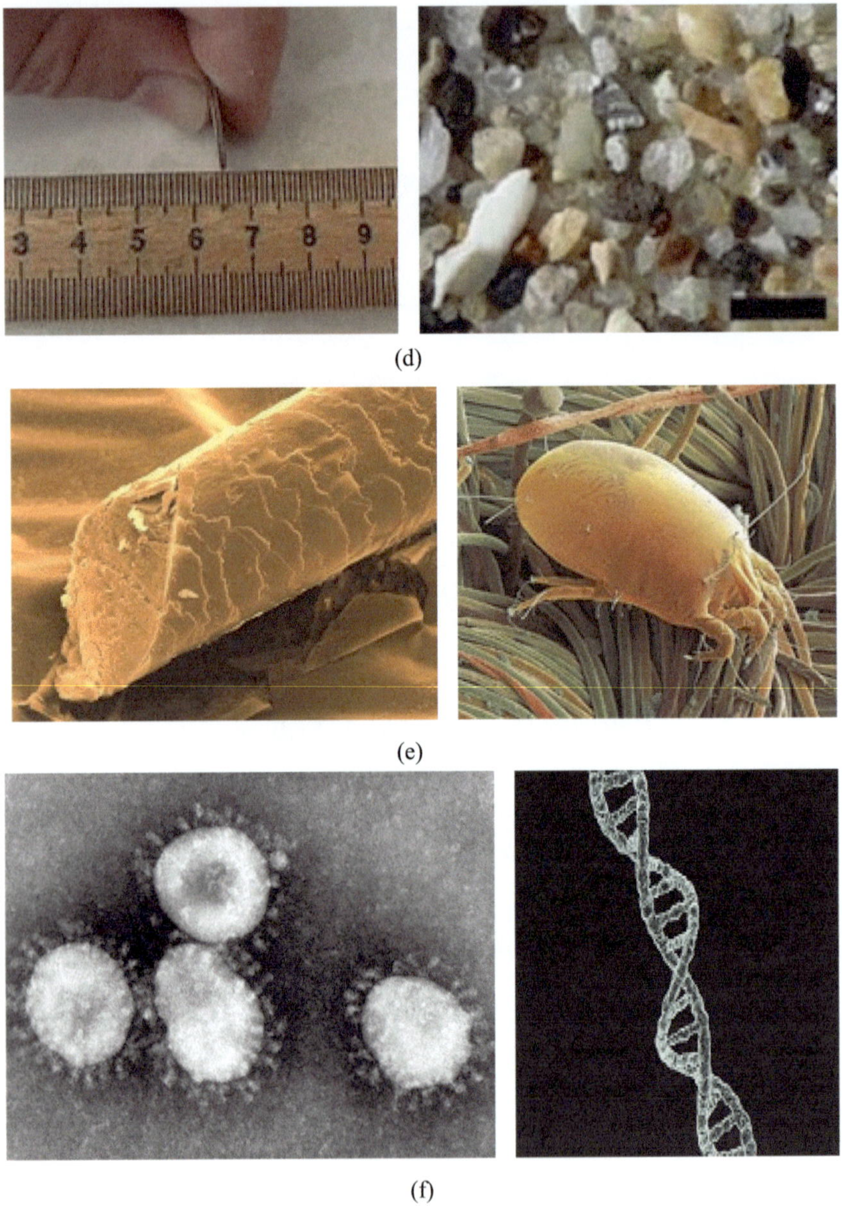

Fig. 1.1 (continued)

printer that takes up less space in the work environment, laboratories such as blood plasma analyzers that are located on an electrical chip and are called lab-on-a-chip, flow detection fluids, and nanoelectronics in which the cooling of electronic circuits is done with nano currents, they all occupy very little space and have made many uses possible with a very small device.

micro- and nano-electro-mechanical systems are a combination of mechanical components, sensors, actuators, and electrical components that have numerous applications in a wide range of today's industries, including medical sciences, the production of biological sensors, and the design and manufacturing of micro-propulsions for microsatellites and similar applications. Figure 1.2a shows a microsensor and a microactuator. As an important application for micro- and nano-electro-mechanical equipment, we can use micropumps together with micro-diffusers (Fig. 1.2b), which are used in medical sciences for the controlled transfer and injection of very small amounts of drugs and chemicals in the range of nanoliters.

Another application of microsystems is in the flow expansion process, which can occur gradually or suddenly. This phenomenon is primarily observed in microsystems, including microturbines, microcompressors, microactuators, and microbirds. An example of the applications of such systems is shown in Fig. 1.2c. Other applications of flow in micro- and nano-channels include purification and separation, especially in the oil and gas industries. In cases where the high sensitivity of flow quality is crucial, the use of nanofilters enables pure fluid flow in channels with micro- and nano-scale dimensions. An example of these filters is shown in Fig. 1.2d.

Another example of an application for micro- and nano-flows is in computer equipment and hard disks. Hard disks are composed of an electronic board that is in charge of controlling reading and writing information by converting the magnetic field into bytes (reading) and converting bytes into the magnetic field (writing), as shown in Fig. 1.3a. This device consists of a slider that stores information in the form of zeros and ones by moving on a screen called a platter. The faster the relative movement of the slider and the screen, the faster the data can be saved or read. Due to the high-density information in advanced hard disks, such as 1 TB disks, it is approximately 625 billion bytes per square inch of the platter. Therefore, the slider should be positioned at a distance of 5–10 nm above the platter. The distance between the slider and the disk is less than the thickness of a fingerprint (Fig. 1.3b) and practically this space could be considered a nanochannel (Fig. 1.3c). The slider and the plate, despite the fluid layer between them, form a model of a sliding bearing. The operating conditions are such that both the Reynolds number and Mach number are very small. For example, in a Winchester-type hard disk, where the distance between the slider and the plate is approximately 50 nm, the Reynolds number is around 0.6, and the Mach number is approximately 0.3.

In the old generation of disks, air was used in the compartment containing the plate and slider, and there were up to five plates and sliders in this space. To rotate the disk faster, the friction between it and its surroundings must be overcome, resulting in a mechanism similar to Couette flow. According to the performed investigations, it was concluded that using helium gas instead of air can reduce the space required and increase the number of plates. The advantage of helium compared to air is that the

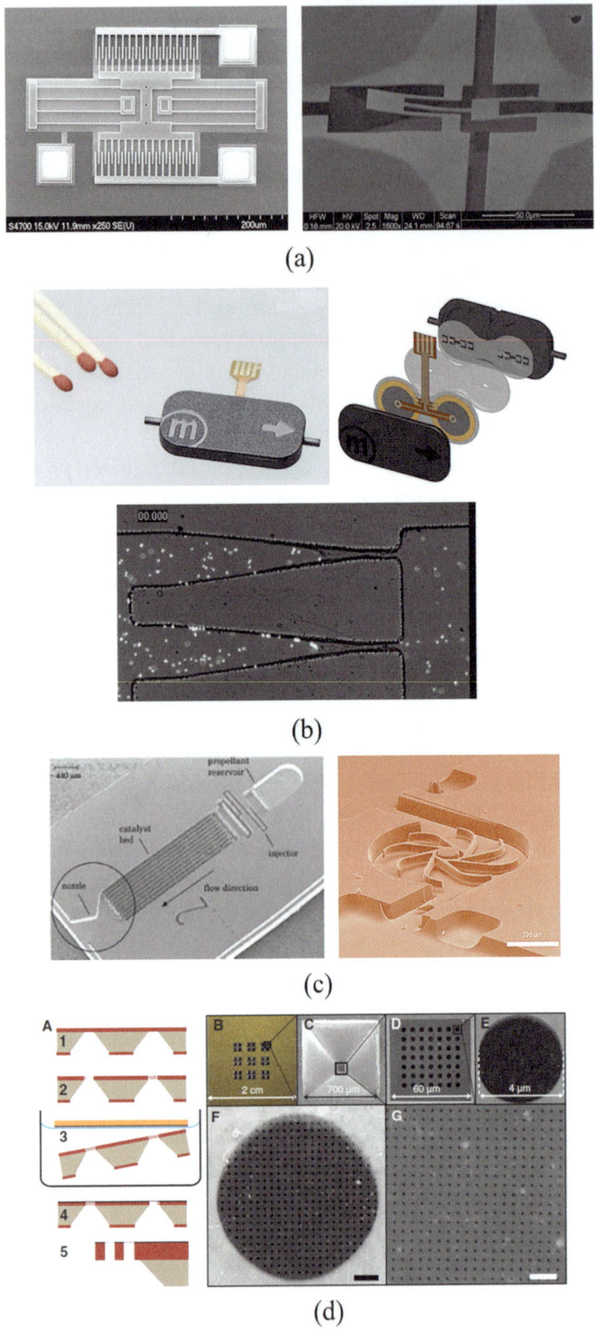

Fig. 1.2 Examples of micro- and nano-technology applications **a** Right: microsensor, left: microactuator **b** small-scale pumps and nozzle diffusers **c** Right: microturbine, left: micronozzle (Reproduced from Ref. [2]) **d** nanofilter for gas separation (Copied from open access Ref. [3])

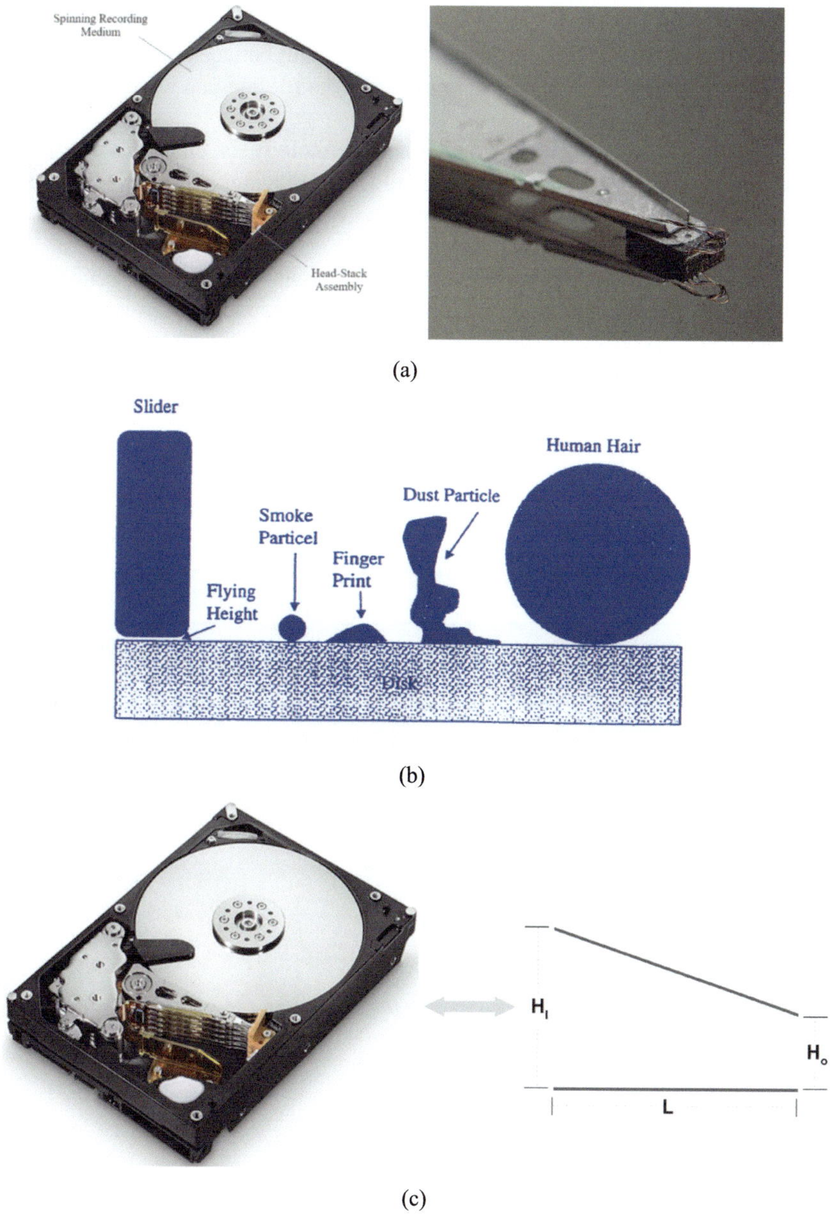

Fig. 1.3 Hard disk components **a** slider and spinning disk **b** comparison of the slider-disk distance with different length values **c** modeling of the slider-disk gap by a channel

coefficient of gas-surface momentum and energy exchange for Helium is much lower than that for air. Therefore, the exchange of momentum between helium molecules and the plate is extremely low, and consequently, the friction and heat produced by helium gas are less compared to the air (Fig. 1.4).

Another application of micro- and nano-technology is changing the DNA sequence using a lab-on-a-chip, as shown in Fig. 1.5. Changing the DNA sequence encompasses any process that alters the position of each nucleotide (adenine, guanine, cytosine, and thymine) within the DNA molecule.

In recent years, researchers have focused on the design and manufacturing of lab-on-a-chip vacuum pumps. Lab-on-a-chip vacuum pumps are utilized in resonators with high-quality factors, such as micro gyroscopes, gas chromatography, and mass spectrometry. As shown in Fig. 1.6, lab-on-a-chip vacuum pumps have become smaller over the past two years, and the number of pump stages has increased by approximately four times, resulting in a higher pressure drop in lab-on-a-chip devices. The working mechanism of these pumps is based on the implementation of a triangular temperature distribution over the walls of the pump components and

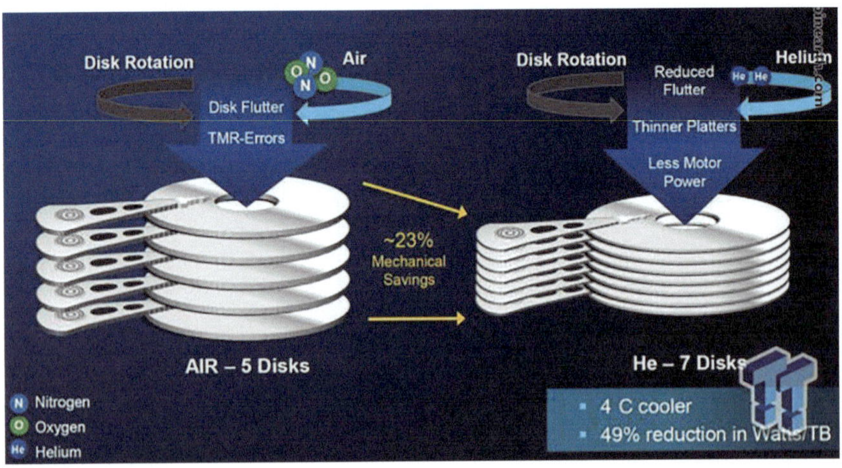

Fig. 1.4 Comparison of helium and air in filling the gap space between disks

Fig. 1.5 Right: DNA schematic, left: lab-on-a-chip for changing DNA sequence. Copied from open access Ref. [4]

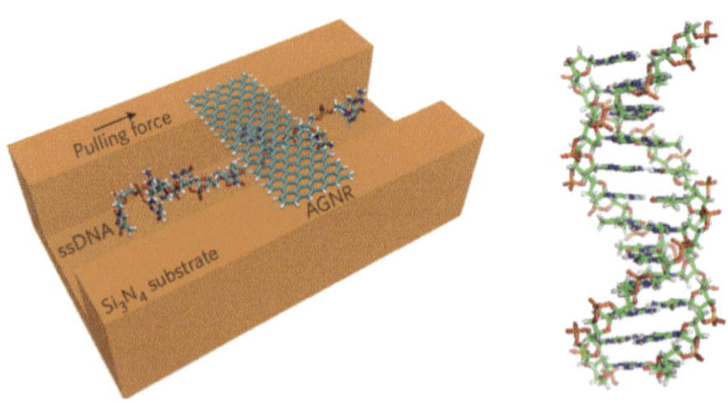

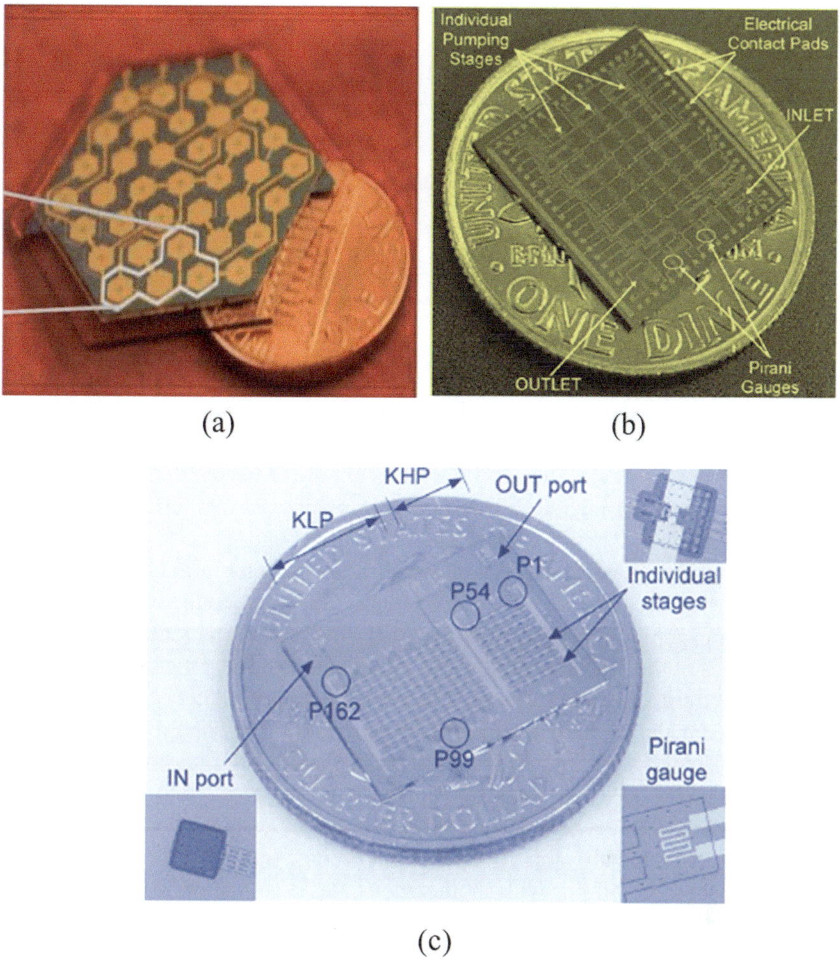

Fig. 1.6 The evolution of lab-on-a-chip manufacturing over two years **a** 24-stages vacuum pump [5] **b** 48-stages vacuum pump (Repro-duced from Ref. [6]) **c** 162-stages vacuum pump (Reproduced from Ref. [7])

the thermal creep phenomenon, which will be examined in detail in Chap. 6 of this book.

1.3 Basic Fluid Concepts

According to the classical definition, a fluid is a substance that continuously changes its shape as long as shear stress, however small, is applied to it. A Newtonian fluid is one in which the shear stress is directly related to the strain rate. Gases and liquids are

Fig. 1.7 The comparison between molecular diameter, molecular spacing and mean free path

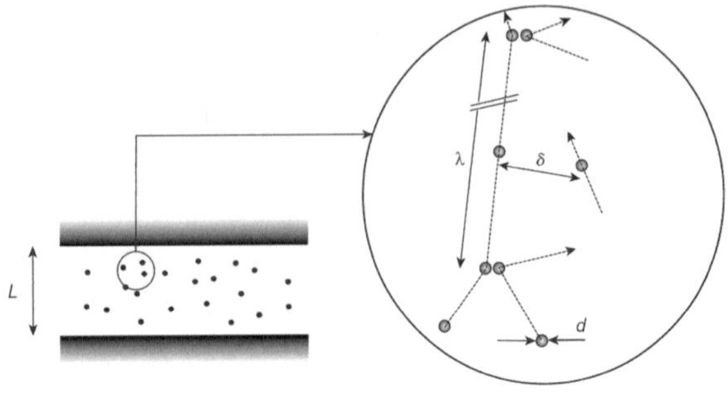

both classified as fluids, but they exhibit completely different behaviors. For example, the viscosity of liquids decreases with increasing temperature, and the viscosity of gases usually increases with increasing temperature. Additionally, the fluid flow can be described in terms of its intrinsic features and flow characteristics. The dynamic characteristics of fluid include linear velocity, angular velocity, circulation, acceleration, and strain rate. From a thermodynamic viewpoint, three key indicators — pressure, temperature, and density—are worth mentioning for introducing fluid flow. The intrinsic transfer characteristics can be expressed by the conductive and diffusion heat transfer capability as well as the viscosity coefficient. The remaining inherent characteristics of the fluid are classified as miscellaneous. Properties such as surface tension, vapor pressure, and surface exchange coefficients fall under this category.

In the gas analysis, there are important parameters that should be considered. The main gas molecular characteristics are molecular diameter (d), average distance between molecules (δ), and mean free path (λ). All the mentioned parameters are shown in Fig. 1.7. The average distance between molecules varies with the mean free path. The mean free path value represents the average distance that each molecule travels freely before colliding with another molecule. This average distance is defined as the inverse of the square root of the numerical density of the gas $\delta = n^{-1/3}$).

The fluid flow is typically modeled in two ways: the first method treats the fluid as a collection of individual molecules, and the second method considers the fluid as a continuous medium, assuming all its properties are continuous and differentiable in space. The second method is the well-known method of classical fluid mechanics. There is a question: Is the continuity assumption still valid when reducing the flow length scale to micro- and nano-scales? The answer to this question is negative, and it is worth noting that the continuity assumption in the mentioned situation yields completely incorrect answers. In general, as long as the conditions for deriving the Navier–Stokes equations are not met, the answers obtained from this set of equations are not reliable. Another assumption made in fluid mechanics for large dimensions is that the fluid does not slip in contact with the wall.

The ideal gas state equation is expressed as

$$P = \rho RT = nkT, \tag{1.1}$$

where k is the Boltzmann constant (1.38×10^{-23}). It could be shown that the number of molecules in the volume decreases as the volume of the space occupied by the fluid decreases

$$N = \frac{P}{kT}\forall. \tag{1.2}$$

The number density is defined as

$$n = \frac{N}{\forall}. \tag{1.3}$$

The behavior of all three types of substances in nature—solids, liquids, and gases—depends on their intermolecular forces. An almost accurate model to express potential and intermolecular forces, known as the Lennard–Jones potential, is described by the following equation: (1.4) and (1.5) [8]. Figure 1.8 shows the distribution of Lennard–Jones potential in terms of intermolecular distance.

$$V_{ij} = 4\varepsilon\left[c_{ij}\left(\frac{r}{\sigma}\right)^{-12} - d_{ij}\left(\frac{r}{\sigma}\right)^{-6}\right] \tag{1.4}$$

$$F_{ij} = -\frac{\partial V_{ij}(r)}{\partial r} = \frac{48\varepsilon}{\sigma}\left[c_{ij}\left(\frac{r}{\sigma}\right)^{-13} - \frac{d_{ij}}{2}\left(\frac{r}{\sigma}\right)^{-7}\right], \tag{1.5}$$

where, r is the intermolecular distance, c_{ij} and d_{ij} are the interaction parameters of molecules, ε is the energy characteristic and σ is the characteristic length where the energy potential between two particles is zero. Actually, ε represents the minimum energy level.

Equation (1.5) shows that when two particles are far apart, the force between them tends to zero and can be neglected as this distance decreases to $r = 2\sigma$. With decreasing distance between two particles distance the attraction force increases and

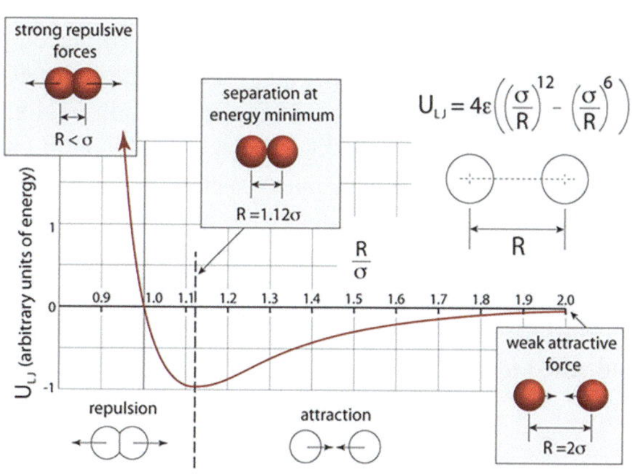

Fig. 1.8 Lennard–Jones potential as a function of intermolecular distance

the surface energy tends to zero. If the distance decreases drastically, the repulsive force becomes dominant. When $r = 1.12\sigma$, there is the lowest energy level between them. As the distance decreases, the dominant intermolecular force is the repulsive force, and for $r < \sigma$ this force is very strong and the energy level tends to infinity.

1.4 Flow Features at Micro- and Nano-Scales

1.4.1 Surface and Body Forces

In classical fluid mechanics analysis, the assumption of continuous flow is considered. For continuum flow analysis, instead of considering the behavior of individual molecules, a set of fluid molecules within a control volume is considered as a continuum medium, and the conservation laws are applied to them. The boundary condition used on the walls in continuum flow is the no-slip condition. According to Eqs. (1.1)–(1.6), the no-slip condition indicates that the velocity of the fluid near the wall is equal to the velocity of the wall.

$$U_{fluid@wall} = U_{wall} \tag{1.6}$$

Investigating micro- and nano-scale flows differs from investigating continuum flows. One of the most important differences is the behavior of the fluid near the wall. Another main difference comes from the small Reynolds number in micro- and nano-scale flows. The Reynolds number in macroscopic flows is inherently very large. For example, the Reynolds number for a one-meter-long airplane wing flying at a speed of 2 m/s in the air is of the order of one million. Micro- and nano-flows have a very small Reynolds number due to their small characteristic length (e.g. a channel with a height of one micrometer) and are typically in the creep flow regime. In other words, due to the small Reynolds number in such flows, the flow is typically laminar at the micro- and nano-scale levels.

In micro- and nano-systems, the discrete molecular structure must be accounted for; the gas flow is rarefied, and the fluid near the walls exhibits velocity slip and a temperature jump relative to the wall. Additionally, peculiar phenomena such as thermal creep, electrokinetic effects, viscous heating in the vicinity of the wall, flow diffusion effects, and even quantum effects can become significant. It should be noted that the surface forces acting on the surface of the flow channels are proportional to the square of the characteristic length of the geometry. On the other hand, volumetric or body forces are related to the third power of the characteristic length. Hence, the surface-to-body forces ratio is very high for flows in micro- and nano-scales, as indicated in Eq. (1.7).

$$\frac{surface\,force}{body\,force} \propto \frac{l^2}{l^3} = \frac{1}{l} \tag{1.7}$$

Therefore, volume forces can be ignored compared to the surface forces in these types of flows. The effects of surface tension on such scales are dominant. This feature is the basis of the design, manufacturing, and implementation of micropumps and microvalves.

1.4.2 Continuity Assumption Breakdown

In solids, the molecules are packed together with a special structure, and they are related to all their neighbors according to the Lennard–Jones formula (1.4). In a solid structure, a molecule needs a certain level of energy to separate from the structure. This energy could be provided by impact or adding heat to the solid. As soon as the temperature of a solid body reaches its melting temperature, all the molecules commonly gain a level of energy that allows them to separate from their neighbors. At this time, the solid object turns into a liquid. If the temperature of the liquid increases, the molecular vibrations increase greatly, and when it reaches the boiling temperature of the liquid, the liquid gradually turns into a gas. Due to the significant intermolecular distance, continuum breakdown occurs faster in gases than in liquids. Table 1.2 makes a comparison between the properties of the two phases of liquid and gas under standard atmospheric conditions.

In the discussions related to micro- and nano-scale flows, an important issue is under what conditions the fluid can be considered continuous. To establish the condition of fluid continuity, the microscopic fluctuations of molecules should not cause significant changes in the average properties of the fluid. Therefore, the following conditions should be met for fluid to be considered continuous [8]:

(1) The point properties of the fluid should be continuous. The point properties include kinematic (velocity and speed of sound) and thermodynamic (pressure and temperature) properties of the fluid.
(2) The transport properties of the fluid should be continuous. Transport properties include the conductivity coefficient and the viscosity coefficient of the fluid.

To establish the first condition, there must be at least 10,000 molecules within a very small control volume. According to Table 1.2, for continuous air at standard conditions, the number density (n) is equal to 3×10^{25}. Hence, the dimensions of the control volume ($\forall$) that meet the condition of continuous flow for the gas phase are as follows

Table 1.2 Sample comparison between gas and liquid phases

Fluid type	Gas (Air)	Liquid (Water)
Molecular diameter (m)	3×10^{-10}	3×10^{-10}
Number density (m^{-3})	3×10^{25}	2×10^{28}
Molecular spacing (m)	3×10^{-9}	0.4×10^{-9}
Mean free path (m)	100×10^{-9}	0.3×10^{-9}

$$\forall = \frac{N}{n} = \frac{10000}{3 \times 10^{25}} = 0.33 \times 10^{-21}\,\mathrm{m}^3 \Rightarrow l_{\text{gas-point}} \simeq 70 \times 10^{-9}\,\mathrm{m}$$

Similarly, the calculation for the liquid phase in Table 1.2 results

$$\forall = \frac{N}{n} = \frac{10000}{2 \times 10^{28}} = 0.50 \times 10^{-24}\,\mathrm{m}^3 \Rightarrow l_{\text{liquid-point}} \simeq 8 \times 10^{-9}\,\mathrm{m}$$

According to this condition, the airflow in a local equilibrium is continuous up to a characteristic length of 70 nm, as determined by point properties. If the characteristic length is less than this number, it is outside the continuum condition. Liquids are usually continuous up to a characteristic length of 8 nm.

It should be noted that the mentioned condition is one of the continuity conditions; another condition, namely, the continuity of transport properties, must also be observed. For this purpose, the control volume should be sufficiently large so that the molecules collide with each other more frequently than they collide with the wall. Therefore, the distance of each side of the control volume must be at least 10 times the length of the mean free path, i.e. $l = 10\,\lambda$. For liquids, however, the mean free path is approximately equal to the diameter of a molecule. This is because the liquid molecules are always in close proximity to one another and are constantly colliding. Hence, the second condition for gases in Table 1.2 can be calculated as follows

$$l_{\text{gas-tr}} = 10\lambda = 10 \times \left(100 \times 10^{-9}\right) = 10^{-6}\,\mathrm{m}$$

For the liquid, the value of $l_{\text{liquid-tr}}$ is as

$$l_{\text{liquid-tr}} = 10\lambda = 10 \times \left(3 \times 10^{-10}\right) = 3 \times 10^{-9}\,\mathrm{m}$$

The conclusion about the continuity condition for the liquids and gases is expressed as follows

$$L_{\text{gas-continuum}} = \max\left(l_{\text{gas - point}}, l_{\text{gas - tr}}\right) = 10^{-6}\,\mathrm{m}$$

$$L_{\text{liquid-continuum}} = \max\left(l_{\text{liquid - point}}, l_{\text{liquid - tr}}\right) = 8 \times 10^{-9}\,\mathrm{m} \simeq 10^{-8}\,\mathrm{m}$$

In general, it can be said that for liquids, microscales are still in the continuum region, but for gases, the micro range marks the beginning of the range of departure from continuous conditions.

Briefly, one can count the main reasons for the difference between micro- and nano-scale flows with macro-scale continuum flows as:

- The surface-to-volume ratio is large in micro- and nano-systems
- The characteristic length of the geometry is comparable to the size of the fluid molecules
- Density fluctuations can be important

- Transport properties, such as the diffusion coefficient and viscosity, differ from those for continuum flows.
- The fluid-surface interaction can have a significant effect on the flow.

1.5 Flow Rarefaction

Rarefaction is one of the special phenomena of gas flows in micro- and nano-scales. The degree of rarefaction in gas flows is measured by Knudsen dimensionless number (Kn), which is the ratio of the molecular mean free path (λ) to the characteristic length scale (L). Equation (1.8) presents the definition of Knudsen number as

$$\text{Kn} = \frac{\lambda}{L} \tag{1.8}$$

Rarefaction becomes noticeable when the mean free path is no longer negligible compared to the flow characteristic length scale; in other words, the Knudsen number is a universal measure of gas rarefaction or characteristic length. The characteristic length varies depending on the specific problem. For example, the characteristic length of a long channel is the height of its entrance opening. A problem can have multiple or variable flow length scales.

For air at standard atmospheric conditions, the molecular mean free path is about 64 nm. Therefore, the flow in microchannels is faced with rarefaction effects. In this condition, the molecules have non-equilibrium behavior. The wall is the most important factor which leads to non-equilibrium conditions. Due to the presence of the wall, boundary conditions are applied to the flow velocity and temperature, which differ from the bulk flow movement. If the distance between gas molecules is significant compared to the mean free path of the flow, the discontinuity applied to the fluid in the vicinity of the walls is not completely damped in the flow, because the probability of gas-surface molecular collisions is substantial compared to inter-molecular collisions. As a result, phenomena such as velocity slip and temperature jump occur. In other words, the rarefied flow is in a thermodynamic non-equilibrium condition.

The Knudsen number can also be defined as the ratio of the collision frequency of the gas-surface molecules with the collision frequency of the gas–gas molecules as

$$\text{Kn} = \frac{\vartheta_{wall}}{\vartheta_{mol}}, \tag{1.9}$$

where, ϑ_{mol} is the intermolecular collision frequency and ϑ_{wall} is the frequency of gas-surface molecular collisions. In porous media, due to the random structure of the medium, the characteristic length of the flow cannot be calculated. This is because the distance between porous particles is completely random, and no specific value can be considered for the characteristic length of the flow. Instead, in such geometries, the frequency-based definition is used. In general, the existence of boundaries leads to non-equilibrium effects in the gas near the boundary, where, the molecules are reflected from the wall and are affected by the wall temperature and other surface properties. If the gas-surface molecular collision frequency is significantly lower than the gas intermolecular collision frequency, a large number of binary collisions in the gas near the walls immediately eliminate any non-equilibrium effects, and the flow can be considered a continuum. The definition given by Eq. (1.9) is also consistent with the definition of the Knudsen number in the dimensionless Boltzmann equation [9]. According to this definition, the Knudsen number is the ratio of the molecular movement due to intermolecular collision (diffusion) to the ballistic movement (convection) [9].

It is worth noting that rarefaction is a flow feature, distinct from the definition of a dilute gas. A dilute gas is a gas in which intermolecular forces are negligible and the ideal gas law is applicable to it. In a dilute gas, the average intermolecular distance (δ) is at least seven times greater than the diameter of the molecules. Air is a dilute gas under standard atmospheric conditions, but it does not exhibit rarefied gas behavior until it enters a microchannel. The definition of dilute gas is the opposite of dense gas. To better understand how air is dilute, consider the following example:

Example 1.1 The number density of air at 1 atmosphere and 20 °C is $2.5 \times 10^{25}\ \text{m}^{-3}$. The radius of the nitrogen molecule as the dominant molecule of air is equal to 0.2 nm. What percentage of the volume of air contains air molecules? Approximate air with nitrogen.

Solution Using the given molecular radius, the volume occupied by the nitrogen molecule is obtained as follows

$$V = \frac{4}{3}\pi r^3 = \frac{4}{3}\pi \left(0.2 \times 10^{-9}\right)^3 = 4 \times 10^{-29}\ \text{m}^3$$

To find the amount of air occupied by the molecules, one should multiply the number of molecules per unit volume by the volume of one molecule as

$$n \times V = 2.5 \times 10^{25}\ \text{m}^{-3} \times 4 \times 10^{-29}\ \text{m}^3 = 0.001 = 0.1\%$$

Example 1.1 illustrates that only 0.1% of the air volume comprises molecules at atmospheric conditions. The remaining 99.99% is empty space. This example concretely shows the concept of dilute gas.

Another situation in which rarefied air occurs is at high altitudes. The effect of increasing the height on the air density is similar to the effect of reducing the dimensions. In other words, both high-altitude and very small-scale conditions lead to an increase in the Knudsen number and, respectively, the existence of rarefied flow. The interesting point here is that many molecular solution methods such as Direct Simulation Monte Carlo (DSMC) were initially proposed for the analysis of hypersonic flow at high altitudes and in the 1990s after observing the similarity of

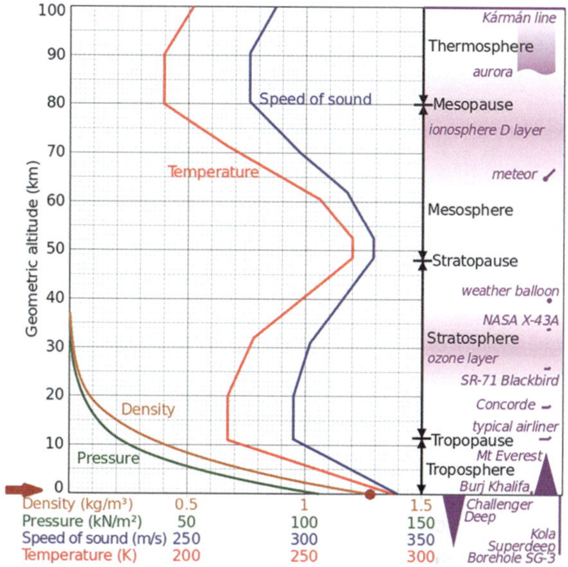

Fig. 1.9 The air pressure, temperature, density, and speed of sound as a function of altitude

the behavior of the flow in small dimensions with the high-altitude flows (from the rarefaction viewpoint), these tools (with modifications) were used to analyze the flow in small scales. Figure 1.9 illustrates the impact of altitude rise on temperature, density, sound speed, and air pressure. On the right side of the diagram, the different layers of the atmosphere are marked.

1.5.1 Gas Kinetic Theory and Collision Cylinder

In this section, a summary of the kinetic theory of gases is presented to explain the phenomena of micro- and nano-flows. Gas molecules experience fewer collisions compared to liquid molecules which are in continuum flow. According to the kinetic theory of gases, molecules are in constant motion, moving in a specific direction with a constant speed, which changes due to intermolecular collisions or collisions with the gas surface. According to Fig. 1.10, if a molecule with a diameter of d is moving at a speed of v, a region around this molecule is defined as a cylinder with a circular cross-section of diameter $2d$, centered at a distance d from the top and bottom, which is produced over time t. Any other molecule inside this cylinder collides with the desired molecule. In other words, the impact cylinder is the volume swept by the impact cross-section that is created at time t.

The definition of this quantity is used to calculate the molecular mean free path. By definition, the average length traveled between two successive collisions is the molecular mean free path as

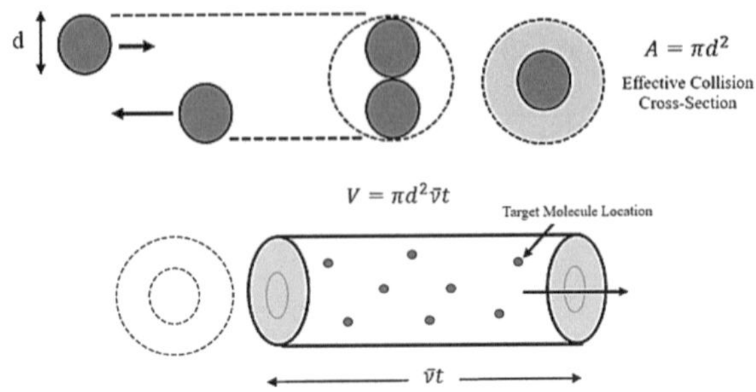

Fig. 1.10 Collision cylinder model

$$\lambda = \frac{\text{traveled distance}}{\text{number of collisions}} = \frac{\overline{v}t}{\pi d^2 \overline{v} t n} = \frac{1}{n\pi d^2} \tag{1.10}$$

The number of collisions is related to the number of particle pairs in the collision cylinder and depends on the relative velocity between them. Since all molecules do not necessarily move in the collision cylinder with the speed of $\overline{v}$, the Eq. (1.10) is not an exact relation. It is more appropriate to employ v_{rel} it in the denominator. The relative velocity is defined as

$$v_{rel} = \sqrt{(\overline{v}_1 - \overline{v}_2)^2} = \sqrt{\overline{v}_1^2 - 2\overline{v}_1\overline{v}_2 + \overline{v}_2^2} = \sqrt{2}\overline{v} \tag{1.11}$$

Therefore, Eq. (1.10) could be modified to Eq. (1.12) as

$$\lambda = \frac{1}{\sqrt{2} n\pi d^2} \tag{1.12}$$

The root mean square velocity of the gas at a specified condition could be calculated as

$$\overline{v} = \sqrt{3RT} = \sqrt{3kT/m}, \tag{1.13}$$

where, k is the Boltzmann constant. For air in the standard atmospheric conditions, the root mean square velocity is equal to 486 m/s.

The mean collision time between intermolecular collisions is defined as the ratio of the mean free path to the average velocity of the gas, as indicated in Eq. (1.14)

$$\overline{t}_c = \lambda/\overline{v} \tag{1.14}$$

For air at standard atmospheric conditions, the average collision time is about 10^{-10} s. According to Eq. (1.14), it can be inferred that when the temperature of the gas increases, the kinetic energy of its molecules increases, the number of collisions increases, and the collision time decreases.

Figure 1.11a shows the changes in the dimensionless number of molecules in terms of molecular speed for different gases. As can be seen, with an increase in molecular mass, the graph shifts to the left, and the maximum speed of the molecules decreases; however, the percentage of molecules at the maximum speed increases. Figure 1.11b shows the changes in the equilibrium distribution function of molecular speeds as a function of temperature. As the temperature increases, the maximum speed of the molecules increases, but the percentage of molecules that are at the maximum temperature decreases.

The average molecular speed can be calculated from the following formula

$$\overline{c'} = \sqrt{\frac{8kT}{\pi m}}, \tag{1.15}$$

where, $c' = c - c_0$.

According to the simplified gas kinetic theory, the viscosity of gas could be defined as

$$\mu = \frac{1}{2}\rho\lambda\overline{c'} \tag{1.16}$$

It should be noted that the Eq. (1.16) has been widely used before. However, it is not an exact formula. By using the Chapman-Enskog expansion of the molecular distribution function, Sonne extracted the stress tensor of the Navier–Stokes equations, compared it with the standard stress tensor of these equations, and derived the Eq. (1.17) for the viscosity coefficient [10, 9]:

$$\mu = \gamma_1\sqrt{\frac{\pi}{8RT}}P\lambda = \gamma_1\sqrt{\frac{\pi RT}{8}}\rho\lambda, \tag{1.17}$$

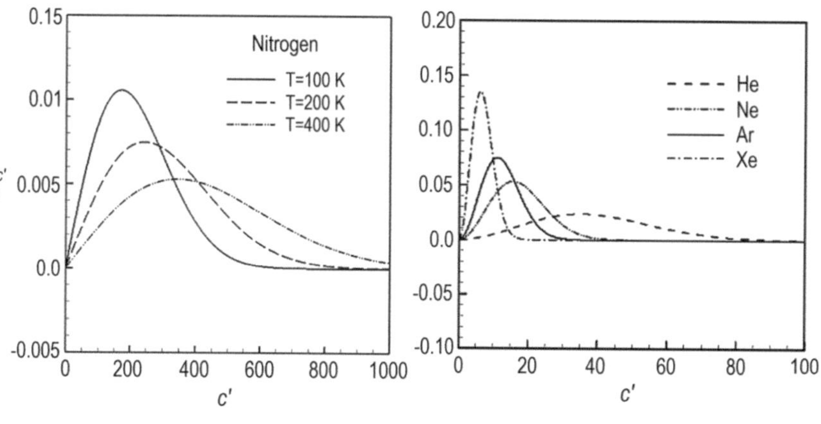

Fig. 1.11 Distribution functions of the velocity of molecules at different temperatures (right) and velocity of molecules of different gases (left)

where, γ_1 is a constant that depends on the collision model. For the hard-sphere and BGK models the γ_1 values are equal to 1.27 and 1, respectively.

In fluid flow investigations, dimensionless numbers play a major role. One of these numbers is the Mach number which is defined in Eq. (1.18) as the ratio of the flow speed to the speed of sound at the flow temperature.

$$\mathrm{M} = u/c_s \tag{1.18}$$

The sound speed corresponds to the speed of spread of information in the gas and is indicated as

$$c_s = \sqrt{\gamma RT} = \sqrt{\frac{\gamma kT}{m}} \tag{1.19}$$

With a combination of Eqs. (1.15) and (1.16) one can drive a new relation for mean free path as

$$\lambda = \frac{\mu}{\rho}\sqrt{\frac{m\pi}{2kT}} \tag{1.20}$$

and dividing by characteristic length a formula for Knudsen number could be obtained

$$\mathrm{Kn} = \frac{\lambda}{L} = \frac{\mu}{\rho L}\sqrt{\frac{m\pi}{2kT}} \tag{1.21}$$

By presenting the definition for the Reynolds number in Eq. (1.22), one can derive Eq. (1.23) which indicates the relation between Mach, Reynolds, and Knudsen numbers.

$$\mathrm{Re} = \frac{\rho U_{\infty L}}{\mu} \tag{1.22}$$

$$\frac{\mathrm{M}}{\mathrm{Re}} = \frac{U_\infty/c_s}{\rho U_\infty L/\mu} = \frac{\mu}{\rho L c_s} = \frac{\mu}{\rho L\sqrt{\frac{\gamma kT}{m}}} = \frac{\mu}{\rho L}\sqrt{\frac{m}{\gamma kT}} \tag{1.23}$$

Finally, Eq. (1.23) could be ordered as

$$\mathrm{Kn} = \sqrt{\frac{\gamma\pi}{2}}\frac{\mathrm{M}}{\mathrm{Re}} \tag{1.24}$$

If viscosity is substituted from Sone's formula Eq. (1.17), instead Eq. (1.16), then Eq. (1.24) will be modified to

$$\mathrm{Kn} = \frac{1}{\gamma_1}\sqrt{\frac{40}{3\pi}}\frac{\mathrm{M}}{\mathrm{Re}} \tag{1.25}$$

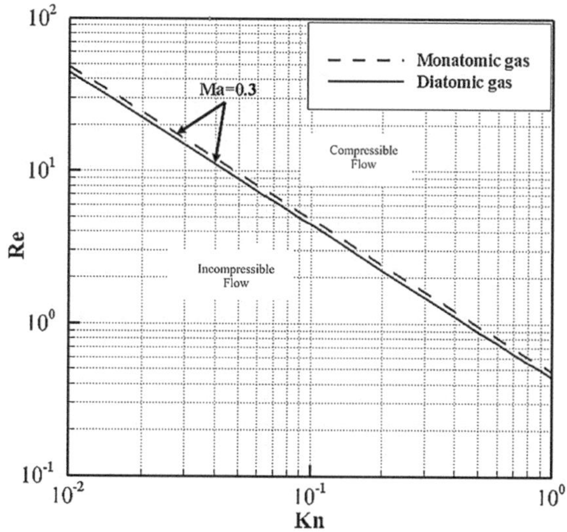

Fig. 1.12 Reynolds number versus Knudsen number graph

In Fig. 1.12, the diagram of Reynolds number versus Knudsen number, Eq. (1.24) is plotted. This indicates that the Reynolds number decreases with an increase in the Knudsen number while maintaining a constant Mach number. The limit of compressibility is considered to be the Mach number of 0.3. The molecular mean free path determines the dependence of the Knudsen and Reynolds numbers on whether the molecule is monoatomic or polyatomic. As it could be seen, at small Knudsen numbers, i.e. $Kn = 10^{-2}$, the flow requires a high Reynolds number (more than 300) to become compressible. On the other hand, the limit of compressible flow shifts to the lower limit of the diagram as the Knudsen number increases. At low Reynolds numbers, i.e. $Re = 10$, the flow velocity can be very low. However, as the Knudsen number increases, the flow enters into the compressible flow range. This compressibility is closely related to density changes in the gas flow field.

1.6 Rarefied Gas Flow Classification

The flow is divided into different regimes based on the Knudsen number. The classification of flow regimes based on the Knudsen number was first presented by Tsien [11]. The mentioned classification and the governing equations in each range are shown in Fig. 1.13.

According to the Fig. 1.13, different flow regimes based on the Knudsen number are as [8]

- Continuum flow ($Kn < 0.001$)
- Slip flow ($0.001 < Kn < 0.1$)
- Transitional flow ($0.1 < Kn < 10$)

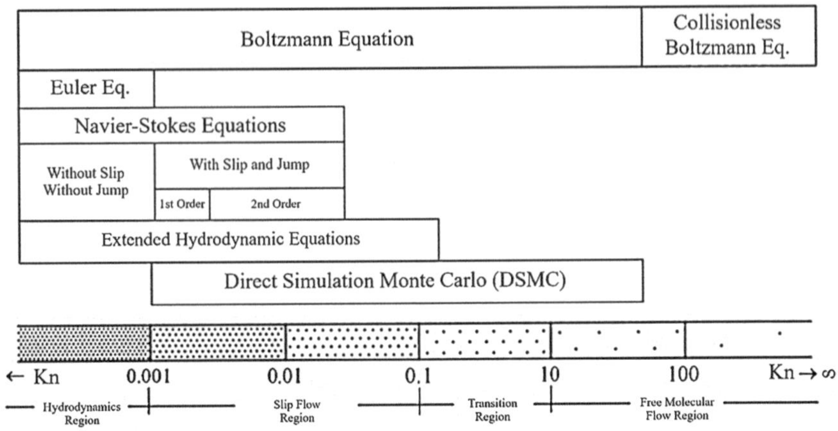

Fig. 1.13 Flow classifications based on Knudsen number

- Free-molecular flow (Kn > 10).

When the Knudsen number approaches zero, Navier–Stokes equations are simplified to Euler equations. In the continuum flow limit, Navier–Stokes equations are valid. As the Knudsen number increases, the continuum regime goes into the so-called slip flow limit. The Navier–Stokes equations are not valid on the boundaries, and the boundary conditions must be modified to account for velocity slip and temperature jump effects. As the Knudsen number continues to increase, the Navier–Stokes equations lose their accuracy in the bulk, and the flow moves to the transition regime. In fact, with the increase in the Knudsen number, the effects of rarefaction are detected throughout the entire flow domain, and continuum-based models fail to accurately model the flow. In this situation, the direct simulation Monte Carlo (DSMC) method is one of the algorithms that directly simulate the behavior of individual molecules, ensuring the solution result converges to the solution of the Boltzmann equation [12]. Boltzmann equation is the governing equation for fluid flow in all flow regimes.

It should be noted that the flow classification presented in Fig. 1.13 has several problems. First, these boundaries are for the geometry of the channel with a length of L and height of h where $L \gg h$, assuming isothermal flow. It should be noted that the classification is not entirely accurate for other geometries. For example, in the case of cavity flow, the slip condition and transition to the transitional flow occur more rapidly. The second point is that these classifications do not consider the Mach number, while the dimensionless parameter should be defined in such a way as to be a function of both the Knudsen number and the Mach number. Additionally, since the local Knudsen number may be larger than the global Knudsen number, it is necessary to determine exactly which location the Knudsen number is defined for. Therefore, the gradient-length Knudsen number (Kn_{GL}) is defined as

$$\mathrm{Kn}_{GL} = \frac{\lambda}{l_{GL}} \tag{1.26}$$

$$l_{GL} = \max\left(\frac{\rho}{\nabla\rho}, \frac{P}{\nabla P}, \frac{T}{\nabla T}, \frac{v}{\nabla v}\right), \tag{1.27}$$

where v is the velocity.

In other words, gradient length shows the ratio of a quantity to its gradient. For example, in the case of the shock wave occurrence, the gradient-length Knudsen number is the parameter that determines the flow condition. Figure 1.14 shows the contours of temperature and the gradient-length Knudsen number around a micro-cylinder with a diameter of 0.3 μm at a Knudsen number of 0.01 and a Mach number of 10. According to Kn_{GL} in the contour, there are significant non-equilibrium conditions in different regions, namely, the shock wave area, the separated flow region behind the cylinder, and the boundary layer. The temperature contour also highlights the limitations of the continuous solution method in accurately predicting the thickness of the shock wave and the extent of the separated flow area.

The graph of characteristic length in terms of number density provides a good guide for determining the range of rarefied gases, as shown in Fig. 1.15. The diagram is plotted at constant Knudsen numbers. The vertical dashed line $n/n_0 = 1$ represents the boundary between a dilute and a dense gas. Diagonal lines are the lines of constant

Fig. 1.14 Flow around a micro cylinder **a** temperature contours obtained by continuum-based (up) and molecular-based (low) ap-proaches **b** contour of gradient-length Knudsen number Reproduced from Ref. [13]

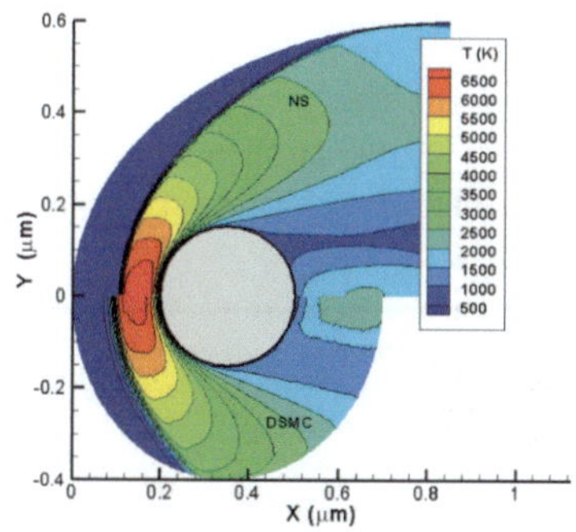

(a) Temperature contours obtained by continuum-based (up) and molecular-based (low) approaches

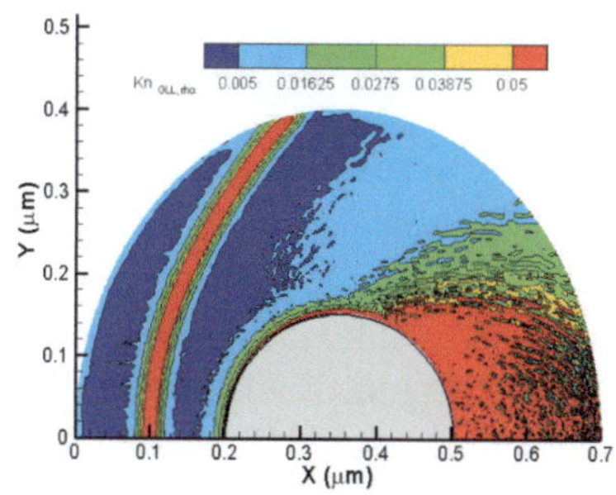

(b) Contour of gradient-length Knudsen number

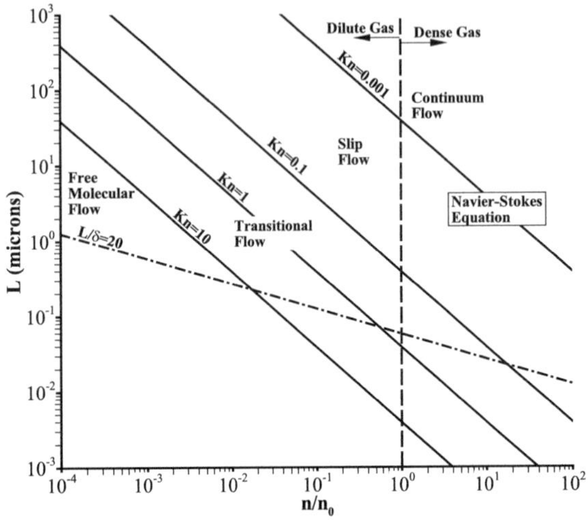

Fig. 1.15 Characteristic length versus number density. Reproduced from Ref. [8]

Knudsen number. It can be seen that with the increase in number density, the characteristic length decreases at a constant Knudsen number. Below the $L/\delta = 20$ line, significant molecular oscillations begin to form. In the situation where the Knudsen number approaches zero, the flow regime is continuum. In this case, the length of the molecular mean free path is much smaller than the characteristic length of the flow. In other words, the number of intermolecular collisions exceeds the number of gas-surface collisions. For this reason, the molecules and the wall reach a state of thermodynamic equilibrium.

When investigating gas flow, one can disregard the molecular structure of the gas and consider only the macroscopic parameters, such as density, velocity, and temperature. However, when the mean free path is no longer negligible compared to the flow characteristic length, the number of intermolecular collisions is significantly reduced compared to gas-surface molecular collisions. Therefore, gas molecules have a lower chance of reaching local thermodynamic equilibrium with the wall. In this condition, the flow regime can be considered as near-free or free molecular. In this case, gas-surface molecular collisions play a major role. Between the continuum and free molecular flow regimes lie the slip and transitional flow regimes, where both intermolecular and gas-surface molecular collisions become significant. The flow rarefaction results in a velocity slip and a temperature jump relative to the wall. In other words, in rarefied gas flows, the no-slip assumption is no longer valid, and the velocity and temperature of the near-wall flow differ from those of the wall itself, including its speed and temperature. Therefore, to predict the flow behavior in the slip flow regime using the Navier–Stokes equations, the no-slip boundary condition must be applied in a modified manner. However, in molecular methods, such as the DSMC method, it is not required to consider flow rarefaction effects (velocity slip and temperature jump) by applying special boundary conditions to the governing equations, since these phenomena are inherently predicted by the molecular behavior of the flow.

By introducing the concept of flow regimes in terms of the Knudsen number, we can visualize the location of micro- and nano-technologies in terms of the Knudsen number (Fig. 1.16). According to the figure, hard disks are positioned in the middle of the transitional flow regime, while micronozzles and microvalves are located at the beginning of the transitional flow regime. On the other hand, micro gyroscopes and some microsensors are placed in the slip flow regime.

Example 1.2 Calculate all the characteristic parameters of nitrogen (N_2) gas flow at a temperature of 350 K, a pressure of 200 kPa, and a speed of 100 m/s in a microchannel with an inlet opening of 10 μm.

Solution Knowing the molecular mass (28.013 kg/kmol) and diameter (3.75×10^{-10}m) of the Nitrogen, we calculate the gas constant and numerical density as

$$R = \frac{\overline{R}}{M} = \frac{8.3145\,(\text{kJ/kmol.K})}{28.013\,(\text{kg/kmol})} = 0.2968\,(\text{kJ/kg.K})$$

$$n = \frac{P}{kT} = \frac{2 \times 10^5\,(\text{N/m}^2)}{1.3805 \times 10^{-23}(\text{J/K} \cdot 350\,\text{K})} = 4.14 \times 10^{25}\,\text{m}^{-3}$$

$$\delta = n^{-1/3} = (4.14 \times 10^{25}\text{m}^{-3})^{-1/3} = 2.89 \times 10^{-9}\ \text{m}$$

The mean free path, the mean square root of the velocity, the collision time, the speed of sound, and the viscosity coefficient of Nitrogen can be calculated as follows

$$\lambda = (\sqrt{2}\pi d^2 n)^{-1} = \left(\sqrt{2}\pi(3.75 \times 10^{-10}\text{m})^2 \times 4.14 \times 10^{25}\text{m}^{-3}\right)^{-1}$$

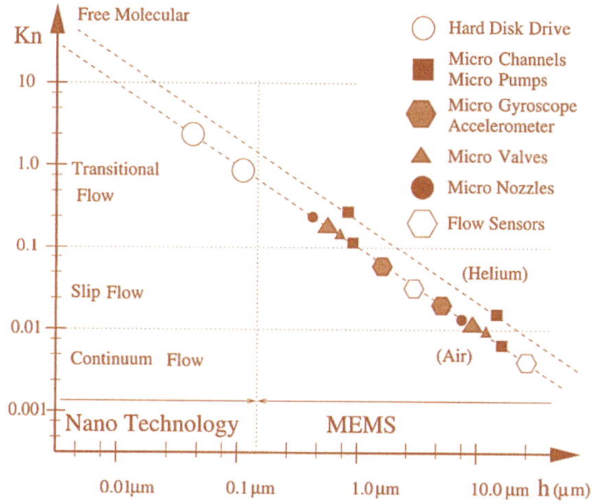

Fig. 1.16 Conventional MEMS applications and nanotechnology at standard atmospheric conditions, *h* indicates the flow characteristic length. Reporduced from Ref. [8]

$$= 3.9 \times 10^{-8}\,\text{m}$$

$$\overline{v} = \sqrt{3RT} = \sqrt{3 \times 296.8\ (\text{J/kg.K}) \times 350\text{K}} = 558\ \text{m/s}$$

$$t_c = \frac{\lambda}{\overline{v}} = \frac{3.9 \times 10^{-8}\text{m}}{558\ (\text{m/s})} = 7.0 \times 10^{-11}\text{s}$$

$$c = \sqrt{\gamma RT} = \sqrt{1.4 \times 296.8\ (\text{J/kg.K}) \times 350\ \text{K}} = 381\ \ \text{m/s}$$

$$\nu = \frac{1}{2}\lambda\overline{v} = \frac{1}{2} \times 3.9 \times 10^{-8}\text{m} \times 558\ (\text{m/s}) = 1.09 \times 10^{-5}\left(\text{m}^2/\text{s}\right)$$

It should be noted that the viscosity has been calculated based on the $\overline{v}$. Table 1.3 presents the nitrogen gas characteristics employed for the abovementioned calculations.

Example 1.3 Consider a 100 nm diameter sphere placed in the middle of a very long microchannel with a height of 1 μm. Argon gas enters this channel with a numerical density of $1 \times 10^{25}\ \text{m}^{-3}$ and a diameter of 4 nm.

(a) Calculate the mean free path of Argon gas
(b) Suppose the sphere and the walls of the channel are at the same temperature, and the flow is driven by the pressure difference between the channel inlet and outlet. What is the characteristic length affecting the problem? Calculate the Knudsen number.
(c) Suppose that the pressure difference between the inlet and outlet of the channel is not applied and the temperature of the channel walls is constant and equal to 300 K. But, the temperature of the sphere is linearly increasing from 300 K (the left side, i.e. at 0°) to 600 K (the right side, i.e. at 180°). Is there a flow inside the channel? What is the characteristic length affecting the problem? Calculate the Knudsen number.

Solution

Table 1.3 Nitrogen gas characteristic parameters

Flow regime	Criteria	Parameter
Rarefied gas	$\frac{\delta}{d} \gg 1$	$\frac{\delta}{d} = \frac{2.89\times10^{-9}\,\text{m}}{3.75\times10^{-10}\,\text{m}} = 7.7$
Incompressible flow	M < 0.3	$M = \frac{u}{c} = \frac{100\,\text{m/s}}{381\,\text{m/s}} = 0.26$
Early slip flow	Kn < 0.01	$Kn = \frac{\lambda}{L} = \frac{3.9\times10^{-8}\,\text{m}}{10^{-5}\,\text{m}} = 0.004$
Laminar flow	Re < 1500	$\text{Re} = \frac{uL}{\nu} = \frac{100\ \text{m/s}\times10^{-5}\,\text{m}}{1.09\times10^{-5}\,\text{m}^2/\text{s}} = 92$
Non-molecular vibrating flow	$L/\delta > 20$	$\frac{L}{\delta} = \frac{10.\times10^{-6}\,\text{m}}{2.89\times10^{-9}\,\text{m}} = 346$

(a) $n = \dfrac{1}{\sqrt{2}\pi d^2 \lambda} \Rightarrow \lambda = \dfrac{1}{\sqrt{2}\pi d^2 n} = 1.4 \times 10^{-9}\text{m} = 1.4\,\text{nm}$

(b) The channel entrance height is the effective characteristic length, hence

$$\text{Kn} = \frac{\lambda}{L} = \frac{1.4 \times 10^{-9}}{1.0 \times 10^{-6}} = 0.0014$$

(c) Due to the temperature gradient on the cylinder, thermal creep occurs, which will be discussed in details in next chapters, resulting in a flow from cold to hot. As a result, the characteristic length is the diameter of the cylinder.

$$\text{Kn} = \frac{\lambda}{L} = \frac{1.4 \times 10^{-9}}{100 \times 10^{-9}} = 0.014$$

The difference in the behavior of the flow in the free molecular regime and the continuum (or near-continuum) flow limit can be described as follows: from a molecular perspective, the behavior of molecules is influenced by intermolecular collisions and motion without collisions. In the free molecular regime, gas molecules move in a straight path, except when they encounter a solid boundary that alters their trajectory. In this situation, any disturbance in the field is quickly felt at all points. In other words, fast information transfer and field gradients are negligible. Therefore, the molecular resolution of such a flow is fast. However, in continuum and near-continuum flows, the transfer of information through the free movement of molecules is ineffective because any disturbance in the field is dampened by numerous intermolecular collisions, and it takes a considerable amount of time to transmit.

1.7 Peculiar Physical Phenomena of Micro- and Nano-Gas Flows

1.7.1 Cold-To-Hot Heat Transfer

One of the famous problems in fluid mechanics is the flow inside a cavity with a moving lid. To better understand the differences between solutions based on classic fluid mechanics and micro- and nano-scale gas flow analysis, Fig. 1.17 provides a suitable comparison. In this figure, the temperature contour and heat flux lines in three different Knudsen numbers, i.e. 0.005, 0.05, and 0.1, from the solution of the Navier–Stokes equations (top row) along with the velocity slip boundary condition and temperature jump and molecular solution (DSMC) flow is shown in the bottom row. The first noteworthy point is that with the increase of the Knudsen number, the temperature contours obtained by a continuum-based approach begin to differ from the molecular solution contours. Considering that the assumption of Fourier's law is valid in the Navier–Stokes solution, heat transfer always occurs from the hot to the cold region. In other words, the heat flows originate from hot areas and move to cold areas. However, in the temperature contours of the molecular-based solution (bottom row), we can observe the phenomenon of cold transfer, also known as "anti-Fourier" heat transfer, which is one of the fundamental differences between micro- and nanoscale gas flows and large-scale (macroscopic) flows. In other words, Fourier law is violated in rarefied gas flow.

Sonne's fundamental equations and higher-order equations, known as Regularized 13 Moments Equations (R13), are able to accurately predict this phenomenon (see Fig. 1.18 and compare with Fig. 1.17). Sonne's fundamental equations [9] state that, in addition to being proportional to the temperature gradient in the first order of the Knudsen number, the heat flux is also proportional to the second gradient of the speed in the second order of accuracy according to the Knudsen number. Therefore, if the Knudsen number increases and a strong turning in the flow velocity field occurs (such as at a cavity corner), there is a possibility of heat transfer occurring under the influence of the second velocity gradient rather than the temperature gradient. Similarly, this phenomenon also occurs in the step flow problem when the temperature of the bottom and side walls of the step differs from the inlet temperature (Fig. 1.19). The cause of this phenomenon, similar to the cavity problem, is the strengthening of the heat flux originating from the second gradient of the velocity.

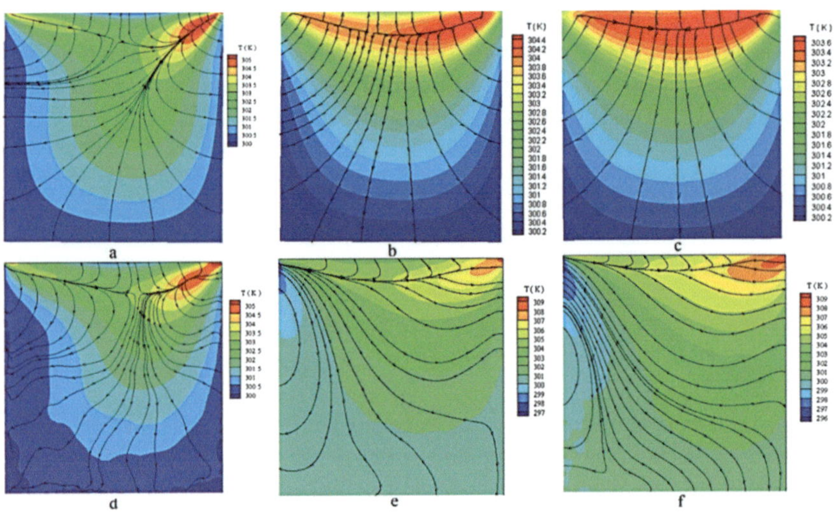

Fig. 1.17 Comparison of the temperature contours and heat flux lines obtained by Navier–Stokes solution (top row) and DSMC solution (bottom row) in the cavity problem with the lid velocity of 100 m/s in three Knudsen numbers of 0.005 (left) and 0.05 (middle), and 0.1 (right side). Copied with permission from Ref. [14]

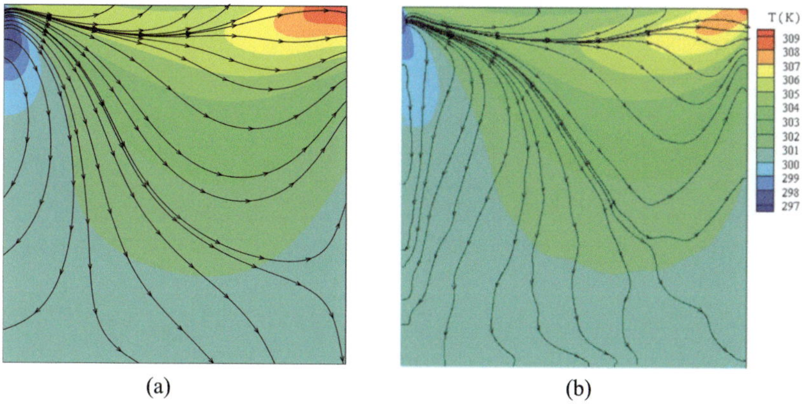

Fig. 1.18 Temperature contours and heat flux lines based solution of R13 equations and Sone equations at Knudsen numbers of 0.05. **a** R13 equations **b** Sone equations Copied with permission from Ref. [14]

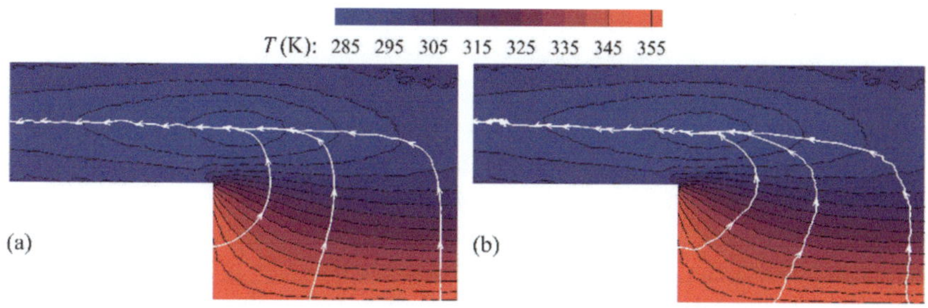

Fig. 1.19 Temperature contours and heat flux lines for the step flow problem at a Knudsen number of 0.05 and a wall temperature of 360 K. **a** DSMC simulation **b** Sone formula Reproduced from Ref. [15]

1.7.2 Knudsen Paradox or Knudsen Minimum

If the mass flow rate through a microchannel is considered, it is expected that when the Knudsen number increases, the mass flow rate decreases. With the increase in the Knudsen number, the mean free path becomes higher, and as a result, the flow density decreases, leading to greater rarefaction and a reduction in the number of gas molecules. Consequently, the mass flow rate is lower. This decrease in mass flow rate is maintained up to the Knudsen number equal to one but near the $Kn = 1$, the flow rate increases, this phenomenon is known as the Knudsen paradox or the Knudsen minimum.

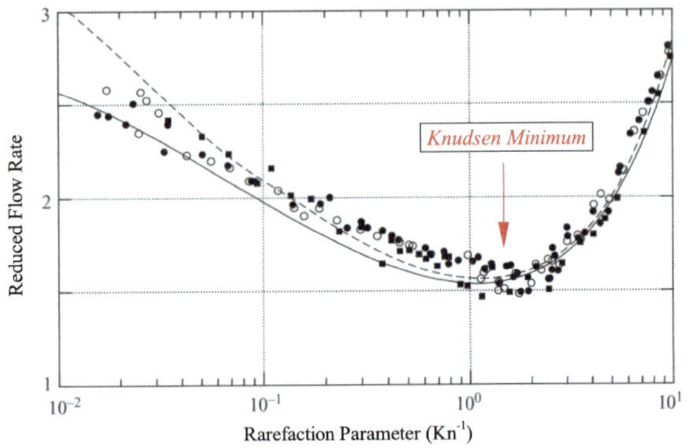

Fig. 1.20 Knudsen minimum phenomenon at Kn = 1. Reproduced from Ref. [19]

According to Fig. 1.20, the Knudsen paradox can be seen in a microchannel.[1] Various reasons have been mentioned in references to explain the phenomenon of the Knudsens paradox [16]. One of these reasons is the activation of the molecular diffusion phenomenon in mass transfer at middle Knudsen numbers in the transitional flow regime [17]. According to the theory of Dongari et al. [17], the term mass diffusion, which depends on the density (or pressure) gradient, is present on the right side of the continuity equation as a source term. With the increase in the Knudsen number in the transitional flow regime, the effect of this source term becomes more noticeable compared to the contribution from the channel pressure difference on mass transfer (or continuum mass transfer). Dongari and his colleagues observed the Knudsen minimum phenomenon by combining the flow rates of the diffusion term, free molecular flow, and continuum flow. In other words, the Knudsen minimum occurs due to a change in mass transfer mechanisms as the Knudsen number increases. Myong [18] demonstrated that the reason for this phenomenon is the increase in the average temperature of the flow at intermediate Knudsen numbers, and the dependence of the flow rate on temperature is also noted.

Another explanation for the Knudsen minimum phenomenon from a molecular perspective is presented by Tatsios et al. [20]. The molecular movement of particles can be divided into two stages: ballistic movement and binary collisions of particles. According to Fig. 1.21, which shows the mass flow rate through the channel in terms of Knudsen number, the behavior of the flow rate related to each of these two phenomena is opposite to each other in terms of Knudsen number. At small Knudsen numbers, most of the flow rate originates from the part related to the collision of particles, but as the Knudsen number approaches one, the flow rate caused by collisionless particles increases, and the contribution of the flow rate from the collision particles starts to decrease. The flow rate for both groups is equal at the Knudsen number of 7. After Kn = 1, despite the decrease in flow rate related to intermolecular

[1] The exact Knudsen number corresponding to the minimum mass flow depends on various parameters such as gas type, pressure ratio, and the gas-surface momentum accommodation coefficient. For example, Ohwada et al. reported that the minimum Knudsen number is about 0.7 [9].

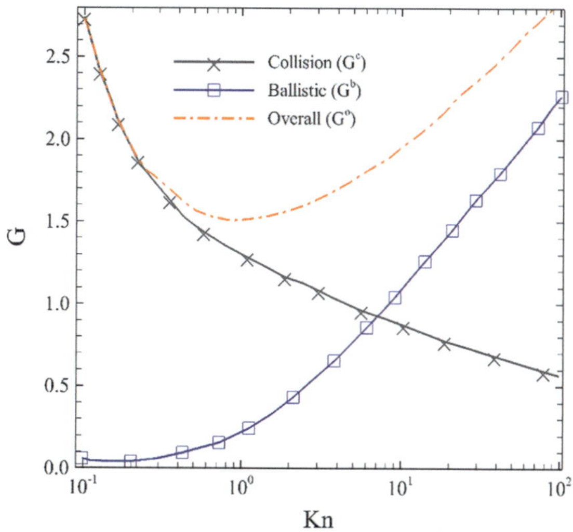

Fig. 1.21 Flow rate decomposition into (1) due to the ballistic motion of the particles and (2) due to the motion of collision particles. Reproduced from Ref. [20]

collisions, the flow rate increases again due to the increase in flow rate caused by ballistic molecules.

It should be noted that the Knudsen minimum value is dependent on the gas type, surface roughness, and gas-surface momentum accommodation. It has also been demonstrated that the value of the Knudsen minimum can be altered by the divergence angle in a divergent channel.

1.7.3 Nonlinear Pressure Distribution

Another phenomenon that occurs in micro- and nano-flows is the simultaneous effect of compressibility and rarefaction. In an incompressible flow inside the channel, the pressure distribution is linear; however, laboratory data show a nonlinear pressure distribution in microchannels. Until the detailed investigations of micro- and nano-flows, it was believed that compressible flow was a high-speed phenomenon, but investigations have shown that compressibility should also be considered for micro- and nano-scale flows. According to the definition, if the change in relative density is equal to 5% or more, the flow is considered compressible. In microchannels, pressure changes are usually noticeable. Therefore, changes in the flow density place it in the compressible regime. Figure 1.22 shows the pressure distributions for different Knudsen numbers. It can be observed that in the case of incompressible flow, the pressure distribution is linear. Increasing the Knudsen number leads to the nonlinearization of the pressure distribution.

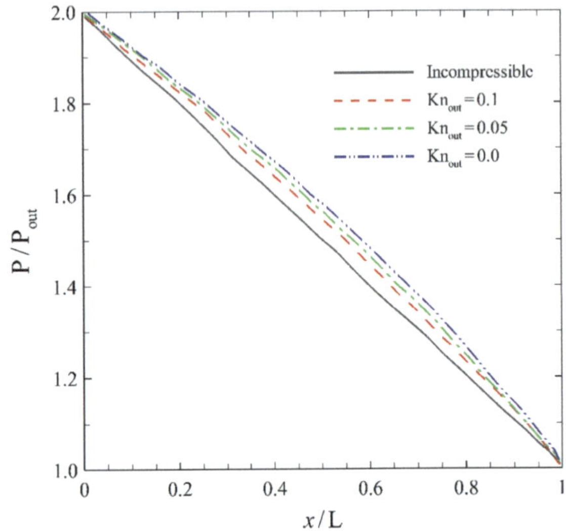

Fig. 1.22 Pressure distributions along the microchannel for different Knudsen numbers. Reproduced from Ref. [21]

1.7.4 Poiseuille Number Deviation

In fluid mechanics, the Poisuille number is defined as the product of the friction coefficient and the Reynolds number. The Poiseuille number indicates the pressure drop resulting from friction in the flow. The ratio of the Poiseuille number to the theoretical value is represented by C^*.

$$\mathrm{Po} = C_f \mathrm{Re} \tag{1.28}$$

$$C^* = \frac{\mathrm{Po}}{\mathrm{Po}_{theo}} \tag{1.29}$$

In Fig. 1.23, the trend for parameter C^* is shown in terms of Reynolds number for two gases of helium and nitrogen. It can be seen that for large Reynolds numbers (small Knudsen) this ratio is about one. With the increase of the Knudsen number (moving to the left side of the horizontal axis), this difference increases, which is due to the reduction of flow friction with the increase of the Knudsen number. As the flow progresses towards rarefaction, the effect of wall surface stress decreases (because basically, the stress on the wall and the friction obtained are mainly due to the fluid sticking to the wall), while in the high Knudsen numbers, the zeroing of the fluid velocity in the vicinity of the wall is irrelevant. In other words, in the rarefied flow, the velocity gradient in the vicinity of the wall is lower than the velocity gradient in the continuum flow due to velocity slip over the wall.

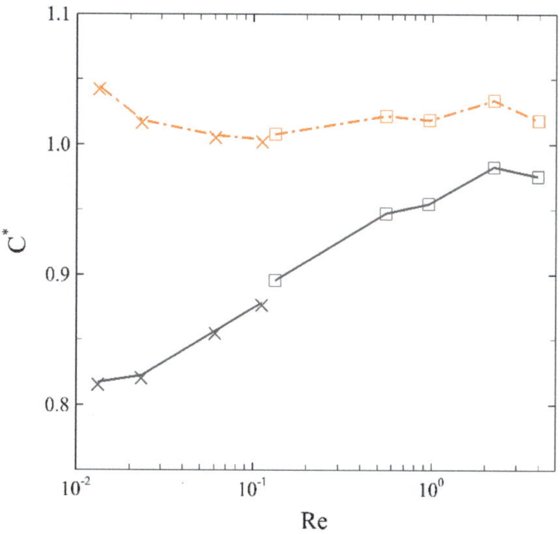

Fig. 1.23 Poiseuille number variations versus Reynolds number. The solid line corresponds to normalization with no-slip theory and the dashed-dotted line to slip theory. Here squares and crosses correspond to the nitrogen and helium flows, respectively. Reproduced from Ref. [8]

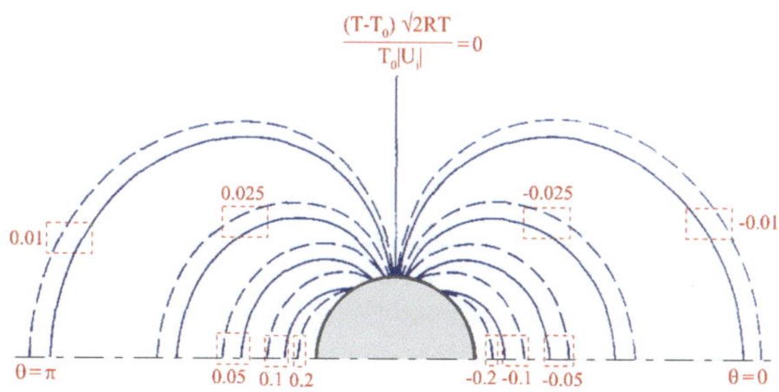

Fig. 1.24 Dimensionless temperature contours around an isothermal sphere within a free molecular flow. Reproduced from Ref. [22]

1.7.5 Thermal Polarization Around a Constant Temperature Sphere

In classical fluid mechanics, the slow motion of an isothermal sphere in the flow does not affect the temperature field. In the rarefied gas flow, the slow motion of the gas induces a non-uniform temperature field. If this phenomenon occurs around a sphere, it is called thermal polarization in rarefied gas flow. This phenomenon is attributed to the presence of a term that depends on the normal gradient of the normal stress in the second-order temperature jump boundary condition. Figure 1.24 shows the dimensionless temperature contours around an isothermal sphere in free molecular flow. The speed of the sphere is negligible compared to the average molecular speed, but it is opposite to zero. The temperature contour indicates that heating occurs in front of the sphere, while cooling occurs behind it. The same temperature polarization occurs in the slip flow regime; however, in this regime, cooling occurs in front of the sphere, and heating occurs behind it. In the slip flow regime, characterized by low flow velocity, the equations governing the velocity and temperature fields are the Stokes and Laplace equations, respectively. These two equations are independent of each other, but the temperature jump boundary condition is a function of the gradient of the gas normal stress, and therefore, the temperature field becomes a function of the velocity field, and the temperature distribution on the surface of the sphere is created. In general, the correlation and influence of temperature and velocity field in rarefied gas flow is much stronger compared to continuum flow [9].

References

1. Feynman, R.P.: There's plenty of room at the bottom. Eng. Sci. **23**(5) (1959)
2. Fekete, Z., Sinkovics, B., Rajta, I., Gál, G.A.B., Fürjes, P.: Characterization of the end-of-range geometric effects in complex 3D silicon micro-components formed by proton beam writing. J. Micromech. Microeng. **20**(6), 064015 (2010)
3. Celebi, K., Buchheim, J., Wyss, R.M., Droudian, A., Gasser, P., Shorubalko, I., Kye, J.I., Lee, C., Park, H.G.: Ultimate permeation across atomically thin porous graphene. Science **344**(6181), 289–292 (2014)
4. Min, S.K., Kim, W.Y., Cho, Y., Kim, K.S.: Fast DNA sequencing with a graphene-based nanochannel device. Nat. Nanotechnol. **6**(3), 162–165 (2011)
5. Besharatian, A., Kumar, K., Peterson, R.L., Bernal, L.P., Najafi, K.: A scalable, modular, multi-stage, peristaltic, electrostatic gas micro-pump. In: IEEE 25th International Conference on Micro Electro Mechanical Systems (MEMS), pp. 1001–1004 (2012)
6. Gupta, N.K., An, S., Gianchandani, Y.B.: A Si-micromachined 48-stage Knudsen pump for on-chip vacuum. J. Micromech. Microeng. **22**, 105026 (2012)
7. An, S., Gupta, N.K., Gianchandani, Y.B.: A Si-micromachined 162-stage two-part Knudsen pump for on-chip vacuum. J. Microelectromech. Syst. **23**, 406–416 (2014)
8. Karniadakis, G., Beskok, A., Aluru, N.: Microflows and Nanoflows: Fundamentals and Simulation. Springer Science & Business Media (2005)
9. Sone, Y.: Molecular Gas Dynamics: Theory, Techniques, and Applications. Birkhäuser (2007)
10. Aoki, K., Baranger, C., Hattori, M., Kosuge, S., Martalo, G., Mathiaud, J., Mieussens, L.: Slip boundary conditions for the compressible Navier–Stokes equations. J. Stat. Phys. **169**(5), 744–781 (2017)
11. Tsien, H.S.: Superaerodynamics, mechanics of rarefied gases. J. Aeronaut. Sci. **13**(12), 653–664 (1946)
12. Wagner, W.: A convergence proof for Bird's direct simulation Monte Carlo method for the Boltzmann equation. J. Stat. Phys. **66**, 1011–1044 (1992)
13. Darbandi, M., Roohi, E.: A hybrid DSMC/Navier-Stokes frame to solve mixed rarefied/nonrarefied hypersonic flows over nano-plate and micro-cylinder. Int. J. Numer. Meth. Fluids **72**(9), 937–966 (2013)
14. Mohammadzadeh, A., Roohi, E., Niazmand, H., Stefanov, S., Myong, R.S.: Thermal and second-law analysis of a micro- or nano-cavity using direct-simulation Monte Carlo. Phys. Rev. E **85**, 056305 (2012)
15. Mahdavi, A.M., Roohi, E.: Investigation of cold-to-hot transfer and thermal separation zone through nano step geometries. Phys. Fluids **27**(7), 051507 (2015)
16. Akhlaghi, H., Roohi, E., Stefanov, S.: A comprehensive review on micro- and nano-scale gas flow effects: slip-jump phenomena, Knudsen paradox, thermally-driven flows, and Knudsen pumps. Sci. Rep. **997**, 1–60 (2023)
17. Dongari, N., Sharma, A., Durst, F.: Pressure-driven diffusive gas flows in micro-channels: from the Knudsen to the continuum regimes. Microfluid. Nanofluid. **6**(5), 679–692 (2009)

18. Myong, R.S.: A full analytical solution for the force-driven compressible Poiseuille gas flow based on a nonlinear coupled constitutive relation. Phys. Fluids **23**(1), 012002 (2011)
19. Ewart, T., Perrier, P., Graur, I.A., Méolans, J.G.: Mass flow rate measurements in a microchannel, from hydrodynamic to near free molecular regimes. J. Fluid Mech. **584**, 337–356 (2007)
20. Tatsios, G., Stefanov, S.K., Valougeorgis, D.: Predicting the Knudsen paradox in long capillaries by decomposing the flow into ballistic and collision parts. Phys. Rev. E **91**(6), 061001 (2015)
21. Kandlikar, S., Garimella, S., Li, D., Colin, S., King, M.R.: Heat Transfer and fluid flow in Minichannels and Microchannels. Elsevier (2005)
22. Aoki, K., Sone, Y.: Temperature field induced around a sphere in a uniform flow of a rarefied gas. Phys. Fluids **30**(7), 2286–2288 (1987)

Chapter 2
A Brief Review of Kinetic Theory

Abstract This chapter provides a foundational review of the kinetic theory of gases, highlighting its importance in modeling non-continuum and rarefied gas flows. It begins by contrasting classical fluid mechanics with molecular approaches and outlines the limitations of continuum assumptions in applications such as microscale devices and high-altitude aerospace flows. The historical evolution of matter theory, from ancient philosophy to atomic theory, sets the stage for exploring kinetic and statistical mechanics. The chapter introduces flow regimes based on the Knudsen number and explains the transition from continuum to free molecular flow. Key components, including molecular spacing, binary collisions, intermolecular forces, and potentials, are discussed in detail using models such as the Lennard–Jones, Hard-Sphere, and Variable Hard-Sphere models. The derivation of pressure and viscosity from molecular principles is demonstrated, and statistical distributions such as the Poisson and Gaussian are used to describe molecular number fluctuations. This comprehensive overview prepares readers for advanced topics in non-equilibrium gas dynamics and methods such as the Direct Simulation Monte Carlo (DSMC) technique.

2.1 Introduction

Classical fluid mechanics considers a gas as a substance composed of infinite components. As a result, assuming a continuous environment, each fluid characteristic can be calculated as a certain (limited) value at any point in space. Density, temperature, velocity, and other gas properties are all considered as a function of space and time. For most common processes in everyday life under standard conditions, i.e., a pressure of one atmosphere, a temperature of approximately 298 Kelvin, and length dimensions ranging from millimeters to meters, the assumption of a continuous environment is valid. However, some other engineering applications, such as space flight or microdevices like micro- and nano-electromechanical systems (MEMS/NEMS), occur in situations where the hypothesis of continuity is not true, and therefore, there is a need to examine the molecular structure of gases more closely.

E. Roohi et al., *Advances in Direct Simulation Monte Carlo: From Micro-Scale to Rarefied Flow Phenomena*, https://doi.org/10.1007/978-981-96-8200-3_2

The current chapter begins with an introduction to matter molecules and especially gases. Then, by introducing the kinetic theory of gases, the pressure is calculated on a molecular level based on the speed of the molecules. Different intermolecular forces and models are investigated, and then the dynamics of intermolecular double collisions are studied. At the end of the chapter, the viscosity of gases is derived based on molecular theories.

2.2 Basic Classic Theories of Matter and Gas

Ancient Greek philosophers believed that matter consists of four main elements: water, earth, air, and fire. Therefore, Aristotle (322–384 BC) believed that all elements include the four characteristics of heat, cold, moisture, and dryness. For example, fire is dry and hot; water is cold and wet; The air is hot and humid, and the earth is cold and dry. Additionally, because these properties were mutable, it was possible to transform copper into gold (alchemically). Due to the focus of this book on gases, further discussions are presented in the field of this state of matter.

Robert Boyle (1627–1691) proposed a theory about air known as the repulsion theory. He stated that the air consists of moving particles that repel each other when they collide. In other words, according to Boyle's Law, the product of pressure and volume of gases is constant.

$$PV = Const.$$

Boyle's hypothesis established the foundation of the kinetic theory of gases. Later, Daniel Bernoulli (1700–1782) disproved Boyle's law of gas pressure by demonstrating that pressure is directly proportional to the kinetic energy of the particles.

Joseph Louis Gay-Lussac (1778–1850) stated that the composition of gaseous substances is always a simple volume ratio. Figure 2.1 shows an example of the composition of gases.

Amedeo Avogadro (1776–1856) stated that equal volumes of gas at equal pressure and temperature contain the same number of molecules. In other words:

$$1\ \text{mol} = 6.022 \times 10^{23}\ \text{molecules}$$

However, this fundamental idea was not accepted by scientists until the 1860s. The atomic theory of John Dalton (1766–1844) was stated as follows:

- All materials are made up of very small particles called atoms.
- All atoms of an element are the same and have the same size, mass, and chemical properties.
- Atoms cannot be created, destroyed, or divided into smaller particles.

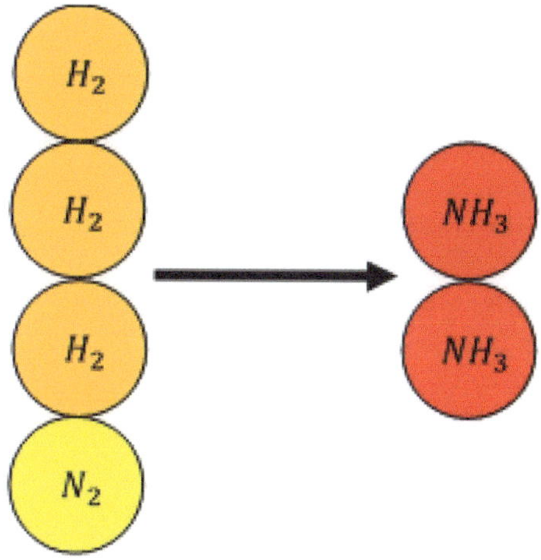

Fig. 2.1 Real examples of different length scale units

- Macroscopic properties, such as density, pressure, and temperature, are defined as averages.
- Different atoms combine in correct numerical ratios and form a new compound (molecule).
- In a chemical reaction, atoms are separated from each other, combined, or their arrangement changes.

2.2.1 Statistical Mechanics and Kinetic Theory

Classical gas dynamics is a macroscopic science that deals with mass and does not deal with microscopic properties. Additionally, fluid mechanics is one of the macroscopic sciences that studies gases and liquids, assuming them as a continuous environment. By applying the principle of conservation of mass, energy, and momentum to a certain mass, all macroscopic properties are assumed and determined.

Although the macroscopic view is suitable for many engineering applications, it cannot accurately predict the physical behavior in certain special conditions where the continuous environmental condition is not met. In branches such as rarefied gases, supersonic flows at high altitudes, multiphase flows, and combustion under special conditions, the assumption of continuity is invalid, and modeling must be conducted at the molecular level.

There are two completely different theories for the molecular behavior of fluids:

- Statistical mechanics
- Kinetic theory

Statistical mechanics is primarily used in analyzing systems in equilibrium, and this theory does not explain the behavior of molecules during the flow and collisions of molecules. Additionally, a series of mathematical relationships is used to determine all possible microscopic structures that contribute to the formation of a system. Then, considering the random movement of the molecules, it is assumed that the system can pass through all possible microscopic states, and each state is in equilibrium. Statistical estimations from the equilibrium state of the system determine the macroscopic properties of the system.

Unlike statistical mechanics, kinetic theory models the movement and collision of molecules using the laws of mechanics and probabilities. Macroscopic properties are also determined by averaging and integrating over the entire range of molecular behavior. Kinetic theory can be applied to both equilibrium and non-equilibrium systems. Additionally, this theory is currently the only available method for analyzing certain non-equilibrium systems. Now the question arises, when does the speed of molecules change? In response, it should be noted that when two molecules with different kinetic energies (speeds) collide, they can exchange their momentum.

2.2.2 Kinetic Theory of Gases

The goal of the kinetic theory [1, 2] describe the relationship between the macroscopic properties of gases (such as pressure, temperature, and velocity) and their microscopic properties (gas mass and molecular speed) as a means to express these properties and their relations. The basic principles of the kinetic theory of gases are as follows:

- A molecule is the smallest component of a chemical substance that still carries the chemical properties of that substance.
- Gases are composed of molecules that are constantly moving randomly.
- Molecules exert force on each other only when they collide, as in an ideal gas.
- The average distance of gas molecules is much larger than their size (dilute gas). As a result, only pairwise collisions of molecules are possible.
- Classical mechanics usually describes intermolecular interaction. In other words, the effects of quantum mechanics and relativity are negligible.

2.2.3 Statistical Origin of Kinetic Theory of Gases

Since it is impossible to follow each molecule and calculate its exact movement path, the statistical description of the behavior of molecules is a logical process. Macroscopic properties, such as density, pressure, and temperature, are defined as the average properties of molecules. By definition, a small macroscopic volume is the smallest volume that satisfies the condition of a continuous medium. To establish the condition of gas continuity, microscopic fluctuations of molecules should not cause significant changes in the average gas properties. Therefore, for the gas to be continuum, the two conditions mentioned in 1.4.2 (which refer to point properties and transport properties) must be met.

The macroscopic analysis examines the flow through macroscopic properties, including velocity, temperature, and density. On the other hand, the analysis of molecular gas dynamics uses the statistical description of the speed distribution functions of molecules. The molecular speed distribution function expresses the changes in the dimensionless number of molecules as a function of molecular speed. The governing equation for the distribution function change is the Boltzmann equation, which will be discussed in detail in Chap. 3 of this book. Macroscopic properties are obtained by averaging (moment calculation) from the velocity distribution functions. The different averaging definitions could be employed as follows:

- Instantaneous average: An average over a small macroscopic volume at a single instant of time.
- Time average: It is obtained from the total properties of molecules inside a cell over a given period.
- Ensemble (cumulative) average: an instantaneous average over the molecules in the desired volume elements for many process trajectories.

2.2.4 *Flow Regimes*

As previously indicated in 1.6, based on the value of the Knudsen number (the ratio of the molecular mean free path to the flow length scale), the flow is classified in different regimes.

Case 1 ($\lambda \ll L$): If a room is considered in a continuous environment, it is clear that the Knudsen number is very small. In this case, the average distance traveled between two molecular collisions is about 64 nm. Therefore, the information a molecule obtains from the wall is transferred to other molecules very quickly, and through these collisions, the molecules approach the equilibrium state rapidly. Since there is no stress or heat transfer in Euler's equation, it can be claimed that Euler's equation corresponds to Kn equal to zero. The limit to continuum conditions is where the length of molecular free motion and the time between two consecutive collisions tend to zero.

Case 2: If the Knudsen number increases slightly and reaches a value less than 0.001, in this case, due to the continuity of the flow, the Navier–Stokes equations can still be used to analyze the flow.

Case 3: The Knudsen number increases slightly and approaches 0.1, in the range of $0.001 < \text{Kn} < 0.1$, considering that the distance between the molecules is significant, the analysis of the flow using the Navier–Stokes equations in the vicinity of the walls is inaccurate. Actually, the interaction of gas molecules with walls leads to the non-equilibrium flow in the vicinity of the walls. Therefore, Navier–Stokes equations should be used with corrected wall conditions, considering the non-equilibrium effects, i.e., using the conditions of velocity slip and temperature jump at the boundaries.

Case 4: For the Knudsen numbers in the range of $0.1 < \text{Kn} < 10$, the number of intermolecular collisions is neither too large to consider the flow in continuum conditions, nor too small to ignore them. This regime is referred to as the "transition regime." For this reason, simulation in the transition regime is relatively complicated. In fact, with the increase of the Knudsen number, the rarefaction effects become apparent throughout the entire solution domain, and the continuous models lose their effectiveness in accurately modeling the flow. In this situation, the DSMC method is one of the candidate approaches, rather than the classical continuum ones.

Case 5 ($\lambda \gg L$): If the Knudsen number becomes very large, the distance between molecules becomes so large that they usually do not collide with each other; in other words, the number of intermolecular collisions is significantly reduced compared to collisions with the surface. In this case, the Boltzmann equation becomes simpler because it enables the skipping of the collision term. In this situation, some analytical solutions exist that can be used to check the flow. The flow regime is referred to as free molecular, where the collision of particles with the object's surface plays a significant role in determining the behavior of the gas.

2.2.5 *Pressure in Stationary Gas*

According to Fig. 2.2, a cube of length L (assuming that $L < \lambda$, consequently there are more molecule-wall collisions than intermolecular collisions) is considered. Let us consider a molecule that hits a wall within a small time interval dt without interacting with other molecules.

From the classical definition, pressure, as the force per unit area, reads:

$$F_x = \frac{\Delta P_x}{2dt}, P_x = \frac{F_x}{A} = \frac{\Delta P_x}{2dtA}, \tag{2.1}$$

Fig. 2.2 Schematic of the assumed 2D cube

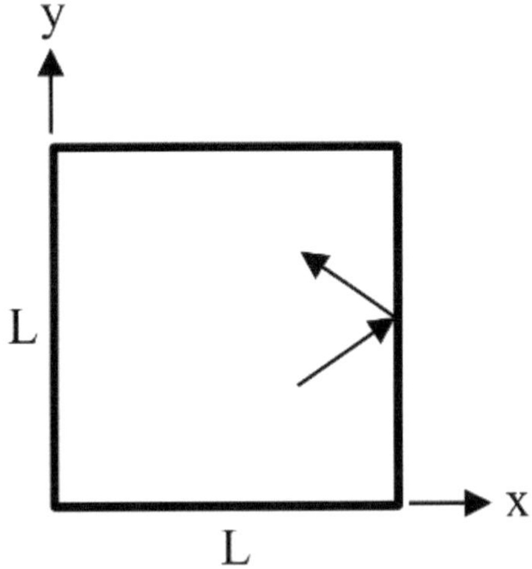

where dt is the traveling time for distance L. The value $2dt$ refers to the time it takes for the molecule to hit the wall and return to its original location. Now, if a molecule with speed c hits a specular surface (a surface where the vertical speed of the molecule is reversed after hitting it and the tangential speed of the molecule does not change) and returns to its original position, the change in the momentum of that molecule is $2mc_n$, where m is the mass of the molecule. and c_n is the component of velocity perpendicular to the surface, which is $c_n = c_x$ for Fig. 2.2. In other words:

$$\Delta P_x = m\vec{c}_x - (-m\vec{c}_x) = 2m\vec{c}_x \tag{2.2}$$

In a time interval dt, all the molecules with the speed c_n that are located at the distance $c_n dt$, hit this wall. Now, the momentum change of all molecules inside the control volume is obtained from the following relationship:

$$\Delta P_x = \text{No. of molecules inside the CV} \times \text{momentum exchange for one molecule}$$

On the other hand:

$$\text{No. of molecules} = \text{number density} \times \text{swept volume} = (nc_x dtds)$$

Therefore, the net rate of change of momentum perpendicular to the surface ds for molecules with speed c_n is equal to:

$$\Delta P_x = \frac{(n_c c_n dtds)2mc_n}{2dt}, \tag{2.3}$$

where n_c is the number density of molecules that have speed c_n. Finally, according to the pressure formula, we have:

$$P_x = \frac{F_x}{A} = \frac{\Delta P_x}{2dtA} = \frac{(c_x dtds) \times n \times (2mc_x)}{2dtds} = nm\bar{c}_x^2 = \rho\bar{c}_x^2 \tag{2.4}$$

Since $\rho = nm$ and considering three molecular velocity components:

$$\begin{cases} P_x = nm\overline{c_x}^2 \\ P_y = nm\overline{c_y}^2 \\ P_z = nm\overline{c_z}^2 \end{cases}$$

Assuming isotropic flow:

$$c_x = c_y = c_z$$

$$c^2 = c_x{}^2 + c_y{}^2 + c_z{}^2$$

Therefore:

$$P = \frac{1}{3}\left(P_x + P_y + P_z\right) = \frac{nm}{3}c^2$$

Generally, the pressure is equal to:

$$P = \frac{n}{3}\overline{mc_n^2\vec{c}}\,,\, c_n > 0 \tag{2.5}$$

The over-bar symbol indicates the average value. In equilibrium conditions, all directions of velocity have the same probability of occurrence. Hence:

$$\overline{c_n^2} = \overline{c_x^2} = \overline{c_y^2} = \overline{c_z^2} = \frac{\overline{c^2}}{3} \tag{2.6}$$

Considering $c^2 = {c_x}^2 + {c_y}^2 + {c_z}^2$:

$$P = \frac{n\overline{mc^2}}{3} = \frac{\overline{\rho c^2}}{3} \tag{2.7}$$

Therefore, the pressure is proportional to the molecular average kinetic energy. In summary, the proof of the pressure relationship can be expressed as follows.

According to the momentum formula, i.e. force × time = mass × velocity, and considering Fig. 2.2, the force in the x-axis for each molecule is defined as

$$F_x = \frac{2mc_x}{\left(\frac{2L}{c_x}\right)} = \frac{mc_x^2}{L} \tag{2.8}$$

Equation (2.8) is written for one molecule, and $2L/c_x$ represents the time it takes for a molecule traveling with velocity c_x to hit the wall and return to its original location. Therefore, this force for the number of N molecules can be expressed by the following relation:

$$F_x = \frac{\sum_N mc_x^2}{L} \tag{2.9}$$

Now, having the force and using the definition, the pressure can be calculated from the following relationship:

$$P_x = \frac{F_x}{A} = \frac{F_x}{L^2} = \frac{\sum_N mc_x^2}{L^3} = \frac{\sum_N mc_x^2}{V}, \tag{2.10}$$

where V is the volume. Then, subsequently, pressures in the y and z directions are similarly obtained from the following relations:

$$P_y = \frac{F_y}{A} = \frac{F_y}{L^2} = \frac{\sum_N mc_y^2}{L^3} = \frac{\sum_N mc_y^2}{V} \tag{2.11}$$

$$P_z = \frac{F_z}{A} = \frac{F_z}{L^2} = \frac{\sum_N mc_z^2}{L^3} = \frac{\sum_N mc_z^2}{V} \tag{2.12}$$

The total (or mechanical) pressure is equal to the average value of the pressures in all directions:

$$P = \frac{1}{3}\left(P_x + P_y + P_z\right) = \frac{\sum_N mc^2}{3V} = \frac{\overline{\rho c^2}}{3} \tag{2.13}$$

It is the same as that obtained in (2.7).

2.2.6 *Molecular Velocity*

The molecular speed (c) is the average speed of the molecules of a system ($\overline{c}$) plus the thermal (or random) speed of each molecule (c'). In other words, $c = \overline{c} + c'$. Each of the mentioned components is calculated as follows:

$$\overrightarrow{c_0} = \left(\overline{c_x}, \overline{c_y}, \overline{c_z}\right), \overline{c_x} = \frac{\sum_{i=1}^{N} c_{x,i}}{N} \tag{2.14}$$

$$\overrightarrow{c'} = \vec{c} - \overrightarrow{c_0}, \tag{2.15}$$

where N is the number of molecules in the field. Two important results come from the above relationships:

- In a gas at rest, due to the fact that there is no flow in it, the average speed of its molecules, which is also the speed of the flow, is equal to zero, i.e. $\overrightarrow{\overline{c}} = 0$.
- The average thermal speed of molecules is always zero; because these speeds are completely random and in different directions.

$$\overline{c'_x} = \overline{c'_y} = \overline{c'_z} = 0 \tag{2.16}$$

In this case, the pressure of a moving gas is defined as the normal momentum exchange rate of the gas to a surface that is moving at the speed of the flow.

$$P = \frac{\overline{nm(c - c_0)^2}}{3} = \frac{\overline{nmc'^2}}{3} \tag{2.17}$$

Equation (2.17) indicates that pressure is directly proportional to the kinetic energy resulting from the thermal molecular speeds.

Now, we consider the molecular translational energy as follows

$$E_{tr} = \frac{\sum_N mc^2}{2} \tag{2.18}$$

Therefore, the pressure could be expressed as

$$P = \frac{2}{3}\frac{E_{tr}}{V} = \frac{2}{3}\rho e_{tr} \tag{2.19}$$

$e_{tr} = E_{tr}/M$ is the translational energy per unit mass. According to Eq. (2.19) and considering ideal gas relation, i.e. $p = \rho RT$, the temperature could be obtained as

$$T = \frac{\frac{2}{3}\rho e_{tr}}{\rho R} = \frac{2}{3}\frac{\hat{e}_{tr}}{k}, \tag{2.20}$$

where $M = m \times N$ is the total mass, $R = k/m$ is the gas constant, k is the Boltzmann constant, and $\hat{e}_{tr} = \frac{E_{tr}}{N}$. Therefore, transfer energy per unit mass and transfer energy per unit number of molecules can be calculated as:

$$e_{tr} = \frac{3}{2}RT = \frac{3}{2}kT/m \tag{2.21}$$

$$\hat{e}_{tr} = \frac{3}{2}kT \tag{2.22}$$

2.2.7 Molecular Energies

In general, for a gas with a value of ξ as a number of degrees of freedom, the kinetic energy is obtained as:

$$e_{tr} = \xi_{tr}\left(\frac{RT}{2}\right) \tag{2.23}$$

Since the monoatomic gases have three degrees of freedom related to translational movement in three directions of coordinates, $\xi_{tr} = 3$. Diatomic molecules are more complex and, as seen in Fig. 2.3, they are usually dumbbell-shaped. These molecules have three degrees of freedom related to translational motion, two degrees of freedom related to rotational motion, and two degrees of freedom related to vibrational motion. In general, the specific heat capacity of constant volume is defined as the following relationship:

Fig. 2.3 Structure of monoatomic, diatomic, and polyatomic molecules

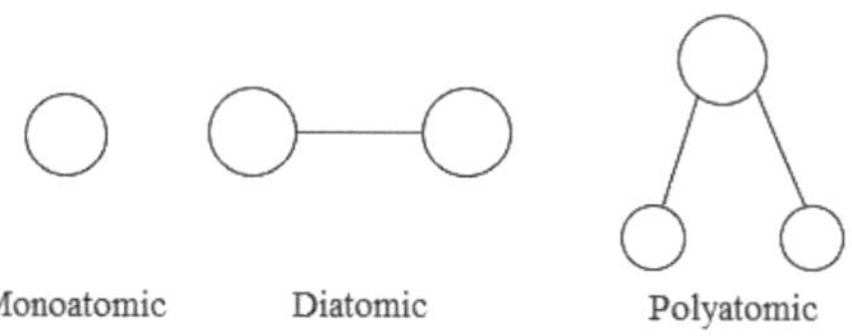

$$c_v = \left(\frac{\partial e}{\partial T}\right)_v = (\xi_{tr} + \xi_{rot} + \xi_{vib})\frac{R}{2} \tag{2.24}$$

As mentioned earlier, monoatomic molecules have only three degrees of freedom related to the translational motion, and therefore they have $\xi_{rot} = 0$, $\xi_{vib} = 0$ and $\xi_{tr} = 3$, and as a result, the specific heat capacity of constant volume will be equal to $c_v = \frac{3}{2}R$. This result is in good agreement with the results of laboratory measurements.

Only three degrees of freedom are active for diatomic molecules at temperatures up to 50 K. Therefore, the specific heat capacity of constant volume for them, like monoatomic molecules, is equal to $\frac{3}{2}R$. But as the temperature exceeds 50 K, their two rotational degrees of freedom are also activated $\xi_{rot} = 2$, and as a result, $c_v = \frac{5}{2}R$. In addition, in diatomic molecules at very high temperatures, the vibrational degrees of freedom are also activated and $\xi_{vib} = 2$, therefore, c_v reaches $\frac{7}{2}R$. All these results are in good agreement with the results obtained from the laboratory measurements shown for hydrogen (Fig. 2.4).

It can be seen that statistical mechanics provides us with accurate results regarding the specific heat capacity at constant volume. But for polyatomic gases in a certain temperature range, $\xi_{vib} = 3$ and $\xi_{rot} = 3$, the existence of vibrational degrees of freedom complicates the calculation of $\frac{c_v}{R}$. In general, the more complex structure of the molecule along with the increase in its degrees of freedom causes an increase in $\frac{c_v}{R}$.

Fig. 2.4 The measured values of $\frac{c_v}{R}$ for Hydrogen

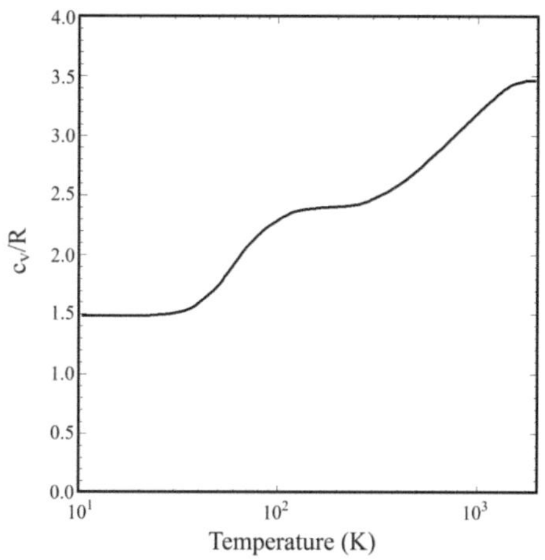

Also, this theory is a good approximation of the specific heat capacity ratio $\gamma = \frac{c_p}{c_v}$. Considering that for an ideal gas $c_p - c_v = R$, therefore:

$$\gamma = \frac{c_p}{c_v} = \frac{\xi + 2}{\xi} \tag{2.25}$$

In this theory, $\gamma = \frac{5}{3}$ for monoatomic gases, which is in good agreement with the results of laboratory measurements. For diatomic gases in a certain temperature range, $\gamma = \frac{7}{5}$ which decreases to $\gamma = \frac{7}{9}$ at higher temperatures (for example, the air is mainly a mixture of two diatomic gases for which $\gamma = \frac{7}{5}$). For polyatomic gases, the values of γ depend on the number of active vibrational degrees of freedom of the gas.

2.2.8 *Poisson Distribution Function*

Number density refers to the number of molecules present in a unit volume. This quantity can be obtained from the ideal gas law. For example, the numerical density of air at a pressure of 101.3 kPa and a temperature of 288 K is equal to $2.7 \times 10^{25}\ \text{m}^{-3}$. Now, consider a volume element of one cubic unit inside a container or a stream. The number of molecules in this volume changes with time at any moment. Because at each moment, a number of molecules enter or leave this volume, the numerical density is the average value of a random variable such as X and is equal to the number of molecules per unit volume. Next, we will calculate the X distribution function.

Suppose N molecules are uniformly distributed inside the volume V_N. Now, we divide the volume of V_N into cells with equal volumes and the same size as V. In this case, due to the uniform distribution of molecules, the probability of having one molecule in volume V is equal to V/V_N and the average number of molecules in each cell in volume V is equal to

$$N\frac{V}{V_N} = nV \tag{2.26}$$

Now, it is possible that there is a group of k molecules in a cell of volume V, and N–k molecules in the rest of the cells are as follows:

$$\left(\frac{V}{V_N}\right)^k \left(1 - \frac{V}{V_N}\right)^{N-k} \tag{2.27}$$

where, $\left(\frac{V}{V_N}\right)^k$ indicates the chance of success and $\left(1 - \frac{V}{V_N}\right)^{N-k}$ shows the chance of failure. Considering the combination of k molecules from N molecules which is represented by C_N^k and is defined as

$$C_N^k = \frac{N!}{k!(N-k)!} \tag{2.28}$$

So, if N molecules are uniformly distributed in the volume V_N, the probability of existing exactly k molecules in a cell of volume V can be calculated as follows

$$P(X = k) = \frac{N!}{k!(N-k)!}\left(\frac{V}{V_N}\right)^k\left(1 - \frac{V}{V_N}\right)^{N-k} \tag{2.29}$$

In the limiting condition $N \gg 1$ and $\frac{V}{V_N} \ll 1$ (indicating a very large number of molecules and very small cells):

$$\frac{N!}{(N-k)!} = N \times (N-1)\ldots(N-k+1) \approx N^k \tag{2.30}$$

and

$$\ln\left(1 - \frac{V}{V_N}\right)^{N-k} = (N-k)\ln\left(1 - \frac{V}{V_N}\right) \approx (N-k)\left(-\frac{V}{V_N}\right) \approx -N\frac{V}{V_N} = -nV \tag{2.31}$$

Considering Taylor series expansion for $\ln(1+x)$ and $x \to 0$

$$\mathrm{Ln}(1+x) = \sum_{j=0}^{\infty}(-1)^j\frac{x^{j+1}}{j+1} = x - \frac{x^2}{2} + \frac{x^3}{3} - \cdots \approx x \tag{2.32}$$

therefore

$$P(X = k) = \frac{(nV)^k e^{-nV}}{k!} \tag{2.33}$$

This discrete distribution is known as the Poisson distribution [3]. In general, phenomena follow the Poisson distribution, and their probability of success is relatively infrequent. In dilute gases, since the collision probability is low, the particle collision probability and the time between particle collisions follow the Poisson distribution. The average value of parameter X is as

$$\begin{aligned}\bar{X} &= \sum_{k=0}^{\infty} kP(X=k) = \sum_{k=1}^{\infty} kP(X=k) = \sum_{k=1}^{\infty}\frac{k(nV)^k e^{-nV}}{k!} \\ &= e^{-nV} \times nV\sum_{k=0}^{\infty}\frac{(nV)^k}{k!} = nVe^{-nV}e^{nV} = nV\end{aligned} \tag{2.34}$$

According to the standard deviation definition

$$\sigma(X) = \left(\overline{X^2} - (\bar{X})^2\right)^{1/2} \tag{2.35}$$

and

$$\begin{aligned}
\overline{X^2} &= \sum_{k=0}^{\infty} k^2 P(X = k) = \sum_{k=1}^{\infty} k^2 P(X = k) = \sum_{k=1}^{\infty} \frac{k^2 (nV)^k e^{-nV}}{k!} \\
&= nVe^{-nV} \sum_{k=0}^{\infty} \frac{(k+1)(nV)^k}{k!} = nVe^{-nV} \left(\sum_{k=0}^{\infty} \frac{k(nV)^k}{k!} + \sum_{k=0}^{\infty} \frac{(nV)^k}{k!} \right) \\
&= nVe^{-nV} \left(nV \sum_{k=0}^{\infty} \frac{(nV)^k}{k!} + \sum_{k=0}^{\infty} \frac{(nV)^k}{k!} \right) = nVe^{-nV} \left(nVe^{nV} + e^{nV} \right) = (nV)^2 + nV
\end{aligned} \tag{2.36}$$

Therefore, the standard deviation is as

$$\sigma(X) = \left(\overline{X^2} - (\bar{X})^2\right)^{1/2} = \sqrt{(nV)^2 + nV - (nV)^2} = \sqrt{nV} \tag{2.37}$$

Finally, if nV itself is large and we are only interested in the values of X that are close to nV (the probability of the rest of the values is specifically zero), the expression $P(X = k)$ can be simplified using Stirling's formula. Sterling's estimate for $k!$ It is as follows

$$k! = \sqrt{2\pi k} \left(\frac{k}{e} \right)^k \textit{ for large } k \tag{2.38}$$

Therefore,

$$\begin{aligned}
\ln P(X = k) &= \ln \left(\frac{(nV)^k e^{-nV}}{k!} \right) \\
&= k \ln(nV) - nV - \frac{1}{2} \ln(2\pi k) - k \ln k + k \\
&= -k \ln \left(\frac{k}{nV} \right) - \frac{1}{2} \ln(2\pi k) + k - nV
\end{aligned} \tag{2.39}$$

Employing Taylor series expansion for small values of $\frac{k-nV}{nV}$ and considering the smallest powers of k–nV

$$\begin{aligned}-k\ln\left(\frac{k}{nV}\right) &= -k\ln\left(1+\frac{k-nV}{nV}\right) = -k\left(\frac{k-nV}{nV}-\frac{1}{2}\left(\frac{k-nV}{nV}\right)^2+\ldots\right)\\ &= -(k-nV)\left(\frac{k-nV}{nV}-\frac{1}{2}\left(\frac{k-nV}{nV}\right)^2+\ldots\right)\\ &\quad -nV\left(\frac{k-nV}{nV}-\frac{1}{2}\left(\frac{k-nV}{nV}\right)^2+\ldots\right)\approx-\frac{1}{2}\frac{(k-nV)^2}{nV}-(k-nV)\end{aligned} \tag{2.40}$$

consequently

$$lnP(X=k) = -\frac{1}{2}\frac{(k-nV)^2}{nV}-\frac{1}{2}\ln(2\pi k) \tag{2.41}$$

and finally

$$P(X=k) = \frac{1}{\sqrt{2\pi k}}e^{\frac{-(k-nV)^2}{2nV}} \tag{2.42}$$

Equation (2.42) is actually is the Gaussian distribution function.

For standard temperature and pressure, i.e. 1 atm, 273 K, the air numerical density is equal to 2.7×10^{25}. Therefore, on average, there are only 27 molecules in a cube whose length is 10 nm. Additionally, the standard deviation of the number of molecules is approximately 20% of the average value, or 2.5. Fluctuations in the number of molecules around the average value can be observed at nano dimensions and at very low densities, corresponding to very high altitudes. Additionally, Fig. 2.5 presents the results obtained from the distribution of the number of particles in a cell via direct simulation of Monte Carlo (DSMC) [4], along with a comparison to the Poisson distribution function. As shown in Fig. 2.5, for a problem with 1000 computational simulator particles and 100 cells, there will be an average of 10 particles per cell. Therefore, as the numerical solution also shows, the probability of having 10 particles in the cell is the highest. Additionally, these results have been compared with the analytical solution, specifically the Poisson distribution.

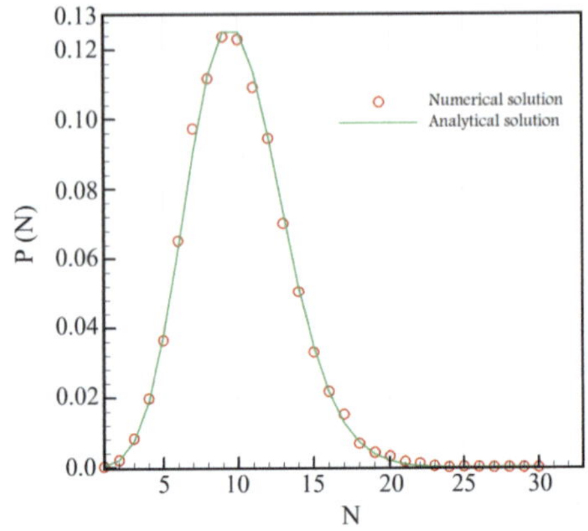

Fig. 2.5 The distribution diagram of the number of particles inside a cell, with an average of 10 particles, and a comparison with the analytical solution

2.2.9 *Dilute Gas and Rarefied Gas*

In molecular gas dynamics, two terms are used to describe a dilute gas, and these terms differ significantly in their physical properties. Dilute gas is a gas in which the distance between gas molecules (δ) is at least seven times the molecular diameter of that gas. In other words, $\frac{\delta}{d} \geq 7$. Therefore, according to this definition, the majority of common gases, such as air, are dilute gases. One of the main characteristics of dilute gas is that because most of the volume of the gas is empty space, intermolecular collisions in these gases are assumed to be binary.

On the other hand, a rarefied gas flow is defined as a flow that is not in the continuum regime and is characterized by a Knudsen number (Kn) of at least 0.001. In other words, non-equilibrium effects are significant in these gases, and the assumption of a continuous environment is not applicable to them. As a result, the Navier–Stokes equations are not sufficiently accurate for analyzing such flows. Therefore, a gas can be diluted; however, due to the small Knudsen number ($\text{Kn} \leq 0.001$), it can be considered a continuum and analyzed using the Navier–Stokes equations. For example, air is a dilute gas under standard atmospheric conditions, but it does not exhibit rarefaction dilution behavior until it enters into a microchannel. The definition of a dilute gas is contrasted with that of a dense gas.

A similar situation exists in classical fluid mechanics for compressible fluid and compressible flow. That is, a gas like air, which is an inherently compressible fluid, exhibits incompressible flow behavior in flows with a Mach number less than 0.3, where the Mach number is the ratio of the flow speed to the speed of sound. That is, the effect of compressibility is insignificant, whereas at Mach numbers greater than 0.3, density changes within the flow become significant, and as a result, the flow should be considered compressible.

2.2.10 Intermolecular Force

The intermolecular force is usually a very strong force, but it is effective over a very short distance. This force includes repulsive force and a very small attractive force. For simple molecules, the force field is symmetric and spherical; however, for more complex molecules, the force field becomes asymmetric. For this reason, the kinetic theory for liquids is very complicated; because in liquids, the molecules are vibrating freely and are very close to each other and are always under the influence of the force fields of the molecules around them. In gases, molecules have more space, but assuming the existence of a force field that extends to infinity, they are always under the influence of the force field. It is often acceptable to ignore intermolecular forces in gases, except during collisions. The closer the behavior of a gas is to that of an ideal gas, the more acceptable this assumption becomes. However, as the density increases, the behavior of the gas deviates from the ideal gas law, and the intermolecular forces become more significant.

2.2.11 Intermolecular Potential

Intermolecular potential characterizes the forces of attraction and repulsion of particles during collision. The intermolecular potential is divided into two categories: long-range and short-range forces. Short-range forces are created when two particles are close enough to each other that their electron clouds interact. In this situation, a strong repulsive force pushes the two particles apart. On the other hand, the long-range forces that cause particle attraction are electrostatic, induction, and dispersion forces. Figure 2.6 shows the types of long-range forces between two particles.

The electrostatic force is created due to the interaction between two charged particles, such as the interaction between ions and electrons. The model used for this force is the Coulomb potential model.

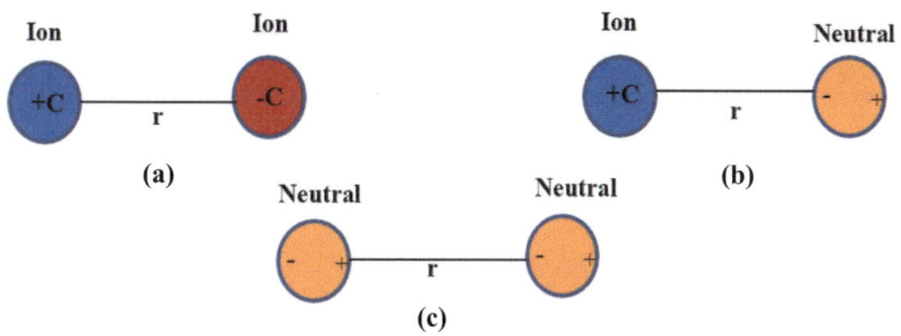

Fig. 2.6 Long-range forces causing intermolecular attraction **a** Electrostatic **b** Induced **c** Dispersion

$$\phi(r) = k_e \frac{(+C)(-C)}{r} \tag{2.43}$$

In this formula, $+C$ and $-C$ represent positive and negative charges, respectively, and Coulomb's constant k_e. In permanent dipole or quadrupole molecules, there is also an electrostatic force, but with a shorter range. In this condition, the power of the potential force in terms of the intermolecular distance is proportional to r^{-3} and r^{-5}, respectively.

The induced force is created by the interaction of a charged particle with a neutral particle. The charged particle creates a polarity in the neutral particle due to the force of attraction between the two. This potential energy depends on the polarity of the neutral particle (α), the amount of charge of the charged particle (C), and the distance between these two particles (r). Permanent dipole molecules can also cause polarity in the neutral particle, in which case the power of the potential relationship will be r^{-6}.

$$\phi(r) = -\frac{C^2\alpha}{2r^{-4}} \tag{2.44}$$

The last type of long-range force is the dispersion force, also known as dipole–dipole induction. This force occurs when two neutral particles come into contact, and a dipole moment is created between them. This effect depends on the polarity and characteristic energy of two particles, and the potential energy in this case is as follows. Also, in the conditions of quadrupole induction and quadrupole induction, the dependence of the potential is r^{-8} and r^{-10}, respectively.

$$\phi(r) = -\frac{3}{2}\left(\frac{h\nu_1 h\nu_2}{h\nu_1 + h\nu_2}\right)\frac{\alpha_1\alpha_2}{r^6}, \tag{2.45}$$

where h is Planck's constant. The terms ν_1 and ν_2 represent the characteristic absorption frequencies of the two interacting neutral particles or atoms. Specifically:

ν_1: characteristic frequency of particle 1
ν_2: characteristic frequency of particle 2

These frequencies correspond to the dominant electronic transition energies of the particles. In general, the most famous models for intermolecular forces are the Lennard–Jones model, the Hard-Sphere model, the Inverse Power Law model, and the Variable Hard-Sphere model.

Lennard–Jones (LJ) Model:

An almost exact model to express potential and intermolecular forces, known as the Lennard–Jones potential, was introduced in Chap. 1.

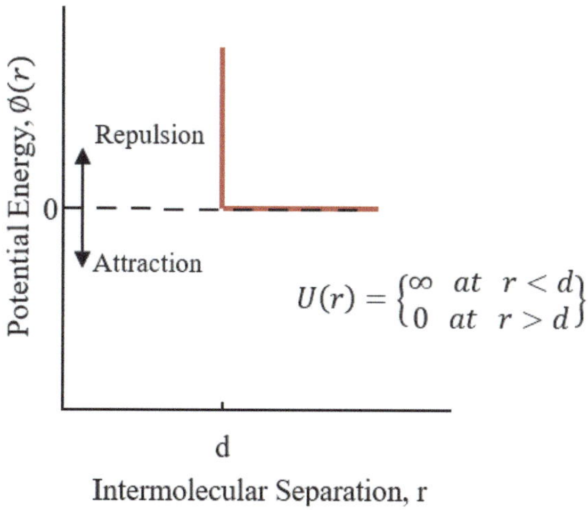

Fig. 2.7 Potential distribution for HS model versus intermolecular distance

Hard-Sphere (HS) Model:

The simplest existing molecular model is the Hard-Sphere model, whose potential is shown in Fig. 2.7. This model treats the molecule as a sphere with a diameter of d and a radius of r_0. In the intermolecular distance $r < d$, this model considers the repulsive force to be infinite and for $r > d$, the intermolecular attractive force is equal to zero. The HS model predicts a power coefficient of 0.5 for the viscosity-temperature relationship, which is not experimentally correct. Therefore, the application of this model is limited to gas flow with very small temperature changes.

Inverse Power-Law (IPL) Model:

In this model, the attraction force is completely ignored and the intermolecular potential and force are defined as the following relationships:

$$U = \frac{k}{(\eta - 1)r^{\eta-1}} \tag{2.46}$$

$$F = \frac{k}{r^{\eta}} \tag{2.47}$$

The coefficient k is a constant, and η is the power of the molecule's hardness. Since the integral of the propagation angle for this model (which will be described later) can be calculated analytically, applying this model in numerical codes is relatively straightforward. Note that if η tends to infinity, this model becomes similar to the HS model. However, usually, for real monoatomic gases, η is equal to 10. Figure 2.8 shows this model.

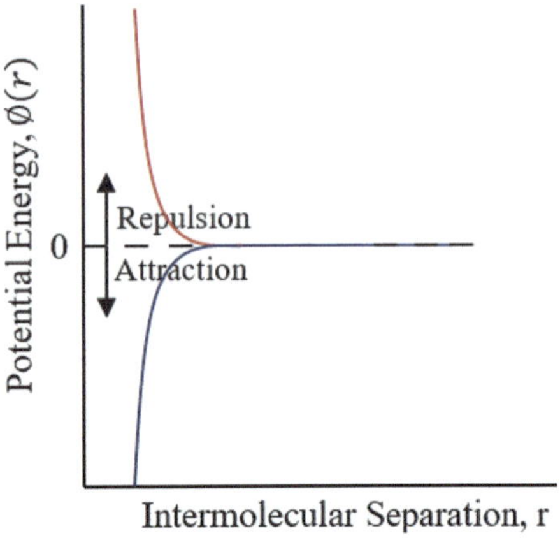

Fig. 2.8 Potential distribution for IPL model versus intermolecular distance

Variable Hard-Sphere (VHS) Model [4]:

In this model, the diameter of the molecule is d. The molecular diameter is not assumed to be constant but is a function of the relative speed of the molecule. For molecular diameter and cross-sectional area (σ_T) in this model, we have:

$$d = d_{ref}\left(\frac{c_{r,ref}}{c_r}\right)^{\alpha} \tag{2.48}$$

$$\sigma_T = \sigma_{ref}\left(\frac{c_r^2}{c_{r,ref}^2}\right)^{-\alpha}, \tag{2.49}$$

the subscript r means relative quantity and α is constant. Since the reference state (ref) is equal to the equilibrium state, its velocity value can be obtained from the following relation

$$c_{r,ref}^2 = \frac{2(2-\alpha)kT}{m_r}, \tag{2.50}$$

where, k is Boltzmann's constant and m_r is the reduced mass, which is defined later. Therefore, the collision cross-sectional area can be calculated as

$$\sigma_T = \sigma_{ref}\left(\frac{m_r c_r^2}{2(2-\alpha)kT_{ref}}\right)^{-\alpha}, \tag{2.51}$$

In Table 2.1, α values for different gases are presented. Figure 2.9 shows the potential distribution of the variable hard sphere model. In this figure, T_1 represents the reference condition and T_2 represents the actual gas condition.

Table 2.1 Conventional length scale units

Gas	α	$d_{ref} \times 10^{10}$ m
Hydrogen (H_2)	0.17	2.92
Helium (He)	0.16	2.33
Nitrogen (N_2)	0.24	4.17
Oxygen (O_2)	0.27	4.07
Argon (Ar)	0.31	4.17
Carbon Dioxide (CO_2)	0.43	5.62

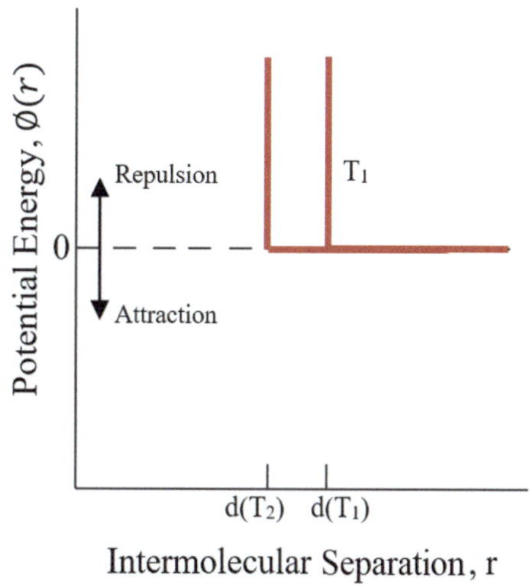

Fig. 2.9 Potential distribution for VHS model versus intermolecular distance

The VHS model results in a temperature paower law formula for the viscosity as

$$\mu_{VHS} = \frac{\left(\frac{15}{8}\right)\sqrt{\pi mk}\left(\frac{4k}{m}\right)^{\alpha} T^{\omega}}{\Gamma(4-\alpha)\sigma_T \xi_{r,ref}{}^{2\alpha}}, \tag{2.52}$$

where, $\omega = 0.5 + \alpha$ and Γ is the gamma function.

The deflection angle (χ), which is defined as the angle between the relative speed before and after the collision, is obtained for this model as

$$\chi = 2\cos^{-1}(b/d), \tag{2.53}$$

Parameter b is the minimum intermolecular distance before the collision, and it is referred to as the impact parameter.

2.2.12 Generalized Hard-Sphere (GHS) Model

The GHS collision model, introduced by Hassan and Hash [6]. It is used to consider the accurate variation of the viscosity with temperature. This model is an extension of the VHS model, incorporating terms that enable the modeling of molecules with both repulsive and attractive potentials. For the GHS model, the total collision cross section could be written as follows:

$$\sigma = \sigma_0\big(\phi(g_0/g)^{2\nu_1} + (1-\phi)(g_0/g)^{2\nu_2}\big) \tag{2.54}$$

where $\sigma_0 = (\pi d_0^2)$ is the reference cross section, $g = \sqrt{4RT}$ and the parameters with subscript zero are calculated at the reference temperature (T_0).

The exponents ν_1 and ν_2 represent velocity-dependent parameters that control how the collision cross section varies with relative speed g. They correspond to:

- ν_1: the exponent associated with the repulsive component of the interaction potential (similar to the exponent in the VHS model),
- ν_2: the exponent associated with the attractive component of the interaction potential.

These parameters enable the GHS model to interpolate between two different interaction types (repulsive and attractive), allowing it to model real gas behavior more accurately over a range of temperatures. The mixing factor governs the interpolation $\phi \in [0,1]$. GHS parameters (ν_1, ν_2, ϕ) can be tuned using experimental viscosity data over a wide range of temperatures.

There are other models, such as GHS, Larsen-Bergnak (LBP), and Variable Soft-Sphere (VSS), which have fewer applications than the previous models and are therefore not reviewed in this book. More details about these models can be found in Ref. [4].

2.2.13 Molecular Spacing

According to Fig. 2.10, the volume V is considered as a cube, which contains N molecules. So, the volume of each small cube contains one molecule equal to V/N, which is equal to $(1/n)$ by definition, where n is the numerical density. So the length of each side of these small cubes, which is the average distance between two molecules, is equal to $\delta = n^{-\frac{1}{3}}$. According to the experimental results, it can be said that if $\frac{\delta}{d} \geq 7$, the assumption of dilute gas is correct.

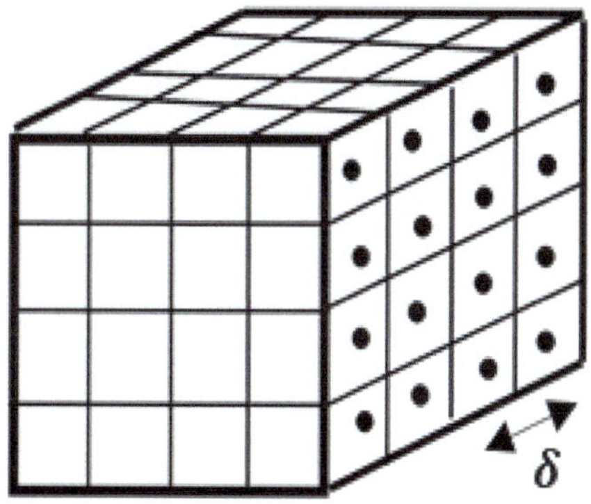

Fig. 2.10 Intermolecular distance concept

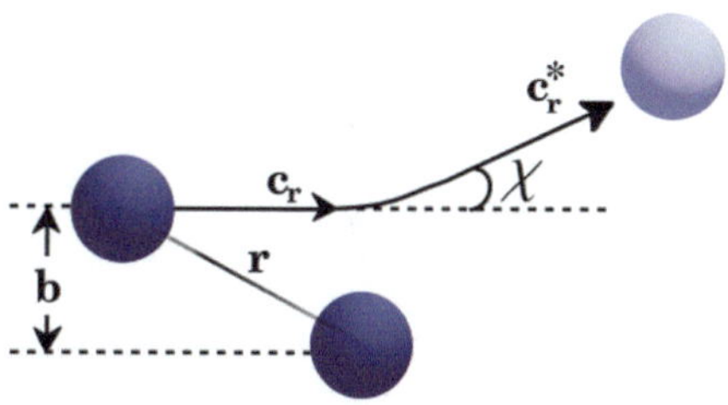

Fig. 2.11 Binary scattering in a fixed reference frame

2.2.14 Binary Collision Concept

The binary collision theory is derived based on classical mechanics. As shown in Fig. 2.11, two particles collide with velocities c_1 and c_2 and move away from each other with the deflection angle χ, and as a result, the velocities after the collision are c_1^* and c_2^* respectively. In this collision, the relative velocity of c_r particles is calculated as $c_r = c_1 - c_2$, remains unchanged ($c_r = c_r^*$). Also, the impact parameter b shows the distance between two particles, and this distance is calculated perpendicular to the relative velocity. In addition to the relative speed and the impact parameter, the scattering angle also depends on the intermolecular potential model.

2.2.15 Scattering Cross-Section Area

The hypothetical area that expresses the probability that light or other radiation is scattered around the scattering center is called the scattering cross-section (see Fig. 2.12). This definition refers to a type of measurement that quantifies the interaction between a scattered particle or wave and one or more scattering centers. The scattering of a particle or wave from a single scattering center is the simplest type of definition of the differential scattering cross-section. The impact parameter (b) is

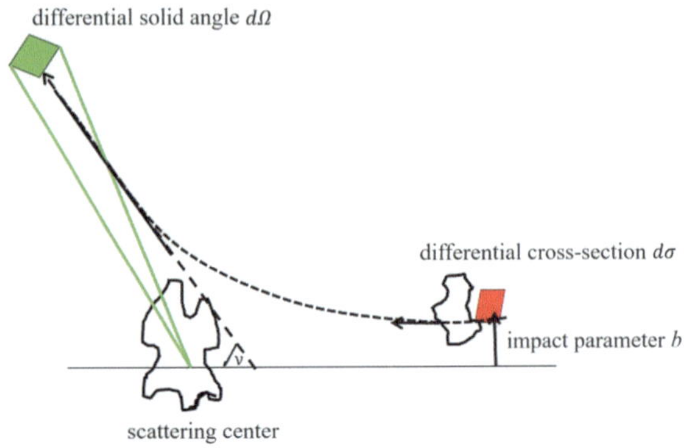

Fig. 2.12 Impact parameter and solid angle

the vertical distance from the trajectory of the incoming particle to the cross-section. The differential cross-sectional area ($d\sigma$) is the area element in the impact parameter plane.

So, we have $d\sigma = d^2b$. The differential cross-sectional area is the result of dividing the differential element of the area by the differential element of the spatial angle in the direction of the exit path of the particles. In other words, it is equal to $|d\sigma/d\Omega|$. This fraction describes the changes in the impact parameter required for changes in the degree of the exit path. The differential cross-section must always be positive. In problems with rotational symmetry, the lateral angle φ does not change in the scattering process and the differential cross-section becomes

$$\frac{d\sigma}{d\Omega} = \frac{b}{sin\vartheta}\left|\frac{db}{d\vartheta}\right| \tag{2.55}$$

As shown in the Fig. 2.12, ϑ is the angle between the incoming and outgoing beam directions. The total cross-sectional area is the integral of the differential cross-sectional area over the entire solid angle as follows

$$\sigma_{tot} = \oint \frac{d\sigma}{d\Omega} d\Omega \tag{2.56}$$

The elastic collision of two hard spheres is an example that gives us a good understanding of the scattering cross-section. R and r are the radii of the scattering center and the scattered sphere, respectively, b is the impact parameter and ϑ is the polar angle of the exit path, as mentioned above. So the differential cross-sectional area will be equal to

$$\left|\frac{d\sigma}{d\Omega}\right| = \frac{1}{4}(r + R)^2 \tag{2.57}$$

and the total cross-section is as

$$\sigma_{tot} = \pi(r + R)^2 \tag{2.58}$$

Therefore, in this case, the cross-sectional area of the total scattering is proportional to the area of a circle with a radius of $(r + R)$, where the center of mass of the incoming sphere is deflected, and it then continues on its way with a fixed scattering center.

2.2.16 *Point Sources of Light*

All problems related to a light source that emits light uniformly in all directions are solved in the same way. A shining metal ball, such as the sun, is an example of such a source, as if it emits light uniformly in all directions. In this case, the luminous flux is uniformly distributed in all directions. This means that the effect of the spring in the light receiver depends only on the distance between the receiver and the center of the luminous sphere, not on the direction of the radius drawn from the center of the bullet to the receiver. In this case, the destination of the light is considered to be at a distance r, which is much larger than the negligible radius of the bright sphere. In this case, it can be assumed that the light is emitted from a single point, i.e. the center of the luminous sphere. In such cases, the source of light is referred to as a point source. A fountain that is considered a point fountain should not always be small, because the ratio between the size of the spring and the properties examined in it is important. For example, stars are excellent examples of point sources in practical applications. Although they are huge in size, their distance from Earth is many times greater than their size. Therefore, a light source that emits light unevenly in different directions is not a point source; Although it is very small compared to the distance to the point of view. Let us clarify more precisely what is meant by uniform radiation in all directions. For this purpose, we need the solid angle Ω, which is equal to the ratio of the area σ, the surface cut-off from the sphere by the cone whose apex is at the point S, to the square of the radius of the sphere r (see Fig. 2.13) as

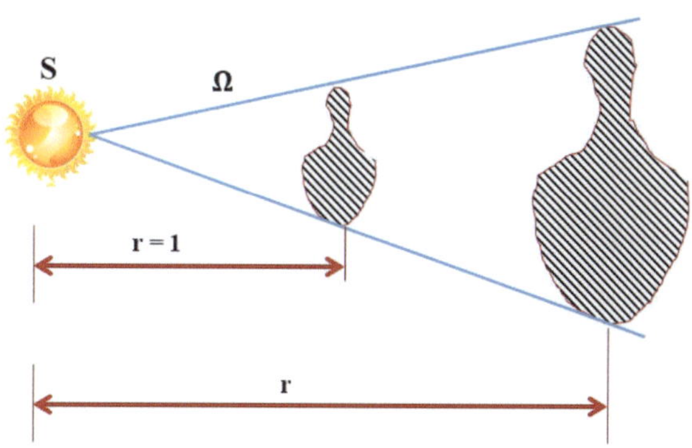

Fig. 2.13 Light point source and solid angle

$$\Omega = \frac{\sigma}{r^2} \tag{2.59}$$

This ratio does not depend on r, because when r increases, the area σ of the surface cut by the cone increases by r^2. If $r = 1$, Ω is numerically equal to σ, that is, the solid angle is measured by a surface that is cut by a cone from a sphere with a radius of 1. The unit of solid angle Stradian (Sr), that is, the solid angle corresponds to a unit area of a sphere with a radius of one. The solid angle that includes the entire space around the fountain is equal to 4π Stradians (Sr); Because the area of the entire sphere whose radius is one is equal to 4π. The total radiation of a source is distributed in a solid angle of 4π Stradians, and the radiation is called uniform or isotropic if it emits the same power in equal space angles separated in each direction. So, a point source is a source whose size is small compared to the distance to the point of observation and emits a uniform luminous flux in all directions.

2.3 Binary Collision

In dilute gases, intermolecular interactions are considered to be binary collisions. Therefore, in this section, binary collisions have been investigated. The governing assumptions in the analysis of binary collisions are:

- Molecules are like points on which the intermolecular force acts. In other words, the size of the molecules is small compared to the distance between them.
- The intermolecular force acts from the middle of the molecules. That is, the force is in the direction that connects the midpoint of one molecule to the midpoint of the other molecule.
- The forces are conservative.
- The assumption of elastic collision is established. In other words, there is no energy loss, and the molecules remain unchanged in shape.

The governing equations in intermolecular interaction are:

- Linear momentum conservation equation
- Angular momentum conservation equation
- Total energy conservation equation
- Kinetic energy conservation equation
- Newton's second law.

Since the intermolecular force is proportional to the distance, in other words, if the distance between two molecules is large, the force between them is zero. While the molecules are close to a certain distance, an attractive force exists between them; however, when the distance is reduced, a strong repulsive force is present. If two molecules of mass m_1 and m_2 are assumed in position r_1 and r_2 as shown in

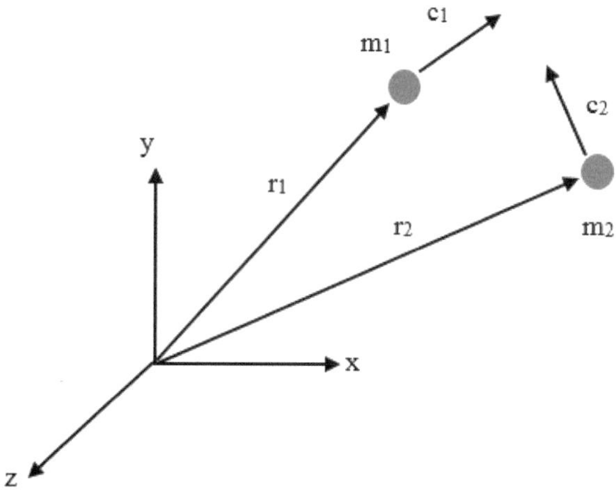

Fig. 2.14 Molecular locations for a binary collision

Fig. 2.14, using the definition of the center of mass, the coordinates and velocity of the center of mass are obtained according to the following relations

$$\boldsymbol{r}_m = \frac{m_1 \boldsymbol{r}_1 + m_2 \boldsymbol{r}_2}{m_1 + m_2} \tag{2.60}$$

$$\boldsymbol{c}_m = \frac{m_1 \boldsymbol{c}_1 + m_2 \boldsymbol{c}_2}{m_1 + m_2} \tag{2.61}$$

Since the collisions are considered elastic in the binary collision, the following relation can be written for the conservation of linear momentum:

$$m_1 \boldsymbol{c}_1 + m_2 \boldsymbol{c}_2 = m_1 \boldsymbol{c}_1^* + m_2 \boldsymbol{c}_2^* = (m_1 + m_2)\boldsymbol{c}_m = (m_1 + m_2)\boldsymbol{c}_m^* \tag{2.62}$$

where the * stands for post-collision properties.

According to Eq. (2.62), it can be seen that the velocity of the center of mass remains constant before and after the collision, as the mass does not change and therefore the velocity of the center of mass remains unchanged. By writing the energy conservation equation before and after the collision, we have

$$\frac{1}{2} m_1 c_1^2 + \frac{1}{2} m_2 c_2^2 = \frac{1}{2} m_1 c_1^{*2} + \frac{1}{2} m_2 c_2^{*2} \tag{2.63}$$

Now, to have better analysis, relative coordinate and relative velocities before and after the collision are defined as follow

$$\boldsymbol{r} = \boldsymbol{r}_1 - \boldsymbol{r}_2 \tag{2.64}$$

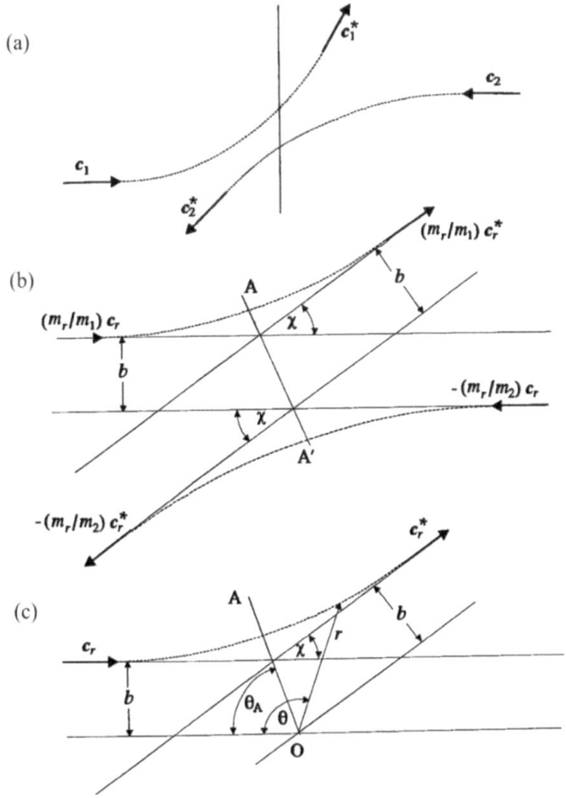

Fig. 2.15 **a** The real collision plane, **b** the collision plane for reduced mass with relative velocity, and **c** the collision plane employed in the present analysis. Adapted from [4]

$$v = \dot{r}e_r + r\dot{\theta}e_\theta \tag{2.83}$$

Now, the angular momentum conservation could be expressed as

$$m_r c_r b = m_r\left(\frac{\partial r}{\partial t} \times r\right) + m_r\left(r\frac{\partial \theta}{\partial t} \times r\right) \tag{2.84}$$

Since, the value of $\left(\frac{\partial r}{\partial t} \times r\right)$ is equal to zero, as angle between them is zero. Equation (2.84) is simplified to

$$m_r c_r b = m_r r^2 \frac{\partial \theta}{\partial t} \tag{2.85}$$

Consider the total energy conservation equation

$$\frac{1}{2}m_r\dot{r}^2 + \frac{1}{2}m_r(c_r sin\theta)^2 + U(r) = \frac{1}{2}m_r c_r^2 \tag{2.86}$$

where $U(r)$ is the potential energy. It should be noted that the value of $(c_r sin\theta)$ is equivalent to tangential velocity (θ direction) before the collision and it is substituted via the relation $sin\theta = \frac{b}{r}$.

According to Eq. (2.85)

$$\dot{r}^2 = \left(\frac{\partial r}{\partial\theta}\cdot\frac{\partial\theta}{\partial t}\right)^2 = \left(\frac{\partial r}{\partial\theta}\right)^2\cdot\left(\frac{c_r b}{r^2}\right)^2 \tag{2.87}$$

Substituting Eqs. (2.87) in (2.86) leads to

$$\frac{1}{2}m_r\left(\frac{\partial r}{\partial\theta}\right)^2\cdot\left(\frac{c_r b}{r^2}\right)^2 + \frac{1}{2}m_r(c_r sin\theta)^2 + U(r) = \frac{1}{2}m_r c_r^2 \rightarrow$$

$$\frac{1}{2}m_r\left(\frac{\partial r}{\partial\theta}\right)^2\cdot\left(\frac{c_r b}{r^2}\right)^2 = \frac{1}{2}m_r c_r^2 - \frac{1}{2}m_r(c_r sin\theta)^2 - U(r), \left(sin\theta = \frac{b}{r}\right)$$

$$\frac{1}{2}m_r\left(\frac{\partial r}{\partial\theta}\right)^2\cdot\left(\frac{c_r b}{r^2}\right)^2 = \frac{1}{2}m_r c_r^2 - \frac{1}{2}m_r\left(c_r\frac{b}{r}\right)^2 - U(r)$$

$$\left(\frac{\partial r}{\partial\theta}\right)^2 = \frac{r^4}{b^2} - r^2 - \frac{2r^4U(r)}{b^2m_rc_r^2} \rightarrow \left(\frac{\partial r}{\partial\theta}\right)^2 = \frac{r^2}{b^2}\left[r^2 - b^2 - \frac{2r^2U(r)}{m_rc_r^2}\right]$$

$$\frac{dr}{d\theta} = \pm\frac{r}{b}\sqrt{r^2 - b^2 - \frac{2r^2U(r)}{m_rc_r^2}} \tag{2.88}$$

Relation (2.88) between r and θ represents the equation of the path to find r_A (the shortest distance from the coordinate origin) which is proportional to θ_A (the angle corresponding to r_A). From Fig. 2.15, the value of $dr/d\theta$ should be zero. Therefore, a relation for r_A is obtained as follows

$$r_A^2 = b^2 + \frac{2r_A^2U(r_A)}{m_rc_r^2} \tag{2.89}$$

Now, to obtain θ_A, the path equation with respect to the r should be determined

$$\frac{d\theta}{dr} = \pm\frac{b}{r}\left(r^2 - b^2 - \frac{2r^2U(r)}{m_rc_r^2}\right)^{-\frac{1}{2}} \tag{2.90}$$

$$\theta - \theta_A = \pm\int_{r_A}^{r}\frac{b}{r}\left(r^2 - b^2 - \frac{2r^2U(r)}{m_rc_r^2}\right)^{-\frac{1}{2}}dr \tag{2.91}$$

Since, $(r \rightarrow \infty)$ then $(\theta \rightarrow o)$, we have

Fig. 2.16 Intermolecular distance for collision in HS model

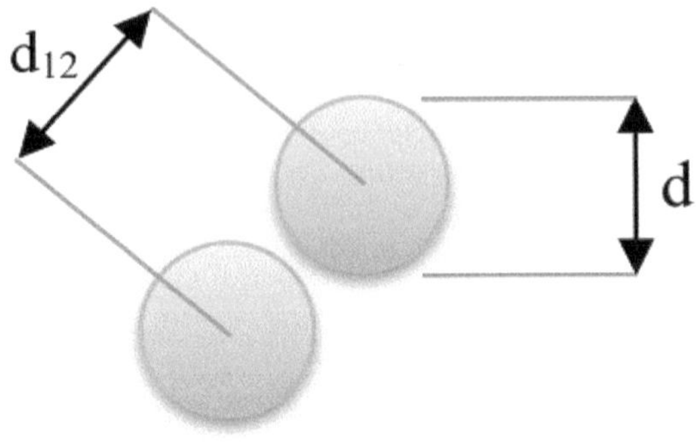

$$\theta_A = \int_{r_A}^{\infty} \frac{b}{r}\left(r^2 - b^2 - \frac{2r^2 U(r)}{m_r c_r^2}\right)^{-\frac{1}{2}} dr \tag{2.92}$$

Due to symmetric collision, the angle between c_r and r_A is equal to θ_A. Then, the angle between and r_A is also equal to θ_A. According to ($\chi = \pi - 2\theta_A$), we have

$$X = \pi - 2\int_{r_A}^{\infty} \frac{b}{r}\left(r^2 - b^2 - \frac{2r^2 U(r)}{m_r c_r^2}\right)^{-\frac{1}{2}} dr \tag{2.93}$$

The meaning of Eq. (2.93) is that if the intermolecular potential and c_r are known, the angle χ can be calculated in terms of parameter b. The parameter b is referred to as the impact parameter.

Considering the HS model (the intermolecular potential is zero at distances up to d, but when the distance between molecules becomes equal to the diameter of the molecule, the repulsive force becomes infinite), the smallest distance where the intermolecular potential can be assumed to be zero is equal to is d_{12} (see Fig. 2.16). The intermolecular force is zero in the interval d to infinity.

$$X = \pi - 2\int_{d_{12}}^{\infty} \frac{b}{r}\left(r^2 - b^2\right)^{-\frac{1}{2}} dr = \pi - \left(2\cos^{-1}\frac{b}{r}\right)_{d_{12}}^{\infty} \tag{2.94}$$

$$X = \pi - \left(\pi - \left(2\cos^{-1}\frac{b}{d_{12}}\right)\right) = 2\cos^{-1}\frac{b}{d_{12}} \tag{2.95}$$

Hence, the impact parameter for the HS model is as

$$b = d_{12} cos\frac{\chi}{2} \tag{2.96}$$

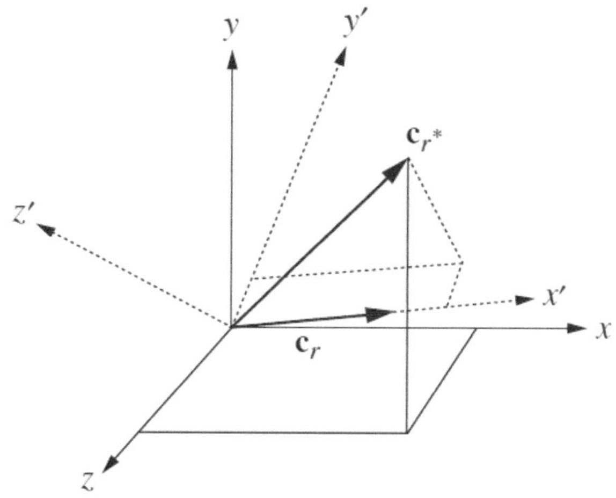

Fig. 2.17 *xyz*- and *x′y′z′*-coordinates

By knowing the angle χ and magnitude of c_r^*, and projecting c_r^* in the $x'y'z'$-coordinate (x' is aligned with c_r), and finally converting the $x'y'z'$-coordinate to *xyz*-coordinate, post-collision velocities will be determined (see Fig. 2.17).

The velocity c_r^* is projected in $x'y'z'$-coordinate via the following relation

$$c_r^{*'} = \left(c_r^* \cos\chi,\, c_r^* \sin\chi \cos\varepsilon,\, c_r^* \sin\chi \sin\varepsilon\right) \tag{2.97}$$

Returning to the main coordinate, i.e. *xyz*-coordinate, the directional cosines between two coordinates is equal to

$$c_r^* = \boldsymbol{X} c_r^{*'} \tag{2.98}$$

$$\boldsymbol{X}_{i,j} = \cos\left(x_i, x_j'\right) \tag{2.99}$$

$$\boldsymbol{X}_{11} = \frac{u_r}{c_r} \quad \boldsymbol{X}_{21} = \frac{v_r}{c_r} \quad \boldsymbol{X}_{31} = \frac{w_r}{c_r} \tag{2.100}$$

$$\boldsymbol{X}_{12} = 0 \quad \boldsymbol{X}_{22} = \frac{w_r}{\sqrt{v_r^2 + w_r^2}} \quad \boldsymbol{X}_{32} = \frac{-v_r}{\sqrt{v_r^2 + w_r^2}} \tag{2.101}$$

$$\boldsymbol{X}_{13} = \frac{-\sqrt{v_r^2 + w_r^2}}{c_r} \quad \boldsymbol{X}_{23} = \frac{u_r v_r}{c_r\sqrt{v_r^2 + w_r^2}} \quad \boldsymbol{X}_{33} = \frac{u_r w_r}{c_r\sqrt{v_r^2 + w_r^2}} \tag{2.102}$$

Here is a summary of the steps in a sample calculation code to obtain c_r^*:

(1) Getting the speed of the center of mass
(2) Obtaining relative speed
(3) Calculating the angle χ

(4) Obtaining the velocities in the $x'y'z'$-coordinate and then transfer to the main coordinate (xyz) using the conductor cosines
(5) Calculation of speeds c_1^* and c_2^* using c_r^* and c_m according to the following formulas.

$$c_1^* = c_m + \frac{m_2}{m_1 + m_2} c_r^* \tag{2.103}$$

$$c_2^* = c_m - \frac{m_1}{m_1 + m_2} c_r^* \tag{2.104}$$

As seen in Fig. 2.18, an angle element is defined instead of the area element. Based on the specified elements the solid angle ($d\Omega$) could be obtained as follows:

$$-b.d\varepsilon.db = \sigma d\Omega \tag{2.105}$$

The right-hand side of the above equation is in the plane perpendicular to c_r^* and the left-hand side is in the plane perpendicular to c_r. It should be noted that r changes between zero and π in the z plane and $r\text{Sin}\theta$ is in the x–y plane. So, for the area element, we need the r component in the x–y plane. In Eq. (2.105), σ is the cross-section differential of the collision surface in the c_r direction, and its product with $d\Omega$ gives the result of the collision element. On the other hand, $d\Omega$ is calculated in the direction of c_r^*. The radial distance to the collision element in the c_r^* direction is represented by r. According to the definition of solid angle, we have:

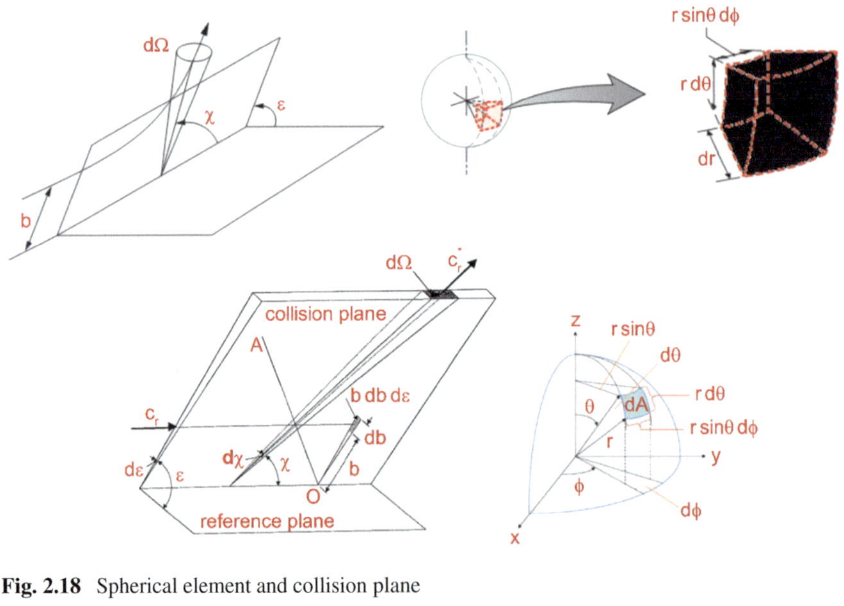

Fig. 2.18 Spherical element and collision plane

$$d\Omega = \frac{(r\sin\chi)d\mathcal{E}(r.d\chi)}{r^2} = \sin\chi d\chi d\varepsilon \tag{2.106}$$

Considering Eq. (2.105)

$$-b.d\varepsilon.db = \sigma\sin\chi d\chi d\varepsilon \tag{2.107}$$

and then

$$\sigma = \frac{-bdb}{\sin\chi d\chi} \tag{2.108}$$

Now, to calculate collision cross-section, one should integrate $\sigma d\Omega$ over the all ranges of solid angle as follow

$$\sigma_T = \int_o^{4\pi}\sigma d\Omega = \int_o^{2\pi}\int_o^{\pi}\sigma\sin\chi d\chi d\varepsilon = 2\pi\int_o^{\pi}\sigma\sin\chi d\chi \tag{2.109}$$

Substituting σ from Eq. (2.108), and also calculating differential impact parameter (db) via derivation from Eq. (2.96), one can calculate the collision cross-section for the HS model as following

$$db = -\frac{d_{12}}{2}\sin\frac{\chi}{2} \tag{2.110}$$

$$\sigma = \frac{\left(-d_{12}\cos\frac{\chi}{2}\right)\left(-\frac{d_{12}}{2}\sin\frac{\chi}{2}\right)}{\sin\chi d\chi} = \frac{d_{12}^2}{4} \tag{2.111}$$

$$\sigma_T = 2\pi\int_o^{\pi}\frac{d_{12}^2}{4}\sin\chi d\chi = \pi\frac{d_{12}^2}{2}(-\cos\chi)_o^{\pi} = \pi d_{12}^2 \tag{2.112}$$

The equality $\sin\chi = 2\sin\frac{\chi}{2}\cos\frac{\chi}{2}$ is employed in Eq. (2.111). collision cross-section is equal to πd^2 for the HS model. In other words, every molecule that moves has a circle with an area of πd^2 which is affected by the collision.

2.4 Collision Cylinder

The initial definition of a collision cylinder was presented in Chap. 1. Each molecule of the c_1 speed group that moves has a cylinder around it as shown in Fig. 2.19, and if a molecule is placed inside this cylinder, it will collide with this molecule.

Therefore, the number of collisions per unit of time is equal to the number of molecules of the arbitrary speed group c_2 that are placed inside the cylinder of the first molecule ($\Delta n c_r \sigma_T$).

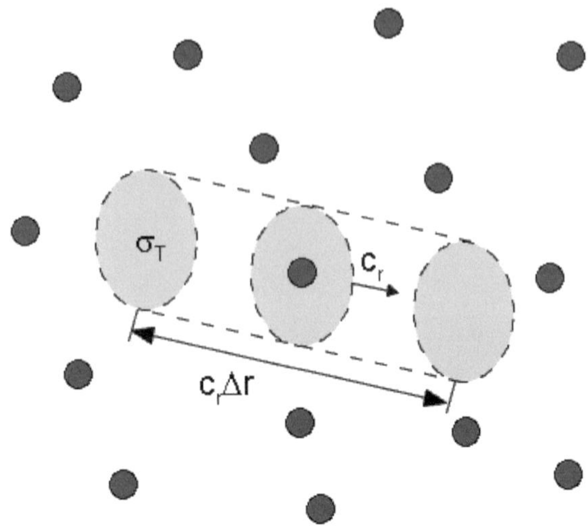

Fig. 2.19 Collision cylinder

2.4.1 Collision Frequency

The number of collisions per unit of time is called the “collision frequency” and is represented by the symbol v. The collision frequency of class 1 molecules with class 2 molecules is calculated according to the following formula

$$\nu = \frac{No.\,of\ collisions}{time\ interval} = \frac{\pi d^2 (c_r \delta t) \Delta n}{\delta t} = \pi d^2 c_r \Delta n \tag{2.113}$$

Δn is a percentage of the total n that is related to the speed category c_2, or in other words, Δn is the percentage of particles in a certain frequency, and c_r is the relative speed of the two speed groups. The total collision frequency is the sum of the mentioned ratio over all speed ranges.

$$\begin{aligned} \nu &= \sum \pi d^2 c_r \Delta n = \pi d^2 \sum c_r \Delta n \frac{n}{n} = \pi d^2 n \sum \frac{\Delta n}{n} c_r \\ &= \pi d^2 n \int_{-\infty}^{+\infty} f(c) c_r dc = \pi d^2 n \overline{c_r} \end{aligned} \tag{2.114}$$

$$\nu = n\overline{c_r}\sigma_T = n\pi d^2 \overline{c_r} \tag{2.115}$$

To calculate the average of any quantity, it is multiplied by the distribution function f, and then the integral is taken. In deriving Eq. (2.115), this definition has been used.

We know that

$$\vec{c}_r = \vec{c}_1 - c_2$$

$$|\overline{c_r}| = \sqrt{\overline{(\vec{c}_1 - \vec{c}_2).(\vec{c}_1 - \vec{c}_2)}} = \sqrt{\overline{c_1^2 + c_2^2 - 2c_1.c_2}} = \sqrt{\overline{c_1^2} + \overline{c_2^2} - \overline{2c_1.c_2}}$$

On the other hand, $\overline{c_1.c_2}$ is equal to zero due to random angles of c_1 and c_2 coming from random movement. In other words, they are parallel at a moment and are anti-parallel at another moment. It could be written that

$$|\overline{c_r}| = \sqrt{2}|\overline{c_1}|$$

$$\nu = \sqrt{2}n\pi d^2\overline{c'}$$

where, $|\overline{c_1}| = |\overline{c'}| = \overline{c'}$.

The interval time for each intermolecular collision is equal to $\frac{1}{\nu}$. Also, knowing that λ (molecular mean free path) is the length that must be traveled by the molecule to make a collision, the following formula is obtained

$$\lambda = \left(\frac{1}{\nu}\right)\overline{c'} = \frac{\overline{c'}}{\nu} = \frac{1}{n\pi d^2}\frac{\overline{c'}}{\overline{c_r}}, \tag{2.116}$$

where $\overline{c'}$ is the average thermal speed of all the molecules. Now, to obtain order of λ, the ratio of $\frac{\lambda}{\delta}$ could be calculated as follows

$$\lambda_{HS} = \frac{1}{\sqrt{2}n\pi d^2} = \frac{1}{\sqrt{2}\left(\frac{1}{\delta^3}\right)\pi d^2} = \frac{\delta^3}{\sqrt{2}\pi d^2} \tag{2.117}$$

$$\frac{\lambda}{\delta} = \frac{\delta^2}{\sqrt{2}\pi d^2} = \frac{1}{\pi}\frac{1}{\sqrt{2}}\frac{\delta^2}{d^2} = \frac{1}{\pi}\frac{\overline{c'}}{\overline{c_r}}\frac{\delta^2}{d^2} \tag{2.118}$$

Knowing the fact that $\delta \gg d$, we can conclude that $\lambda \gg \delta$ holds. Therefore, $\lambda \gg \delta \gg d$ and as a result, λ is usually one or two orders of magnitude larger than d.

The range of effect of intermolecular potential forces is approximately equal to the diameter of molecules. For example, according to the hard-sphere model, if the molecules are very close together, the repulsive force between them is infinite. The range of effect of intermolecular forces for most intermolecular models is about $2d$, where the importance of $\lambda \gg \delta \gg d$ is determined here. In fact, the intermolecular force (in dilute gases) appears during collisions and has a short duration. Considering that n is a large number, the collision frequency is also noticeable, and as a result, the collision time (1/ν) is a very short period. intermolecular forces are active within very short times. By considering the relationship between the collision frequency and the number of molecules, it can be seen that the collision time is short.

$$\frac{v}{N} = \frac{n\pi d^2\sqrt{2}\,\overline{c'}}{nV} = \frac{\pi d^2\sqrt{2}\,\overline{c'}}{V} \tag{2.119}$$

Therefore, we can conclude that the number of collisions is small compared to the total number of molecules. However, the following points should be considered:

- The collisions are random events. That is, a single molecule experiences very few collisions, but because the total number of molecules is large, the number of collisions is considerable.
- The collision between the molecules is binary. Considering that the probability of a single molecule colliding is low, the probability of three molecules colliding simultaneously becomes even lower. Therefore, most collisions in gases are binary in nature. Ternary collisions may occur for liquids.

2.4.2 Collision Probability

The probability of a collision between two molecules is equal to the probability of a molecule from the c_2 speed category being placed in the collision cylinder of a c_1 speed molecule. In the time interval δt, the volume of the collision cylinder swept by the first molecule compared to the second molecule is equal to $\pi d^2(c_r\delta t)$. If the number of molecules of the second group that are in the collision cylinder of the first group is equal to Δn, the probability of collisions between the molecules of the first and second group is equal to the number of molecules of group c_2 that are in the collision cylinder of group c_1, i.e. $\Delta n\pi d^2 c_r \delta t$.

The collision probability is obtained by summing over all molecules as follows

$$\begin{aligned} P(\delta t) &= \sum \pi d^2 c_r n \delta t = \sum \pi d^2 c_r n \frac{n}{n} \delta t = \\ &= n\pi d^2 \delta t \int_{-\infty}^{+\infty} f(c) c_r dc = n\pi d^2 \delta t \overline{c_r}, \end{aligned} \tag{2.120}$$

where, $\frac{\Delta n}{n}$ has been replaced by $f(c)dc$, where $f(c)$ is the velocity distribution function. Consider that the probability of no collision between molecules is expressed by $P(\delta t)$. We know from introductory mathematics that

$$P(t + \delta t) = P(t) + \delta t \frac{dP(t)}{dt} \tag{2.121}$$

On the other hand, we know that the probability value of $P(t + \delta t)$ is equal to the probability P_1 of the particle not colliding at time t multiplied by the probability of not colliding at the next time δt, i.e.

$$P(t + \delta t) = P(t)P_1(\delta t) = P(t)(1 - P(\delta t)) = P(t)\left(1 - n\pi d^2 \delta t \overline{c_r}\right) \tag{2.122}$$

According to Eqs. (2.121) and (2.122)

$$\frac{1}{P(t)}\frac{dP(t)}{dt} = -n\pi d^2\overline{c_r} \tag{2.123}$$

Assuming $P(0) = 1$, then

$$P(t) = \exp\left(-n\pi d^2\overline{c_r}t\right) \tag{2.124}$$

Now, the probability of no collision up to time t and collision during δt is equal to

$$\exp\left(-n\pi d^2\overline{c_r}t\right)n\pi d^2\delta t\overline{c_r} \tag{2.125}$$

The average time for each collision could be calculated as follows

$$\tau = \int_0^\infty t\exp\left(-n\pi d^2\overline{c_r}t\right)n\pi d^2\delta t\overline{c_r} \tag{2.126}$$

$$\begin{aligned}\tau &= \frac{1}{n\pi d^2\overline{c_r}}\int_0^\infty (n\pi d^2\overline{c_r}t)\exp(-n\pi d^2\overline{c_r}t)d(n\pi d^2\delta t\overline{c_r}) \\ &= \frac{1}{n\pi d^2\overline{c_r}}\int_0^\infty (x)\exp(-x)d(x) = \frac{1}{n\pi d^2\overline{c_r}}\end{aligned} \tag{2.127}$$

Therefore, the mean collision time is equal to the inverse of collision frequency.

2.5 Viscosity

Viscosity is the property of the fluid's resistance against the forces acting on it that creates shear stress. Viscosity is the main factor of momentum transfer in fluid layers and appears when there is relative movement between fluid layers. Liquids are more viscous than gases. The origin of viscosity can be attributed to two primary factors.

- Molecular attraction forces that arise in the liquids.
- Molecular momentum exchange forces that arise in the gases.

The following points should be considered:

- Viscosity changes with temperature in liquids are more than gases.
- If the pressure increase is not too high, the viscosity is independent of the pressure; but at high pressure gradient conditions, the viscosity of liquids and gases increases with increasing pressure. Note that the ineffectiveness of pressure on

the viscosity of gases is because, with increasing pressure, the speed of molecular collision increases but the length of collision decreases. The ideal fluid is one whose viscosity is zero and is incompressible. The ideal gas is one whose viscosity is non-zero and it is compressible.

2.5.1 Dynamic Viscosity

Viscosity can be measured by placing a fluid between two plates separated by a distance d. Let us assume according to Fig. 2.20 that we keep the lower plate fixed and the upper plate moves with velocity u in the direction of x. If you want the upper plate to move at a constant speed, you need to apply a constant force F. To derive the viscosity coefficient for a base case, we examine the flow between two parallel plates, for which we move the upper plate with velocity u. This movement causes the molecules near the upper plate to move with the speed of the plate and this kinetic energy is transferred layer by layer inside the fluid.

Near the upper surface, a frictional force causes the fluid to drag along with it at the same speed u. However, near the bottom plate, the fluid is stationary (we considered the flow without slip). Experimentally, it can be seen that the force per unit area that must be applied to the upper surface is proportional to the velocity gradient as

$$\frac{F}{A} = \mu \frac{du_x}{dz} \approx \mu \frac{u}{d} \tag{2.128}$$

The second approximate equality holds for small distances d. The proportionality factor μ is called viscosity (or to be more precise, dynamic viscosity). The goal is to calculate the momentum changes between fluid layers. To calculate the momentum transferred from an arbitrary cross-section in the field, we first need to know the number of particles moved from that cross-section. A cross-section of the fluid could be considered as shown in Fig. 2.21.

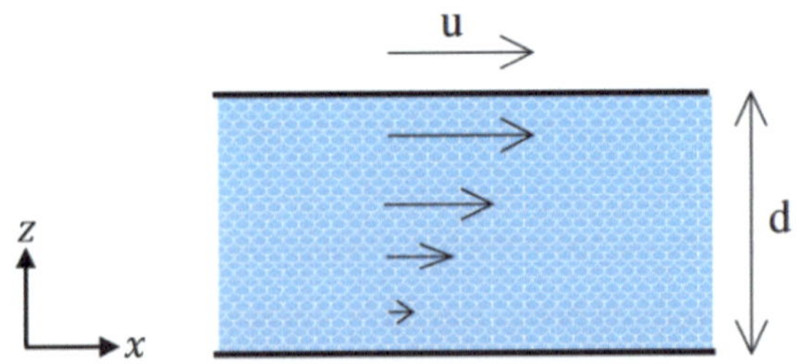

Fig. 2.20 Schematics of fluid between two parallel plates (one moving and one stationary)

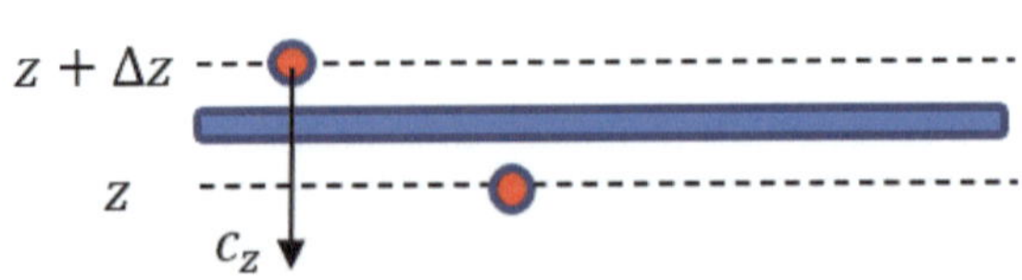

Fig. 2.21 A fluid control volume moving with speed of c_z in z-direction

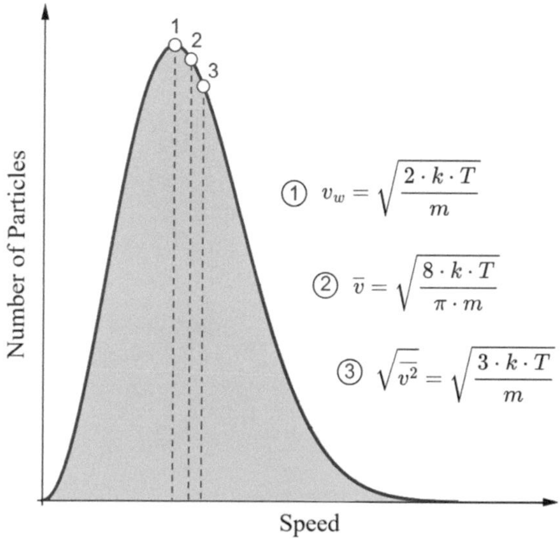

Fig. 2.22 Maxwel-Boltzmann velocity distribution function

It should be noted that the flow here is one-dimensional and horizontal, but the molecules move in three directions, and c_z is the thermal velocity molecules in the vertical direction. If we consider the velocity distribution function as shown in Fig. 2.22, the number of molecules located in an element $d\overrightarrow{c}$ is as

$$nf(c)d\overrightarrow{c} \tag{2.129}$$

It should be noted that employing $d\overrightarrow{c}$ in Eq. (2.129) indicates that a velocity vector distribution function (see Fig. 2.23) is needed to calculate the number of molecules.

Therefore, the flux of the number of molecules, i.e. number of molecules per unit area, within the moving (with c_z) element $d\overrightarrow{c}$ is as

$$\left(nf(c)d\overrightarrow{c}\right)c_z \tag{2.130}$$

It should be considered that the c_z is the molecular speeds in z-direction. Momentum exchange in z-direction for a transfer molecule from $z + \Delta z$ to z is equal to

$$\Delta p = m(u_x(z + \Delta z) - u_x(z)) \approx m\frac{du_x}{dz}\Delta z \tag{2.131}$$

The distance Δz is dependent on the molecule inclination angle during the movement. Before the velocity change, a molecule moves freely a distance of mean free

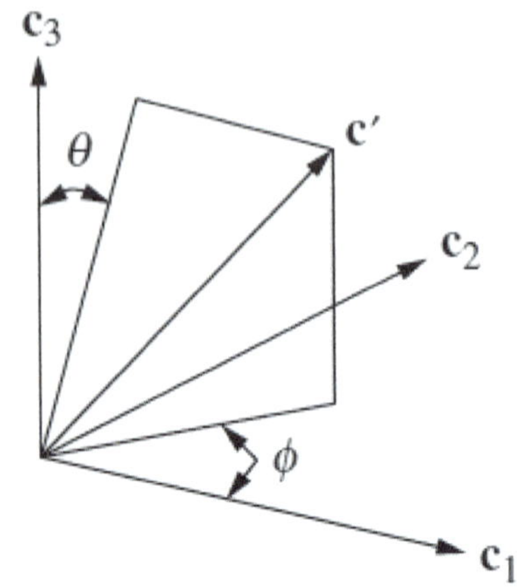

Fig. 2.23 Velocity vector distribution function

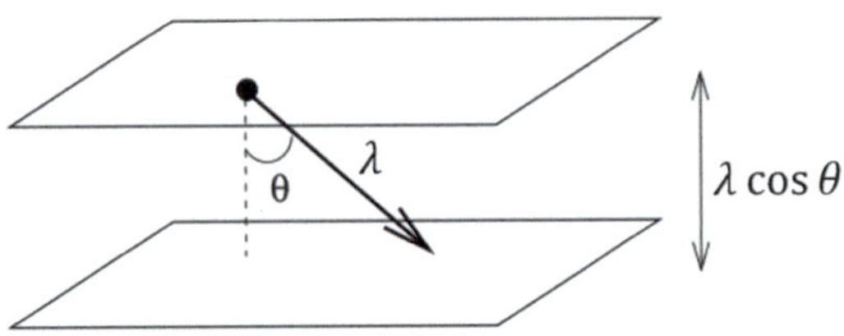

Fig. 2.24 Molecular movement up to mean free path under inclination angle of θ

path in average. Therefore, Δz could be calculated as

$$\Delta z = \lambda \cos\theta \tag{2.132}$$

The movement of the molecule up to is illustrated in Fig. 2.24. If one molecule goes from top to bottom, $0 \leq \theta < \frac{\pi}{2}$. If one molecule moves from bottom to top, we have $\pi < \theta \leq \frac{\pi}{2}$ and $\Delta z < 0$ (see Fig. 2.25).

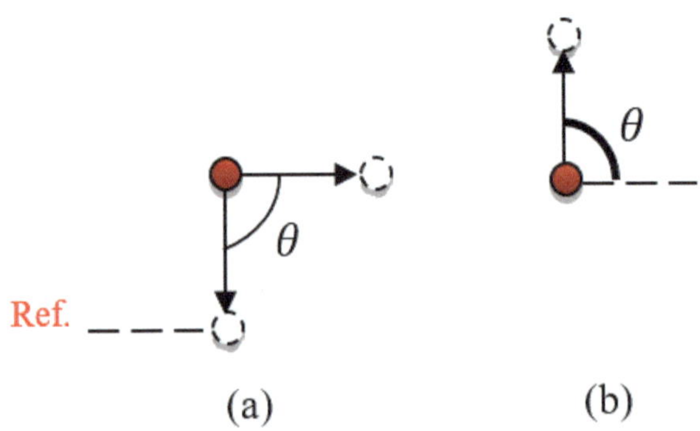

Fig. 2.25 Molecule movement between plates; **a** $\boldsymbol{\theta} \in \left[0, \frac{\pi}{2}\right)$ and **b** $\boldsymbol{\theta} \in \left(\frac{\pi}{2}, \boldsymbol{\pi}\right]$

These particles are usually absorbed by the z layer. The goal is to consider the force per unit area for each layer at height z. This leads to the rate of change of the momentum exchange as follows

$$\frac{F}{A} = \frac{1}{A}\frac{\Delta p}{\Delta t} \tag{2.133}$$

It could be written that

$$\frac{F}{A} = n\int \Delta p c_z f(\vec{c})d\vec{c} \tag{2.134}$$

The integration in Eq. (2.134) is due the fact that the flux for all possible velocity elements should be considered. By substituting Eqs. (2.131) in (2.134) and employing Maxwellian velocity distribution, i.e. $f(\vec{c}) = \left(\frac{m}{2\pi k_B T}\right)^{\frac{3}{2}} e^{\frac{-mc^2}{2k_B T}}$, we have

$$\frac{F}{A} = mn\frac{du_x}{dz}\int c_z\left(\frac{m}{2\pi k_B T}\right)^{\frac{3}{2}} e^{-\frac{mc^2}{2k_B T}} d\vec{c}\,\lambda\cos\theta \tag{2.135}$$

It is assumed that the fluid is moving with $\langle v_x \rangle = u_x$ in x-direction. Usually, the flow speed u is much less than the average molecular speed $\langle c_x \rangle$.

The integral of Eq. (2.135) in the polar form could be expressed as

$$\frac{F}{A} = mn\frac{du_x}{dz}\int dcc^2\int_0^{\pi} d\theta\sin\theta\int_0^{2\pi} d\phi(c\cos\theta)\lambda\cos\theta\left(\frac{m}{2\pi k_B T}\right)^{\frac{3}{2}} e^{-\frac{mc^2}{2k_B T}} \tag{2.136}$$

To obtain Eq. (2.136), the differential $d\vec{c}$ is substituted from the below formula (see Fig. 2.26)

$$d\vec{c} = c^2 dc d\theta \sin\theta d\phi \tag{2.137}$$

$$c_z = c\cos\theta \tag{2.138}$$

It should be noted that in the case of Couette flow, the term $\frac{du_x}{dz}$ in Eq. (2.136) is constant. In Eq. (2.136), $c = |\boldsymbol{c}|$. Based on the following obvious relations

$$\int_0^{2\pi} d\phi = 2\pi$$

$$\int_0^{\pi} d\theta \cos^2\theta \sin\theta = \frac{2}{3}$$

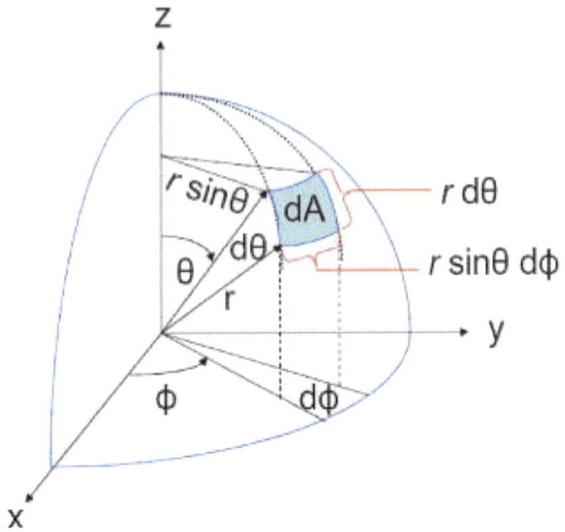

Fig. 2.26 3-D element in polar coordinate

We have

$$\frac{F}{A} = \frac{mn\lambda}{3}\frac{du_x}{dz}\int dc 4\pi c^3 \left(\frac{m}{2\pi k_B T}\right)^{\frac{3}{2}} e^{-\frac{mc^2}{2k_B T}} \tag{2.139}$$

The integral term in Ea. (2.138) is equal to $\overline{c} = \langle c \rangle$, then

$$\frac{F}{A} = \frac{1}{3} mn\lambda\overline{c}\frac{du_x}{dz} \tag{2.140}$$

Combining Eqs. (2.139) and (2.128) leads to the below formula for the viscosity

$$\mu = \frac{1}{3} mn\lambda\overline{c} \tag{2.141}$$

Equation (2.141) was first derived by Maxwell and it is accepted as a relation with low accuracy. According to Eq. (2.141) and considering the mean free path formula $\lambda_{HS} = \frac{1}{\sqrt{2}n\pi d^2}$ one can conclude that the viscosity is not proportional to the density (n). Viscosity is always dependent on the temperature and not pressure. Since it was assumed that the velocity profile is completely linear, it follows that the flow regime is continuum.

Now, we obtain the exact expression for the integral of Eq. (2.139). We know that

$$\int_0^\infty x^3 exp(-a^2x^2)dx = \frac{1}{2a^4} \tag{2.142}$$

Therefore,

$$\int_0^\infty dc4\pi c^3\left(\frac{m}{2\pi k_B T}\right)^{\frac{3}{2}} e^{-\frac{mc^2}{2k_BT}} = \left(\frac{m}{2\pi k_B T}\right)^{\frac{3}{2}} 4\pi \int c^3 e^{-\frac{mc^2}{2k_BT}} dc$$
$$= \left(\frac{m}{2\pi k_B T}\right)^{\frac{3}{2}} 4\pi \left(\frac{2k_B T}{m}\right)^2 = \sqrt{\frac{8kT}{\pi m}} \tag{2.143}$$

The average velocity in the Maxwellian condition reads

$$\overline{c} = \sqrt{\frac{8kT}{\pi m}} \tag{2.144}$$

By substitution Eq. (2.144) and considering $\lambda_{HS} = \frac{1}{\sqrt{2}n\pi d^2}$ in Eq. (2.141), we have

$$\mu = \frac{1}{3}mn\left(\frac{1}{\sqrt{2}n\pi d^2}\right)\left(\sqrt{\frac{8kT}{\pi m}}\right) = \frac{2}{3d^2}\sqrt{\frac{mkT}{\pi^3}} \tag{2.145}$$

Generally, the gas viscosity could be described as

$$\frac{\mu}{\mu_0} = \left(\frac{T}{T_0}\right)^n \tag{2.146}$$

μ_0 is a reference which corresponds to $T_0 = 273\text{K}$. and it is dependent on the gas type. In general, the Sutherland law is employed to capture variations of the gas viscosity with respect to the temperature.

2.5.2 Viscosity Coefficient from Molecular Viewpoint

Consider a velocity profile that varies in one direction, i.e. it is a function of y, or on the other word $\overline{c_x}(y)$ (see Fig. 2.27) [5]. Consider a molecule which has a x-momentum equal to mc'_x. This molecule is displaced in y-direction via c'_y. Therefore, the momentum flux in y-direction is equal to $mc'_x c'_y$. Now, the average x-momentum flux transferred in y-direction for N molecule is equal to $-nm\overline{c'_x c'_y}$. The negative sign here indicates that if $c'_y > 0$, momentum has been decreased with respect to the previous position. In the other word, the y-momentum decreased with respect to the initial state.

$$\sum_{i=1}^{N} mc'_x c'_y = nm\left[\frac{\Delta n(c'_x c'_y)_1 + \Delta n(c'_x c'_y)_2 + \ldots}{n}\right] = nm\overline{c'_x c'_y} \tag{2.147}$$

The term $\Delta n(c'_x c'_y)_1$ indicates the number of molecules existing in the first speed category. This term should be integrated over all ranges of speed to obtain the average value of momentum flux. On the other hand, the momentum change for each molecule between successive collisions is as

$$m\left(\overline{c}_{x(y=o)} - \overline{c}_{x(y=-\lambda)}\right) \tag{2.148}$$

By employing the Taylor series expansion, we have

$$m\left(\overline{c}_{x(y=o)} - \overline{c}_{x(y=-\lambda)}\right) = m\left(\lambda \frac{d\overline{c}_x}{dy} + \frac{\lambda^2}{2}\frac{d^2\overline{c}_x}{dy^2} + \ldots\right) = m\lambda \frac{d\overline{c}_x}{dy} + O(\lambda^2) \tag{2.149}$$

The number of collisions per unit area and time is equal to nc'. Therefore, by multiplying this term by Eq. (2.149) and setting the result equal to Eq. (2.147)

$$\beta_\mu n\overline{c'} m\lambda \frac{d\overline{c_x}}{dy} = -nm\overline{c'_x c'_y} \tag{2.150}$$

The right-hand side of the above equation is the stress from a microscopic point of view, which must be multiplied by a constant to be equal to its corresponding value from a macroscopic point of view, where β_μ is a proportionality constant and for monoatomic gases $\beta_\mu \approx 0.5$. The above relation can be rewritten as follows

$$\beta_\mu \rho \overline{c'} \lambda \frac{d\overline{c_x}}{dy} = -\rho\overline{c'_x c'_y} \tag{2.151}$$

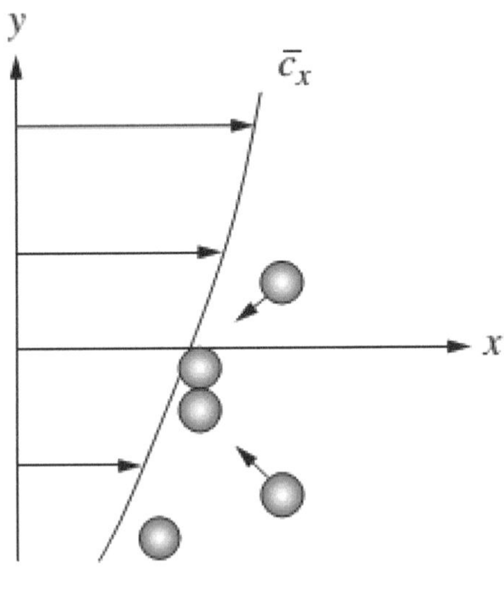

Fig. 2.27 Velocity profile and molecular behavior

According to the Eq. (2.151), the viscosity could be calculated as

$$\mu = \beta_\mu \rho \overline{c'} \lambda \tag{2.152}$$

Using Eq. (2.152), one can calculate the viscosity based on the thermal molecular speed. It is shown that the viscosity is a function of the molecular mean free path (λ). The parameter λ is dependent on the collision frequency (ν) which itself is a function of collision cross-section (σ_T). Collision cross-section is also relies on deflection angle (χ). Therefore, the deflection angle (χ) and intermolecular potential model play major roles in the viscosity.

References

1. Cercignani, C.: Rarefied Gas Dynamics: From Basic Concepts to Actual Calculations, vol. 21. Cambridge university Press (2000)
2. Hirshfelder, J.O., Curtiss, C.F., Bird, R.B.: Molecular Theory of Gases and Liquids (1964)
3. Feller, W.: An introduction to Probability Theory and Its Applications, vol. 2. Wiley, New York (1991)
4. Bird, G.A.: Molecular Gas Dynamics and the Direct Simulation of Gas Flows. Clarendon Press, Oxford (1994)
5. Liou, W.W., Fang, Y.: Microfluid Mechanics: Principles and Modeling (2006)
6. Hassan, H.A., Hash, D.B.: A Generalized Hard-Sphere Model for Monte Carlo Simulation. Phys. Fluids A 5 738–744 (1993)

Chapter 3
Direct Simulation Monte Carlo Method

Abstract The DSMC technique uses a finite set of model particles (simulators) that move and collide in a computational domain to perform a stochastic simulation of real molecular gas dynamics. The basic concept of the method is based on the discretization of time and space in the real gas dynamics process, and it splits the motion into two successive stages: free molecular motion and binary intermolecular collisions within the grid cells at each time step. In this chapter, the novel N-particle equation governing the DSMC technique is presented, and the principles of the DSMC method are described, with potential applications to MEMS filled with gas in mind. A preview of classic collision algorithms, including Majorant Collision Frequency, Bernoulli Trials, and No Time Counter, will be provided, with particular attention to their relationship to the Boltzmann equation, their corresponding accuracy, limitations, and restrictions. The application of different boundary conditions will be discussed. The simulation of near-continuum gas flows using time-relaxed and hybrid schemes will be presented. The simulation of low-speed gas flows is an essential microfluidic issue. Some new DSMC schemes for multiscale simulation will be presented. Finally, the simulation of gas flow in complex two and three domains will be commented on, and new DSMC algorithms using a small number of particles in cells will be described.

3.1 Introduction

Direct Simulation Monte Carlo (DSMC) [1, 2] has been widely used for many years to investigate problems related to rarefied gas flows. It was proposed and demonstrated for the first time in 1963, when the originator Prof. G. Bird published a short paper in Physics of Fluids [3]. Formally, the evolution of the method could be divided into three periods. Initially, the DSMC was considered a numerical technique that simulates real molecular gas dynamics by using a limited number of model particles with specific locations and speeds $\{\mathbf{r}_i, \mathbf{v}_i\}$, $i = 1, \ldots, N$, which move and collide randomly within a simulation area. The method works by dividing the real gas dynamics process into small units of time and space, and then simulating two steps in each unit: first, the free movement of molecules, and second, the pairwise collisions

E. Roohi et al., *Advances in Direct Simulation Monte Carlo: From Micro-Scale to Rarefied Flow Phenomena*, https://doi.org/10.1007/978-981-96-8200-3_3

of molecules inside each grid cell. Modeling the pairwise collisions in cells is the more difficult step, and many researchers have tried to enhance the original "Time Counter" (TC) collision scheme that Graeme Bird [1] first suggested. The original scheme had a problem with calculating the pairwise collision frequency, which was affected by the inverse number of model particles in each grid cell per time unit. In practice, this meant that at least 20–30 particles per cell (depending on the simulated gas flow) were needed to obtain accurate values for the macroscopic flow features. Within this period the first modifications of the TC collision scheme were proposed by Koura [4] and Yanitskiy [5]. The former suggested the time interval between particle collisions in a cell to be calculated from an exponential distribution instead of from the formula defined by the binary collisions in equilibrium used in the TC scheme. The latter proposed a different approach to counting binary collisions by using a strict Markov process based on Bernoulli trials with a variable probability proportional to the product of the particle's relative velocity and the total collision cross-section. The new schemes improved the accuracy of the collision algorithm, but their efficiency was less compared to TC one and, at that time, remained unpopular. Wagner [6] provided a strict mathematical demonstration that the method approaches the Boltzmann equation when the number of particles in cells is very large. Sone's monograph [7] contains a review of the method's theoretical basis.

In the second period, continuing theoretical studies have suggested several collision schemes with improved features: "Null-Collision" [8], "Modified-Nanbu" [9, 10], "Majorant Collision Frequency" [11], and "No Time Counter (NTC)" [2]. The Bird's NTC scheme has become the most common scheme, and we will refer to it as the "standard collision scheme" later in the text. All these methods employ the same principle of identifying the highest collision frequency in cells to determine the number of particle pairs to test for collision. Then they use the acceptance-rejection procedure to find the actual number of collisions in cells. The primary difference between these schemes lies in their approach to determining the highest collision frequency. When simulating flows with strong non-equilibrium phenomena, such as shock waves, a large number of particles per cell (at least 20–30) is required to obtain reliable results. During this period, an alternative approach was proposed by Yanitskiy in the "Ballot-box" collision scheme [12]. Instead of calculating the maximum collision number per time step first, he suggested subdividing the time step into subintervals, in which only one collision of a randomly chosen pair occurs with a given probability. This scheme is built on a strict Markov random process and does not use the "acceptance-rejection" method. The "Ballot-box" scheme demonstrates similar properties to those of the "No Time Counter".

The third period of the DSMC method development is associated with the emergence of more sophisticated modifications of the method to meet the increasing computational requirements of DSMC applications in complex, rarefied gas dynamics and microfluidic problems. The DSMC calculations revealed two fundamental problems, in addition to the inherent discretization issues of deterministic numerical approaches, namely, the presence of statistical noise in the numerical solution and the dependence of the results on the particle number per cell. A promising way to overcome the first one can be found in the proposed Monte Carlo schemes for near-continuum [13, 14, 15, 16] and low speed [17, 18, 19] gas flows, that consider the related asymptotic features of the Boltzmann equation. Especially in the last papers, the authors propose different interesting and promising ways to reduce variation in the numerical solution by simulating only the difference from the local equilibrium state. However, precise simulation appears to require a large number of particles per cell to prevent the random errors that occur during the particle creation and removal process, which is a fundamental part of the suggested algorithms. The second problem is related to the generation of random values from a small group of particles that represent the local velocity distribution function in a cell at each time unit. This causes a systematic error in estimating important parameters in the simulation. Similar thoughts highlight the limitations of the standard DSMC method when applied to small numbers of model particles (simulators) per cell. Little attention was paid to the mentioned problem until recently. To ensure reliable results, all collision schemes that utilize the principle of maximizing the number of collision pairs require a specified number of particles per cell $N \sim 10 - 20$. The reason why the algorithms that use acceptance-rejection method reflect the mentioned behavior is that they allow the same pair of particles to collide multiple times when calculating the actual number of collisions. In cells with a small number of particles, the effect of repeat collisions is dominant and leads to distortion of the collision process (see [20]). Bird suggested a new version of the DSMC method with better details of some of the processes (see the materials of his short course presented at the DSMC conference in Santa Fe, 2007, and also the book [21]). In the enhanced collision NTC algorithm, the problem of the low number of unrealized possible collisions between particles in nearby cells was partially addressed in the subcell areas of the simulation grid by utilizing a complex algorithm to find partners for collisions among single particles from different nearby subcells. However, the occurrence of repeat collisions stayed as a source of random error. The convergence of the complex Bird's algorithm is examined by Gallis et. al. [22]. Their analysis revealed that, in general, a lower number of particles than 10–20 in cells reduces the accuracy of the complex method. Su et al. [23] proposed a version of the transient adaptive sub-cell (TAS) approach in the DSMC method, which represents an interesting modification of the sophisticated method for simulating unstructured grids.

Yanitskiy [12] proposed an alternative way of construction of collision algorithms on the basis of the Kac stochastic model of the collision process in a homogenous rarefied gas [24] (see also [25]). Following Yanitskiy's approach, a simplified version (SBT) of the Bernoulli trials (BT) scheme with a linear dependence on the number of particles is proposed in [26, 27] that prevents repeated collisions and reveals improved accuracy under a small number of particles per cell. The algorithms BT or SBT have been suggested for use in combination with a universal two-step collision procedure to reduce the number of unrealized potential collisions between particles situated in neighboring cells. This leads to more accurate calculations with a smaller mean number of particles in the grid cells. From the perspective of the general simulation algorithm, the two-step collision procedure is an essential component of a symmetric Strang splitting scheme [28] that enhances the accuracy of the splitting method with respect to the time step.

The third phase of DSMC development is not yet complete. The DSMC parallel code implementations [29, 30, 31, 23, 32, 33, 34] on computational systems with different parallel organization of the computations such as supercomputers, clusters, and GPU systems can also be related to this period; however.

A subsequent fourth period of the DSMC development and its application began to emerge, which could be associated with the growing need for multiscale and multiphysics computational modeling and simulation. Some hybrid schemes combining molecular dynamics and the DSMC method, as well as DSMC and continuum gas dynamic models, have emerged, showing promising results (see Section *Multiscale flow simulation*).

In the present chapter, the basic elements of the DSMC method are described mostly for single monoatomic gas. Presented are different models of binary molecular interactions and boundary conditions. Finally, some examples of micro gas flow simulations taken from the authors' work are provided. Where necessary, the consideration is extended to a neutral gas mixture. The simulation of a chemical reacting gas mixture required a lot of additional information and was not included in the chapter.

3.2 Basic Elements of the DSMC Method

3.2.1 *The Splitting Scheme*

In general, a complete mathematical explanation of the movement of a rarefied gas system can be written by an evolving kinetic equation in this open form with respect to the velocity distribution function $f(t, \mathbf{r}_i, \mathbf{v}_i)$:

$$\frac{\partial}{\partial t} f(t, \mathbf{r}_i, \mathbf{v}_i) = -D\big[f(t, \mathbf{r}_i, \mathbf{v}_i)\big] + Q\big[f^{(2)}\big(t, r_i, \mathbf{v}_i, \mathbf{r}_j, \mathbf{v}_j\big)\big], \tag{3.1}$$

where $f(t, \mathbf{r}_i, \mathbf{v}_i) = f^{(1)}(t, \mathbf{r}_i, \mathbf{v}_i)$ and $f^{(2)}\big(t, \mathbf{r}_i, \mathbf{v}_i, \mathbf{r}_j, \mathbf{v}_j\big)$ are one-particle and two-particle distribution functions corresponding to particle velocities of $\mathbf{v}_i$ and $\mathbf{v}_j$ at time t and spacial coordinate $\mathbf{r}_i$. D and Q are the linear differential operator describing the free particle motion and the nonlinear integral operator describing the particle binary interactions, respectively. Equation (3.1) is just the final equation in the known BBGKY-hierarchy for an N-particle system with short-range intermolecular potential [35] (the generalized form of Eq. (3.1) governing the local N-particle distribution is derived in [36]). Considering the collision cross-section σ with a finite diameter d of a cut-off intermolecular potential (for a hard-sphere gas $\sigma = \pi d^2$, where d is the molecular diameter), in the Grad-Boltzmann limit ($N \to \infty, \sigma \to 0, \sigma N$ bounded) and under the assumption of molecular chaos, then

$$f^{(2)}(t, \mathbf{r}_i, \mathbf{v}_i, \mathbf{r}_j, \mathbf{v}_j) = f^{(1)}(t, \mathbf{r}_i, \mathbf{v}_i) f^{(1)}(t, \mathbf{r}_j, \mathbf{v}_j), \tag{3.2}$$

Transforming the Eq. (3.1) into the well-known Boltzmann equation

$$\frac{\partial}{\partial t} f(t, \mathbf{r}_i, \mathbf{v}_i) = D\big[f(t, \mathbf{r}_i, \mathbf{v}_i)\big] + Q\big[f(t, \mathbf{r}_i, \mathbf{v}_i), f(t, \mathbf{r}_j, \mathbf{v}_j)\big], \tag{3.3}$$

where operators D and Q take the form

$$D = -\mathbf{v}\frac{\partial}{\partial \mathbf{r}}, \quad Q = \int\limits_{\Re^3 \times \mathbb{S}^2} \left(f_i' f_j' - f_i f_j\right) B(g, \boldsymbol{\theta}) d\Omega(\boldsymbol{\theta}) d\mathbf{r}. \tag{3.4}$$

The external force has been diminished in Eq. (3.3) for the simplicity. In Eq. (3.4), $f_i = f(t, \mathbf{r}_i, \mathbf{v}_i)$, $f_j = f(t, \mathbf{r}_j, \mathbf{v}_j)$, $f_i' = f(t, \mathbf{r}_i, \mathbf{v}'_i)$ and $f_j' = f(t, \mathbf{r}, \mathbf{v}'_j)$. The post-collision particle velocities v_i' and v_j' are as:

$$\begin{aligned} \mathbf{v}'_i &= \mathbf{v}_i + \boldsymbol{\theta}[(\mathbf{v}_i - \mathbf{v}_j) \cdot \boldsymbol{\theta}] \\ \mathbf{v}'_j &= \mathbf{v}_j - \boldsymbol{\theta}[(\mathbf{v}_i - \mathbf{v}_j) \cdot \boldsymbol{\theta}] \end{aligned} \tag{3.5}$$

The function $B(g, \boldsymbol{\theta})$ is dependent on the intermolecular potential, $\boldsymbol{\theta}$ is a unit vector that expresses the variation of the particle velocity owing to a particle collision [7], $g = |\mathbf{v}_i - \mathbf{v}_j|$ is the relative molecular velocity, and $d\Omega$ is a solid angle element in the direction of $\boldsymbol{\theta}$. In general, the function B is connected to the collision cross-section σ in a straightforward manner; see Refs. [35, 7]. A mathematical expression of a rarefied gas problem (3.3) must be specified by related initial and boundary conditions for the velocity distribution function defined within the volume and at the edges of the simulation area $\mathfrak{D}$.

A computational domain $\mathfrak{D}$ divided into M cells by step $h = \{\Delta x, \Delta y, \Delta z\}$ could be considered. Also, the time t could be divided into small intervals such as $\tau = \Delta t$ ($t_k = k\tau$, $k = 1, ..., K$). Let the real gas in the area be considered by a limited set of model particles shown by their locations and speeds $\{\mathbb{R}(t_k), \mathbb{V}(t_k)\} = \{\mathbf{r}_i(t_k), \mathbf{v}_i(t_k)\}$, $i = 1, \ldots, N$ which are changing over the simulation time. The model particles are distributed into the grid cells so that $N = \sum_{l=1}^{M} N^{(l)}$ ($N^{(l)}$ is the number of particles in cell l) and $\{\mathbf{r}_j^{(l)}(t_k), \mathbf{v}_j^{(l)}(t_k)\}$, $j = 1, \ldots, N^{(l)}$. An approximation for the velocity distribution function $f(t, \mathbf{r}, \mathbf{v})$ is as follows (see Sone [7])

$$\begin{aligned} f(t, \mathbf{r}, \mathbf{v}) &= w_N \sum_{l=1}^{M} \sum_{j=1}^{N^{(l)}} \delta(\mathbf{r} - \mathbf{r}_j^{(l)}(t)) \delta(\mathbf{v} - \mathbf{v}_j^{(l)}(t)), \quad (\mathbf{r} \in \mathfrak{D}), \\ f(t_k, \mathbb{R}(t_k), \mathbb{V}(t_k)) &\equiv f(t_k, \mathbf{r}, \mathbf{v}) \end{aligned} \tag{3.6}$$

where the Dirac delta function is indicated as δ and W_N is a scaling factor. Calculating the mean value of f from (3.6) over the cell volume $V^{(l)} = |\mathbf{dr}^{(l)}| = h^\alpha$ ($\alpha = 1, 2,$ or 3 is the volume dimension) results in a rough approximation of the velocity distribution function in cell (l)

$$f(t, \mathbf{r}^{(l)}, \mathbf{v}) = \frac{w_N}{\left|\mathbf{dr}^{(l)}\right|} \sum_{j=1}^{N^{(l)}} \delta\left[\mathbf{v} - \mathbf{v}_j^{(l)}(t)\right] = \frac{\Delta}{\left|\mathbf{dr}^{(l)}\right|} \int_{\mathbf{dr}^{(l)}} f(t, \mathbf{r}, \mathbf{v})\mathbf{dr}, \tag{3.7}$$

where $\mathbf{r}^{(l)}$ is a reference point in cell (l), e.g. the geometric center of cell (l).

The standard DSMC method uses the following splitting scheme, applied to the discrete velocity distribution function known at time t_k, to obtain an approximation of the solution of Eq. (3.1) at t_{k+1}:

$$\begin{aligned} f[t_k + \tau, \mathbb{R}(t_k), \mathbb{V}(t_k + \tau)] &= S_Q^{\tau,h}\{f[t_k, \mathbb{R}(t_k), \mathbb{V}(t_k)]\}, \\ f(t_0, \mathbb{R}(t_0), \mathbb{V}(t_0)) &\equiv f(0, \mathbf{r}, \mathbf{v}) \\ f[t_k + \tau, \mathbb{R}(t_k + \tau), \mathbb{V}(t_k + \tau)] &= S_D^{\tau}\{f[t_k + \tau, \mathbb{R}(t_k), \mathbb{V}(t_k + \tau)]\}. \end{aligned} \tag{3.8}$$

For the sake of brevity from now on, it is reasonable to assume that all necessary mathematical conditions for proper consideration are met. In splitting Eqs. (3.8), we indicate the numerical algorithms approximating the action of the collision and convective terms in (3.1) by operators $S_Q^{\tau,h}$ and S_D^{τ}, respectively. Operator $S_Q^{\tau,h}$ acts locally on the particle subset $N^{(l)}$ in each cell $\mathbf{dr}^{(l)}$, while operator S_D^{τ} acts on the total particle set within the entire computational domain $\mathfrak{D}$. A special feature of $S_Q^{\tau,h}$ is that the particle locations are ignored and a random even distribution of the particles in each cell is assumed. The mentioned assumption is equivalent to the transformation to a course-grained distribution function indicated by (3.7). A strict consideration should consider the fact that the DSMC collision algorithm $S_Q^{\tau,h}$ does not suppose that the particle speeds are statistically independent, i.e. equality (3.2) must be met. Hence, instead of acting on a one-particle distribution function given by (3.7), operator $S_Q^{\tau,h}$ acts on a course-grained two-particle velocity distribution function $f^{(2)}(t, \mathbf{r}^{(l)}, \mathbf{v}_i, \mathbf{v}_j)$. On the other hand, the action of operator S_D^{τ} is invariant to both one- and two-particle distribution functions. This indicates that the DSMC method can estimate the evolution of a course-grained two-particle velocity distribution function, and consequently, calculate corresponding local correlations and fluctuations of macroscopic variables; see [37].

Considering $S_{Q+D}^{\tau,h}$ as the operator evaluating the solution of (3.3) at $t_k + 1$ from the state at t_k, then the splitting method could be approximated as

$$S_{Q+D}^{\tau,h} \approx S_D^{\tau} S_Q^{\tau,h} \approx S_Q^{\tau,h} S_D^{\tau}. \tag{3.9}$$

A conventional viewpoint is that the DSMC is a method that statistically solves the Boltzmann equation. However, as recently shown [2, 37, 25, 38, 39], DSMC does not necessarily model the Boltzmann equation. DSMC actually solves an equation with less restriction. There are two observed features that prove the above-mentioned statement. First, DSMC could be simulated under a limited number of particles in spite of the fact that (was proved mathematically) the DSMC solution converges to the Boltzmann equation under an infinite number of simulators [25]. Second, in DSMC, the post-collision velocities of colliding particles are strongly correlated. This indicates that the assumption of molecular chaos, embodied in the Boltzmann equation, is not generally fulfilled in the DSMC collision process. It also should be noted that DSMC has been widely used to obtain correct numerical estimations of fluctuations and correlations of macroscopic variables [2, 37, 38, 39]. The aforementioned remarks suggest that the Boltzmann equation may not be the governing equation of the DSMC method. Stefanov [40] introduced a novel nonhomogeneous, local N-particle kinetic equation, which can be considered the governing equation that DSMC statistically solves. The mentioned equation is derived via the spatial and temporal discretization of the splitting form of kinetic Eqs. (3.11)–(3.13) and applying to the N-particle distribution function ($\tilde{F}_N$). The function $\tilde{F}_N$ is randomized at instant time t over the number of particles in the cell l, i.e. $\overline{N}^{(l)}$ and velocities $\mathbb{V} = \{\mathbf{v}_1, \ldots, \mathbf{v}_i, \ldots, \mathbf{v}_j, \ldots, \mathbf{v}_N\}$ in a small volume of V around point $\mathbf{r}$. The Stefanov's governing equation has the following form:

$$\frac{\partial \tilde{F}_N(t,\mathbf{r},\mathbb{V})}{\partial t} + \sum_{i=1}^{N} \mathbf{v}_i \frac{\partial \tilde{F}_N(t,\mathbf{r},\mathbb{V})}{\partial \mathbf{r}} = \sum_{1 \le i \le j \le N} \left\{ \int g_{i,j} \left[\tilde{F}_N\left(t,\mathbf{r},\mathbb{V}'_{\mathrm{i,j}}\right) - \tilde{F}_N(t,\mathbf{r},\mathbb{V}) \right] d\sigma_{i,j} \right\}, \tag{3.10}$$

where $d\sigma_{i,j}$ is the differential collision cross-section and $\mathbb{V}'_{\mathrm{i,j}} = \{\mathbf{v}_1, \ldots, \mathbf{v}'_{\mathrm{i}}, \ldots, \mathbf{v}'_{\mathrm{j}}, \ldots, \mathbf{v}_{\mathrm{N}}\}$ is the velocity vector update after the collision occurring between particles i and j, and $g_{i,j} = |\mathbf{v}_i - \mathbf{v}_j|$. This equation describes the evolution of the local N-particle distribution in the vicinity of the physical point $\mathbf{r} \in \Re^3$ which is defined in 3D physical space and $3N$-dimensional velocity space. The discretized for of Eq. (3.10) is as:

$$t < \tau \le t + \Delta t,\ l = 1, M$$

$$\left| \begin{array}{l} \tilde{F}^*_{N^{(l)}}\left(t+0, r^{(l)}, \mathbb{V}^{(l)}\right) = \tilde{F}_{N^{(l)}}\left(t, r, \mathbb{V}^{(l)}\right), \quad r \in \mathfrak{D}^{(l)} \subset \Re^3 \\ \dfrac{\partial \tilde{F}^*_{N^{(l)}}\left(t, r^{(l)}, \mathbb{V}^{(l)}\right)}{\partial t} = \dfrac{1}{V^{(l)}} \displaystyle\sum_{1 \le i \le j \le N^{(l)}} \left\{ \int g_{i,j} \left[\tilde{F}^*_{N^{(l)}}\left(t, r^{(l)}, \mathbb{V}'^{(l)}_{i,j}\right) - \tilde{F}^*_{N^{(l)}}\left(t, r^{(l)}, \mathbb{V}^{(l)}\right) \right] d\sigma_{i,j} \right\} \end{array} \right. \tag{3.11}$$

$$\left| \begin{array}{l} \tilde{F}^{**}_{N^{(l)}}\left(t+0, \mathbf{r}^{(l)}, \mathbb{V}^{(l)}\right) = \tilde{F}^*_{N^{(l)}}\left(t+\Delta t, \mathbf{r}^{(l)}, \mathbb{V}^{(l)}\right), \\ \dfrac{\partial \tilde{F}^{**}_{N^{(l)}}\left(t, \mathbf{r}^{(l)}, \mathbb{V}^{(l)}\right)}{\partial t} = -\displaystyle\sum_{i=1}^{N^{(l)}} \mathbf{v}_{\mathrm{i}} \dfrac{\partial \tilde{F}^{**}_{N^{(l)}}\left(t, \mathbf{r}^{(l)}, \mathbb{V}^{(l)}\right)}{\partial \mathbf{r}} \quad \mathbf{r} \in \mathfrak{D}^{(l)} \end{array} \right. \tag{3.12}$$

$$\left| \begin{array}{l} F_N(t+\Delta t, \mathbf{r}, \mathbb{V}) = \sum_{l=1}^{M} \tilde{F}^{**}_{N^{(l)}}\left(t+\Delta t, \mathbf{r}^{(l)}, \mathbb{V}^{(l)}\right) \\ \tilde{F}_{N^{(l)}}\left(t+\Delta t, \mathbf{r}^{(l)}, \mathbb{V}^{(l)}\right) = \int\limits_{D^{(l)}} F_N(t+\Delta t, \mathbf{r}, \mathbb{V}) d\mathbf{r}, \quad \mathbf{r} \in \mathfrak{D} \subset \Re^3 \end{array} \right. , \tag{3.13}$$

where Δt is the time step and M represents the number of subdomains $\mathfrak{D}^{(l)}$. Actually, $\mathfrak{D}^{(l)}$ is analogous to computational cells in the DSMC method. $\tilde{F}_N$ represents the N-particle distribution function for particle velocities around the spatial point $\mathbf{r} \in \mathfrak{D} \subset \Re^3$. Equations (3.11)–(3.13) correspond to algorithmic interpretations in the DSMC method by employing three successive procedures during each time interval. Equations (3.12) and (3.13) correspond to the ballistic motion and indexation of particles, respectively. Equation (3.11) indicates the mathematical interpretation of the binary collision relaxation process in the DSMC. Equation (3.11) could be written in the operator form as follows:

$$\frac{\partial \tilde{F}^{*}_{N^{(l)}}\left(t, \mathbf{r}^{(l)}, \mathbb{V}^{(l)}\right)}{\partial t} = \Omega \tilde{F}^{*}_{N^{(l)}}\left(t, \mathbf{r}^{(l)}, \mathbb{V}^{(l)}\right), \tag{3.14}$$

$$\begin{aligned} &\Omega \tilde{F}^{*}_{N^{(l)}}\left(t, r^{(l)}, \mathbb{V}^{(l)}\right) \\ &= \frac{1}{V^{(l)}} \sum_{1 \le i \le j \le N^{(l)}} \left\{ \int g_{i,j} \left[\tilde{F}^{*}_{N^{(l)}}\left(t, r^{(l)}, \mathbb{V}'^{(l)}_{i,j}\right) - \tilde{F}^{*}_{N^{(l)}}\left(t, r^{(l)}, \mathbb{V}^{(l)}\right) \right] d\sigma_{i,j} \right\} \end{aligned} \tag{3.15}$$

Operator Ω produces an update $\tilde{F}^{*}_{N^{(l)}}$ for infinitesimally small time step $\Delta t \to 0$ that comes from a single collision (defined as an instantaneous event). The solution of Eq. (3.14) at time $t + \Delta t$ can be described based on the below series of the solution at time t considering the number of collisions k:

$$\tilde{F}^{*}_{N^{(l)}}\left(t+\Delta t, \mathbf{r}^{(l)}, \mathbb{V}^{(l)}\right) = \sum_{k=0}^{\infty} \frac{\Omega^k \tilde{F}^{*}_{N^{(l)}}\left(t, \mathbf{r}^{(l)}, \mathbb{V}^{(l)}\right)}{k!} (\nu \Delta t)^k, \tag{3.16}$$

where ν is the binary collision frequency, which is generally unknown and depends on the solution.

The exponential form of the collision transition operator for a time step $\Delta t \to 0$, which gives the solution of the discrete collision Eq. (3.16) at instant $t + \Delta t$ from an initial state t is given by

$$G(\Delta t) = \exp[\Delta t \nu (\mathrm{T} - \mathrm{I})], \tag{3.17}$$

where the new operators I and T decompose the exponential operator into identity and 3D-velocity rotation operators

$$\begin{aligned}
I\psi &= \psi \\
T_{i,j}\psi &= \frac{1}{\sigma_{i,j}} \int\limits_{4\pi} \psi(\mathbf{v}_i, \mathbf{v}_j)\sigma(g_{i,j}, \theta) d\theta d\varepsilon \\
T\psi &= \sum_{1 \le i < j \le N^{(l)}} \omega_{i,j} T_{i,j}\psi
\end{aligned} \tag{3.18}$$

The unknown binary collision frequency ν is defined as:

$$\begin{aligned}
\nu &= \sum_{1 \le i < j \le N^{(l)}} \omega_{i,j} \\
\omega_{i,j} &= \frac{\sigma_{i,j} g_{i,j}}{V^{(l)}}
\end{aligned} \tag{3.19}$$

The transition operator $G(\Delta t)$ serves for the derivation of the series of Bernoulli trial schemes [36].

3.2.2 General Flowchart

The splitting approximation (3.9) illustrates nothing else but the traditional two-stage splitting scheme used in the DSMC method. The algorithmic interpretation of the basic DSMC elements is illustrated in the following generalized flowchart (Fig. 3.1):

3.3 Basic Collision Schemes

Over the years, a number of collision schemes (see the chapter introduction) realizing operator $S_Q^{(\tau,h)}$ have appeared and been used for solving a wide variety of rarefied gas dynamic problems. In this chapter, four essential parts of the modern realizations of the DSMC method are chosen to represent the basic ideas included in all binary collision algorithms.

3.3.1 No Time Counter (NTC) Scheme

The most popular and frequently used is the Bird's No Time Counter collision (NTC) scheme [2]. The latest version of the NTC collision procedure (see the materials of the short course delivered by Prof. Bird at the DSMC conference in Santa Fe, 2007 and his book [21]) consists of the following steps, performed in each cell l, $l = 1, ..., M$:

NTC Algorithm:

Fig. 3.1 General flowchart of the DSMC algorithm

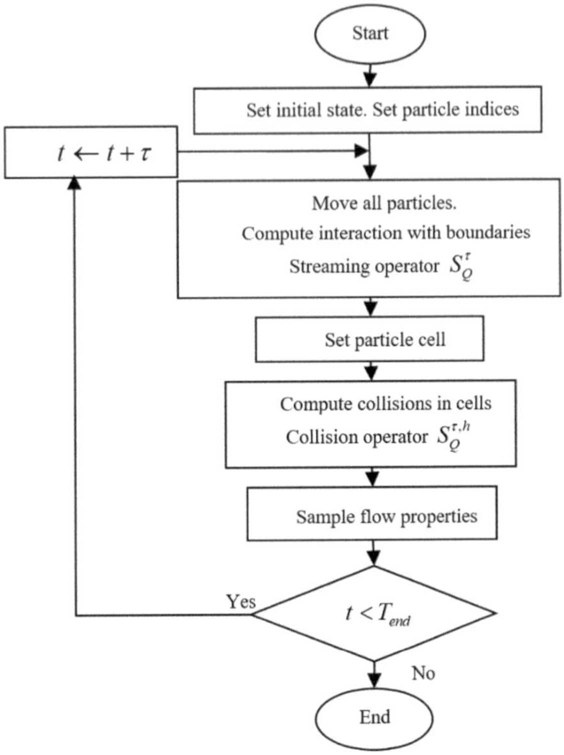

Step 1. Calculate the number of possible collision pairs N_c to be checked for the collision using

$$N_c = \mathrm{w_N} \frac{N^{(l)}\left(N^{(l)} - 1\right)}{2V^{(l)}} (\sigma g)^{(l)}_{\max} \tau, \tag{3.20}$$

where $(\sigma g)^{(l)}_{\max}$ is a superior parameter, which is updated continuously during the simulation, factor $\mathrm{w_N}$ is the number of real molecules presented by one simulated particle, τ is the time step and N^l and V^l represent the number of particle and volume of the l-th cell.

Step 2. Choose at random a pair (i, j), $1 \le i < j \le N^{(l)}$ from the particle subset $N^{(l)}$;

Step 3. Check pair (i, j) for collision with probability

$$p_{i,j} = \frac{\sigma_{i,j} g_{i,j}}{(\sigma g)_{\max}}, \tag{3.21}$$

where σ_{ij} is the effective collision cross-section of pair (i, j) and $g_{i,j} = \left|\mathbf{v}_i - \mathbf{v}_j\right|$;

Step 4. If the collision is accepted then the particle velocities are changed to their post-collision values

$$\mathbf{v}'_i = \frac{1}{2}\left[\left(\mathbf{v}_i + \mathbf{v}_j\right) + \left|\mathbf{v}_i - \mathbf{v}_j\right|\boldsymbol{\omega}\right]$$
$$\mathbf{v}'_j = \frac{1}{2}\left[\left(\mathbf{v}_i + \mathbf{v}_j\right) - \left|\mathbf{v}_i - \mathbf{v}_j\right|\omega\right], \tag{3.22}$$

otherwise, the velocities remain unchanged.

Step 5. If the number of checked particle pairs is less then N_c transition to step 2.

The system of Eqs. (3.22) is an alternative of (3.5), where $\boldsymbol{\omega}$ is a unity vector pointing out from the new direction of relative velocity $\left(\boldsymbol{v}'_i - \boldsymbol{v}'_j\right)$. The end of vector $\boldsymbol{\omega}$ is uniformly distributed on a sphere of unit radius. Operator $S_Q^{\tau,h}$ does not act on the particle positions and they remain unchanged.

The algorithm's performance is proportional to the number of particles in the cell, with a factor that is usually well below 1, making the NTC collision algorithm very efficient in cases where the number of particles is large, $N^{(l)} > 1$. However, the method of random selection of particle pairs in a cell allows repeatedly the same pair to be checked for collision and, consequently, multiple duplicate collisions to be realized. For $N^{(l)} > 1$, this occurs rarely; however, when a few particles are present in a cell, the probability of repeated collisions increases rapidly and becomes unacceptable. In practice, numerous calculations demonstrate that, in most cases of DSMC calculations, using the NTC scheme, an average number of particles is sufficient to obtain reliable numerical data with sufficient accuracy.

There is an important detail related to the number of checked-for collision pairs N_c specified by the formula (3.20). The right-hand side value is not an integer, and in NTC, it is rounded to the nearest lower integer. In this way, N_c remains systematically underestimated. To overcome this effect, the residual from the rounding operation (the remainder) is retained and added to the right-hand side value, which is calculated in the same cell at the next time step. The use of remainders is avoided in the Majorant Collision Frequency scheme considered in the next subsection.

3.3.2 Majorant Collision Frequency (MCF) Scheme

The MFC scheme [11, 41] consists of the following steps, performed in each cell l, $l = 1, \ldots, M$:

MCF algorithm:

Step 1. Compute the majorant collision frequency for cell l

$$\nu^{(l)} = \mathrm{w_N}\frac{N^{(l)}\left(N^{(l)} - 1\right)}{2V^{(l)}}(\sigma g)^{(l)}_{\max}, \tag{3.23}$$

where $(\sigma g)^{(l)}_{\max}$ is a superior parameter, which is updated continuously during the simulation.

Set the collision number $m = 0$;

Step 2. Sample a random time interval δt to the next collision from the probability density $\nu^{(l)} \exp(-\nu^{(l)}\delta t)$ and increase $t_{m+1} = t_m + \delta t$;

Step 3. If $t_{m+1} > \tau$ stop MCF procedure and proceed to the next cell. If $t_{m+1} \le \tau$ choose at random a pair (i, j), $1 \le i < j \le N^{(l)}$ from the particle subset $N^{(l)}$;

Step 4. Check pair (i, j) for collision with probability

$$p_{i,j} = \frac{\sigma_{i,j} g_{i,j}}{(\sigma g)_{\max}}, \tag{3.24}$$

where σ_{ij} is the total collision cross-section of pair (i, j) and $g_{i,j} = \left|\mathbf{v}_i - \mathbf{v}_j\right|$;

Step 5. If the collision is accepted, then the particle velocities are changed to their post-collision values

$$\begin{aligned} \mathbf{v}'_i &= \frac{1}{2}\left[\left(\mathbf{v}_i + \mathbf{v}_j\right) + \left|\mathbf{v}_i - \mathbf{v}_j\right|\boldsymbol{\omega}\right] \\ \mathbf{v}'_j &= \frac{1}{2}\left[\left(\mathbf{v}_i + \mathbf{v}_j\right) - \left|\mathbf{v}_i - \mathbf{v}_j\right|\boldsymbol{\omega}\right], \end{aligned} \tag{3.25}$$

otherwise, the velocities remain unchanged.

As one can see, the MCF scheme differs from NTC in the way it calculates the number of selected pairs for collision within a cell during a time step. First, the time interval to the next collision is computed from the exponential distribution by using the majorant collision frequency. The time intervals are accumulated until the sum exceeds the time step, which is the condition that stops the iteration process. As a result of many realizations of the stochastic process, the number of selected collision pairs in each cell m is a random variable with a Poisson distribution

$$P(m, \lambda) = \frac{\lambda^m e^{-\lambda}}{m!}, \tag{3.26}$$

where the mean value, λ of m is exactly equal to the real value of the right-hand side of Eq. (3.20). Thus, the MCF algorithm evaluates the correct mean values of the number of selected collision pairs $\overline{N}_c$ in cells without using remainders, which is an advantage of this scheme. The algorithm's performance is proportional to the number of particles in the cell, and, respectively, the MCF scheme has similar efficiency to the NTC scheme.

3.3.3 Bernoulli Trials (BT) Scheme

Yanitskiy [5, 12] proposed an alternative of the NTC and MCF schemes by using the Bernoulli trials process of checking for collision of all possible particles $N^{(l)}(N^{(l)}-1)/2$ in cell l with probability, which is derived from the Kac stochastic collision model [24]. Unlike NTC and MCF the BT collision scheme does not use an "acceptance-rejection" procedure and avoids repeat collisions. The BT algorithm consists of the following steps performed for all combinations of particle pairs (i, j) chosen sequentially from $N^{(l)}$ particles in cell l.

BT Algorithm:

Step 1. The first particle of each pair is the particle with index $i = 1, \ldots, (N^{(l)} - 1)$ chosen in strict order from the particle list created for cell l;

Step 2. The second particle is the particle with index $j \in \left[i+1, N^{(l)}\right]$ chosen at random from $k = (N^{(l)} - i)$ particles taking place in the list after the particle i.

Step 3. The particle pair (i, j) with velocities $(\mathbf{v}_i, \mathbf{v}_j)$, $i < j = 1, \ldots, N^{(l)}$ is checked for collision with probability

$$\mathrm{W}_{i,j} = \mathrm{w_N}\frac{\sigma_{i,j}g_{i,j}\tau}{V^{(l)}}, \tag{3.27}$$

where $\mathrm{W}_{i,j}$ must satisfy the condition

$$Prob\{\mathrm{W}_{i,j} \geq 1\} \rightarrow 0. \tag{3.28}$$

Step 4. If the collision is accepted, the velocities $(\mathbf{v}_i, \mathbf{v}_j)$ are changed to their post-collision velocities $(\boldsymbol{v}'_i, \boldsymbol{v}'_j)$ according to (3.22), otherwise the velocities are left unchanged.

The number of Bernoulli trials is exactly equal to $N^{(l)}(N^{(l)}-1)/2$. The algorithm BT requires number of operations of order $O(N^{(l)2})$ in cell l and, respectively, is less efficient than the NTC algorithm. When the number of particles in cell is small $N^{(l)} \rightarrow 1$, the efficiency of algorithm BT can be improved by reducing the quadratic dependence on the particle number. The low computational efficiency was the reason for avoiding the use of Bernoulli trials for a long time. However, the BT algorithm has an advantage whose importance has only recently been realized: it does not produce repeat collisions within a cell during a time step. As shown in the new simplified algorithm [26, 27], with a number of operations of order, the avoidance of repeated collisions is a crucial requirement for accurately simulating the collision process in cells with a small number of particles.

3.3.4 Intermolecular Potentials

There are several collision models that describe different kinds of interactions between particles. The inverse power law model is based on the assumption that the repulsive force between the particles is inversely proportional to $1/r^{\eta}$. The dependence of the viscosity and thermal conductivity with temperature is proportional to T^{ω}, where

$$\omega = \frac{1}{2} + \nu = \frac{1}{2}\frac{\eta + 3}{\eta - 1} \tag{3.29}$$

According to this, the dependence is proportional to $T^{1/2}$ for hard sphere (HS), ($\eta \rightarrow \infty$) and to T for Maxwellian interaction ($\eta = 5$), respectively. The HS model has a limitation that the collision cross-section does not depend on the relative translational energy (E_t) of the colliding particles, which is not realistic. To match the powers of the order of 0.75, which are typical for real gases, a cross-section that varies with the conditions is needed. This is achieved by introducing the Variable Hard Sphere (VHS) model [2], which represents a hard sphere molecule whose diameter d is a function of the relative molecular velocity g.

$$\frac{\sigma_{i,j}}{\sigma_{ref}} = \left(\frac{d_{i,j}}{d_{ref}}\right)^2 = \left(\frac{g_{i,j}}{g_{ref}}\right)^{-2\nu}, \tag{3.30}$$

where σ_{ref}, d_{ref} and g_{ref} are reference values. In the VHS model, the ratio of the momentum to the viscosity cross-section follows the HS value. However, this is a deficiency because the ratio varies according to the power law in the case of the inverse power law model, and the real gas values differ from the HS values.

This deficiency is remedied by the Variable Soft Sphere (VSS) model [42], where the deflection angle χ is given by

$$\chi = 2\cos^{-1}\left[\left({}^{b}/_{d}\right)^{1/\alpha}\right], \tag{3.31}$$

where b is impact parameter.

When α is equal to 1 and v is equal to 0 ($\omega = 1/2$), the VSS model and the HS model are equivalent. The reference molecular diameter is related to the reference viscosity for all the cases

$$d_{ref}^2 = \frac{5(\alpha + 1)(\alpha + 2)\left(\frac{mkT_{ref}}{\pi}\right)^{1/2}}{4\alpha(5 - 2\omega)(7 - 2\omega)\mu_{ref}} \tag{3.32}$$

The expressions for the components of the post-collision relative velocity are given by:

$$g'_x = \cos\chi\, g_x + \sin\chi\, \sin\varepsilon \left(g_y^2 + g_z^2\right)^{1/2}$$
$$g'_y = \cos\chi\, g_y - \sin\chi \left(g g_z \cos\varepsilon + g_x g_y \sin\varepsilon\right) / \left(g_y^2 + g_z^2\right)^{1/2}$$
$$g'_z = \cos\chi\, g_z + \sin\chi \left(g g_y \cos\varepsilon - g_x g_z \sin\varepsilon\right) / \left(g_y^2 + g_z^2\right)^{1/2} \tag{3.33}$$

Further progress on the collision models has been achieved by Hassan and Hash [43] by introducing the Generalized Hard Sphere (GHS) model, introduced in Chap. 2. The GHS model is capable of reproducing the analytic viscosity coefficients available in the literature, which are derived for a variety of interaction potentials that incorporate both attraction and repulsion. Following this line [44] proposed a Generalized Soft Sphere (GSS) scheme and, recently, Sharipov and Strapasson [45] implemented an arbitrary intermolecular potential into the DSMC method by using the tabulated dependence of the deflection angle χ on the impact parameter b and the energy of the relative motion of molecules E.

Collision models, treating the internal degree of freedom of gas molecules, neutral and chemical reactive gas mixtures, are considered in detail in Bird's monograph [2]. A special note should be added for the recently published quantum-kinetic (Q–K) chemistry model [46] that substantially expands the scope of the DSMC method with respect to chemical reaction simulation.

3.4 Free Particle Movement

The streaming operator S_D^τ acts on the total particle set within the entire computational domain $\mathfrak{D}$. It calculates the new coordinates and velocities of all particles obtained as a result of their free collisionless movement in the computational domain. If the initial set of coordinates and velocities at time t is $\{\mathbb{R}(t), \mathbb{V}(t)\} = \{\mathbf{r}_i(t), \mathbf{v}_i(t)\}, i = 1, \ldots, N$ then the new coordinates and velocities are calculated following the laws of classical mechanics. Below several examples of particle movement are presented.

3.4.1 No External Force Acting on the Particles

The particles are moved to the new positions proportionally to their velocities, which remain unchanged

$$\mathbb{R}(t+\tau) = \mathbb{R}(t) + \mathbb{V}(t)\tau$$
$$\mathbb{V}(t+\tau) = \mathbb{V}(t). \tag{3.34}$$

3.4.2 A Constant External Force Acting on the Particles

The particles are moved with acceleration $\mathbf{F}/m$ and their new positions and velocities are given by

$$\mathbb{R}(t+\tau) = \mathbb{R}(t) + \mathbb{V}(t)\tau + \frac{\mathbf{F}}{m}\frac{\tau^2}{2}$$
$$\mathbb{V}(t+\tau) = \mathbb{V}(t) + \frac{\mathbf{F}}{m}\tau, \tag{3.35}$$

where m is the particle mass.

3.4.3 Flows with Axial Symmetry

The particle motion presented in an appropriate coordinate system can be used to reduce the number of dimensions required for simulating gas flow with a certain symmetry. The procedure for axially symmetric flows, proposed by Bird [2], is described below. The axial symmetry is assumed with respect to axis Ox. At the beginning of this stage, all particles have the same polar coordinate. Thus, only two distance coordinates (axial and radial) and three motion components are kept for each particle. Then, the molecules are moved according to their velocities and acquire certain coordinates x', y' and z' given by

$$x' = x + v_x\tau, \quad y' = y + v_y\tau, \quad z' = v_z\tau,$$
$$x^+ = x', \quad y^+ = \sqrt{y'^2 + z'^2} \tag{3.36}$$

while x^+, y^+ are taken to be the new axial and radial ordinates. The new velocity components, v_x^+, v_y^+ and v_z^+, are computed by taking the components of the previous velocity in x- direction and in the two directions parallel and orthogonal to the radius vector with components y', z'. Following this procedure, the new velocities read

$$v_x^+ = v_x, \quad v_y^+ = \frac{v_y y' - v_z z'}{y^+}, \quad v_z^+ = \frac{v_y z' - v_z y'}{y^+}. \tag{3.37}$$

3.4.4 Rotating Coordinate System

For the simulation of gas flow around rotating solid bodies, a very useful approach is to present the particle movement in a framework rotating with the angular velocity $\boldsymbol{\omega} = \left(\omega_x, \omega_y, \omega_z\right)$ of the rotating body [47]. In this framework, the body is stationary, and the relative movement of the particles concerning the body is calculated

$$\mathbf{r}'(t+\tau) = \mathbf{r}(t) + \mathbf{v}(t)\tau,$$
$$\mathbf{r}(t+\tau) = \mathbf{A}(\boldsymbol{\omega}\tau)\mathbf{r}'(t+\tau), \tag{3.38}$$

where $\mathbf{A}(\boldsymbol{\omega}\tau)$ is a 3×3 matrix transforming the new particle coordinates with respect to the position in the rotating frame at time $(t+\tau)$. For a rotating framework with respect to z-axis the matrix, i.e. $\boldsymbol{\omega} = (0, 0, \omega)$, $\mathbf{A}(\boldsymbol{\omega}\tau)$ takes the form

$$\mathbf{A}(\omega\tau) = \begin{pmatrix} \cos\omega\tau & \sin\omega\tau & 0 \\ -\sin\omega\tau & \cos\omega\tau & 0 \\ 0 & 0 & 1 \end{pmatrix} \tag{3.39}$$

A good approximation of the new particle velocity in the rotating framework is given by

$$\mathbf{v}(t+\tau) = \mathbf{v}(t) + \{2\boldsymbol{\omega} \times \mathbf{v}(t) + \boldsymbol{\omega} \times [\boldsymbol{\omega} \times \mathbf{r}(t)]\}\tau. \tag{3.40}$$

It is worth noting that the particle movement is mesh-free and the streaming operator realization does not contribute to the error of approximation defined by the splitting scheme.

3.5 Boundary Conditions

The action of the streaming operator S_D^τ includes the implementation of the corresponding initial and boundary conditions.

Most frequently, the initial condition for the particle initial positions and velocities is taken in the form of an equilibrium homogenous state $p_0 = n_0 k_B T_0$ at a given bulk velocity $\mathbf{V}_0$, pressure p_0 and temperature T_0, which is described by the Maxwell velocity distribution function with

$$f_M(0, \mathbf{r}, \mathbf{v}) = \frac{n_0}{(2\pi R T_0)^{3/2}} e^{-\frac{(v-V_0)^2}{2RT_0}}. \quad (\mathbf{r} \in \mathfrak{D}) \tag{3.41}$$

The velocity distribution function of reflected and emitted particles $f(t, \mathbf{r}, \mathbf{v}')$ and $f_e(t, \mathbf{r}, \mathbf{v})$ at a boundary can be presented in the following form:

$$|\mathbf{v} \cdot \mathbf{n}| f_w(t, \mathbf{r}, \mathbf{v}) = \int_{\mathbf{v}' \cdot \mathbf{n} < 0} \mathrm{R}\big(\mathbf{v}' \to \mathbf{v}, \mathbf{r}, t\big) \big|\mathbf{v}' \cdot \mathbf{n}\big| f\big(t, \mathbf{r}, \mathbf{v}'\big) d\mathbf{v}' +$$
$$+ |\mathbf{v} \cdot \mathbf{n}| f_e(t, \mathbf{r}, \mathbf{v}) \quad (\mathbf{r} \in \partial\mathfrak{D};\, \mathbf{v} \cdot \mathbf{n} > 0), \tag{3.42}$$

where $R(\mathbf{v}' \rightarrow \mathbf{v}, \mathbf{r}, t)$ is a scattering kernel expressed in terms of the probability density for an incident particle with velocity after reflection, to obtain the velocity $\mathbf{v}$. In general, one can classify the boundary conditions as follows:

- solid (liquid) surface-gas boundary conditions;
- free stream boundary conditions;
- inlet and outlet boundary conditions;
- symmetry and periodicity boundary conditions.

3.5.1 Solid (Liquid) Surface-Gas Boundary Conditions

A complete description of gas-surface interaction requires knowledge of both the kinetic theory of gases and condensed matter physics; therefore, a multiphysics approach should be applied. Instead, in kinetic theory the problem is simplified by using model boundary conditions in the form of Eq. (3.42) for the velocity distribution function with emitting term $f_e(t, \mathbf{r}, \mathbf{v}) = 0$ for the case of particle reflections. The following two models encompass the most frequently used surface-gas boundary conditions in kinetic theory.

- Specular-diffuse reflection

In the case of not moving solid wall, this model is presented by kernel $R(\mathbf{v}' \rightarrow \mathbf{v}, \mathbf{r}, t)$ in the simplest form [48]

$$R(\mathbf{v}' \rightarrow \mathbf{v}, \mathbf{r}, t) = (1-\alpha)\delta(\mathbf{v} - \mathbf{v}' + 2n(\mathbf{v} \cdot \mathbf{v}')) + \alpha f_0(t, \mathbf{r}, \mathbf{v})|\mathbf{v} \cdot \mathbf{n}|, \quad (3.43)$$

where

$$f_0 = \frac{2\beta^4}{\pi} e^{-\beta^2 \mathbf{v}^2} \quad (3.44)$$

$\beta = 1/\sqrt{2RT_w}$ and T_w is the surface temperature. The kernel describes the well-known Maxwell's specular-diffuse model according to which a fraction $(1-\alpha)$ of particles is reflected specularly and the remaining fraction α is reflected diffusely in accordance with the Maxwellian velocity distribution. For $\alpha = 0$ the particle reflection is pure specular and for $\alpha = 1$ a complete diffuse reflection is realized. In the case of diffuse reflection with $\alpha = 1$ the algorithmic interpretation of the stochastic kernel gives the following statements for the post-collision normal, tangential and bi-normal components of velocity of the reflected particle with respect to a local coordinate system $(\mathbf{n}, \mathbf{t}, \mathbf{b})$ attached to the wall in the incident point:

$$v_n = \left[-\ln(R_{f_1})\right]^{1/2} \Big/ \beta,$$
$$v_t = \left[-\ln(R_{f_2})\right]^{1/2} \sin(2\pi R_{f_3}) \Big/ \beta,$$

$$v_b = \left[-\ln(R_{f_2})\right]^{1/2} \cos(2\pi R_{f_3}) \Big/ \beta, \tag{3.45}$$

where $R_{f_1}, R_{f_2}, R_{f_3}$ are three random numbers uniformly distributed in the interval (0, 1).

- Cercignani-Lampis model

The Cercignani-Lampis (CL) model [49] is considered more realistic than the Maxwell model (which has only one adjustable parameter) because it preserves the reciprocity relation and generates realistic distributions of direction and energy for the molecules that are re-emitted. According to the CL model, the normal and tangential components of the gas molecule velocity scatter independently. The scattering kernel comprises three parts, each corresponding to one velocity component. The CL model contains two adjustable parameters: the normal energy accommodation coefficient α_n and the tangential momentum accommodation coefficient α_t. The first one α_n can range from 0 to 1 and is related to the part of kinetic energy corresponding to the normal velocity, while the latter α_t can range from 0 to 2 and is related to the tangential momentum. The scattering kernel [49] reads as follows

$$\begin{aligned} R(\mathbf{v}' \to \mathbf{v}, \mathbf{r}, t) = {} & \frac{[\alpha_n \alpha_t (2-\alpha_t)]^{-1}}{\pi/2} \beta^4 v_n \exp\left[-\beta^2 \frac{v_n^2 + (1-\alpha_n) v_n'^2}{\alpha_n} \right] \\ & \times \exp\left[-\beta^2 \frac{|v_t - (1-\alpha_t) v_t\prime|^2}{\alpha_t (2-\alpha_t)} \right] I_0\left(2\beta^2 \frac{(1-\alpha_n)^{1/2} v_n v_n\prime}{\alpha_n} \right), \\ & \big((\mathbf{n}\cdot\mathbf{v}) > 0,\ (\mathbf{n}\cdot\mathbf{v}') < 0 \big), \end{aligned} \tag{3.46}$$

where v_n' and v_t' are the normal and tangential components of the incident velocity in a local coordinate system, in which the incident velocity vector is expressed by $v' = (v_n', v_t', 0)$. Thus, if the surface is parallel to (y, z)-plane of a global Cartesian coordinate system $v_n' = v_x'$ and $v_t' = \sqrt{v_y'^2 + v_z'^2}$. The term I_0 denotes the modified Bessel function of first kind and zeroth order defined by

$$I(y) = (2\pi)^{-1} \int_0^{2\pi} \exp(y \cos\phi) d\phi. \tag{3.47}$$

The Cercignani-Lampis gas surface model can be incorporated into DSMC simulations as proposed by Lord [50, 51]. The values of the post-collision tangential and bi-normal velocity components v_t and v_b are computed in the local coordinate system as follows:

$$\begin{aligned} &\theta = 2\pi R_{f_1} \quad r = \beta^{-1}\left[-\alpha_t(2-\alpha_t)\ln\left(R_{f_2}\right)\right]^{1/2}, \\ &v_t = (1-\alpha_t) v_t\prime + r\cos\theta, \quad v_b = r\sin\theta. \end{aligned} \tag{3.48}$$

The value of the normal velocity component v_n is obtained from:

$$\theta = 2\pi R_{f_3}, \quad r = \beta^{-1}\left[-\alpha_n \ln\left(R_{f_4}\right)\right]^{1/2},$$
$$v_n = \left(r^2 + (1-\alpha_n)v_n'^2 + 2r\sqrt{(1-\alpha_n)}\left|v_n'\right|\cos\theta\right)^{1/2}. \qquad (3.49)$$

Finally, the obtained post-collision velocity components $(v_n, v_t, 0)$ are to be reverted to components (v_x, v_y, v_z) in the global coordinate system i.e. in the case of a scattering surface, which is parallel to (y, z)-plane, the corresponding components are equal to

$$\begin{aligned} v_x &= v_n \\ v_y &= v_t \cos\varphi + v_b \sin\varphi \\ v_z &= -v_t \sin\varphi + v_b \cos\varphi, \end{aligned} \qquad (3.50)$$

where $\cos\varphi = v_y' \big/ v_t'$ and $\sin\varphi = v_z' / v_t'$.

3.5.2 Free Stream Boundary Conditions

When the computational domain has open boundaries that expose the gas flow in the domain to the effect of an external free stream flow with bulk velocity V then the boundary condition (3.42) at these boundaries takes the form:

$$|\mathbf{v}\cdot\mathbf{n}|f_w(t,\mathbf{r},\mathbf{v}) = |\mathbf{v}\cdot\mathbf{n}|f_e(t,\mathbf{r},\mathbf{v}) \quad (\mathbf{r}\epsilon \quad \partial\mathfrak{D}) \qquad (3.51)$$

where $(\mathbf{v}\cdot\mathbf{n})$ corresponds to inward-directed flow. The free stream velocity distribution f_e is given by

$$f_e = \frac{n_e\beta^3}{\pi^{3/2}}e^{-\beta^2(\mathbf{v}-\mathbf{V})^2}, \qquad (3.52)$$

where $\beta = 1\big/\sqrt{2RT_e}$. Using distribution, the number flux of the molecules entering the computational domain can be calculated based on the local temperature and the mean flow velocity as follows:

$$\dot{N} = \frac{n_e}{2\sqrt{\pi}\beta}\left\{\exp\left(-s^2\cos^2\theta\right) + \sqrt{\pi}s\cos\theta[1 + \mathrm{erf}(s\cos\theta)]\right\}, \qquad (3.53)$$

where $s = V\beta$ denotes the molecular speed ratio and θ is the angle between the free stream velocity vector and the normal vector to the open boundary. The coordinates of the entering molecules through the open boundary are randomly distributed over the whole boundary surface A. The number of entering particles within a time step τ is calculated using the Poisson distribution with a mean value $\overline{N} = \dot{N}A\tau$ in order to be consistent with the free-stream equilibrium state. The time instance δt of each particle injection within a time step is calculated from the corresponding exponential distribution as follows

$$\delta t = -\tau \ln R_f \tag{3.54}$$

The algorithmic interpretation of equality gives the following statements for the generation of the normal, tangential, and bi-normal components of velocity of the injected particles with respect to a local coordinate system (**n**, **t**, **b**) attached to point **r**:

The normal velocity component v_n is computed from the distribution

$$f_{v_n} = \pi^{-1/2}\beta v_n \exp\left[-\beta^2 (v_n - V_n)^2\right] \tag{3.55}$$

by using the acceptance-rejection method as suggested by [2, 52]. The maximum $f_{v_n,\ \max}$ of the distribution (3.55) occurs at

$$v_n^* = V\cos\theta + \left[\left(s^2\cos^2\theta + 2\right)^{1/2} - s\cos\theta\right] \Big/ (2\beta). \tag{3.56}$$

According to the acceptance-rejection procedure, the first step is to compute a value

$$v_n = (V\cos\theta + 3\beta^{-1})R_f \tag{3.57}$$

that satisfy inequality ($\mathbf{n}\cdot\mathbf{v} \geq 0$) for the upstream boundary or

$$v_n = (V\cos\theta - 3\beta^{-1})R_f \tag{3.58}$$

for the downstream boundary ($\mathbf{n}\cdot\mathbf{v} \leq 0$). Thus, velocity magnitude is uniformly distributed in a range well extended into the tails of the velocity distribution. In the second step, this value is accepted with probability $P = f_{v_n} \big/ f_{v_n,\max}$ or rejected with probability $1 - P$. In the case of rejection, a new value v_n is computed from (3.57) and checked for acceptance. The iteration process terminates when a value v_n is accepted.

The tangential components are computed, according to, as follows

$$\begin{aligned} v_t &= \left[-\ln(R_{f_1})\right]^{1/2}\sin(2\pi R_{f2})/\beta, \\ v_b &= \left[-\ln(R_{f_1})\right]^{1/2}\cos(2\pi R_{f2})/\beta. \end{aligned} \tag{3.59}$$

3.5.3 Inlet and Outlet Boundary Conditions

The inlets and outlets are typical boundaries for the geometry of many internal flows in tubes, channels, and chambers that connect larger vessels (such as reservoirs) or open areas filled with gas under varying pressures, temperatures, and other conditions. Particularly interesting is the treatment of such boundaries in the DSMC method in the case of gaseous flows through microchannel configurations. The simplest but not very precise way is to assume that the equilibrium conditions existing in the large vessels (reservoirs) are applied directly at the corresponding inlet and outlet boundaries. In this case, the inlet and outlet boundary conditions can be considered as free stream ones (usually with bulk velocity $V = 0$), and formulas (3.51)–(3.59) can be used. The error introduced by the implementation of these conditions might be reduced by considering sufficiently long channels when the end effects can be ignored. The short channel consideration requires more realistic inlet/outlet flow conditions by using large reservoirs with open boundaries attached to the microchannel ends. At the open boundaries, the equilibrium conditions (3.51)–(3.59) with required pressure and temperature can be applied.

In the case of subsonic steady-state gas flow, it is suitable at the micro-configuration ends to implement pressure-driven implicit boundary conditions [53, 52, 54]. According to these conditions, inlet pressure P_{in} and temperature T_{in}, and outlet pressure P_{out} are considered as known macroscopic variables. The velocity profiles $\mathbf{V} = \{u, v, w\}$ at both inlet and outlet and the temperature at the outlet are unknown. Based on the 1-D characteristic equation $du/a = -d\rho/\rho$, where the speed of sound is defined as $a = \sqrt{dp/d\rho}$, the following implicit boundary conditions are obtained for cell j adjacent to the outlet:

$$\begin{aligned}
\rho_{out,j} &= mn_{out,j} = \rho_j + \frac{P_{out} - p_j}{a_j^2} \\
T_{out,j} &= \frac{P_{out}}{\rho_{out,j} R} \\
u_{out,j} &= u_j + \frac{p_j - P_{out}}{\rho_j a_j} \\
v_{out,j} &= v_j, \quad w_{out,j} = w_j,
\end{aligned} \tag{3.60}$$

where u is the streamwise velocity component, $a_j = \sqrt{\gamma R T_j}$ is the sound speed. Subscript j indicates the averaged macroscopic quantities during a given computational time in the boundary cell j. The time-averaging procedure is described in the next section.

The inlet boundary conditions for density and velocity components are computed as follows:

$$\rho_{in,j} = mn_{in,j} = \frac{P_{in}}{RT_{in}}$$

$$u_{in,j} = u_j + \frac{P_{in} - p_j}{\rho_j a_j}$$

$$v_{in,j} = v_j, \quad w_{in,j} = w_j, \tag{3.61}$$

where subscript j represents cells, adjacent to the inlet boundary. The computed macroscopic variables and are used for calculation of the number of injected through the boundary face A_j of cell j particles and generation of their velocity. For this aim, the approach described by equations—has to be applied by replacing the free stream parameters with the corresponding cell quantities and. Thus, the formula for the number flux through cell-surface A_j takes the form:

$$\dot{N}_j = \frac{n_j A_j}{2\sqrt{\pi}\beta_j}\left\{\exp\left(-s_j^2\right) + \sqrt{\pi} s_j\left[1 + \mathrm{erf}(s_j)\right]\right\}, \tag{3.62}$$

where $n_j = n_{out,j}$, $s_j = \beta_j u_{out,j}$, $s_j = \beta_j u_{out,j}$, $\beta_j = 1/\sqrt{2RT_{out,j}}$ for the outlet boundary and $n_j = n_{in,j}$, $s_j = \beta_j u_{in,j}$, $\beta_j = 1/\sqrt{2RT_{in}}$ for the inlet boundary.

3.6 Sampling of Macroscopic Properties

3.6.1 Volume-Based Averaging

The DSMC is a stochastic numerical method that simulates realizations of a random process following the behavior of a model gas consisting of a large number of simulators that collide and move in a computational domain. The evaluation of the flow macroscopic properties is a process of averaging the molecular characteristics of all particles present in each cell of the grid covering the computational domain. Time averaging over a single trajectory of the random process is employed when the gas flow is stationary, allowing for the calculation of the macroscopic properties of the established flow within the grid cells. It is worth noting that time averaging assumes a stationary ergodic process. Thus, if at the instant $t_k = k\tau(k = 1, K_T)$ the number of particles in cell l is equal to $N^{(l)}(t_k)$, then the averaged over K_T time steps magnitudes of the basic macroscopic variables as density, mean velocity components, and translational temperature are computed as follows:

$$N_T^{(l)} = \sum_{k=1}^{K_T} N^{(l)}(t_k), \quad \overline{N^{(l)}} = \frac{N_T^{(l)}}{M_T},$$

$$\rho^{(l)} = mn^{(l)} = m\frac{\mathrm{W_N}}{\left|d\mathbf{x}^{(l)}\right|},$$

$$
\begin{aligned}
&\mathbf{V}_\alpha^{(l)} = \overline{v_{\alpha,i}^{(l)}} = \frac{1}{N_T^{(l)}} \sum_{k=1}^{K_T} \sum_{i=1}^{N^{(l)}(t_k)} v_{\alpha,i}^{(l)},\\
&\overline{\left(v_\alpha'^{(l)}\right)^2} = \overline{\left(v_\alpha^{(l)} - \mathbf{V}_\alpha^{(l)}\right)^2} = \frac{1}{N_T^{(l)}} \sum_{k=1}^{K_T} \sum_{i=1}^{N^{(l)}(t_k)} \left(v_{\alpha,i}^{(l)}\right)^2 - \left(\mathbf{V}_\alpha^{(l)}\right)^2, \quad \alpha = \{x, y, z\},\\
&\frac{1}{2} R T_\alpha^{(l)} = \frac{1}{2} \overline{\left(v_\alpha^{(l)}\right)^2} = \frac{1}{2} \overline{\left(v_\alpha^{(l)} - \mathbf{V}_\alpha^{(l)}\right)^2}\\
&T^{(l)} = \left(T_x^{(l)} + T_y^{(l)} + T_z^{(l)}\right)/3.
\end{aligned}
\tag{3.63}
$$

Similarly, the scalar pressure $p^{(l)}$ and viscous stress tensor $\boldsymbol{\tau}^{(l)}$ in cell l take the forms

$$
p^{(l)} = \frac{1}{3} \rho \overline{\left(v'^{(l)}\right)^2} = \frac{1}{3} \rho \left(\overline{\left(v_x'^{(l)}\right)^2} + \overline{\left(v_y'^{(l)}\right)^2} + \overline{\left(v_z'^{(l)}\right)^2} \right) \tag{3.64}
$$

and

$$
\boldsymbol{\tau}^{(l)} \equiv \tau_{\alpha\beta}^{(l)} = -\left(\rho \overline{v_\alpha'^{(l)} v_\beta'^{(l)}} - \delta_{\alpha\beta} p^{(l)}\right), \alpha, \beta = \{x, y, z\}. \tag{3.65}
$$

The heat flux vector without taking into account the internal energy of molecules results in

$$
\mathbf{q} = \frac{1}{2} \rho \overline{\left(v'^{(l)}\right)^2 \mathbf{v}'^{(l)}} \tag{3.66}
$$

The instantaneous macro-characteristics of an unsteady flow are computed at instant $t_k = k\tau (k = 1, K_T)$ in cell l by using ensemble averaging over a group of process trajectories $m = 1, M_T$ run within a time interval $(0, K_T \tau)$ as follows

$$
\begin{aligned}
&N_T^{(l)}(t_k) = \sum_{m=1}^{M_T} N_m^{(l)}(t_k), \overline{N^{(l)}(t_k)} = \frac{N_T^{(l)}(t_k)}{K_T},\\
&\rho^{(l)}(t_k) = m n^{(l)}(t_k) = m \frac{\mathrm{W_N} \overline{N_T^{(l)}(t_k)}}{\left|d\mathbf{x}^{(l)}\right|},\\
&\mathbf{V}_\alpha^{(l)}(t_k) = \overline{v_\alpha^{(l)}(t_k)} = \frac{1}{N_T^{(l)}(t_k)} \sum_{m=1}^{M_T} \sum_{i=1}^{N^{(l)}(t_k)} v_{\alpha,i,m}^{(l)}(t_k),\\
&\overline{\left[v_\alpha'^{(l)}(t_k)\right]^2} = \overline{\left[v_\alpha^{(l)}(t_k) - \mathbf{V}_\alpha^{(l)}(t_k)\right]^2} = \frac{1}{N_T^{(l)}(t_k)} \sum_{m=1}^{M_T} \sum_{i=1}^{N^{(l)}(t_k)} \left[v_{\alpha,i,m}^{(l)}(t_k)\right]^2 - \left[\mathbf{V}_\alpha^{(l)}(t_k)\right]^2,\\
&\frac{1}{2} R T_\alpha^{(l)} = \frac{1}{2} \overline{\left[v_\alpha^{(l)}(t_k)\right]^2} = \frac{1}{2} \overline{\left[v_\alpha^{(l)}(t_k) - \mathbf{V}_\alpha^{(l)}(t_k)\right]^2}, \quad \alpha = \{x, y, z\}
\end{aligned}
$$

$$T^{(l)}(t_k) = \left[T_x^{(l)}(t_k) + T_y^{(l)}(t_k) + T_z^{(l)}(t_k)\right]/3. \quad (3.67)$$

The instantaneous magnitudes of pressure, shear stress, and heat flux in a cell are calculated in a similar manner.

When the kinetic and macroscopic time scales are well separated the time averaging can be applied to a relatively short time interval on the macroscopic scale, but is large enough from a microscopic viewpoint to obtain a reliable instantaneous estimate of the macroscopic variables in cells. Thus, the single-trajectory simulation advances with large portions of time from a kinetic viewpoint, presenting the macroscopic flow dynamics. Each portion contains a large enough number of kinetic time steps K_T (centered over the sampling time) in order to accumulate a sufficient sample size in cells (see for details [55]). This approach remains the only possible when one wants to simulate a flow with chaotic spatiotemporal behavior (see for example [56, 57, 58]).

3.6.2 Surface-Based Averaging

So far, we have considered volume-based methods for averaging the molecular properties of particles present in a cell at a given moment. There is an alternative approach to averaging the flux of molecular quantities across a surface element. Typically, this is suitable for any flux properties, but can also be used to evaluate basic macroscopic quantities, such as velocity and temperature. We will demonstrate it by calculation of velocity slip and temperature jump at the interface "gas–wall" as described in [59, 60]. According to the surface-based approach, one can obtain the velocity slip and temperature jump by accumulating the impulses of all particles striking a given surface element within a time interval as follows

$$v_{slip} = \frac{\sum (v_t / |v_n|)}{\sum (1/|v_n|)} \quad (3.68)$$

$$T_{gas} - T_{wall} = \frac{1}{3R} \frac{\sum (v^2/|v_n|) - \sum (1/|v_n|)\, v_{slip}^2}{\sum (1/|v_n|)} - T_{wall}, \quad (3.69)$$

where v_n and v_t are the normal and tangential velocity components of each particle striking the regarding surface. It is worth noting that each particle is counted twice—once as incident and again as reflected, with the normal velocity taken with the corresponding sign.

This approach is illustrated in Fig. 3.2, where the velocity slip and temperature jump along the moving lid of a micro-sized cavity are shown (for details see [60]). They are computed by both volume-based (macroscopic) and surface-based (microscopic) DSMC approaches and compared to the solution of Navier–Stokes-Fourier equations with first-order slip boundary conditions for Knudsen number Kn $= 0.05$

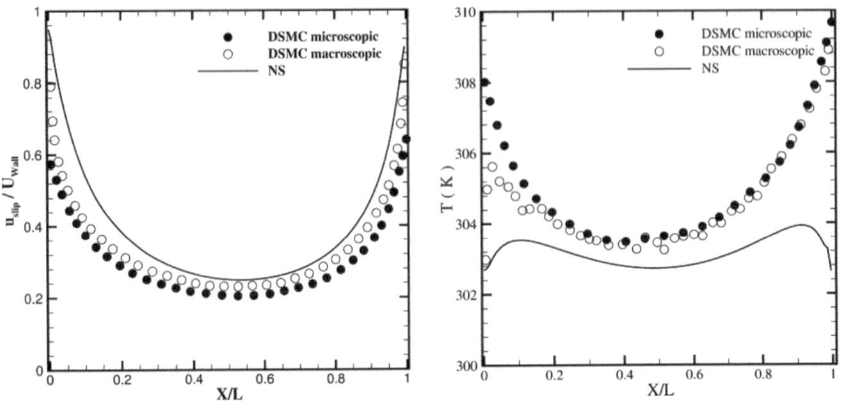

Fig. 3.2 Comparison of velocity slip (left) and temperature jump (right) along the driven lid of the cavity from the DSMC and NS solutions at $\mathrm{Kn} = 0.05$ (Copied with permission from Ref. [60])

and lid velocity U_{wall} [60]. As seen, the macroscopic (volume-based) approach, which accounts for the properties of all particles found in the adjacent cells to the surface, predicts a closer to continuum NSF solution of velocity slip and temperature jump along the driven cavity lid. Additionally, in contrast to the continuumlike slip velocity, microscopic slip is finite and does not diverge at the corner.

3.6.3 *DSMC Data Filtering*

Statistical scatter is an inherent part of the stochastic simulation, and as a result, the sampled DSMC data contain statistical fluctuations whose level depends on the sample size. Besides increasing the sample size, post-processing filtering provides an alternative for obtaining smoother and more reliable estimations of the macroscopic variables. A simple filter is realized by using a diffusive smoothing algorithm for 1 or 2 passes [61, 62]. Directly, it performs simple averaging of data φ_l in each cell l with its surrounding N_s cells as follows

$$\overline{\varphi}_l = \frac{\varphi_l + \sum_{i=1}^{N_s} \varphi_i}{N_s + 1}. \tag{3.70}$$

The number of surrounding cells N_s depends on the number of space dimensions and type of mesh. For rectangular grids in 1D, 2D and 3D N_s is equal to 3, 8 and 26, respectively. The diffusive character of the averaging does not allow the filter to be applied iteratively over many passes.

3.6.4 Flux-Corrected Transport (FCT) Filtering

A nonlinear FCT filtering method was suggested by Kaplan and Oran [62] to improve post-processing performance by reducing statistical fluctuations and CPU runtime. Unlike fully diffuse simple filters, the FCT filter has the advantage of leading the noisy flow to a solution that is already verified to be correct. The FCT filtering procedure includes three parts: diffusion, anti-diffusion flux, and anti-diffusion flux limiting. Considering the parameter φ^0 as a noisy field that should be represented in the filtered form φ^1, we could write [61]

$$\varphi^0 \rightarrow \varphi^d \rightarrow \varphi^{dal} \equiv \varphi^1 \tag{3.71}$$

The steps are as follows:

(a) The initial values φ^0 are diffused, resulting in φ^d. This is achieved by the definition of diffusive flux $f^d_{l+1/2}(\varphi)$ across the surface between neighboring cells l and $l+1$ in the form

$$f^d_{l+1/2}(\varphi^0) = \nu_{l+1/2}(\varphi^0_{i+1} - \varphi^0_i), \tag{3.72}$$

where $\{\nu_{l+1/2}\}$ are diffusion coefficients. Then

$$\varphi^d_l = \varphi^0_i + f^d_{l+1/2}(\varphi^0) - f^d_{l-1/2}(\varphi^0). \tag{3.73}$$

(b) To remove excess diffusion, anti-diffusion fluxes are defined

$$f^{da}_{l+1/2}(\varphi^d) = \mu_{l+1/2}(\varphi^d_{i+1} - \varphi^d_i), \tag{3.74}$$

where $\{\mu_{l+1/2}\}$ are anti-diffusion coefficients. If fluxes (3.74) were used directly to evaluate φ^1 new extrema could be introduced.

(c) To ensure stability and elimination of nonphysical overshoots, the anti-diffusion fluxes f^{da} should be limited before applying to the field. The limiter function, following Boris and Book [111], could be defined as

$$f^{dal}_{l+1/2} = S.max\Big\{0, min\Big[S.(\varphi^d_{l+2} - \varphi^d_{i+1}), \left|f^{da}_{l+1/2}(\varphi^d)\right|, S.(\varphi^d_l - \varphi^d_{l-1})\Big]\Big\} \tag{3.75}$$

where $S = sgn(\varphi^d_{l+1} - \varphi^d_l)$. Then

$$\varphi^1_i \equiv \varphi^{dal}_i = \varphi^d_i - f^{dal}_{l+1/2}(\varphi^d) + f^{dal}_{l-1/2}(\varphi^d). \tag{3.76}$$

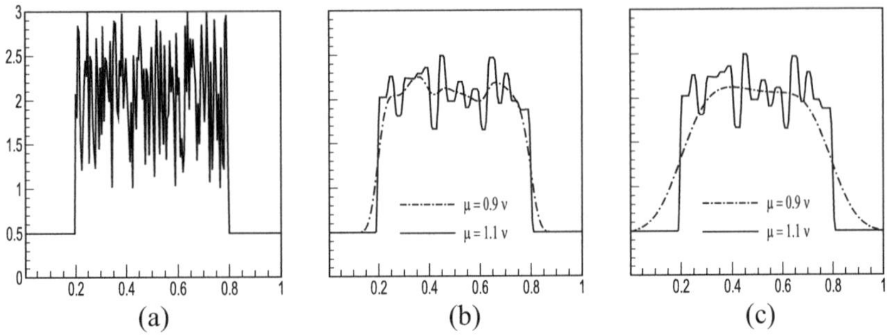

Fig. 3.3 The effect of FCT filter on a plain wave with superimposed random noise (**a**), after 1000 passes (**b**), and after 10,000 passes (**c**) [61]

The procedure smooths the local peaks while keeping the neighbor values nearly unchanged. The FCT filter's performance can be adjusted by varying the diffusion and anti-diffusion coefficients. For $\nu > \mu$, the filter exhibits a diffusive behavior, and the degree of inequality affects how quickly it can smooth the peaks; however, this could make the iteration process unstable. On the other side, if $\nu < \mu$ then it works as a high-frequency filter that removes the sharp peaks and the iteration process of passing multiple times converges to a solution, which essentially stop changing after a certain number of passes. Figure 3.3 illustrates this, where a noisy square wave with zero velocity (Fig. 3.3a) goes through the FCT filter 1000 times (Fig. 3.3b) and 10,000 times (Fig. 3.3c) for each of the previous cases. As can be seen, in the case of $\nu < \mu$, less noisy wave could be achieved by filtration. However, smoothing is stopped for more than 1000 passes through the filter. Also, in the case of $\nu > \mu$, the filter smooths the wave, but it is diffusive and does not reach a solution.

3.7 Accuracy of the DSMC Simulation

3.7.1 Time and Space Discretization Errors

The analysis of the accuracy of the DSMC method requires a complex approach that must take into account issues typical of deterministic numerical methods, as well as those related to the stochastic nature of the particle method. Firstly, to quantify the method errors, one must estimate the approximation error of the splitting scheme for a reference mathematical model. As mentioned before, we prefer to relate the DSMC solution to the evolutionary Eq. (3.1) or Stefanov Eq. (3.10) rather than directly to the Boltzmann Eq. (3.3), which follows from (3.1) under the assumption of molecular chaos (3.2). Using the result obtained in [63] and taking into account (3.7) one can show that the splitting method approximates Eqs. (3.1) and (3.3) with accuracy $O(\tau + h)$.

Instead the standard two-step splitting approximation (3.9) one can apply a symmetric splitting scheme proposed by Strang [28], (see also [63, 56]), which is second-order accurate in time. The splitting scheme employs one of the following operator approximations:

$$S_{Q+D}^{\tau,h} \approx S_Q^{(\tau/2,h)} S_D^{\tau} S_Q^{'(\tau/2,h)}, \tag{3.77}$$

or

$$S_{Q+D}^{\tau,h} \approx S_D^{\tau/2} S_Q^{(\tau,h)} S_D^{\tau/2}, \tag{3.78}$$

where $S_Q^{(\tau/2,h)}$ and $S_D^{\tau/2}$ act within a half-time step. The operator order symmetry requires the sampling procedure to be executed between two half-time steps of $S_Q^{(\tau/2,h)}$ or $S_D^{\tau/2}$. Being of the same order of accuracy both approximations (3.77) and (3.78) are not equivalent. The first should be applied for the simulation of small and moderate Knudsen number gas flows, while the second will be more efficient at high Knudsen numbers.

Time and space discretization has an impact on all characteristics of the particle simulation. The results are illustrated by the analysis of viscosity and thermal conductivity, performed for small and moderate Knudsen numbers where the Chapman-Enskog first-order approximation theory is valid. Using the Green–Kubo theory, Alexander, Garcia and Adler [64] and Hadjiconstantinou [65] have obtained the following estimations of shear viscosity and thermal conductivity, depending on cell size h

$$\mu = \frac{5}{16d^2}\sqrt{\frac{mk_BT}{\pi}}\left(1 + \frac{16}{45\pi}\frac{h^2}{\lambda^2}\right)$$
$$k = \frac{75}{64d^2}\sqrt{\frac{k_B^3T}{m\pi}}\left(1 + \frac{32}{225\pi}\frac{h^2}{\lambda^2}\right) \tag{3.79}$$

and time step τ

$$\mu = \frac{5}{16d^2}\sqrt{\frac{mk_BT}{\pi}}\left(1 + \frac{32}{150\pi}\frac{\left(V_{mp}\tau\right)^2}{\lambda^2}\right)$$
$$k = \frac{75}{64d^2}\sqrt{\frac{k_B^3T}{m\pi}}\left(1 + \frac{64}{675\pi}\frac{\left(V_{mp}\tau\right)^2}{\lambda^2}\right), \tag{3.80}$$

respectively. The second-order deviations from the Chapman-Enskog expressions of shear viscosity and thermal conductivity shown in formulae and are consistent with the first-order accurate splitting scheme. Specifying a maximum desired deviation in (3.79) and (3.80) one can calculate the upper limits of time and space steps that can be used for simulating near-continuum gas flows. The analog of the well-known Courant condition written for the particle simulation of a rarefied gas

$$V_{mp}\frac{\tau}{h} \leq 1 \tag{3.81}$$

can be used as an additional limiting condition on time and space steps. In practice, the empirical conditions, suggested by G. Bird

$$h \leq \lambda/3, \quad \tau \leq \tau_m/3, \tag{3.82}$$

are commonly used to provide a sufficiently accurate simulation of a rarefied gas flow at small or moderate Knudsen numbers. In inequalities, τ_m means the free-flight time of the particles.

3.7.2 Statistical Errors

Another important parameter of the simulation is the average number of particles (simulators) $\overline{N^{(l)}}$ in a cell. The number of particles $N^{(l)}(t_k)$ found in a cell l defines a random velocity point-set representing the velocity space at instance t_k. The estimations—are obtained under the assumption of a very large number of simulators in cell $N^{(l)}(t_k) \to \infty$. However, any real particle simulation is carried out with a finite total number of simulators and, correspondingly, with a much smaller number of random particles in each cell. The finite number of simulators leads to statistical errors that are an inherent part of the overall error of the DSMC method. The mean values (3.63)–(3.67) of the macroscopic variables fluctuate around their mathematical expectations and have a variance depending on the number of particles in the cell. The statistical errors of the basic macroscopic variables are derived in [66] using equilibrium statistics. Following the paper [66] we illustrate the dependence of statistical errors on the average number of particles in a cell. Let N_0 is the average number of particles in the statistical cell, V_0 is the average velocity, T_0 is the average temperature and the statistical mean value is estimated over M_T independent samples. The instantaneous fluctuations of density, velocity components and temperature are equal to

$$\delta N = N - N_0, \quad \delta v_\alpha = v_\alpha - V_\alpha, \quad \delta T = T - T_0. \tag{3.83}$$

The one-standard-deviation error bars for the sample estimates are equal to

$$\sigma_N = \frac{\sqrt{\overline{(\delta N)^2}}}{\sqrt{M_T}}, \quad \sigma_{v,\alpha} = \frac{\sqrt{\overline{(\delta v_\alpha)^2}}}{\sqrt{M_T}}, \quad \sigma_T = \frac{\sqrt{\overline{(\delta T)^2}}}{\sqrt{M_T}}. \tag{3.84}$$

The corresponding fractional errors are

$$
\begin{aligned}
E_N &= \frac{\sigma_N}{N_0} = \frac{1}{\sqrt{M_T N_0}} \\
E_v &= \frac{\sigma_{v,\alpha}}{|v_\alpha|} = \frac{1}{\sqrt{M_T N_0}} \frac{1}{Ma\sqrt{\gamma}} \\
E_T &= \frac{\sigma_T}{T_0} = \frac{1}{\sqrt{M_T N_0}} \sqrt{\frac{k_B}{c_v}}.
\end{aligned}
\tag{3.85}
$$

As seen, the fractional error depends inversely on the square root of the product of the numbers of particles per cell and independent samples.

It is worth noting (see [66]) that the statistical errors (3.83) correspond to the magnitudes of the physical fluctuations of the macroscopic variables in small volumes and can serve to calculate the latter. Here one can find a significant difference in the properties of the Boltzmann solution and the DSMC results. As known the Boltzmann Eq. (3.3) does not describe the physical fluctuations while the equation for the two-particle velocity distribution function (3.1) contains them (see [56]). As already mentioned, the DSMC method enables the estimation of the temporal evolution of a coarse-grained two-particle velocity distribution function and, from that, the calculation of the associated local correlations and fluctuations of macroscopic variables. That is why, we prefer to relate the DSMC solution to the evolutionary Eq. (3.1) or Stefanov Eq. (3.10) rather than the Boltzmann Eq. (3.3).

In practice, numerical experiments performed using the traditional DSMC method reveal that the average number of simulators per cell, which yields accurate values of the macroscopic flow characteristics, depends strongly on the simulated flow regime. Typically, simulating gas flow with small gradients of macroscopic variables requires no more than a few particles per cell to obtain an accurate enough solution. The solution dependence on the number of simulators becomes more evident for high-speed gas flows with large gradients. The simulation of a strong shock wave is a typical example that requires the particle number per cell in the shock wave front to be larger than 30. Commonly accepted is the empirical rule $N > 20 - 30$ that provides a good solution in the general case.

3.7.3 Stochastic Errors Caused by Insufficient Number of Simulators

As the previous remark indicates, the collision procedure is stochastic and causes stochastic errors that differ from statistical errors, i.e. Eq. (3.85). To analyze the stochastic errors more deeply, one needs to apply strict mathematical methods (see, for example, [67]), which may be beyond the present scope and need additional studies. The focus here is on the stochastic errors that arise from unsuitable spatial and temporal discretization and the occurrence of regions of computational cells with an inadequate number of model particles. The case of simulation, where the entire computational domain has an inadequate total number of particles, is also important to consider. The above situations are common in large two- and three-dimensional DSMC calculations or simulations that need grids with local subcell areas. When the grid has a low number of particles per cell, the conditions for correct simulation using the standard DSMC scheme $S_D^{\tau} S_Q^{\tau,h}$ are violated in the following items:

- A strongly degenerated probability distribution is formed by the relative velocities between the particle pairs in a cell, which are few in number. This distribution is used repeatedly in the acceptance-rejection procedure within each time step;
- Many pairs of particles that are close enough to collide in different neighboring cells are ignored. These pairs could significantly impact the local collision rate. The standard DSMC scheme cannot handle these potential collisions.

One can demonstrate that the first situation causes an increase in the number of collisions that occur repeatedly within cells, while the second situation causes an increase in the number of potential collisions that are missed between particles in neighboring cells. Both actions together cause a local decrease in potential collisions. This is clearly evident in pairs that consist of particles in neighboring cells. For example, some particles could switch their positions in stage 2 before checking for collisions. However, the idea that repeated collisions have a reducing effect on the number of effective collisions requires further elaboration. The assertion can be proved easily by using the elastic collision model with central symmetry of the interaction potential. Let us estimate the effect of n repeated collisions between particle pairs (i, j) belonging to a given cell. Let particles have initial velocities $(\mathbf{v}_i^{(0)}, \mathbf{v}_j^{(0)})$. Applying Eq. (3.22) n times we obtain the post-collision velocities after n repeated collisions:

$$
\begin{aligned}
\mathbf{v}_i'^{(n)} &= \frac{1}{2}\left[\left(\mathbf{v}_i^{(n-1)} + \mathbf{v}_j^{(n-1)}\right) + \left|\mathbf{v}_i^{(n-1)} - \mathbf{v}_j^{(n-1)}\right| \boldsymbol{\omega}^{(n)}\right] \\
\mathbf{v}_j'^{(n)} &= \frac{1}{2}\left[\left(\mathbf{v}_i^{(n-1)} + \mathbf{v}_j^{(n-1)}\right) - \left|\mathbf{v}_i^{(n-1)} - \mathbf{v}_j^{(n-1)}\right| \boldsymbol{\omega}^{(n)}\right]
\end{aligned}
\tag{3.86}
$$

Taking into account that $(\mathbf{r}_i^{(n-1)} + \mathbf{r}_j^{(n-1)}) = (\mathbf{r}_i^{(0)} + \mathbf{r}_j^{(0)})$ and $\left|\mathbf{r}_i^{(n-1)} - \mathbf{r}_j^{(n-1)}\right| = \left|\mathbf{r}_i^{(0)} - \mathbf{r}_j^{(0)}\right|$ we obtain

$$
\begin{aligned}
\mathbf{r}_i'^{(n)} &= \frac{1}{2}\left[\left(\mathbf{r}_i^{(0)} + \mathbf{r}_j^{(0)}\right) + \left|\mathbf{r}_i^{(0)} - \mathbf{r}_j^{(0)}\right| \boldsymbol{\omega}^{(n)}\right] \\
\mathbf{r}_j'^{(n)} &= \frac{1}{2}\left[\left(\mathbf{r}_i^{(0)} + \mathbf{r}_j^{(0)}\right) - \left|\mathbf{r}_i^{(0)} - \mathbf{r}_j^{(0)}\right| \boldsymbol{\omega}^{(n)}\right].
\end{aligned}
\tag{3.87}
$$

Therefore, the post-collision velocities obtained from repeatedly using elastic collisions are statistically equivalent to those from a single collision of the selected pair. A reduction in the collision effect could also be seen in the case of non-elastic repeated collisions. The primary effect of a standard DSMC simulation with a small number of particles in cells is that the local collision frequency decreases, and it only matches the Boltzmann collision frequency when there are sufficient particles per cell.

3.7.4 Probability Analysis of Repeated Collisions

Let us examine the events of repeated collisions and discover how the acceptance-rejection procedure operates. The paper by Shevyrin et al. [68] was the first to examine numerically how the accuracy of the DSMC method is affected by the number of repeated collisions. They simulated two classical one-dimensional rarefied gas flows: the plane Couette flow and the heat transfer between two parallel plates. The established correlation between the number of repeated collisions and the difference between the computational results and the solution of the Boltzmann equation clearly demonstrated a real effect of repeated collisions on the simulation accuracy. In paper [56] we analyze the probability of repeated collisions of a certain pair (i, j) chosen from $N \geq 2$ particles that occupy a cell within a time step τ. Consider within this time step the maximum number of collisions is $m \geq 2$, based on (3.20), that can be realized in the considered cell. The acceptance-rejection procedure of the standard collision algorithm is used m times to obtain the actual number of collisions.

The result of each iteration of the algorithm with respect to the pair (i, j) can be described by the set of mutually exclusive and exhaustive events:

A, the velocities of particles (i, j) are changed to their post-collision values (3.22) due to a collision between these particles;

B, the velocity of particle i or particle j is changed due to a collision with another particle $k \neq (i, j)$;

C, the velocities of particles i and j are not changed (collisions with particles i and j do not occur).

Let P_A, P_B, P_C be the corresponding probabilities for the occurrence of one of A, B, or C events. The following equality is held after each iteration of the collision algorithm:

$$P_A + P_B + P_C = 1. \tag{3.88}$$

For m iteration we obtain

$$(P_A + P_B + P_C)^m = 1. \tag{3.89}$$

The trinomial (3.89) expands as follows:

$$(P_A + P_B + P_C)^m = \sum_{k_1+k_2+k_3=m} \binom{m}{k_1, k_2, k_3} P_A^{k_1} P_B^{k_2} P_C^{k_3}, \tag{3.90}$$

where each multinomial term of the expansion (3.90)

$$P_m(k_1, k_2, k_3) = \binom{m}{k_1, k_2, k_3} P_A^{k_1} P_B^{k_2} P_C^{k_3}, \tag{3.91}$$

determines the probability of occurrence of a sequence of events with length m containing k_1 events A, k_2 events B and k_3 events C. The multinomial coefficients

$$\binom{m}{k_1, k_2, k_3} = \frac{m}{k_1!k_2!k_3!} \tag{3.92}$$

sum up the number of possible different ways to change the sequence of m events, and k_1, k_2, k_3 are the multiplicities of each of the distinct events.

To calculate the probability of having one or more repeated collisions of a specific pair (i, j), we need to consider all sequences that have at least two events $A(k_1 \geq 2)$ and exclude from them the sequences in which every two events A are separated by at least one event B. Thus, sequences of type $'AABC...'$, $'ACCAB...'$, $'BACAB...'$ belong to the set of sequences containing repeated collisions while sequences of type $'ABABABA't$, $'ACCBACBA'$ must be excluded from this set. We use the standard (NTC) collision algorithm to study short event sequences that involve cell collisions with few particles. To find the likelihood of multiple collisions, we first need to estimate the probabilities P_A, P_B and P_C. Calculation of a collision using acceptance-rejection procedure consists of two steps: first, choosing a random pair (i, j) from $N_p = N(N-1)/2$ possible pairs with probability $p_1 = 1/N_p = 2/N(N-1)$ and second, colliding the pair (i, j) with probability $p_{i,j}$ (see (3.21)). Considering two mentioned steps, the probability P_A for collision of the pair (i, j) is equal to $P_A = p_1 p_{i,j} = 2p_{i,j}/N(N-1)$. Similar reasonings are used for calculation of P_B. Finally, we obtain

$$\begin{aligned} P_A &= \frac{2p_{i,j}}{N(N-1)}; \\ P_B &= \frac{1}{N(N-2)}\Big(\sum_{k\neq i,j} p_{i,k} + \sum_{k\neq i,j} p_{j,k}\Big); \\ P_C &= 1 - (P_A + P_B). \end{aligned} \tag{3.93}$$

For a qualitative analysis of the repeated collision probability, it is enough to consider event sequences with maximum number of collisions $m = 2$ and $m = 3$. For $m = 2$ equality (3.90) is reduced to

$$(P_A + P_B + P_C)^2 = P_A^2 + P_B^2 + P_C^2 + 2(P_A P_B + P_A P_C + P_B P_C). \tag{3.94}$$

Only the sequence of events $'AA'$ leads to a repeated collision of pair (i, j). This means that the probability P_R that a collision is repeated is equal to

$$P_R = P_2(2,0,0) = P_A^2. \tag{3.95}$$

For $m = 3$ equality (3.90) reads

$$\begin{aligned}(P_A + P_B + P_C)^3 = (P_A^3 + 3P_A^2P_B + 3P_AP_B^2 + P_B^3 + 3P_A^2P_C+ \\ 6P_AP_BP_C + 3P_B^2P_C + 3P_AP_C^2 + 3P_BP_C^2 + P_C^3).\end{aligned} \tag{3.96}$$

Now the probability that at least one repeated collision occurs is equal to

$$P_R = \begin{cases} P_A^3 + 3P_A^2P_C = P_A^2(1+2P_C), \; P_B = 0, \quad N = 2, \\ P_A^3 + 2P_A^2P_B + 3P_A^2P_C = P_A^2(1 + P_B + 2P_C), \; N \geq 3. \end{cases} \tag{3.97}$$

The probability of repeated collisions is derived according to the probabilistic description of multiple iterations of the acceptance-rejection procedure used in the standard collision algorithm. Based on the analysis of Eqs. (3.95), (3.97), the probability of repeated collisions is of order $O(p_{i,j}^2/N_p^2)$. When the number of particles in cell decreases, the probability value increases sharply. Due to the proportionality of $p_{i,j}$ to $g_{i,j}$ (see equiation (3.24)), the value P_R is quadratically dependent on the relative velocity. As a result, pairs with larger relative velocities reflect more contributions to the repeated collisions. Thus, high relative velocities cause an additional distortion in the collision process. The mentioned feature could be observed in the DSMC calculations of strong shock waves. The problem is also majority remained even with choosing smaller time steps. Due to remaining many particles in the same cell within multiple successive time steps, the realized repeated collisions may increase. The solution is to employ the grid with more spatial resolution. The implementation of all these conflicting each other requirements for better accuracy reduces the performance of the method and increases the computational resources. The improvement of the DSMC method with respect to accuracy and efficiency is a key objective of the recent developments considered in the next section.

3.8 Multiscale Flow Simulation

Over the past two decades, considerable efforts have been applied to bridge the gap between micro and macro descriptions (models) of multi-scaled processes and phenomena occurring in rarefied gas flows interacting with solid or liquid bodies or boundaries. To a large extent, these problems also concern gas flows in micro- and nano-structured devices, such as MEMS and NEMS. According to the review [112] the existing general strategies of numerical methods for resolving multiscaled problems include:

(a) Domain decomposition methods. In this approach, the macro and microscale models are solved on different domains which may or may not overlap, and the two models are matched either over a handshake region or across an interface;

(b) Adaptive mesh and model refinement. This is a modification of the adaptive mesh refinement method, except for an added option to switch to a more refined model when it is necessary;
(c) The heterogeneous multiscale method. Here, the basic idea is to utilize both macroscopic and microscopic models throughout the entire computational domain, extracting or reconstructing missing data over time and space for the corresponding model by leveraging information from the other model. The important task is how to use effectively the existing time and space separation;
(d) Gap-tooth schemes or patch dynamics. The primary objective is to combine the results of microscale simulations on small spatial and temporal domains through interpolation and extrapolation, thereby capturing the large-scale behavior by exploiting scale separation.

The majority of studies on coupling DSMC with continuum models in hybrid computational methods [32, 69, 70, 71, 72, 73, 74, 75, 76, 77] could be related to the first strategy (a) of domain-decomposition methods. Strategies of adaptive mesh and model refinement (b) have been also widely used by many authors [2, 78, 79, 80, 81]. For more details on strategies (a) and (b) we refer the reader to the corresponding articles cited above.

In this section, we will focus primarily on the application of the DSMC method within the framework of strategy c). First of all, the information preservation IP method [82, 83, 84, 85, 86, 87], which has been created as a method reducing the statistical fluctuations of the macroscopic variables, could be classified in some sense as a heterogeneous method. Here, we leave the details of the IP method discussed widely in the reference papers to the reader and are going to present some recent developments which use specific properties of gas flows in order to design a heterogeneous multiscale strategy.

The use of the last multiscale strategy (d) will be illustrated in the section on microfluidic applications with an example of a DSMC application to simulate the dynamics of a rotating micro-actuator driven by gas thermal effects. In this case, the time scales of gas kinetic processes and rotation dynamics are widely separated, and a dynamic patch strategy in the form of a predictor–corrector scheme can be applied.

3.8.1 Multiscale Methods Based on Algebraic Decomposition of the Distribution Function

The continuum limits [35, 7] of the Boltzmann Eq. (3.3), namely, the Euler and Navier–Stokes-Fourier equations, when the Knudsen number tends to zero, yield a fundamental relation between the microscopic and macroscopic descriptions in kinetic theory. Another important limit is the local equilibrium state when the distribution function perturbations around the local equilibrium are small. Examples include low-speed gas flows and small temperature difference heat transfer. In all these cases, the useful macroscopic information is diminishing compared to the equilibrium thermal noise. In both limits, the straightforward application of the DSMC method including any of the collision schemes described in this chapter, becomes increasingly computationally expensive because of the existence of different scales. Recently, two promising methods have been proposed to resolve these multiscale simulation problems and both are based on the idea of algebraic decomposition of the local distribution function into equilibrium and non-equilibrium parts

$$f(t, \mathbf{r}, \mathbf{v}) = f_{eq}(t, \mathbf{r}, \mathbf{v}) + f_d(t, \mathbf{r}, \mathbf{v}), \tag{3.98}$$

where f_{eq} is a Maxwellian distribution

$$f_{eq}(t, \mathbf{r}, \mathbf{v}) = \frac{n(t, \mathbf{r})}{[2\pi RT(t, \mathbf{r})]^{3/2}} e^{-\frac{[v - V(t,r)]^2}{2RT(t,r)}} \tag{3.99}$$

and f_d is a function that measures the deviation from equilibrium in time and phase space. In general, the deviational part may be smaller, comparable, or larger to the equilibrium part. The key point here is that for the Maxwellian function (3.99), the collision integral Q (see (3.4)) is equal to zero and the equilibrium part can be evaluated by using knowledge of macroscopic variables n, v, and T only. When the Knudsen number becomes smaller the gas flow approaches more and more closely the equilibrium state and respectively, the distribution function (3.98) is relegated more and more to the equilibrium part (3.99). Thus for small Knudsen numbers, the presentation (3.98) may be associated with the Champan-Enskog expansion. The decomposition of the distribution function with respect to a small Knudsen number is a crucial aspect of the first approach considered in this subsection. In a sub-sequence of papers [88, 89, 14, 16] the method has evolved to the following computational scheme [71, 90, 91].

The starting point is that the functions f and f_d satisfy the following coupled system of equations [71]

$$\partial_t \boldsymbol{\rho} + \nabla_x \cdot F(\boldsymbol{\rho}) + \nabla_x \cdot \langle \mathbf{v}\psi(\mathbf{v}) f_d \rangle = 0, \tag{3.100}$$

$$\partial_t f + \mathbf{v} \cdot \nabla_x f = \frac{1}{\varepsilon} Q(f, f), \tag{3.101}$$

where $\psi(\mathbf{v}) = \left(1, \mathbf{v}, |\mathbf{v}|^2\right)$ are collision invariants, $\boldsymbol{\rho} = (\rho, \rho\mathbf{V}, \rho e)$ with density $\rho = mn$ and energy $e = |\mathbf{V}|^2/2 + (3/2)RT$, and $F(\boldsymbol{\rho}) = \left\langle \mathbf{v}\psi f_{eq} \right\rangle$; ε is the Knudsen number. The weighted integrals of f over velocity space are denoted by

$$\langle \phi f \rangle = \int \phi(\mathbf{v}) f d\mathbf{v} \tag{3.102}$$

with $\phi(\mathbf{v})$ any function of velocity.

Formally, the basic idea behind the first approach becomes clear from the proposed author's algorithm for the numerical solution of the equation system (3.100) and (3.101) at each time step:

- Solve the kinetic Eq. (3.101) by DSMC method and compute a first set of moments along with the correction term $\nabla_x \cdot \langle \mathbf{v}\psi(\mathbf{v})f_d \rangle$ from the particle data set;
- Solve the fluid Eq. (3.100) with any type of finite volume scheme and obtain a second set of moments;
- Match the moments of kinetic and fluid solutions through a transformation of the sample values;
- Proceed to the next time step and repeat the steps above.

In order to take full advantage of the distribution function decomposition at small Knudsen numbers, the authors [71] have introduced a transition function $w(t, \mathbf{r}) \in [0, 1]$ defined on three disjoined sets $\Omega_1, \Omega_2, \Omega_3$ belonging to the physical space. The first set Ω_1 is assumed a domain where the flow is far from equilibrium. In Ω_2 the flow is supposed to be in local equilibrium ($f_{\mathrm{d}} = 0$). In the handshake zone Ω_3 the flow is supposed to be close to equilibrium $\left(f_{\mathrm{d}} \ll f_{\mathrm{eq}}\right)$. The function $w(t, \mathbf{r})$ is defined by:

$$w(t, \mathbf{r}) = \begin{cases} 1 & , \mathit{for}\ \mathbf{r} \in \Omega_1 \\ 0 & , \mathit{for}\ \mathbf{r} \in \Omega_2 \\ 0 \le w(t, \mathbf{r}) \le 1 & , \mathit{for}\ \mathbf{r} \in \Omega_3 \end{cases} \tag{3.103}$$

Multiplying (3.101) successively by w and $1 - w$, after some mathematical manipulations, the authors get the final form of their kinetic-fluid model

$$\partial_t \boldsymbol{\rho} + \nabla_x \cdot F(\boldsymbol{\rho}) + \nabla_x \cdot \langle \mathbf{v}\psi(\mathbf{v}) g_K \rangle = 0, \tag{3.104}$$

$$\partial_t f_K + \mathbf{v} \cdot \nabla_x f_K = \frac{1}{\varepsilon} Q\left(f_K + E[\boldsymbol{\rho}_F], f_K + E[\boldsymbol{\rho}_F]\right) + \left(\frac{\partial_t w + \mathbf{v} \cdot \nabla_x w}{w}\right) f_K, \tag{3.105}$$

where $f_K = wf$, $f_F = (1 - w)f$, $g_K = hf_{\mathrm{d}}$, and $g_F = (1 - w)f_{\mathrm{d}}$. The distribution $E[\boldsymbol{\rho}_K]$ is defined as the Maxwellian whose moments are given by $\langle \psi(\mathbf{v}) f_K \rangle$, while $E[\boldsymbol{\rho}_F]$ is the Maxwellian whose moments are given by $\boldsymbol{\rho} - \langle \psi(\mathbf{v}) f_K \rangle$. For small Knudsen number, $g_K = f_K - E[\boldsymbol{\rho}_K]$. For zone Ω_2 the system and reduces to the well-known continuum Euler equations and for zone Ω_1 reduces to the system and, which is equivalent to the Boltzmann equation. A detailed description of the method and DSMC application can be found in paper [71].

As shown, the proposed multiscale approach at small Knudsen numbers utilizes elements from both domain decomposition (a) and heterogeneous (c) multiscale strategies. A razor edge for the method is the evaluation of the transition function $w(t, \mathbf{r})$, which must be computed from a precisely chosen breakdown parameter.

The second multiscale approach is also based on the decomposition of the distribution function into equilibrium and non-equilibrium parts. However, instead small Knudsen number flows, the authors have assumed a small deviation from equilibrium as the result of a small bulk velocity (the Mach number smaller than 1) or small temperature difference without requiring Kn < 1. In a series of articles [17, 92, 93, 18, 19, 94] the authors develop a low-variance deviational particle Monte Carlo (LVDSMC) method, which achieves variance reduction by simulating only the difference between the velocity distribution and an underlying equilibrium distribution. Briefly, the author's idea comprises several major steps [94]. By using decomposition, the collision operator Q in the Boltzmann equation is split into linear and nonlinear parts

$$Q(f,f) = P(f_{\mathrm{d}}) + Q(f_{\mathrm{d}},f_{\mathrm{d}}), \tag{3.106}$$

where f_{eq} is supposed to be a known equilibrium distribution function. The key point here is a special presentation [95] of the linear operator P in a gain term, depending on kernel functions k_1 and k_2, and a lost term depending on the mean collision frequency of a molecule belonging to the non-equilibrium part f_d with molecules belonging to the equilibrium part f_{eq} as shown below

$$P(f_d) = \int_{\mathbb{R}^3} d^3\mathbf{v}_*[2K_1(\mathbf{v},\mathbf{v}_*) - K_2(\mathbf{v},\mathbf{v}_*)]f_d(\mathbf{v}_*) - \nu f_d, \tag{3.107}$$

where ν is mean collision frequency and ν, K_1K_2 depends on f_{eq}. The specific forms of ν, K_1K_2 are given in [95]. It is worth noting that for the hard sphere model, functions K_1 and ν are found analytically (see [93, 18, 95]). For VHS collisions, Wagner [95] has proposed a series of Markov processes for the creation and deletion of particles to simulate the gain and lost terms in. The nonlinear part of expresses the collision operator between deviational particles. It is simulated by the procedure, proposed originally by Baker and Hadjiconstantinou [92].

The streaming operator D of the Boltzmann equation is simulated for the deviational particles in the same manner as the free motion of physical particles. The boundary conditions, however, are modified. Additional particles must be generated at the boundary to account for the differences in fluxes between equilibrium and boundary distributions [93, 18].

The proposed variance-reduced particle method reduces significantly the level of statistical noise in comparison with the standard DSMC method for problems where the departure from equilibrium is small. The method is very suitable for the simulation of gas flows in small-scale devices (MEMS and NEMS).

A good critical review of both multiscale approaches presented above is given in paper [96].

A quite different and interesting approach for variance reduction by using the same decomposition form has been used in [97, 98]. The major advantage of the proposed variance-reduced direct simulation Monte Carlo (VRDSMC) compared to the LVDSMC (described above) is that it can be incorporated into the standard DSMC with relatively small modifications. The basic idea behind the proposed approach is very simple. Instead of using a particle scheme that differs from DSMC, the method achieves variance reduction by combining a standard DSMC calculation with an auxiliary equilibrium simulation (correlated sampling), utilizing the same set of particles in cells as obtained in the DSMC calculations. The key element is the introduction of a set of important weights for all particles involved in the simulation. Each weight W_i measures the relative frequency of finding a particle i with velocity v_i in the equilibrium state relative to the non-equilibrium state. Practically, a separate equilibrium simulation is not performed, and its actual evolution over time is replaced by a procedure for updating the weights. The details of the VRDSMC method can be found in [99, 97].

According to the author's remarks in the reference papers, all presented in this section, multiscale methods require auxiliary procedures to ensure the stability of the computational process, which increases the complexity of the corresponding numerical algorithms.

3.8.2 Internal Flow Multiscale Method (IMM)

In a recent paper [100], the authors present a novel multiscale approach, termed the Internal Flow Multiscale Method (IMM), for simulating streamwise micro/nano flows in high aspect ratio geometries, such as within long ducts, tubes, or channels, of varying sections. The scheme combines the standard hydrodynamic equations for mass and momentum flux conservation with molecular dynamics (MD) within a single framework. It is worth noting that a DSMC algorithm could efficiently replace the molecular dynamics method in cases of gas flow. The proposed approach employs strategy c) by leveraging the specific features of low-speed internal gas flows in domains with channel geometry. It is assumed that in the longitudinal direction, the length L_x is much larger than L_y and L_z in the transversal direction. There is a scale separation between hydrodynamic variation along streamlines and microscopic processes transverse to the flow direction (in y and z) and it is effectively used here by applying a simple multiscale strategy. The method applies a simple hydrodynamic model over the entire domain with coordinates aligned with the local flow

$$\frac{\partial p}{\partial s} = \frac{\partial \tau_{sy}}{\partial y} + \frac{\partial \tau_{sz}}{\partial z} + F_s, \tag{3.108}$$

where τ_{sy} and τ_{sz} are normal components of shear stress, p is the scalar pressure and F_s is a body force per unit volume in the flow direction. An additional simplification here is the assumption for a local parallel flow. At the micro-scale, a set of micro-subdomains covering the full cross-section of the channel is distributed along the longitudinal axis with a spacing that allows for capturing the flow variations along the channel. Within each micro sub-domain, a constant body force Φ acts on gas molecules in such a way that

$$\Phi \equiv \frac{\partial p}{\partial s}. \tag{3.109}$$

The coupling between micro and macro domains is performed by an iteration procedure with respect to Φ. The iteration procedure converges to a distribution of Φ along the channel length to ensure two constraints, namely, an imposed pressure difference $p_A - p_B$ of both channel ends and a constant mass flow rate through each of the micro subdomains. Initially, a constant pressure gradient $\Phi = \Delta p/L_s$ is imposed in each micro subdomain. In the first iteration, the simulation in each microdomain generates different mass flow rates throughout the corresponding cross-section of the channel. The corresponding Φ_i in the sub-domains are accordingly corrected toward producing equal mass flow rates and the second iteration is realized. This process is repeated until both conditions for pressure difference and mass flow rate are fulfilled with the desired accuracy. The details of the iteration procedure can be found in the paper [100].

3.8.3 Multiscale Methods Based on Fokker-Plank (FP) Equation

The stiffness of the collision operator, which results in massive computational costs in the near-continuum regimes for DSMC, can be treated by passing to the diffusion limit of the Boltzmann equation. In this context, the Fokker–Planck (FP) equation naturally arises as a diffusion approximation of the underlying Master equation [101]. Although there is acceptable conformity between the FP and other approaches such as DSMC, there are some drawbacks, such as incorrect Prandtl numbers [102]. Gorji and co-workers [103] developed an algorithm to solve this problem by introducing a cubic drift term. Later, this algorithm was optimized and extended for diatomic molecules [104, 105, 106]. Combining DSMC and FP has become more popular as an efficient particle method over the last few years. Some researchers proposed criteria for switching between FP and DSMC based on the number of collisions between gas molecules [105, 107, 108, 109, 110].

Gorji and Jenny [105] generalized a simple linear model with the objective to fix inconsistencies regarding the Prandtl number. The generalized form which is actually equivalent to the term Q in Eq. (3.4) is as:

$$S^{FP} = -\frac{\partial (A_i F)}{\partial v_i} + \frac{D^2}{2}\frac{\partial^2 F}{\partial v_i \partial v_j}, \tag{3.110}$$

where the $A_i(t, \mathbf{r}, \mathbf{v}')$ and $D(t, \mathbf{r})$ correspond to the drift and diffusion terms, respectively. The pecular velocity is defined as $\mathbf{v}' = \mathbf{v} - \mathbf{V}$, where $\mathbf{V}$ is the mean (or gas) velocity. Note that the mass density function F is related to the probability density function f as:

$$F(t, \mathbf{r}, \mathbf{v}) = \rho(t, \mathbf{r})f(t, \mathbf{r}, \mathbf{v}), \tag{3.111}$$

For the drift coefficient, a cubic model as a polynomial function of the fluctuating velocity has been proposed as [103]:

$$A_i \tilde{c}_{i,j} v'_j + \gamma_i \left(v'_j v'_j - \overline{u'_j u'_j} \right) + \Lambda \left(v'_i v'_k v'_k - \overline{u'_i u'_k u'_k} \right), \tag{3.112}$$

where $\mathbf{u}'$ is a realization from sample space $\mathbf{v}'$ and therefore:

$$\overline{\Psi(\mathbf{u}')} = \int \Psi(\mathbf{v}') f d\mathbf{v}'. \tag{3.113}$$

The scalar is defined as:

$$\Lambda = -\frac{1}{\alpha \rho^3} \left| \det(\pi_{i,j}) \right|, \tag{3.114}$$

where $\det(\pi_{i,j})$ is the determinant of the stress tensor and $\alpha = \tau \left(\overline{u'_i u'_i} \right)^4$ a scaling factor with $\tau = 2\mu / p$ the relaxation time. The diffusion term in Eq. (3.110) has been considered to be as:

$$D = \sqrt{\frac{4e_s}{3\tau}}, \tag{3.115}$$

where $e_s = 2kT / 2m$ is the sensible energy.

3.9 Conclusion

The development of the DSMC method has a history spanning more than 60 years. Over the years, numerous DSMC schemes and improvements have been proposed and applied to a wide range of problems in various areas, including microfluidics, which treats the gas structure on the particle level and considers rarefied gas dynamics as an appropriate model for gas flow description. In this chapter, we provided a description of the basic principles and present mostly well-established elements of the DSMC method, as viewed by the authors. At the same time, we have not addressed many interesting and important issues in particle simulation, such as gas mixtures, chemical reactions, and quantum effects considerations. Other important omissions are related to the DSMC analysis of macroscopic variable fluctuations, flow instability, and self-organization in microgas flows. For these flows, as well as for details on other topics we refer to relevant monographs [2, 21], papers [56, 57, 58], and the literature quoted therein.

References

1. Bird, G.A.: Molecular Gas Dynamics. Oxford University Press, Oxford (1976)
2. Bird, G.A.: Molecular Gas Dynamics and the Direct Simulation of Gas Flows. Oxford University Press, Oxford (1994)
3. Bird, G.A.: Approach to translational equilibrium in a rigid sphere gas. Phys. Fluids **6**(10), 1518–1519 (1963)
4. Koura, K.: Transient couette flow of rarefied binary gas mixture. Phys. Fluids **13**(6), 1457–2000 (1970). https://doi.org/10.1063/1.1693102
5. Belotserkovskii, O.M., Yanitskii, V.E.: The statistical particles-in-cells method for solving rarefied gas dynamics problems. USSR Comput. Math. Math. Phys.Comput. Math. Math. Phys. **15**, 101–114 (1975)
6. Wagner, W.: A convergence proof for Bird direct simulation Monte-Carlo method for the Boltzmann equation. J. Stat. Phys. **66**(3–4), 1011–1044 (1992). https://doi.org/10.1007/bf01055714
7. Sone, Y.: Molecular Gas Dynamics: Theory, Techniques, and Applications. Birkhäuser, Boston (2007)
8. Koura, K.: Null-collision technique in the direct-simulation Monte-Carlo method. Phys. Fluids **29**(11), 3509–3511 (1986)
9. Babovsky, H., Neunzert, H.: On a simulation scheme for the Boltzmann equation. Math. Methods Appl. Sci. **8**, 223–233 (1986)
10. Nanbu, K.: Direct simulation scheme derived from the Boltzmann-Equation.1. Monocomponent gases. J. Phys. Soc. Jpn. **49**(5), 2042–2049 (1980). https://doi.org/10.1143/Jpsj.49.2042
11. Ivanov, M., Rogasinsky, S.: Theoretical analysis of traditional and modern schemes of the DSMC method. In: Beylich, A. (ed.) 17th Symposium on Rarefied Gas Dynamics 1990, pp. 629–642. VCH, New York (1991)
12. Yanitskiy, V.E.: Operator approach to direct monte-carlo simulation theory in rarefied-gas dynamics. In: Beylich, A.E. (ed.) Rarefied Gas Dynamics, pp. 770–777. VCH Aahen, Germany (1991)
13. Fan, J., Shen, C.: Statistical simulation of low-speed rarefied gas flows. J. Comput. Phys.Comput. Phys. **167**, 393–412 (2001)
14. Pareschi, L., Russo, G.: Time relaxed Monte Carlo methods for the Boltzmann equation. Siam J. Sci. Comput. **23**(4), 1252–1272 (2001)
15. Rjasanow, S., Wagner, W.: Stochastic weighted particle method for the Boltzmann equation. J. Comput. Phys.Comput. Phys. **124**(2), 243–253 (1996)
16. Trazzi, S., Pareschi, L., Wennberg, B.: Adaptive and recursive time relaxed Monte Carlo methods for rarefied gas dynamics. Siam J. Sci. Comput. **31**(2), 1379–1398 (2009). https://doi.org/10.1137/07069119x
17. Baker, L.L., Hadjiconstantinou, N.G.: Variance reduction for Monte Carlo solutions of the Boltzmann equation. Phys. Fluids **17**(5) (2005)
18. Homolle, T.M.M., Hadjiconstantinou, N.G.: A low-variance deviational simulation Monte Carlo for the Boltzmann equation. J. Comput. Phys.Comput. Phys. **226**(2), 2341–2358 (2007). https://doi.org/10.1016/j.jcp.2007.07.006

19. Radtke, G.A., Hadjiconstantinou, N.G., Wagner, W.: Low-noise Monte Carlo simulation of the variable hard sphere gas. Phys. Fluids **23**(3), Artn 030606 (2011a). https://doi.org/10.1063/1.3558887
20. Shterev, K.S., Stefanov, S.K.: Influence of reservoirs on pressure driven gas flow in a microchannel. Aip Conf. Proc. **1404** (2011). https://doi.org/10.1063/1.3659911
21. Bird, G.A.: The DSMC Method, Version 1'1 edition ed. CreateSpace Independent Publishing Platform (2013)
22. Gallis, M.A., Torczynski, J.R., Rader, D.J., Bird, G.A.: Convergence behavior of a new DSMC algorithm. J. Comput. Phys.Comput. Phys. **228**(12), 4532–4548 (2009). https://doi.org/10.1016/j.jcp.2009.03.021
23. Su, C.C., Tseng, K.C., Cave, H.M., Wu, J.S., Lian, Y.Y., Kuo, T.C., Jermy, M.C.: Implementation of a transient adaptive sub-cell module for the parallel-DSMC code using unstructured grids. Comput. Fluids. Fluids **39**(7), 1136–1145 (2010). https://doi.org/10.1016/j.compfluid.2010.02.003
24. Kac, M.: Probability and Related Topics in Physical Sciences. Interscience Publishers Ltd., London-New York (1959)
25. Wagner, W.: A convergence proof for bird direct simulation Monte-Carlo method for the Boltzmann-equation. J. Stat. Phys. **66**(3–4), 1011–1044 (1992)
26. Stefanov, S.K.: On DSMC calculations of rarefied gas flows with small number of particles in cells. Siam J. Sci. Comput. **33**(2), 677–702 (2011). https://doi.org/10.1137/090751864
27. Stefanov, S.K.: Particle Monte Carlo algorithms with small number of particles in grid cells. Lect. Notes Comput. Sci.Comput. Sci. **6046**, 110–117 (2011)
28. Strang, G.: On construction and comparison of difference schemes. Siam J. Numer. Anal. **5**(3), 506–+ (1968). https://doi.org/10.1137/0705041
29. Dietrich, S., Boyd, I.D.: Scalar and parallel optimized implementation of the direct simulation Monte Carlo method. J. Comput. Phys.Comput. Phys. **126**(2), 328–342 (1996)
30. Roohi, E., Darbandi, M.: Recommendations on performance of parallel DSMC algorithm in solving subsonic nanoflows. Appl. Math. Model. **36**(5), 2314–2321 (2012). https://doi.org/10.1016/j.apm.2011.08.036
31. Scanlon, T.J., Roohi, E., White, C., Darbandi, M., Reese, J.M.: An open source, parallel DSMC code for rarefied gas flows in arbitrary geometries. Comput. Fluids. Fluids **39**(10), 2078–2089 (2010). https://doi.org/10.1016/j.compfluid.2010.07.014
32. Wu, J.S., Tseng, K.C.: Parallel DSMC method using dynamic domain decomposition. Int. J. Numer. Meth. Eng.Numer. Meth. Eng. **63**(1), 37–76 (2005). https://doi.org/10.1002/Nme.1232
33. Wu, J.S., Tseng, K.C., Wu, F.Y.: Parallel three-dimensional DSMC method using mesh refinement and variable time-step scheme. Comput. Phys. Commun.. Phys. Commun. **162**(3), 166–187 (2004). https://doi.org/10.1016/j.cpc.2004.07.004
34. Wu, J.S., Tseng, K.C., Yang, T.J.: Parallel implementation of DSMC using unstructured mesh. Int. J. Comput. Fluid D **17**(5), 405–422 (2003). https://doi.org/10.1080/1061856031000081497
35. Cercignani, C.: The Boltzmann Equation and its Applications. Springer, New York (1988)
36. Stefanov, S., Roohi, E., Shoja-Sani, A.: A novel transient-adaptive subcell algorithm with a hybrid application of different collision techniques in direct simulation Monte Carlo (DSMC). Phys. Fluids **34**(9), Artn 092003 (2022). https://doi.org/10.1063/5.0104613
37. Stefanov, S.K., Boyd, I.D., Cai, C.P.: Monte Carlo analysis of macroscopic fluctuations in a rarefied hypersonic flow around a cylinder. Phys. Fluids **12**(5), 1226–1239 (2000)
38. Bruno, D., Frezzotti, A., Ghiroldi, G. P.: Rayleigh-Brillouin scattering in molecular Oxygen by CT-DSMC simulations. In Aip Conf. Proc. **1786**(1), 160001. AIP Publishing LLC (2017)
39. Ladiges, D., Gallis, M. A., Torczynski, J. R.: A direct simulation Monte Carlo study of thermal fluctuations in a rarefied gas. Phys. Fluids **31**(7), 077104 (2019)
40. Stefanov, S. K.: On the basic concepts of the direct simulation Monte Carlo method. Phys. Fluids **31**(6) (2019). https://doi.org/10.1063/1.5099042

41. Ivanov, M.S., Gimelshein, S.F.: Computational hypersonic rarefied flows. Annu. Rev. Fluid Mech.. Rev. Fluid Mech. **30**, 469–505 (1998)
42. Koura, K., Matsumoto, H.: Variable soft sphere molecular-model for air species. Phys. Fluids a-Fluid **4**(5), 1083–1085 (1992)
43. Hassan, H.A., Hash, D.B.: A generalized hard-sphere model for Monte-Carlo simulation. Phys. Fluids a-Fluid **5**(3), 738–744 (1993)
44. Fan, J.: A generalized soft-sphere model for Monte Carlo simulation. Phys. Fluids **14**(12), 4399–4405 (2002). https://doi.org/10.1063/1.1521123
45. Sharipov, F., Strapasson, J.L.: Direct simulation Monte Carlo method for an arbitrary intermolecular potential. Phys. Fluids **24**(1), Artn 011703 (2012). https://doi.org/10.1063/1.3676060
46. Bird, G.A.: The Q-K model for gas-phase chemical reaction rates. Phys. Fluids **23**(10), Artn 106101 (2011). https://doi.org/10.1063/1.3650424
47. He, Y., Stefanov, S.K., Ota, M.: The Monte Carlo simulation of a model microactuator driven by rarefied gas thermal effects. Rarefied Gas Dyn. **1084**, 281–286 (2009)
48. Cercignani, C.: Rarefied Gas Dynimics: From Basic Concepts to Actual Calculations. Cambridge University Press., (2000)
49. Cercignani, C., Lampis, M.: Kinetic Models for Gas-Surface Interactions. Transp. Theor. Stat. **1**(2), 101–110 (1971)
50. Lord, R.G.: Application of the Cercignani-Lampis scattering kernel to direct simulation Monte-Carlo calculations. Rarefied Gas Dyn. 1427–1433 (1991a)
51. Lord, R.G.: Some extensions to the Cercignani-Lampis gas-surface scattering kernel. Phys. Fluids a-Fluid **3**(4), 706–710 (1991)
52. Liou, W.W., Fang, Y.C.: Implicit boundary conditions for direct simulation Monte Carlo method in MEMS flow predictions. Cmes-Comp. Model Eng. **1**(4), 119–128 (2000)
53. Fang, Y.C., Liou, W.W.: Computations of the flow and heat transfer in microdevices using DSMC with implicit boundary conditions. J. Heat Trans.-T Asme **124**(2), 338–345 (2002)
54. Wang, M.R., Li, Z.X.: Simulations for gas flows in microgeometries using the direct simulation Monte Carlo method. Int. J. Heat Fluid Fl **25**(6), 975–985 (2004)
55. Cave, H.M., Tseng, K.C., Wu, J.S., Jermy, M.C., Huang, J.C., Krumdieck, S.P.: Implementation of unsteady sampling procedures for the parallel direct simulation Monte Carlo method. J. Comput. Phys.Comput. Phys. **227**(12), 6249–6271 (2008). https://doi.org/10.1016/j.jcp.2008.03.015
56. Stefanov, S., Roussinov, V., Cercignani, C.: Rayleigh-Benard flow of a rarefied gas and its attractors. I. Convection regime. Phys. Fluids **14**(7), 2255–2269 (2002a). https://doi.org/10.1063/1.1483837
57. Stefanov, S., Roussinov, V., Cercignani, C.: Rayleigh-Benard flow of a rarefied gas and its attractors. II. Chaotic and periodic convective regimes. Phys. Fluids **14**(7), 2270–2288 (2002b). https://doi.org/10.1063/1.1483839
58. Stefanov, S., Roussinov, V., Cercignani, C.: Rayleigh-Benard flow of a rarefied gas and its attractors. III. Three-dimensional computer simulations. Phys. Fluids **19**(12) (2007)
59. Lofthouse, A.J., Scalabrin, L.C., Boydt, L.D.: Velocity slip and temperature jump in hypersonic aerothermodynamics. J. Thermophys. Heat Tr. **22**(1), 38–49 (2008). https://doi.org/10.2514/1.31280
60. Mohammadzadeh, A., Roohi, E., Niazmand, H., Stefanov, S., Myong, R.S.: Thermal and second-law analysis of a micro- or nanocavity using direct-simulation Monte Carlo. Phys. Rev. E **85**(5) (2012)
61. Amiri-Jaghargh, A., Roohi, E., Niazmand, H., Stefanov, S.: Low speed/low rarefaction flow simulation in micro/nano cavity using DSMC method with small number of particles per cell. In: Frijns, A. (ed.) 1st European Conference on Gas Micro Flows, vol. 362; J. Phys. Conf. Ser. (2012)
62. Kaplan, C.R., Oran, E.S.: Nonlinear filtering for low-velocity gaseous microflows. Aiaa J **40**(1), 82–90 (2002)

63. Bobylev, A., Ohwada, T.: The error of the splitting scheme for solving evolutionary equations. Appl. Math. Lett. **14**, 45–48 (2001)
64. Alexander, F.J., Garcia, A.L., Alder, B.J.: Cell size dependence of transport coefficients in stochastic particle algorithms. Phys. Fluids **10**(6), 1540–1542 (1998)
65. Hadjiconstantinou, N.G.: Analysis of discretization in the direct simulation Monte Carlo. Phys. Fluids **12**(10), 2634–2638 (2000)
66. Hadjiconstantinou, N.G., Garcia, A.L., Bazant, M.Z., He, G.: Statistical error in particle simulations of hydrodynamic phenomena. J. Comput. Phys.Comput. Phys. **187**(1), 274–297 (2003). https://doi.org/10.1016/S0021-9991(03)00099-8
67. Dimov, I., Faragó, I., Havasi, Á., Zlatev, Z.: Operator splitting and commutativity analysis in the Danish Eulerian model. Math. Comput. Simul. **67**, 217–233 (2004)
68. Shevyrin, A.A., Bondar, Y.A., Ivanov, M.S.: Analysis of repeated collisions in the DSMC method. In: Capitelli, M. (ed.) 24th Symposium on Rarefied Gas Dynamics, pp. 565–570. Melville, New York (2005)
69. Burt, J.M., Boyd, I.D.: A hybrid particle approach for continuum and rarefied flow simulation. J. Comput. Phys.Comput. Phys. **228**(2), 460–475 (2009)
70. Darbandi, M., Roohi, E.: A hybrid DSMC/Navier-Stokes frame to solve mixed rarefied/ nonrarefied hypersonic flows over nano-plateandmicro-cylinder. Int. J. Numer. Meth. Fl **72**(9), 937–966 (2013)
71. Degond, P., Dimarco, G.: Fluid simulations with localized boltzmann upscaling by direct simulation Monte-Carlo. J. Comput. Phys.Comput. Phys. **231**(6), 2414–2437 (2012)
72. Garcia, A.: DSMC/continuum hybrid methods. Rarefied Gas Dyn. **585**, 371–371 (2001)
73. Liu, H.L., Fan, J., Shen, C.: Validation of a hybrid grid scheme of DSMC in simulating three-dimensional rarefied gas flows. Rarefied Gas Dyn. **663**, 382–389 (2003)
74. Roveda, R., Goldstein, D.B., Varghese, P.L.: Hybrid Euler/particle approach for continuum/ rarefied flows. J. Spacecraft Rockets **35**(3), 258–265 (1998)
75. Sun, Q.H., Boyd, I.D., Candler, G.V.: A hybrid continuum/particle approach for micro-scale gas flows. Rarefied Gas Dyn. **663**, 752–759 (2003)
76. Wang, W.L., Sun, Q.H., Boyd, I.D.: Assessment of a hybrid method for hypersonic flows. Rarefied Gas Dyn. **663**, 923–930 (2003)
77. Wijesinghe, H.S., Hornung, R.D., Garcia, A.L., Hadjiconstantinou, N.G.: Three-dimensional hybrid continuum-atomistic simulations for multiscale hydrodynamics. J. Fluid Eng.-T Asme. **126**(5), 768–777 (2004). https://doi.org/10.1115/1.1792275
78. Garcia, A.L., Bell, J.B., Crutchfield, W.Y., Alder, B.J.: Adaptive mesh and algorithm refinement using direct simulation Monte Carlo. J. Comput. Phys.Comput. Phys. **154**(1), 134–155 (1999)
79. Kolobov, V.I., Arslanbekov, R.R., Aristov, V.V., Frolova, A.A., Zabelok, S.A.: Unified solver for rarefied and continuum flows with adaptive mesh and algorithm refinement. J. Comput. Phys.Comput. Phys. **223**(2), 589–608 (2007)
80. Su, C.C., Tseng, K.C., Wu, J.S., Cave, H.M., Jermy, M.C., Lian, Y.Y.: Two-level virtual mesh refinement algorithm in a parallelized DSMC code using unstructured grids. Comput. Fluids. Fluids **48**(1), 113–124 (2011). https://doi.org/10.1016/j.compfluid.2011.04.002
81. Wu, J.S., Tseng, K.C., Kuo, C.H.: The direct simulation Monte Carlo method using unstructured adaptive mesh and its application. Int. J. Numer. Meth. Fl **38**(4), 351–375 (2002)
82. Roohi, E., Darbandi, M.: Extending the Navier–Stokes solutions to transition regime in two-dimensional micro- and nano-channel flows using information preservation scheme. Phys. Fluids **21**(8), Artn 082001 (2009). https://doi.org/10.1063/1.3177351
83. Shen, C.: Information preservation (IP) method in simulation of internal rarefied gas flows in MEMS. Rarefied Gas Dyn. **762**, 141–149 (2005)
84. Shen, C., Jiang, J.Z., Fan, J.: Information preservation method for the case of temperature variation. Rarefied Gas Dyn. **585**, 185–192 (2001)
85. Sun, Q.H., Boyd, I.D., Fan, J.: Development of an information preservation method for subsonic, micro-scale gas flows. Rarefied Gas Dyn. **585**, 547–553 (2001)

86. Zhang, J., Fan, J.: Information preservation modeling of Rayleigh-Benard transition from thermal conduction to convection. Rarefied Gas Dyn. **1084**, 359–364 (2009)
87. Zhang, J., Fan, J., Jiang, J.Z.: Multiple temperature model for the information preservation method and its application to nonequilibrium gas flows. J. Comput. Phys.Comput. Phys. **230**(19), 7250–7265 (2011). https://doi.org/10.1016/j.jcp.2011.05.025
88. Caflisch, R.E., Pareschi, L.: Towards a hybrid Monte Carlo method for rarefied gas dynamics. Ima V Math. **135**, 57–73 (2004)
89. Pareschi, L., Caflisch, R.E.: An implicit Monte arlo method for rarefied gas dynamics—1. The space homogeneous case. J. Comput. Phys. **154**(1), 90–116 (1999)
90. Degond, P., Dimarco, G., Pareschi, L.: The moment-guided Monte Carlo method. Int. J. Numer. Meth. Fl **67**(2), 189–213 (2011)
91. Dimarco, G., Pareschi, L.: Fluid solver independent hybrid methods for multiscale kinetic equations. Siam J. Sci. Comput. **32**(2), 603–634 (2010)
92. Baker, L.L., Hadjiconstantinou, N.G.: Variance-reduced particle methods for solving the Boltzmann equation. J. Comput. Theor. Nanos **5**(2), 165–174 (2008)
93. Homolle, T.M.M., Hadjiconstantinou, N.G.: Low-variance deviational simulation Monte Carlo. Phys. Fluids **19**(4), Artn 041701 (2007a). https://doi.org/10.1063/1.2717721
94. Radtke, G.A., Hadjiconstantinou, N.G., Wagner, W.: Low variance particle simulations of the Boltzmann transport equation for the variable hard sphere collision model. Aip Conf. Proc. **1333**, 307–312 (2011). https://doi.org/10.1063/1.3562666
95. Wagner, W.: Deviational particle Monte Carlo for the Boltzmann equation. Monte Carlo Methods Appl. **14**(3), 191–268 (2008). https://doi.org/10.1515/MCMA.2008.010
96. Radtke, G.A., Peraud, J.P.M., Hadjiconstantinou, N.G.: On efficient simulations of multiscale kinetic transport. Philos. T. R. Soc. A **371**(1982) (2013)
97. Al-Mohssen, H.A., Hadjiconstantinou, N.G.: Low-variance direct Monte Carlo simulations using importance weights. Esaim-Math. Model Num. **44**(5), 1069–1083 (2010). https://doi.org/10.1051/M2an/2010052
98. Al-Mohssen, H.A., Hadjiconstantinou, N.G.: A practical variance reduced DSMC method. Aip Conf. Proc. **1333**, 219–224 (2011). https://doi.org/10.1063/1.3562651
99. Al-Mohssen, H.A.: An excursion with the Boltzmann equation at low speeds : variance-reduced DSMC. Ph.D., Massachusetts Institute of Technology (2010)
100. Borg, M.K., Lockerby, D.A., Reese, J.M.: A multiscale method for micro/nano flows of high aspect ratio. J. Comput. Phys.Comput. Phys. **233**, 400–413 (2013)
101. Gorji, M.H., Torrilhon, M.: Entropic Fokker-Planck kinetic model. J. Comput. Phys.Comput. Phys. **430**, 110034 (2021)
102. Jenny, P., Torrilhon, M., Heinz, S.: A solution algorithm for the fluid dynamic equations based on a stochastic model for molecular motion. J. Comput. Phys.Comput. Phys. **229**(4), 1077–1098 (2010)
103. Gorji, M.H., Torrilhon, M., Jenny, P.: Fokker-Planck model for computational studies of monatomic rarefied gas flows. J. Fluid Mech. **680**, 574–601 (2011)
104. Gorji, M.H., Jenny, P.: A Fokker-Planck based kinetic model for diatomic rarefied gas flows. Phys. Fluids **25**(6), 62002 (2013)
105. Gorji, M.H., Jenny, P.: Fokker—Planck—DSMC algorithm for simulations of rarefied gas flows. J. Comput. Phys.Comput. Phys. **287**, 110–129 (2015)
106. Sadr, M., Gorji, M.H.: A continuous stochastic model for non-equilibrium dense gases. Phys. Fluids **29**(12), 122007 (2017)
107. Jun, E., Gorji, M.H., Grabe, M., Hannemann, K.: Assessment of the cubic Fokker–Planck–DSMC hybrid method for hypersonic rarefied flows past a cylinder. Comput. Fluids. Fluids **168**, 1–13 (2018)
108. Jun, E.: Cubic Fokker-Planck-DSMC hybrid method for diatomic rarefied gas flow through a slit and an orifice. Vacuum **159**, 125–133 (2019)
109. Küchlin, S., Jenny, P.: Parallel Fokker–Planck-DSMC algorithm for rarefied gas flow simulation in complex domains at all Knudsen numbers. J. Comput. Phys.Comput. Phys. **328**, 258–277 (2017)

110. Mahdavi, A., Roohi, E.: A novel hybrid DSMC-Fokker Planck algorithm implemented to rarefied gas flows. Vacuum **181**, 09736 (2020)
111. Boris, J.P., Book, D.L.: Flux-corrected transport. 1. Shasta, a fluid transport algorithm that works. J. Comput. Phys. **11**(1), 38–69 (1973). https://doi.org/10.1016/0021-9991(73)90147-2
112. Engquist, B., Li, X., Ren, W., Vanden-Eijnden, E.: Heterogeneous multiscale methods: A review. Commun. Comput. Phys. **2**(3), 367–450 (2007)
113. Gorji, M.H., Jenny, P.: An efficient particle Fokker-Planck algorithm for rarefied gas flows. J. Comput. Phys.Comput. Phys. **262**, 325–343 (2014)
114. Liou, W.W., Fang, Y.C.: Heat transfer in microchannel devices using DSMC. J. Microelectromech. S **10**(2), 274–279 (2001)
115. Tseng, K.C., Cave, H.M., Kuo, T.C., Jermy, M.C., Krumdieck, S.P., Wu, J.S.: Implementation of the transient adaptive sub-cell module for the parallel DSMC code. Aip Conf. Proc. **1084**, 287–292 (2009)

Chapter 4
Recent Advances in DSMC Collision Treatments

Abstract This chapter reviews the new collision models that have emerged in the DSMC method over the past few years. The new collision models for the two categories, i.e., models based on the Boltzmann equation and models based on the Kac master equation, are presented. However, the main focus is on the BT family schemes (i.e., SBT, ISBT, GBT, hybrid NTC-GBT-SBT-TAS, SSBT, SGBT schemes), which are based on the Kac master equation (or Stefanov equation). For each collision models, the collision algorithm, the obtained results, a comparison with traditional DSMC, and the achieved improvements will be described. At the end of this chapter, new treatments related to the DSMC collision procedure are introduced and discussed. These techniques primarily focus on eliminating collisions or reducing the time required for the DSMC collision process in near-continuum flows. The considered approaches are DSMC-Fokker-Plank (DSMC-FP), Low Diffusion-DSMC (LD-DSMC), low-variance deviational simulation Monte Carlo (LVDSMC), and collision limiters.

4.1 Introduction

In this section, we present a summary of recent advances in collision models developed within the framework of the Direct Simulation Monte Carlo (DSMC) method. The DSMC schemes can be categorized into two general classes, depending on the approach to collision partner selection about the collision models. In the first category, collision models have been developed based on assumptions that lead to the derivation of the Boltzmann equation. The most popular collision models for this group are No Time Counter (NTC) [1] and Time Counter (TC) [2]. The second group of collision schemes has been derived based on the Kac master equation. For the second group, the traditional collision models are Pseudo-Poisson (collision frequency) [3], Majorant Frequency (MF) [4], Bernoulli Trials (BT) [5], Ballot Box Scheme [5],

---

Supplementary Information The online version contains supplementary material available at https://doi.org/10.1007/978-981-96-8200-3_4.

E. Roohi et al., *Advances in Direct Simulation Monte Carlo: From Micro-Scale to Rarefied Flow Phenomena*, https://doi.org/10.1007/978-981-96-8200-3_4

Null Collision [6]. The NTC and MF schemes are extensively discussed in Chap. 3. The recently proposed collision models from both abovementioned categories will be discussed and reviewed in the present chapter.

4.2 Nearest Neighbor (NN)

LeBeau et al. [7] introduced the "virtual subcell" (VSC) method as an improvement in the particle selection procedure of the NTC, in which the first particle is chosen randomly, while the second partner is selected from available neighbors of the selected particle, i.e., the nearest neighbor (NN) of a given particle is found and selected. The flowchart of the NN collision model is shown in Fig. 4.1. VSC requires the collision separation calculation for all available particles in the cell to determine the nearest neighbor. The NTC-NN algorithm was implemented in new versions of DSMC codes distributed by Gream Bird [8]. Bird et al. [9] and Gallis et al. [10] compared the convergence behavior of the NN collision model compared with the standard NTC collision and showed that NN achieves very high efficiency as it minimizes the mean collision separation between collision partners (Fig. 4.1); however, the new algorithm is very sensitive to the selection of the time step and almost always needs smaller time steps compared to the original NTC. They showed that NN significantly reduces the computational resources required for a DSMC simulation to achieve a particular level of accuracy.

As mentioned, NN requires lower time steps because it overlooks a crucial aspect in pair selection: higher-velocity particles may reach particles at a greater distance than their nearest neighbor; therefore, a sufficiently small time step should be utilized. On the other hand, according to the Boltzmann equation, the working space is phase space with three coordinates (positions) and three velocities variables. From

Fig. 4.1 NN collision procedure

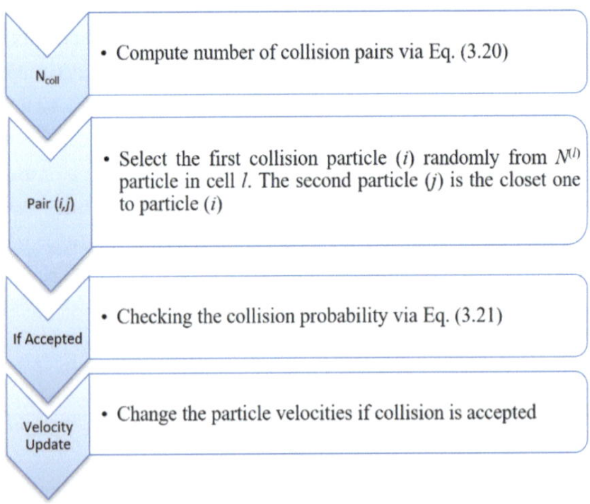

this viewpoint, NN considers only the space of coordinates, using the criterion of minimum distance for colliding pairs. This means that NN works well when the whole velocity distribution function is available at every coordinate point. To achieve this, one must have sufficient particles in all small volumes. Otherwise, there will be negative effects and erroneous solutions. Bird [11] demonstrated that NN requires at least 7 particles per collision cell to predict the equilibrium collision frequency accurately.

It should be noted that Bird states that lower values of the separation of free paths (SOF), which is the ratio of the mean collision separation (MCS) to the mean free path, largely self-validate the accuracy of DSMC solutions. However, it should be mentioned that this is a necessary but not a sufficient condition. It is worth noting that a low SOF does not always guarantee the correct solution, especially if it is not accompanied by adequate particles per cell (PPC) and sufficiently fine cells.

Figure 4.2 shows the convergence behavior of the thermal-conductivity ratio for the NN and NTC schemes, using 60 simulators per cell as a function of cell size and time step, respectively. According to the different trends, one can conclude that limiting the search for the nearest neighbor (i.e., NN scheme) significantly reduces the total error for all combinations of time step and cell size [10].

Figure 4.3 presents the Mach contour for argon gas flow at a Mach number of 10 past a 12-inch-diameter circular cylinder, using the NN scheme with the Bird's DS2V code. The stream temperature was 200 K, and the surface temperature was a uniform 500 K. The case, characterized by a number density of 4.247×10^{20} particles per cubic meter and a Knudsen number of 0.009, has been selected as the subject for benchmark calculations using a variety of DSMC codes. There is an option in the DS2V program to turn off the procedures that cause near-nearest-neighbor collisions. According to Fig. 4.4, with cell adaptation, the deterioration in the results was approximately twice that due to not adapting the cells with nearest neighbors

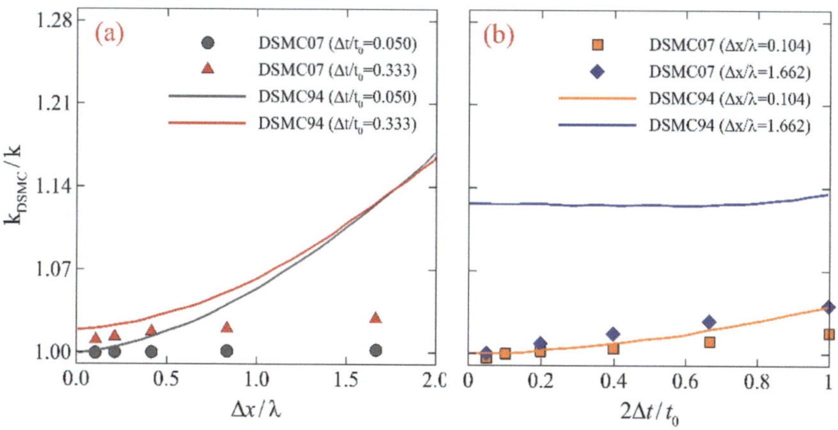

Fig. 4.2 Comparison between the NN (DSMC07) and NTC (DSMC94) schemes for predicting the thermal conductivity of Argon. Reproduced from Ref. [10]

Fig. 4.3 Mach number contour obtained for the benchmark test case using the NN scheme with the DS2V code ([11], open source)

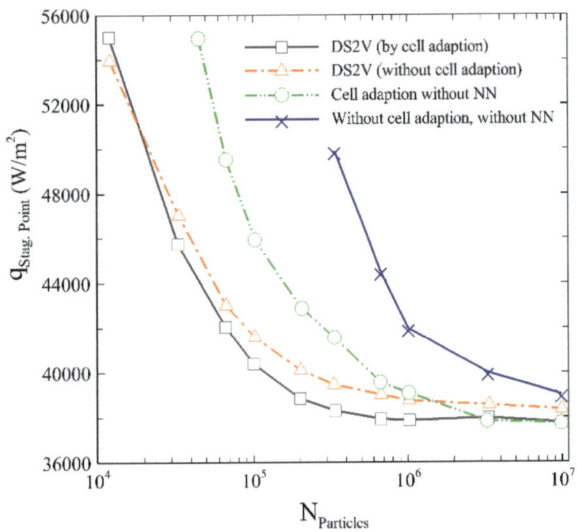

Fig. 4.4 Convergence behavior of DS2V code concerning the number of simulated particles. Reproduced from Ref. [11]

enabled. The deterioration when neither nearest neighbors nor cell adaptation was employed was markedly greater than the sum of the individual deteriorations.

4.3 Near Neighbor

As a modification of the NN model, Macrossan [12] suggested the 'pseudo-subcell' collision method, in which the search for a collision partner discontinues if a 'near-enough' particle is found, i.e., whenever another particle is found within the 'pseudo-subcell' of radius δ centered on the first particle. Macrossan showed that in structured cells, the pseudo-subcell method gives a 5% increase in mean collision separation (MCS) compared to the virtual-subcell method and is up to 20% faster than the virtual-subcell method, see Fig. 4.5.

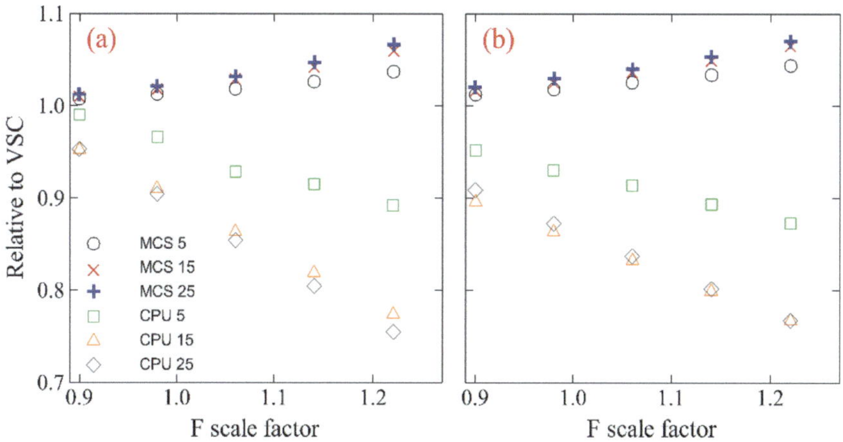

Fig. 4.5 Mean collision separation (MCS) and CPU time for the pseudo-subcell method relative to the virtual-subcell method in the cases of **a** 3D cells with an aspect ratio (AR) of 1:2:3 and **b** 2D cells with an AR of 1:2. Reproduced from Ref. [12]

4.4 Simplified Bernoulli Trial (SBT)

An alternative approach to the collision schemes described so far was developed based on the approximations of the transition collision operator [13] of the Kac equation. As a result, the first widely discussed Bernoulli Trials (BT) scheme [5] checks collisions with a given probability for all available collision pairs, i.e., $N(N-1)/2$, which makes this algorithm quite time-consuming. Stefanov [14, 15] showed that in small time-step intervals, it is possible to simplify the collision process, and instead of checking all available possibilities, for the first selected particle, say i, one can choose randomly the second pair among the next in the list available sorted-particles. Hence, the number of selections would be reduced to $(N-1)$. At the same time, theoretically, the probability of collision for each pair remains unchanged, based on the strict Markov process, equal to the corresponding one in the BT method. This leads to the simplified Bernoulli Trials (SBT) scheme. In SBT, each particle pair can be checked for collision only once per time step, thereby preventing repeated collisions and resulting in significantly reduced computational costs compared to the BT method.

According to the Eqs. (3.17)–(3.19), the final form of transition operator could be obtained as:

$$\boldsymbol{G}(\Delta t) = exp\left[\Delta t \sum\nolimits_{1 \le i < j \le N^{(l)}} \omega_{i,j}\left(T_{i,j} - I\right)\right] \tag{4.1}$$

For small interval Δt, operator G can be expanded concerning Δt degrees and terms of order equal to or higher than $O(\Delta t^2)$ can be neglected. The transition operator $G(\Delta t)$ corresponding to this approximation is obtained as:

$$G(\Delta t) \approx \prod_{1 \le i < j \le N^{(l)}} exp\left[\left(T_{i,j} - I\right)\omega_{i,j}\Delta t\right] = \prod_{i=1}^{N^{(l)}-1} \prod_{j=i+1}^{N^{(l)}} exp\left[\left(T_{i,j} - I\right)\omega_{i,j}\Delta t\right] \quad (4.2)$$

Each cofactor in this presentation, as shown in Eq. (4.2) is replaced by an approximation that is linear in Δt. This gives

$$G_{BT}(\Delta t) = \prod_{i=1}^{N^{(l)}-1} \prod_{j=i+1}^{N^{(l)}} \left[\left(1 - \omega_{ij}\Delta t\right)I + \omega_{i,j}\Delta t T_{i,j}\right] \quad (4.3)$$

The approximation operator G_{BT} has a probability interpretation, which gives the Bernoulli-Trials (BT) scheme. As a simplification, Stefanov demonstrated that it is possible to extend the internal product on the right-hand side of the Eq. (4.3), in a series of j for t to reach a new simplified transition operator G_{SBT} as follows [15]:

$$G_{SBT}(\Delta t) = \prod_{i=1}^{N^{(l)}-1} \left\{\left[1 - \sum_{j=i+1}^{N^{(l)}} \frac{1}{k}\left(k\omega_{i,j}\Delta t\right)\right]I + \sum_{j=i+1}^{N^{(l)}} \frac{1}{k}\left(k\omega_{i,j}\Delta t\right)T_{i,j}\right\}, \quad (4.4)$$

where $k = N^{(l)} - i$. The algorithmic interpretation of the operator $G_{SBT}(\Delta t)$ states that instead of checking $k = \frac{N^{(l)}(N^{(l)}-1)}{2}$, it is possible in each acceptance-rejection event to select the second particle of the collision pair at random and reduce the number of pair selections to $N^{(l)} - 1$. The SBT algorithm enables simulations with a significantly lower average number of particles per cell. When the required conditions are met, it effectively eliminates the algorithm's accuracy dependence on the number of particles, thereby reducing computational costs compared to the BT algorithm. The numerical procedure of SBT is as follows: particles in the l-th cell should be locally indexed to produce a particle list numbered as 1, 2, ..., $N^{(l)}$. The first particle of the collision pair (i, j), say i, is selected in sequence from the particle list: $i = 1, \ldots, N^{(l)} - 1$. The second particle, say j, is chosen randomly with the probability of $\frac{1}{k}$ from $k = N^{(l)} - i$ particles taking place in the list after particle i.

$$j = (i + 1) + int\left(k \times R_{nf1}\right) \quad (4.5)$$

Each pair is then checked for collision with probability.

$$W_{i,j} = \frac{\left(N^{(l)} - i\right)F_{num}\Delta t\sigma_{i,j}g_{i,j}}{V^{(l)}} > R_{nf2} \quad (4.6)$$

It should be noted that the Δt should be adjusted so that $W_{i,j}$ does not exceed unity, say

$$prob\left\{W_{i,j} \ge 1\right\} \to 0 \quad (4.7)$$

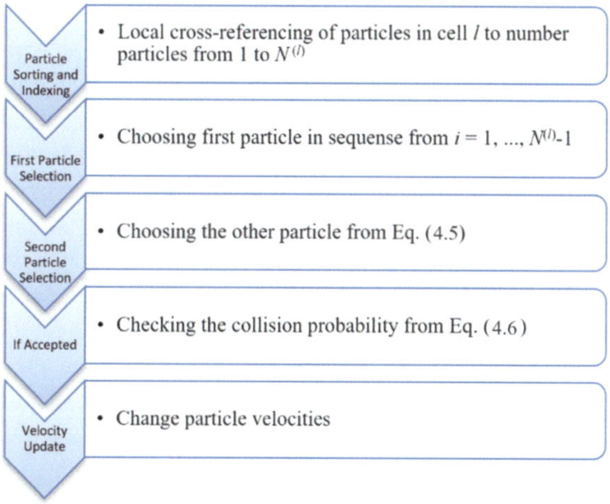

Fig. 4.6 SBT collision procedure

It is worth noting that the probability $W_{i,j} > 1$ should be kept always close to zero by choosing the appropriate time step and cell size. The SBT procedure avoids, within a time step, the production of eventually successively repeated collisions, which occurs in the NTC scheme when it is applied with a small number of particles. More details on the theoretical background of this scheme can be found in Refs. [14, 15]. The sequence of the SBT collision procedure is shown in Fig. 4.6.

4.4.1 SBT and Repeated Collisions

The SBT procedure prevents the possibility of repeated collisions in each time step. However, for possible repeated collisions in successive time steps that can occur in SBT, there are two possible answers: one mathematical and the other physical. The mathematical answer is based on the fact that the SBT algorithm was derived from the Kac collision model, assuming a strict Markov collision process within a time step [16]. The collision process is founded on its initial state, i.e., on a set of particle velocities in a cell at the beginning of each time step and the transition collision probabilities determined for each particle pair in a cell. After completing the collisions within a time step, a new state, i.e., the velocity distribution, is formed and used as the initial state for the collisions in the next step. At this moment, the step of particle-free motion, which changes the set of particles in the considered cell, is neglected. It is not necessary to remember the past of the collision history for a strict Markovian process. That is, the error realized by repeated collisions is included in the error of the Markov approximation of the collision process. Practically, the second answer is that, first, the exchange of particles between cells occurs during the particle-free motion. Then, the probability of a collision of a given pair depends on the time step—the smaller the time step, the smaller the probability of collision. The probability of a repeated collision is proportional to the square of pair probability. Therefore, it can be shown that the order of a repeated event is $O(\delta t^2)$. Since the accuracy of the splitting scheme is $O(\delta t)$, one can neglect the repeated collisions as introducing an error of higher order for very small time steps. Considered in conjunction with the particle-free motion step, this allows us to neglect the effect of the repeated collisions realized in successive time steps.

In the NTC scheme, on the other hand, the situation is different; namely, the decrease in time step results in a reduction of the maximum collision number within a collision cell during a given time step. However, for a given particle pair, the collision probability is determined by the acceptance-rejection rule with the probability given by Eq. (3.21), which does not depend on the time step. Thus, the probability of a repeated collision of a twice-chosen pair in the NTC algorithm does not decrease with a decrease in the time step. That is why NTC codes must introduce a repeated collision-preventing rule within a greater number of time steps. However, direct prevention of repeated collisions introduces extra errors in all schemes using the maximum collision frequency approach (NTC, MFS, Null-collision) when the number of particles is small. These schemes are based on a combination of the acceptance-rejection procedure and the random selection of particle pairs. When the number of particles is too small to complete the time step, the collision subroutine must repeatedly check the same pair that was previously chosen but rejected multiple times. Next time, the routine can choose the same pair and check until a collision is realized. Using this direct prevention procedure, a conditional probability that excludes consideration of certain pairs undermines the correct random selection of pairs, and the non-uniformity of the probability becomes significant. This leads to a deviation from the correct probability, which is given in the Kac stochastic model for the collision of each particle pair. In these cases, the probability of collision is not equal to Kac probability. This explains why the NTC scheme cannot work accurately with a small number of particles in cells.

Figure 4.7a presents the average number of particles per cell decreases from 10 to 0.5 while maintaining a constant total number of particles of 24,240 in the case of the Fourier heat conduction problem. This is equivalent to increasing the number of cells from 2400 cells for PPC = 10 to 48,480 cells for PPC = 0.5. The time step decreases in proportion to the cell size to maintain a constant dx/dt ratio. The results indicate that SBT can predict the correct wall heat flux with high accuracy, even with a low PPC, if appropriate combinations of cell size and time step are employed. However, the NTC results rapidly deviate when the number of particles per cell is decreased under the same grid conditions. The same conclusion for the domain heat flux distribution could be drawn considering Fig. 4.7b. The figure shows variations of the domain heat flux with axial direction from SBT and NTC solutions with PPC = 1, 2, and 5. The figure clearly shows that all SBT solutions coincide, while NTC solutions with PPC = 5 or less falls below the converged value of 1512 W/m^2. This different behavior of the NTC and SBT schemes could be described as follows. The NTC algorithm randomly selects collision pairs. This could allow multiple repeated collisions between the same particle pair if there are a few particles per cell or if the particle set in a cell remains constant over time. This leads to distortion of the

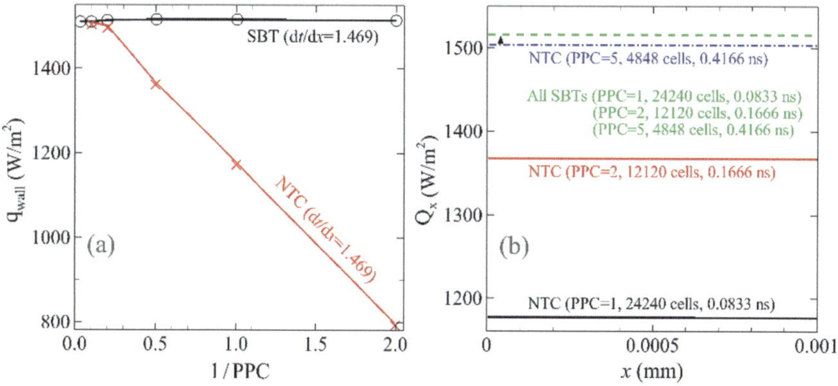

Fig. 4.7 Dependence of the obtained wall heat flux (**a**) and domain heat flux (**b**) on the number of particles per cell (PPC) in the Fourier heat conduction problem for SBT and NTC schemes. Reproduced from Ref. [17]

collision process in cells with a small number of particles. They are the reason for correlations between particle velocity and break the molecular chaos theory, which is the building block in the derivation of the Boltzmann equation. Consequently, the NTC method requires a sufficient number of particles per cell to minimize the probability of repeated collisions.

4.4.2 SBT on Dual Grids

One reason for errors in the DSMC method is that particles at a collision separation distance, which have the potential to be selected as a collision pair, are not checked for collision because they are located in two neighboring cells. This difficulty can be overcome by using a dual grid (also known as a staggered grid), which is created by translating cells (or subcells) [14]. This strategy enhances collision accuracy, particularly when there are a small number of particles per cell. For an orthogonal grid, displaced grids are created by translating cells in each axis direction by a distance equal to half the cell size. Figure 4.8 shows the formation of the displaced grid in structured cells. In this figure, solid lines represent the main grid, and dashed lines represent the displaced grid, which is obtained by displacing the main grid by a magnitude of $\Delta x/2$ in the axial direction and $\Delta y/2$ in the normal direction. It should be noted that the number of cells in the displaced grid is (NCX + 1) × (NCY + 1), where NCX and NCY represent the number of cells in the original grid in the x- and y-directions, respectively. The volume of internal cells in the displaced grid is equal to the volume of internal cells in the main grid. Nevertheless, the volume of boundary cells must be adjusted. Figure 4.8 also depicts two particles that, despite being close, are not selected for collision because they are in different cells in the main grid. After the grid is displaced, these particles belong to the same cell, and the possibility of collision is checked.

SBT on transient adaptive subcells (SBT-TAS)

Previous works [14] have shown that the computational efficiency of the SBT method decreases with an increasing number of particles in a collision cell. Using the TAS technique [16], this problem may be avoided because the transient adaptive subcells are set dynamically in such a way that the number of particles always remains limited in every subcell; see Fig. 4.9. During the simulation, the particle distribution within cells is not uniform; therefore, there may be subcells that are empty or contain only a single particle. In this case, no collisions are possible in such cells. In the NTC-TAS approach proposed by Su et al. [19], if such a case occurs, a search is typically performed among neighboring subcells to find a partner for the collision pair. It could be shown that there is no need to follow this procedure in the SBT-TAS algorithm. Instead, it is suggested to use a two-half-time-step collision algorithm and shift the transient adaptive subcell grid in the second half-time step to prevent a reduction in collision frequency within the cell. By using a dual subcell grid, the single particles in cells within the first half-time step will have the chance to find a collision partner within the second half-time step.

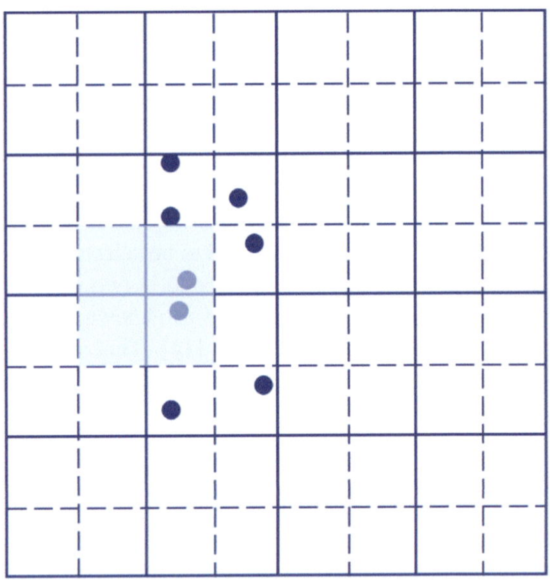

Fig. 4.8 Schematic of the displaced grid (dashed line): solid lines are the main grid; the grey cell is a typical cell created after grid displacement. Reproduced from Ref. [18]

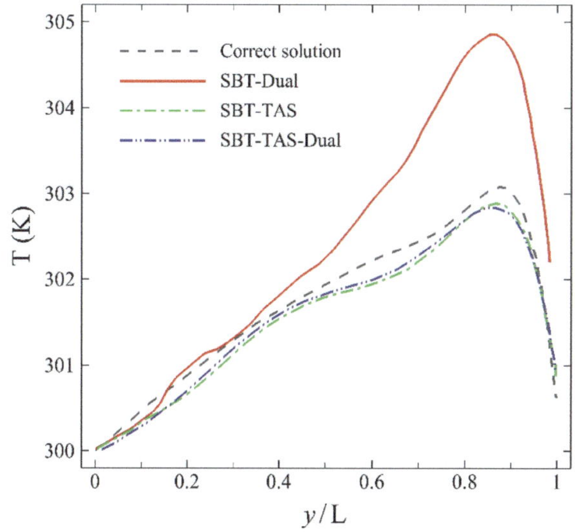

Fig. 4.9 Evaluation of SBT-Dual, SBT–TAS, and SBT–TAS–Dual in predicting the temperature profile in the cavity using a 50 × 50 grid. Reproduced from Ref. [16]

4.4.3 SBT-TAS with Cell Volume Estimation

Unlike the NTC collision scheme, the collision probability function of the SBT scheme, as given in Eq. (4.6), depends on subcell volumes. Therefore, there is a necessity for having an accurate volume computation (approximation) of subcells in the SBT scheme. TAS divides the collision cell into equally spaced rectangular spaces, and the SBT procedure can operate independently within each space. By producing subcell divisions over a generally shaped collision cell, there are three types of subcells, as can be seen in Fig. 4.10a: (1) The first group is those that are out of the borders of the collision cell and are not considered in the flow, (2) The second group is those which are in flow but are vacant; thus, they are assumed inactive in the collision occurrence, and (3) The third group of subcells is the ones that are totally or partially in the flow and at the same time contain at least one particle. These subcells are referred to as 'in-flow active' subcells. They are the subcells that SBT could use as the areas for performing a collision. In the probability function of the SBT scheme, as given in Eq. (4.6) $V^{(l)}$ is the volume assigned to the particles. Since the collision cell is considered a homogeneous space, Goshayeshi et al. [20] suggested that the cell volume is equally distributed among only in-flow active subcells (type 3). Regarding this definition, the probability function could be modified in the form of Eq. (4.8). For the lth subcell, it will be:

$$W_{i,j}^{(l)} = F_{num}\left(N^{(l)} - i\right)(\sigma g)_{i,j}\frac{\Delta t}{V^{(l)}} \tag{4.8}$$

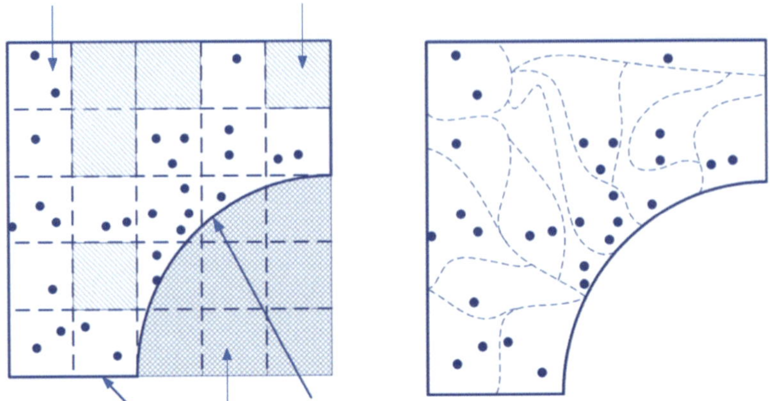

Fig. 4.10 Different subcell types are produced around a circular cylinder: **a** rectangular subcell divisions for indexing particles, and **b** subcell divisions based on the assigned volume in the estimation of the probability function, where divisions are attempted to represent an equivalent volume of V(l). Reproduced from Ref. [20]

$$V^{(l)} = \frac{V^c}{NS_{type3}}, \tag{4.9}$$

where $W_{i,j}^{(l)}$ is the collision probability of the pair (i, j) in the lth subcell, $N^{(l)}$ is the number of indexed particles for this subcell, NS_{type3} is the number of type 3 subcells, and $V^{(l)}$ is calculated in a way that all the collision volume (V^c) is equally distributed among type 3 subcells. Corresponding to this volume propagation, Fig. 4.10b demonstrates how the cell volume is heterogeneously divided among the type 3 subcells. Equation (4.8) could be substituted for the direct calculation of the subcell volume, which could be relatively time-consuming. This equation maintains consistency with the probabilistic nature of the DSMC method and is not limited to boundary cells; it can be applied to all subcells in the simulation domain, thereby providing improved consistency and accuracy.

4.4.4 SBT-TAS with Cell Time Step Control

The investigation described in the previous section revealed a requirement for time-step interval control at the subcell level [20]. There are some restrictive rules on the time step in the DSMC method, applicable to both the motion and collision stages of particles. In this sense, any manipulation that violates these rules will result in errors. The SBT's features require the TAS implementation to be dependent on the local parameters of the subcells. SBT does not allow the use of dependent TAS grids similar to NN or as NTC does. This requirement stems from the nature of SBT, which compels the scheme to determine the unknown collision frequency by straightforwardly investigating the binary collision probabilities of the selected particle pairs, unlike other schemes that fulfill this task by first estimating an upper bound of the total number of collision events within a time step. Consequently, implementing TAS with cell time step control in SBT will produce the so-called "subcells" as smaller collision cells. In contrast, the time step interval employed in these subcells is a controlled ratio of the basic time step used in the regular grid cells. This time step interval is calculated to ensure consistency between the allowed distances for movement (transit time) and the allowed collision frequency based on the mean collision time determined in the main cell. Of course, there is no necessity that the mean collision time of the main cell and each of its subcells be the same.

Figure 4.11 depicts this probable time variation in a series of subcells. In SBT-TAS implementation, all the subcells, S_1,..., S_5 have the same working time step equal to the main cell time-step ($Time_C$). Consequently, subcell S_1 receives a larger time step than its actual value, whereas S_2–S_5, accepts smaller values. In other words, the relation between the working time step and the actual time step of S_1–S_5 subcells relative to the $Time_C$ will be as follows:

$$Time\ step_{actual\ S_1} < Working\ time\ step_{S_1-S_5} = Time_C < Time\ step_{actual\ S_2-S_5}$$

This phenomenon will be intensified if the considered cell is placed in complex regions such as shock interactions. In these situations, implementing SBT-TAS using the collision cell time step will result in a time error, and violations of DSMC's movement and collision decoupling principles will occur. In this state, those subcells that have a time step smaller than their actual value (S_2–S_5 subcells), will underpredict the number of collisions. On the other hand, subcells that have a time-step larger than

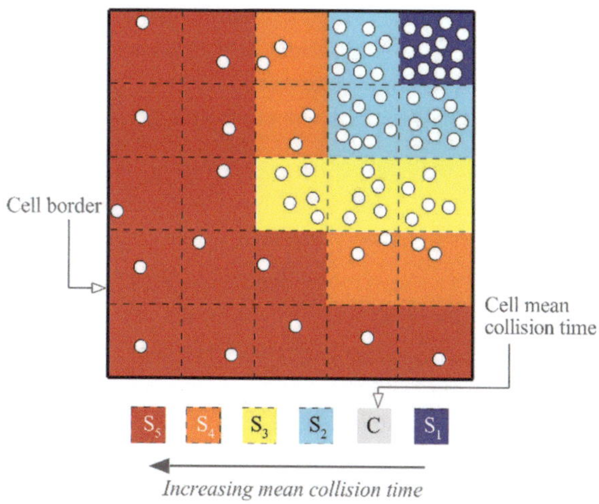

Fig. 4.11 Variation of the mean collision time among the subcells of a typical cell. Reproduced from Ref. [20]

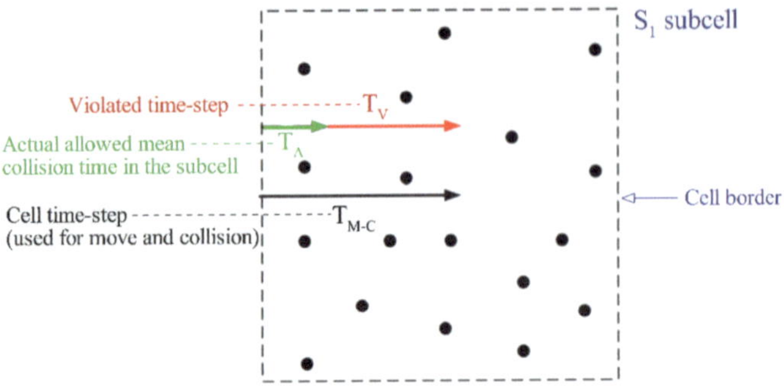

Fig. 4.12 S_l subcell and time-step violation. Reproduced from Ref. [20]

the actual value, subcell S_l, overpredict the number of collisions, while particle-transition paths are larger than the allowed amount. Consequently, violations of movement and collision decoupling will emerge in these subsets, see Fig. 4.12.

The violation time-scale, i.e., as called *Time-Slip*, $S_v = \frac{(T_M - C)}{T_A}$, is a constant that almost preserves its magnitude, even when the time-step of the cell $(T_M - C)$ is controlled in such a way that it becomes equal to the available smallest mean collision time among concurrent subcells of the collision cell. Therefore, any attempt to avoid this error by reducing the working time-step will be unsuitable and it will just decrease the speed of the numerical procedure, while the S_v still exists in the cell space as a time-scaling factor.

In the SBT scheme, the acceptance-rejection depends on the value of the probability function $W_{i,j}^{(l)}$ of the two colliding particle pairs (i, j) in the lth subcell. It is observed that cells likely containing S_l subcells are predominantly located in shock or shock–shock interaction regions. A key characteristic of these cells is their heterogeneous distribution of particles within their space, which is a crucial factor in generating S_l subcells. With increasing the usage of the TAS grid on these cells, the number of S_l subcells will also increase.

Hence, the TAS grid should be fabricated over these cells to such an extent that there would be no S_l subcells. It could be assumed that $W_{i,j}^{(l)}$ is a sign of the presence of S_l subcells. Therefore, the controlling variable for each collision cell is defined as this:

$$CV = \frac{N_E}{N_C} \tag{4.10}$$

where, N_E is the times that $W_{i,j}^{(l)}$ occurs and N_C is the number of collision occurrences. It is a ratio between the times that the probability function $(W_{i,j}^{(l)} > 1)$ exceeds the acceptable limit at all times that a collision occurs in that area, and at this work, it is also referred to as the *P*-exceed ratio or *P_E_R*. Therefore, the implemented solution suggested in Ref. [20] aims to enable the system to sense the degree of the generated error and secure it within a safe zone, where the threat of creating S_l subcells is at its minimum state. At the same time, this solution should enable the TAS system to operate at its maximum capacity for the SBT scheme. This combination of SBT-TAS with this control mechanism is called “SBT-TAS C.M.”.

Figure 4.13 compares the contours of Mach number and SOF obtained by the SBT-TAS scheme, indicated by solid lines, with the results obtained by the NN scheme, indicated by flood contours. Hypersonic rarefied gas flow past a cylinder is considered a well-known 2D benchmark problem. It is a Mach 10 flow of Argon at T = 200 K passing over a 12 inch circular cylinder with a fully diffusive surface at 500 K and a nominal free-stream Knudsen number of 0.01. Since this benchmark problem encompasses a wide range of collision frequencies, from the downstream back-cylinder zone to the upstream stagnation area, it is a suitable choice for challenging any new collision scheme. According to Fig. 4.13, the suitable agreement is observed for results obtained by SBT-TAS and NN schemes.

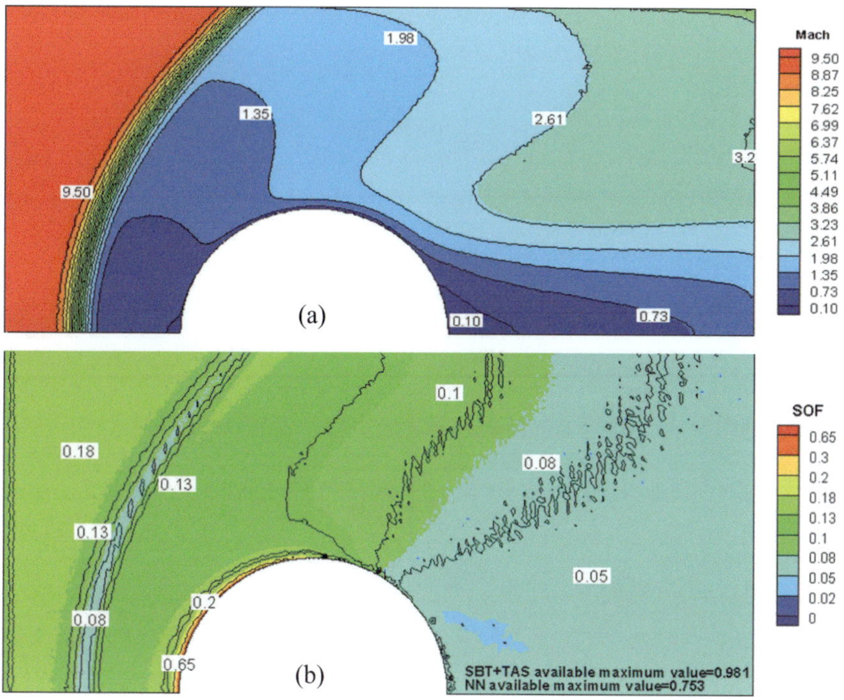

Fig. 4.13 Contours of Mach number (**a**) and SOF (**b**) obtained by SBT-TAS scheme (solid lines) compared with NN scheme (flood). Copied with permission from Ref. [20]

4.5 Intelligent SBT (ISBT)

The SBT collision scheme is designed to obtain correct collision frequencies with the minimum number of particles per cell. A modified version of this scheme, called intelligent SBT (or ISBT), has a semi-perception of inter-particle distances and can prioritize the acceptance of closer pairs, was suggested by Goshayeshi et al. [21]. The aim of developing ISBT is to alter the particle sorting and acceptance-rejection process in the SBT scheme to bias the selections in support of closer partners. ISBT has two main stages. The first step aims to define a smart way of locally indexing particles. The particle indexing should be modified in a way that a semi-perception of distance appears in the indexing process, i.e., in a collision cell with N_c particles, it should be an indexing method that satisfies the condition of ($\delta_i < \delta_j$, $0 < i < j \leq N_C$), where δ_i is the distance of the i-th particle to a reference point®; see Fig. 4.14a.

Investigation over the optimized location of the point® exhibits that the indexed points, relative to their hierarchical order, would be closer to each other if the point® is

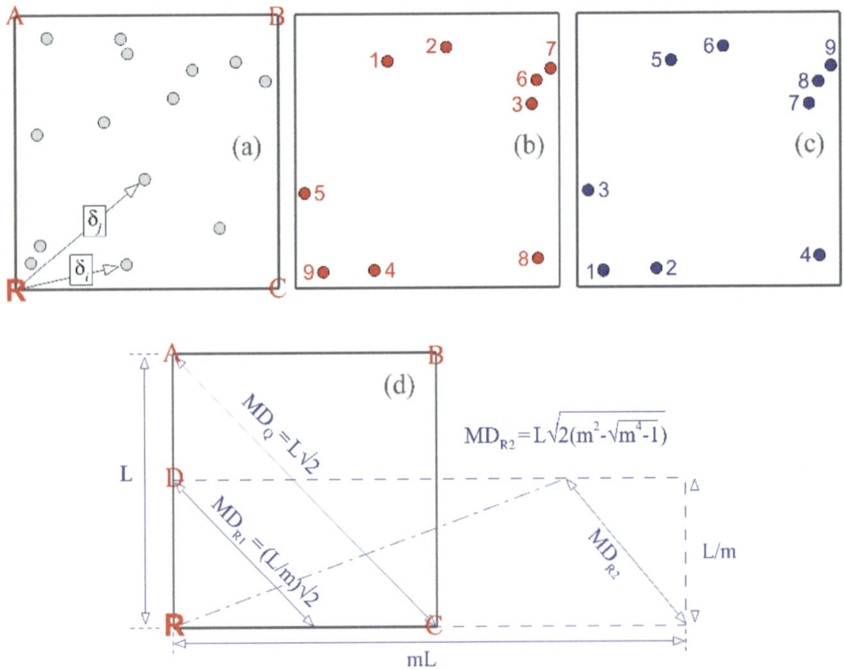

Fig. 4.14 Distribution of some particles in a cell/subcell; **a** a system of particle indexing based on the distance of particles to a definite center point®, **b** the indexing results for a group of nine particles when the point® is located on the center of the cell, **c** when this point is on the left-down corner of the cell, and **d** the geometrical comparison between one quadrangle cell/subcell ('RABC') and its equal-area elongated cell/subcell ('RDEF'), while the calculated values for the maximum distance (MD) between two far-indexed particles are shown; they denote that elongated cell/subcell has smaller MD value. Reproduced from Ref. [21]

located on the corners of a rectangle that surrounds the particles; compare Fig. 4.14b, c. Therefore, Goshayeshi et al. [21] suggested the random selection of cell/subcell corners to efficiently index particles quite close to each other. However, since the δ distance does not consider angular distances, some particles are still indexed successively even when they are far apart, specifically those particles on opposite sides of the diagonal, e.g., particles 4 and 5 in Fig. 4.14c. Statistically, these particles are not the majority and, on the other hand, any other way for particle indexing would unavoidably produce situations in which some indexed particles would be far from each other. It is crucial to note that while along the indexing direction, diagonal RB in Fig. 4.14a, particles are prevented from colliding in opposite corners (because they are indexed at the bottom and the top of the hierarchical order), along the other one, i.e., diagonal AC in Fig. 4.14a, particles might be indexed as neighbors, and their collision, reverses the angular momentum associated with those collision pairs. Since these kinds of collisions might be inevitably supported in the ISBT scheme, as a solution, Goshayeshi et al. [21] suggested using elongated cell/subcells. In a quadrangular cell/subcell, i.e., 'RABC' in Fig. 4.14d, the maximum distance (MD) that two far indexed particles could have is $\sqrt{2}$ times greater than the side length. However, using elongated cell/subcell obliges the time step to be reduced accordingly. Therefore, in cases of hypersonic shock interaction simulations, Goshayeshi et al. [21] suggested using elongated cells/subcells with an elongation factor of two ($m = 2$), while reducing the time step by about one-third. Therefore, the ISBT technique could be implemented alongside the transient adaptive sub-cell technique (TAS) as described before.

The ISBT procedure can be implemented as follows: The first particle (i) is sequentially selected from 1 to $N^{(l)}$, while its pair (j) at each subsequent step is randomly selected among the further hierarchically indexed particles, as described in the previous paragraph and following Eq. (4.5). The main difference with the SBT is that instead of regenerating the second random number (R_{nf2}), which was used in the acceptance-rejection section of the SBT scheme in Eq. (4.6), ISBT does not produce any new random number and implements the first generated one (R_{nf1}):

$$W_{i,j} \geq R_{nf1}, \tag{4.11}$$

where W_{ij} is calculated through Eq. (4.6). Figure 4.15 shows the sequences of the ISBT procedures. As a result of this policy, pseudo-circular subcells (Fig. 4.16) will be created that allow an easier acceptance for closer available pairs and a more difficult condition for further particles; it is because that smaller number for j, as the second particle in the hierarchical list of indexed numbers, is only obtained if the R_{nf1} is closer to zero, and conversely, greater number for j, as the second particle, is merely obtained if the R_{nf1} is further from zero.

Figure 4.17 illustrates the hypersonic Nitrogen flow of Mach 15.6 passing over a biconic geometry, Fig. 4.17a contains quite demanding conditions consisting of a laminar recirculation zone, laminar expanding zone, high-speed low-density region, and low-speed high-density region Fig. 4.17b. The accuracy and performance of the ISBT-TAS and SBT-TAS schemes are compared with those of their validated

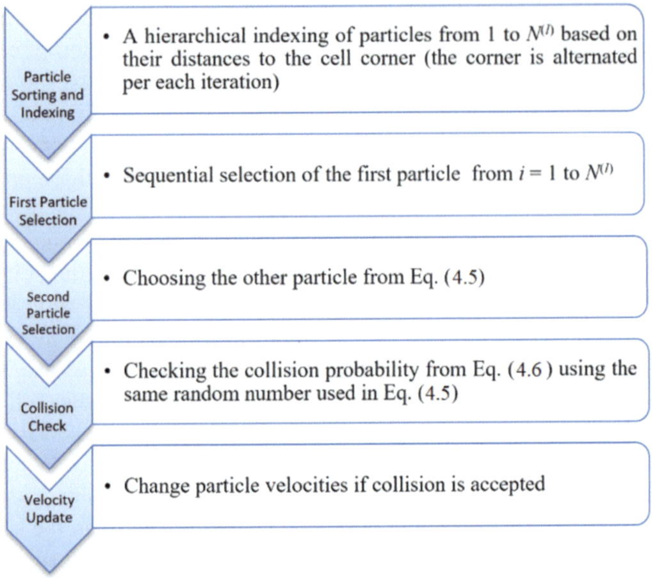

Fig. 4.15 ISBT collision procedure

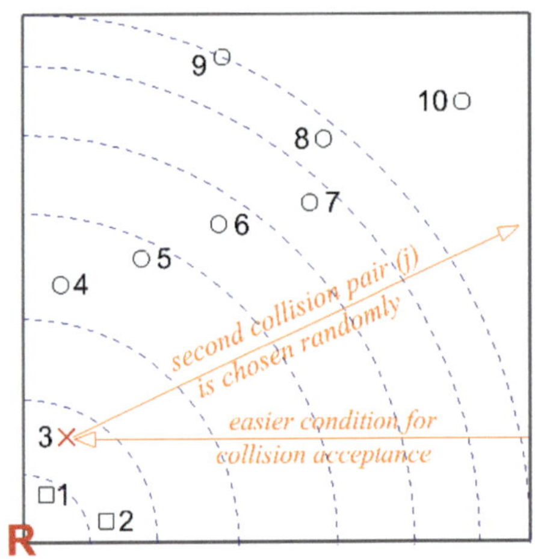

Fig. 4.16 Pseudo-circular subcells in the ISBT scheme lead to a biased selection of closer collision pairs. Reproduced from Ref. [20]

counterparts in the DS2V code, specifically the NN scheme. As could be seen from the Fig. 4.17, the ISBT-TAS scheme follows the profiles obtained by the NN scheme and experimental data.

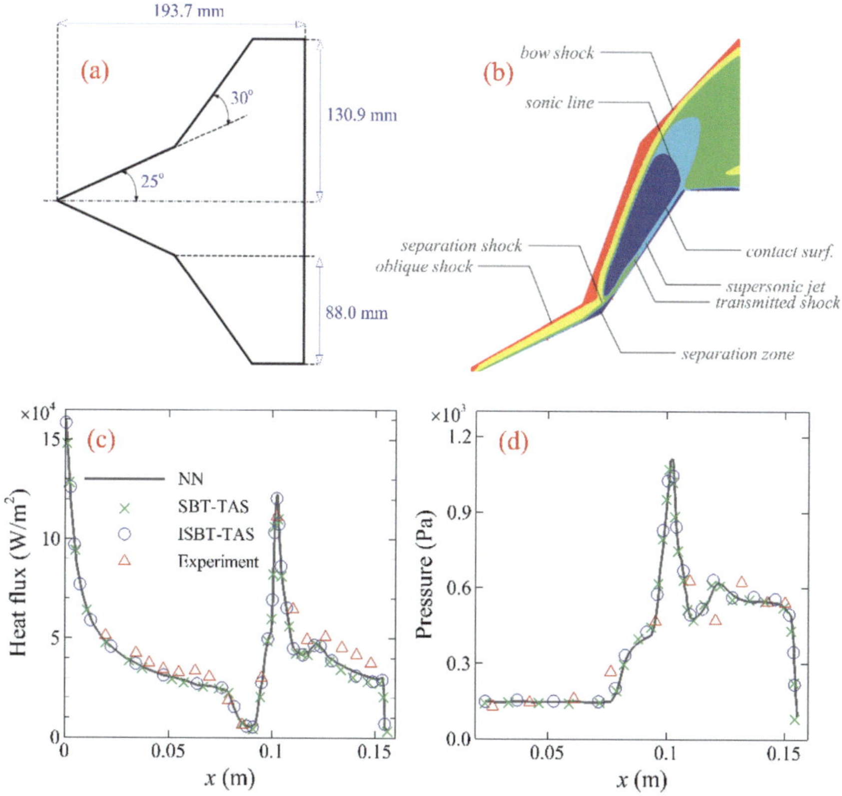

Fig. 4.17 Biconic geometry and dimensions (**a**), numerical flow filed description (**b**), and surface properties obtained by ISBT scheme and comparison with other schemes and experimental data (**c**, **d**). Reproduced from Ref. [21]

4.6 Generalized Bernoulli Trial (GBT)

An additional partial linearization of the SBT operator, Eq. (4.4), with regard to Δt, leads to a generalized form of the approximation operator called the generalized Bernoulli trial (GBT) scheme [13]. The SBT algorithm is more efficient than the original BT scheme, and its efficiency is comparable to the NTC when the number of particles in a cell is small. However, it remains less efficient than the standard NTC in simulations that use a relatively large number of particles per cell and a small time step. This is because a fixed number of pairs should be selected and checked for possible collisions, whereas the NTC typically selects a smaller number of pairs to check for collisions using an estimate of the maximum number of selected pairs for a given time step. The effect in SBT is that most of the selected pairs will not have a significant acceptance chance because their probability ($W_{i,j}$) is low at a small time step. In this case, it is preferable to reduce the number of selected pairs (N_{sel}) to a lower magnitude, while increasing the pair weight and, respectively, the collision probability, in such a way that the collision frequency remains constant. In this regard, a rational option could be considered to reduce the number of selected pairs to $N_{sel} = \frac{N^{(l)}}{2}$ or $N_{sel} = \frac{N^{(l)}}{3}$ and even less in order to be of the same number range as the number of selected pairs in the NTC scheme. To start with, the operator form remains the same as given in (4.4), but the upper bound of the product is reduced to N_{sel} as follows:

$$G_{GBT}(\Delta t) = \prod_{i=1}^{N_{sel}} \left\{ \left[1 - \sum_{j=1}^{N^{(l)}} \frac{1}{k'k} (k'k\omega_{i,j}\Delta t) \right] I + \sum_{j=1}^{N^{(l)}} \frac{1}{k'k} (k'k\omega_{i,j}\Delta t) T_{i,j} \right\} \tag{4.12}$$

where, k' is a correction coefficient to compensate for the reduction of selected pairs and it must be found. The linearization to Δt of the product given in Eq. (4.12) yields the following transition operator:

$$G'_{GBT}(\Delta t) = \left[1 - \sum_{i=1}^{N_{sel}} \sum_{j=i+1}^{N^{(l)}} \frac{1}{k'k} (k'k\omega_{i,j}\Delta t) \right] I + \sum_{i=1}^{N_{sel}} \sum_{j=i+1}^{N^{(l)}} \frac{1}{k'k} (k'k\omega_{i,j}\Delta t) T_{i,j} \tag{4.13}$$

The operator $G'_{GBT}(\Delta t)$ can be realized by a selection of one pair similar to the Ballot Box scheme; however, here we use it to demonstrate how to define the correction coefficient for the selected pairs in the operator with a reduced product. The correction coefficient k' is obtained from the equality:

$$k' \sum_{i=1}^{N_{sel}} \sum_{j=i+1}^{N^{(l)}} 1 = \sum_{i=1}^{N^{(l)}-1} \sum_{j=i+1}^{N^{(l)}} 1 \tag{4.14}$$

$$k' \left[\sum_{i=1}^{N^{(l)}-1} \sum_{j=i+1}^{N^{(l)}} 1 - \sum_{i=1}^{N^{(l)}-N_{sel}-1} \left(\sum_{j=i+1}^{N^{(l)}-N_{sel}} 1 \right) \right] = \sum_{i=1}^{N^{(l)}-1} \sum_{j=i+1}^{N^{(l)}} 1 \tag{4.15}$$

According to (4.15), one can determine the below expression for k':

$$k' = \frac{C(N^{(l)}, 2)}{C(N^{(l)}, 2) - C(N^{(l)} - N_{sel}, 2)} \tag{4.16}$$

In Eq. (4.16), ${}_{\alpha}C_{\beta}$ is the number of combinations of α from β elements, for example:

$$C(N^{(l)}, 2) = \frac{N^{(l)}!}{2!(N^{(l)} - 2)!} = \frac{N^{(l)}(N^{(l)} - 1)}{2} \tag{4.17}$$

Hence, the final expression for k' is as:

$$k' = \frac{\frac{\left[N^{(l)}\left(N^{(l)}-1\right)\right]}{2}}{\frac{\left[N_{sel}\left(2N^{(l)}-N_{sel}-1\right)\right]}{2}} = \frac{N^{(l)}\left(N^{(l)}-1\right)}{N_{sel}\left(2N^{(l)}-N_{sel}-1\right)} \tag{4.18}$$

The inverse value of the correction coefficient $\frac{1}{k'}$ has a simple probabilistic meaning of selection probability of N_{sel} pairs from all available pairs in the cell, whose total number is $\frac{N^{(l)}\left(N^{(l)}-1\right)}{2}$. They are located in the upper triangle of matrix elements of all collision probabilities as:

$$\left\{\begin{array}{ccccccc}
p_{1,N^{(l)}} & p_{2,N^{(l)}} & p_{3,N^{(l)}} & p_{4,N^{(l)}} & \cdots & p_{s-1,N^{(l)}} & p_{s,N^{(l)}} \\
p_{1,N^{(l)}-1} & p_{2,N^{(l)}-1} & p_{3,N^{(l)}-1} & p_{4,N^{(l)}-1} & \cdots & p_{s-1,N^{(l)}-1} & p_{s,N^{(l)}-1} \\
\vdots & \vdots & \vdots & \vdots & \vdots & \vdots & \vdots \\
p_{1,s} & p_{2,s} & p_{3,s} & p_{4,s} & \cdots & p_{s-1,s} & \times \\
\vdots & \vdots & \vdots & \vdots & \vdots & \vdots & \vdots \\
p_{1,3} & p_{1,3} & \times & p_{4,3} & \cdots & p_{s-1,3} & p_{s,3} \\
p_{1,2} & \times & p_{3,2} & p_{4,2} & \cdots & p_{s-1,2} & p_{s,2} \\
\times & p_{2,1} & p_{3,1} & p_{4,1} & \cdots & p_{s-1,1} & p_{s,1}
\end{array}\right|$$
$$\left.\begin{array}{ccccc}
p_{s+1,N^{(l)}} & \cdots & p_{N^{(l)}-2,N^{(l)}} & p_{N^{(l)}-1,N^{(l)}} & \times \\
p_{s+1,N^{(l)}-1} & \cdots & p_{N^{(l)}-2,N^{(l)}-1} & \times & p_{N^{(l)},N^{(l)}-1} \\
\vdots & \vdots & \vdots & \vdots & \vdots \\
p_{s+1,s} & \cdots & p_{N^{(l)}-2,s} & p_{N^{(l)}-1,s} & p_{N^{(l)},s} \\
\vdots & \vdots & \vdots & \vdots & \vdots \\
p_{s+1,3} & \cdots & p_{N^{(l)}-2,3} & p_{N^{(l)}-1,3} & p_{N^{(l)},3} \\
p_{s+1,2} & \cdots & p_{N^{(l)}-2,2} & p_{N^{(l)}-1,2} & p_{N^{(l)},2} \\
p_{s+1,1} & \cdots & p_{N^{(l)}-2,1} & p_{N^{(l)}-1,1} & p_{N^{(l)},1}
\end{array}\right\} \tag{4.19}$$

In the BT scheme, all particle pairs are checked for collision sweeping the matrix elements one by one. In SBT, only one pair from each column is selected at random with probability $\frac{1}{k}$ to be checked for collision and the corresponding collision probability is corrected by coefficient k'. In the transition operator $G_{GBT}(\Delta t)$, N_{sel} pairs are selected by choosing one pair per column at random from the first N_{sel} columns of the matrix. Thus, N_{sel} pairs are selected from $C(N^{(l)}, 2) - C(N^{(l)} - N_{sel}, 2)$ where C is the combination function, pairs and the corresponding selection probability of all N_{sel} pairs from all available pairs is $\frac{1}{k'}$. Respectively, the probability of selection of each pair (i, j) is $\frac{1}{(kk')}$ and consequently, the total collision probability of each selected pair must be corrected by coefficient kk' in accordance with the transition operator (4.12).

It is worth noting that it is assumed that the particle set in each cell within a time step is random. In most cases, the exchange of particles between cells during

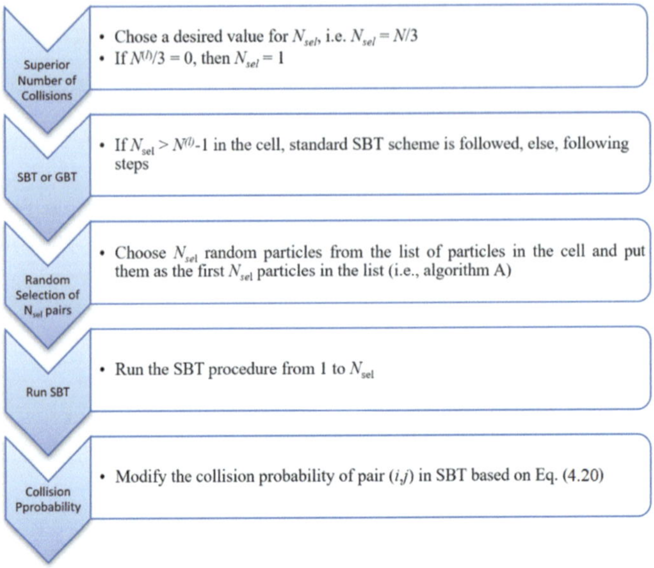

Fig. 4.18 GBT collision procedure

free particle motion is sufficient, and no additional actions are required. However, the universal algorithm should contain an additional procedure of particle rearrangement. In this regard, the suggested procedure for the GBT scheme is illustrated in Fig. 4.18 and described as follows:

I. Select N_{sel}, i.e., assume N_{sel} is a fraction/function of the number of particles in the cell. The following procedure is performed if $N_{sel} < N^{(l)} - 1$ in the cell; otherwise, the standard SBT scheme is followed.
II. Algorithm A. Choose N_{sel} random particles from the list of particles in the cell, and reorder the particle list exchanging the positions of the selected particles with those occupying the first N_{sel} positions in the list, i.e., put the N_{sel} selected particles in the index list ordered as first, second, etc. particle in the list. Place the former first, second, and so on, particles in their respective places in the index list.
III. Run the SBT procedure from $i = 1$ to N_{sel}, but modify the collision probability of every pair with the following correction:

$$W_{i,j} = \frac{k'kF_{num}\Delta t\sigma_{i,j}g_{i,j}}{V^{(l)}} \tag{4.20}$$

where,

$$k'k = \frac{N^{(l)}\left(N^{(l)} - 1\right)}{N_{sel}\left(2N^{(l)} - N_{sel} - 1\right)}\left(N^{(i)} - i\right) \tag{4.21}$$

Similar to SBT, Δt should be adjusted so that $W_{i,j}$ in Eq. (4.20) does not exceed unity, i.e., inequality (4.7) should be satisfied. Instead of (4.21), an alternative collision probability correction in the GBT algorithm is conceivable considering a generalization of the BB and SBT collision formulas, i.e., GBT probability correction should approach to BB and SBT corrections if $N_{sel} = 1$ and $N_{sel} = N^{(l)} - 1$, respectively. To achieve this aim, we slightly modify the BB algorithm to avoid repeated collisions. The original BB algorithm chooses two particles in random from the whole set of particles in the list. However, we change the particle order and put the first selected particle in the BB procedure as the first particle in the list. The second particle is chosen at random from the particles located after the first one ($i = 1$) in the list of particles. However, the probability correction in the BB scheme, either the original one or the modified version presented here, is $C(N^{(l)}, 2)$ (which is the combination of two particles from $N^{(l)}$ ones), as BB considers only one pair on behalf of all available pairs in the cell. If we pick more pairs out to check their collision possibility, i.e., $N_{sel} > 1$ in GBT, a smaller correction should be introduced as:

$$\widehat{kk} = \frac{C\left(N^{(l)}, 2\right)}{C(N_{sel} + 1, 2)} \tag{4.22}$$

Equation (4.22) is the correction incorporating the effects of considering N_{sel} number of selected pairs to be checked for collisions instead of all $C(N^{(l)}, 2)$ pairs.

The correction given by (4.22) reduces to $C(N^{(l)}, 2)$ if $N_{sel} = 1$, i.e., BB scheme, and to 1 if $N_{sel} = N^{(l)} - 1$, i.e., the SBT scheme. However, the GBT probability correction should include another term compensating the selection of just one particle from the particles located after the ith particle in the list. If SBT with $N_{sel} = N^{(l)} - 1$ is considered, the additional term of the probability correction for the i-th particle should be $\left(N^{(l)} - i\right)$, while BB with $N_{sel} = 1$, there is no need for an extra correction in (4.22). In order to capture both SBT and BB limits, the complementing correction term in (4.22) should read as $(N_{sel} - i + 1)$. Consequently, an alternative GBT collision probability correction instead of (4.21) reads as:

$$k'k_2 = \frac{C\left(N^{(l)}, 2\right)}{C(N_{sel} + 1, 2)}(N_{sel} + 1 - i). \tag{4.23}$$

Figure 4.19 compares the temperature fields for argon gas obtained with those obtained by the NN scheme (flood) in the cavity flow problem, with the Knudsen number and lid velocity set to 0.1, 1000 m/s, and the wall temperature at 300 K, respectively. The temperature contours from various $N_{\rm sel}$ of the GBT scheme are plotted against that of the NN algorithm. According to Fig. 4.19, an excellent agreement between the temperature fields calculated by both schemes is observed.

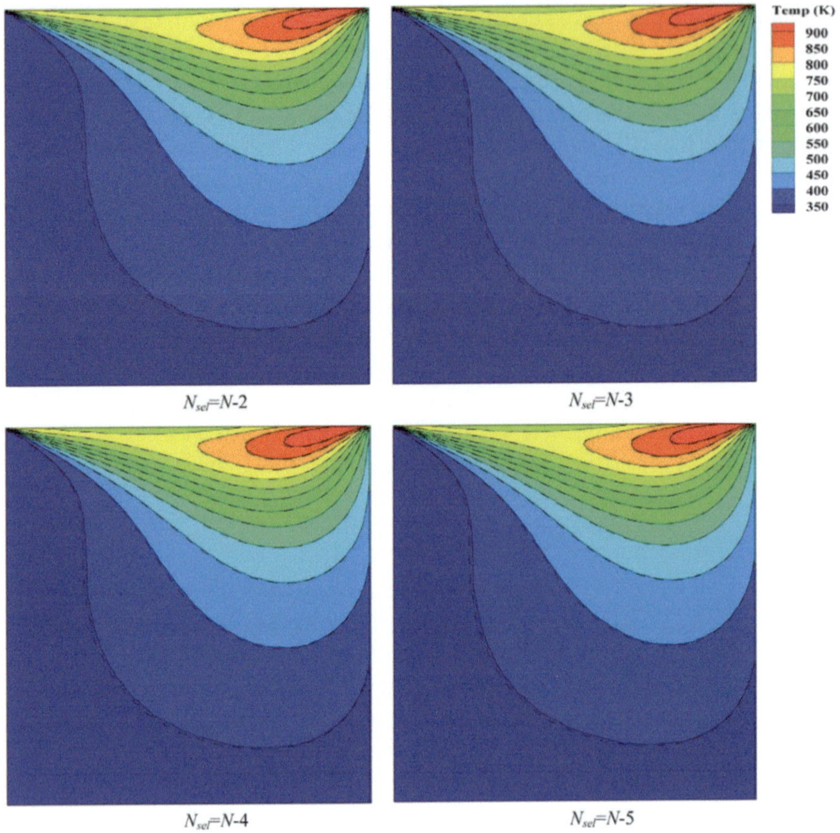

Fig. 4.19 Temperature contours in the 2D cavity problem obtained by the the GBT scheme (line) and NN scheme (flood). Unpublished results from authors

4.7 Hybrid NTC-GBT-SBT-Tas Scheme

Recently, Stefanov et al. [22] proposed a novel hybrid transient adaptive subcell DSMC algorithm to simulate rarefied gas flows in a wide range of Knudsen numbers. The novel algorithm is considered together with the standard and hybrid collision algorithms built on uniform grids. The standard collision algorithm uses only one single scheme (i.e. NTC, GBT, or SBT). The hybrid algorithm employs NTC, GBT, or SBT depending on the instantaneous number of particles in the considered cell. The novel hybrid TAS algorithm benefits from both the hybrid collision approach and the transient adaptive subcell grid, which covers each collision cell, to achieve a uniform accuracy of order $O(\Delta t, \Delta r)$ independently of the number of particles in the cells. To this end, a local time step is defined as coherent with the TAS grid that covers the corresponding collision cell.

NTC (applied to a cell l)	GBT or SBT (applied to a cell l)
Compute $A_{sel} = \frac{N^{(l)}\left(N^{(l)}-1\right)}{2}\frac{F_{num}(\sigma g)_{max}}{V^{(l)}}\Delta t + \Delta A'_{sel}$ $N_{sel} = \lfloor A_{sel} \rfloor$ $\Delta A'_{sel} = A_{sel} - N_{sel}$	Compute $N_{sel} = \begin{cases} N^{(l)}-1, & N^{(l)} < 5 \quad \text{(SBT)} \\ N^{(l)}/2, & 5 \le N^{(l)} \le 15 \quad \text{(GBT)} \\ N^{(l)}/3, & N^{(l)} > 15 \quad \text{(GBT)} \end{cases}$ $k'k = \frac{N^{(l)}\left(N^{(l)}-1\right)}{N_{sel}\left(2N^{(l)}-N_{sel}-1\right)}\left(N^{(l)}-i\right)$
1. **Select a pair (i,j) at random from $N^{(l)}$ particles**	1. Select the first particle in order from the list $i=1,\ldots,N_{sel}$; Select the second particle (j) at random from $\left(N^{(l)}-i\right)$ particles after i_{th} particle in the list $j \in \left\{i+1,\ldots,N^{(l)}\right\}$
2. **Accept collision of (i,j) with probability** $w_{i,j} = \frac{\sigma_{i,j} g_{i,j}}{(\sigma g)_{max}}$ **and change $(\mathbf{c}_i, \mathbf{c}_j)$ to post-collision $(\mathbf{c}'_i, \mathbf{c}'_j)$**	2. Accept collision (i,j) with probability $w_{i,j} = k'k\frac{F_{num}\sigma_{i,j} g_{i,j}}{V^{(l)}}\Delta t$ and change $(\mathbf{c}_i, \mathbf{c}_j)$ to post-collision $(\mathbf{c}'_i, \mathbf{c}'_j)$
Repeat 1and 2 for N_{sel} collision pairs	Repeat 1and 2 for N_{sel} collision pairs

Fig. 4.20 Flowchart of the hybrid algorithm

The analysis of the collision scheme properties [14, 23] and recent comparisons of different collision schemes on a uniform grid [13, 24] shows that on a coarse grid with the number of particles in a cell $N^{(l)} \ge 15 - 20$, the Bird's NTC has good accuracy and faster performance than the simplified Bernoulli-trials scheme. Under these conditions and optimal time and space resolution, the Generalized Bernoulli-trials algorithm GBT keeps the accuracy almost independent of N_{sel} and has almost equal or better performance than NTC if the number of selected pairs N_{sel} is set to appropriate values comparable to the corresponding number N_{sel} calculated in the NTC scheme (see Fig. 4.20).

The accurate simulation of gas flows in areas with large gradients of the macroscopic variables requires a grid with finer resolution. The same applies to shock waves and boundary areas. However, the fine uniform grid often increases the computational memory and time beyond the available resources and the computational cost becomes prohibitively expensive. An optimal solution to this problem is employing the TAS concept on the entire domain D with variable sizes of the cells in the different subdomains $D^{(l)}$ [22]. The operator $G_{GBT\text{-}TAS}$ corresponding to the new TAS hybrid collision algorithm can be written as follows:

$$S_{B+G}^{\Delta t} = \prod_{n=1}^{K} S_{B+G}^{\Delta t/K} \approx \prod_{n=1}^{K} \mathrm{IB}(\Delta t/K)\left(\sum_{l=1}^{M} G_{GBT-TAS}^{(l)}(N_{sel})\right)$$

$$G^{(l)}_{GBT-TAS}(N_{sel}) = \left| \begin{array}{ll} \sum_{l_s=1}^{\eta(l)^d \sum_s(l)} G^{l_s}_{GBT}\left(N_{sel}, \Delta t^* = \tau \frac{\eta_{max}}{\eta(l)(l)}()I^{(\eta(l))}\right) \\ I, t_s(l) \quad > n\tau \end{array} \right. \quad (4.24)$$

where $K = \eta_{\max} = \bar{n}_{\max}(t) + 0.5$, $\tau = \Delta t / \eta_{\max}$, I is the identity operator, $I^{(\eta(l))}$ is a local indexation operator in cell l depending on the TAS level, and $t_s(l) = \sum^{\sum} \tau \frac{\eta_{max}}{\eta(l)}$ is the interim local time after every collision step $\Delta t^* = \tau \frac{\eta_{max}}{\eta(l)}$, which is determined for subcells of the cell l. The number of subcells in cell l is equal to $\eta(l)^d$, where d is the space dimension. In our consideration, the collision generator $G^{(l)}_{GBT-TAS}(N_{sel})$ acting in the collision cells has a pyramidal structure that at each level has a number of subcells in each coordinate direction equal to $\eta(l) = [\bar{n}^{(l)} + 0.5]$. The simulation results (Fig. 4.21) indicated that by reducing the total number of particles to a hundred times, it was demonstrated on grids with different resolutions that the hybrid TAS algorithm retains accuracy and provides a solution in very good agreement with the reference one.

Important remark 1. When the computations are performed on a coarse grid with large dx and dt, significant macroscopic gradients (density, pressure, temperature) within cells can still exist. The random choice of collision pairs (NTC, MFC) causes the collision process to follow these gradients, i.e., more pair selection in areas with higher density, etc. It makes the solution more accurate than the splitting scheme allows, first order O(dx, dt). For the Bernoulli-trials schemes, the situation is different. The pair selections are partially ordered assuming the distribution of particles in a cell is uniform and homogeneous. So, if one uses a coarser grid, NTC and MCF are

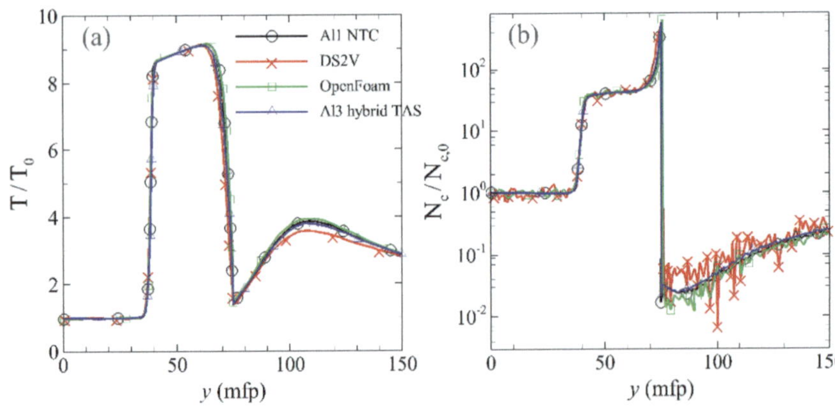

Fig. 4.21 Non-dimensional profiles of temperature (**a**) and normalized binary collision rate (**b**) obtained by four different codes: NTC on the uniform grid 800 × 800 and $N_{tot} = 35 \times 10^6$ (A11), DS2V NN on the adapted grid and $N_{tot} = 35 \times 10^6$, OpenFOAM on a uniform 800 × 800 grid and $N_{tot} = 35 \times 10^6$, hybrid TAS on reference grid 400 × 400 and $N_{tot} = 0.35 \times 10^6$ (A13). Reproduced from Ref. [22]

slightly better than BT ones. The positive quality of BT schemes is that they exhibit uniform accuracy regarding PPC when the accuracy conditions are met [22]. They are achieved when the discretization steps *dx*, *dt* are appropriately reduced. It means the use of finer grids and smaller time steps. Of course, there is a large common area of parameters (*dx* and *dt*) where NTC with large PPC can be compared to BT schemes with small PPC.

In summary, the optimal DSMC solution is the choice of a collision hybrid TAS scheme, which combines NTC, GBT, and SBT depending on the PPC for uniform grids (without TAS) (Algorithm 2 in [22]) or with TAS (Algorithm 3 there).

This ensures uniform accuracy in space and time throughout all areas of the computation domain.

Important remark 2. Simulating flows with complex structures, such as vortices, shock waves, and intricate boundaries, requires a minimum number of simulators throughout the entire computational domain. The phenomena cannot be accurately reproduced if the number is reduced below this threshold. Therefore, the total number of simulators cannot be arbitrarily small, as this can compromise the particle model's self-similarity and its ability to accurately represent real processes.

4.8 Symmetrized and Simplified Bernoulli Trials (SSBT)

A symmetrized SBT method (SSBT scheme) was introduced by [25] to provide less restrictive conditions for selecting pairs based on a probabilistic approach. Unlike the SBT method, the SSBT scheme selects the second particle of a chosen pair from a complete list of particles with equal probability, it prevents repetitive collisions by introducing a procedure to avoid duplicate colliding pairs. Actually, SSBT has been proposed to eliminate the "no-chance collision possibility" for some collision pairs in SBT and GBT methods. Additionally, a mathematical mechanism has been implemented for the SSBT scheme to prevent successive collisions.

In the SSBT scheme, the collision operator should be considered symmetric using the obvious equality of collision probabilities $p_{i,j} = p_{j,i}$. The symmetric form of collision probabilities is clearly seen in the collision matrix (4.19). The symmetry comes from the simple fact of the relativity of the binary collision, i.e., whether particle i hits particle j or particle j hits particle i are considered the same collision. By introducing a symmetric form of probability $\frac{(p_{i,j}+p_{j,i})}{2}$, both upper (above the diagonal elements) and lower (below the diagonal elements) triangular parts of the matrix (4.19) are allowed to be taken into account. This gives freedom for the second particle j to be selected from the elements of both the upper and lower triangular parts. To complete the symmetric form of (4.19), in the operator $G_{SBT}(\Delta t)$, the external product must be extended to $N^{(l)}$ and the internal summation performed for $\left(1, N^{(l)},\ i \neq j\right)$. Then, a new generator is obtained for the SSBT scheme having the following symmetric form:

$$G_{SSBT}(\Delta t) = \prod_{i=1}^{N^{(l)}} \left\{ \left[1 - \sum_{j=1, i \neq j}^{N^{(l)}} \frac{1}{k'} \left[\frac{k'}{2} (p_{i,j} + p_{j,i}) \right] \right] I + \sum_{j=1, i \neq j}^{N^{(l)}} \frac{1}{k'} \left[\frac{k'}{2} (p_{i,j} + p_{j,i}) \right] T_{i,j} \right\}, \tag{4.25}$$

where $k' = N^{(l)} - 1$. The algorithmic interpretation of the operator $G_{SSBT}(\Delta t)$ determines the SSBT scheme. The computational costs of the SSBT are equal to $O\left[\left(N^{(l)}\right)\right]$. The number of selected pairs per time step is equal to $N^{(l)}$, and the collision probability is equal to $P_{i,j} = \frac{k'(p_{i,j}+p_{j,i})}{2}$. An essential requirement for eventual overruns of condition $P_{i,j} \leq 1$ is that they must be sparse. As can be seen from the operator, it allows consideration of duplicate collisions. In the realized algorithm, duplicate collisions are avoided by a procedure described below, which does not alter the result of the operator's action $G_{SSBT}(\Delta t)$. The algorithm of the SSBT method is described in the following steps and illustrated in Fig. 4.22:

- The scheme uses the local indexing of $N^{(l)}$ particles in each cell. The first particle (i) of the selected pair (i, j) is chosen in sequence from the list of indexed particles: $i = 1, 2, \ldots, N^{(l)}$.
- The second particle (j) is selected randomly among the remaining particles of the list. By considering equal chance for each particle of cell l, j can be obtained:

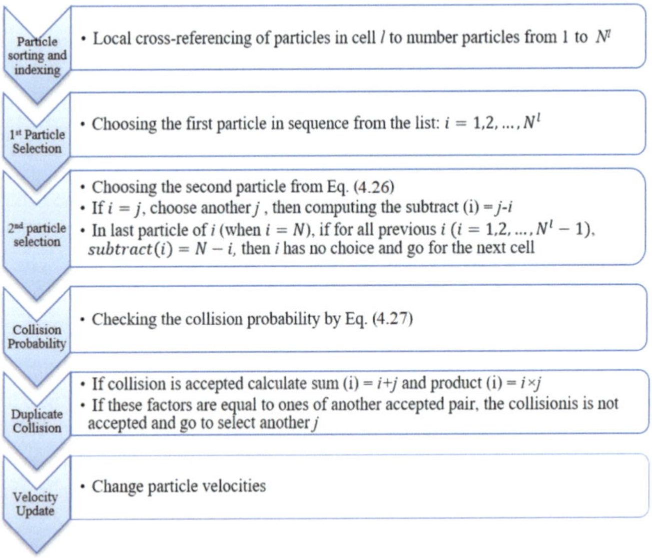

Fig. 4.22 The flowchart for SSBT collision procedure per each cell

$$j = int\left(R_f \times N^{(l)}\right) + 1 \tag{4.26}$$

Thus, $1 \leq j \leq N^{(l)}$, which makes it possible for the selected pair (i, j) to duplicate the previously chosen pair (i, j).

- If $j = i$, (diagonal element of the matrix), then another j will be selected.
- The subtraction of the numbers of each indexed particle of a chosen pair is calculated (Subtract $(i) = j - 1$) to be used in the prevention of an infinite loop for $i = N^{(l)}$, as given below:
- If $i = N^{(l)}$, therefore, the possibility of choosing the second particle is investigated with an internal loop (the first internal loop). If for each i ($i = 1,2, \ldots, N^{(l)} - 1$), the subtract (i) is equal to $N^{(l)} - i$, then the last particle of the list ($i = N^{(l)}$) will not have any choice of the second particle as its selected pair, and the selection procedure goes to the next cell.
- Check the collision probability with the corrected probability equation:

$$W_{i,j} = \frac{\left(N^{(l)} - 1\right) F_{num} \Delta t \sigma_{i,j} g_{i,j}}{2V^{(l)}} \tag{4.27}$$

- If the collision is accepted ($W_{i,j} > R_f$), the possibility of a duplicate collision is investigated, as given below [26]:

  Define a new integer array I_p with a length equal to a total number of particles. If $I_p(i) = j$ and $I_p(j) = i$, the selected pair (i, j) is rejected as a duplicated one, otherwise, the selected pair is approved, and I_p elements (i), (j) are changed to $I_p(i) = j$ and $I_p(j) = i$.
- If the accepted collision is not duplicated, then the particles' velocities will be updated.

Figure 4.23 clarifies the differences between SBT and SSBT selection procedures. Here, each row of circles represents the indexed particles in a given cell. Moreover, each pair of rows, which are separated with a bracket, shows the selection of pair (i, j) in which the first row corresponds to the first particle (i) of the pair, and the second row corresponds to the second particle (j) of that pair. For the SBT method, the first particle of the pairs (i) is selected in sequence from the indexed particles $(1 \leq i \leq N - 1)$ in a given cell (the black circle of the first row in each bracket). The second particle (j) of each pair is selected randomly from the list of particles that are located after the first particle $(i + 1 \leq j)$. In the second row of each bracket, the black circle is a random representative of j, and the crossed circles are the particles that are not allowed to be chosen as the second particle of the pair (i, j). For the SSBT method, the first particle of the pairs (i) is selected in sequence from the indexed particles $(1 \leq i \leq N)$ in a given cell. The second particle (j) of each pair is chosen randomly from the entire list of the particles. As can be seen, the restriction that was imposed on the selection of j in the SBT method has been removed in SSBT.

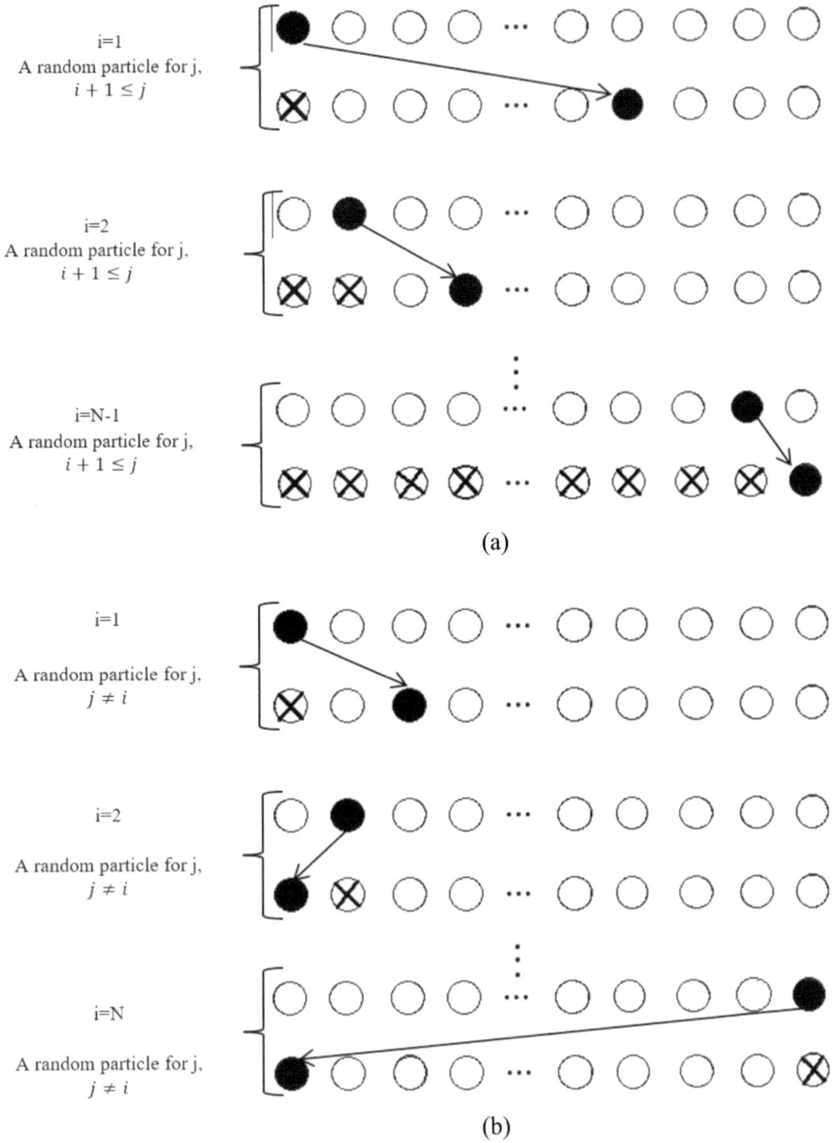

Fig. 4.23 Schematic picture of the selection procedures for **a** SBT and **b** SSBT schemes

4.9 Symmetrized and Generalized Bernoulli Trials (SGBT)

A similar concept employed in the SSBT scheme [25] has been adapted for use in the GBT scheme, resulting in a new collision scheme known as symmetrized and generalized Bernoulli trials (SGBT) [26]. SGBT aims to facilitate the choice of collision pairs by employing a symmetrization of the collision pair selection process. Actually, the SGBT scheme uses a combination of SSBT and GBT schemes to select collision pairs. The fundamental differences between the GBT and SGBT collision schemes are outlined in Table 4.1.

Considering symmetric probability in the SSBT scheme, i.e. $P_{i,j} = \frac{k'(p_{i,j}+p_{j,i})}{2}$, and Eq. (4.25), a coefficient k'' is considered to limit the number of selections (similar to the GBT scheme) as follows:

$$k''\sum_{i=1}^{N_{sel}}\sum_{j=1,j\neq i}^{N^{(l)}} 1 = \sum_{i=1}^{N^{(l)}}\sum_{j=1,j\neq i}^{N^{(l)}} 1 \tag{4.28}$$

$$k''\left[\sum_{i=1}^{N^{(l)}-1}\sum_{j=1,j\neq i}^{N^{(l)}} 1 - \sum_{i=1}^{N^{(l)}-N_{sel}}\left(\sum_{j=1,j\neq i}^{N^{(l)}-N_{sel}} 1\right)\right] = \sum_{i=1}^{N^{(l)}}\sum_{j=1,j\neq i}^{N^{(l)}} 1 \tag{4.29}$$

$$k''\left[\sum_{i=1}^{N^{(l)}} 1 - \sum_{i=1}^{N^{(l)}-N_{sel}} 1\right] = \sum_{i=1}^{N^{(l)}} 1 \tag{4.30}$$

$$k''\left[N^{(l)} - \left(N^{(l)} - N_{sel}\right)\right] = N^{(l)} \tag{4.31}$$

Thus:

$$k'' = \frac{N^{(l)}}{N_{sel}} \tag{4.32}$$

Table 4.1 Conventional length scale units

Statement	GBT Scheme	SGBT Scheme
Basic equation	Kac	Kac
Number of selected pairs in each cell	$N_{sel} \leq N^{(l)} - 1$	$N_{sel} \leq N^{(l)} - 1$
Selection of collision pair	From the given list, start by selecting the initial particle in the sequence and, subsequently, choose a random particle from the remaining $N^{(l)} - i$ particles located after the first one in the list	From the provided list, begin by selecting the starting particle in the sequence and, subsequently, choose a random particle from the remaining $N^{(l)} - 1$ ones in the list
Repeated collisions are possible	No	Yes (with some remedies)

$$P_{i,j} = \frac{k'k''(p_{i,j} + p_{j,i})}{2} \quad (4.33)$$

The following steps outline the SGBT method algorithm:

- Randomly choose N_{sel} particles from the list of particles and position them at the start of the particle list within the cell. Since the SGBT scheme utilizes only part of the particles in the cell, similar to the GBT algorithm, a shuffling procedure must be included. For that, choose N_{sel} random particles from the list of particles in the cell and reorder the particle list, exchanging the positions of the selected particles with those occupying the first N_{sel} positions in the list, i.e., put the N_{sel} selected particles in the index list ordered as first, second, etc., particle in the list. Place the former first, second, and so on, particles in their respective places in the index list.
- Choose the first particle in the list (i). Randomly select the second particle (j) from the initial $N^{(l)} - 1$ particles, ensuring an equal probability for each particle within cell l.

$$j = int\left[R_f \times \left(N^{(l)} - 1\right)\right] + 1 \quad (4.34)$$

- If $j = i$, choose another j.
- As a result, $1 \leq j \leq N^{(l)} - 1$, and it is probable that the chosen pair (i, j) will replicate the previously picked pair (i, j).
- The parameter of (Subtract(i) $= j - i$) is calculated to be employed to prevent a boundless loop for $i = N^{(l)}$ as below:

  For $i = N^{(l)}$, if for each i ($i = 1,2, ..., N^{(l)} - 1$), Subtract($i$) $= N^{(l)} - i$, subsequently, the selection procedure proceeds to the next cell in the sequence since the final particle on the list ($i = N^{(l)}$) is excluded from the random selection process for the second particle.
- Assessing the collision probability using the adjusted probability value:

$$W_{i,j} = \frac{N^{(l)}\left(N^{(l)} - 1\right)F_{num}g_{i,j}\sigma_{i,j}\Delta t}{N_{sel} \times 2V^{(l)}} \quad (4.35)$$

- The following procedure is used to avoid a repeated collision if the collision has been accepted, i.e., $W_{i,j} > R_f$:

  Define a new integer array I_p with a length equal to the total number of particles. If $I_p(i) = j$ and $I_p(j) = i$, the selected pair (i, j) is rejected as a duplicated one, otherwise, the selected pair is approved, and I_p elements (i), (j) are changed to$I_p(i) = j$ and $I_p(j) = i$
- If the collision between particles i and j is approved, the velocities of the particles are updated. The flowchart of the SGBT procedure is illustrated in Fig. 4.24.

The newly proposed SSBT and SGBT schemes symmetrize the collision pair selection process and are tested on two fundamental problems, the normal shock wave and the Fourier heat transfer problem, to evaluate their performance.

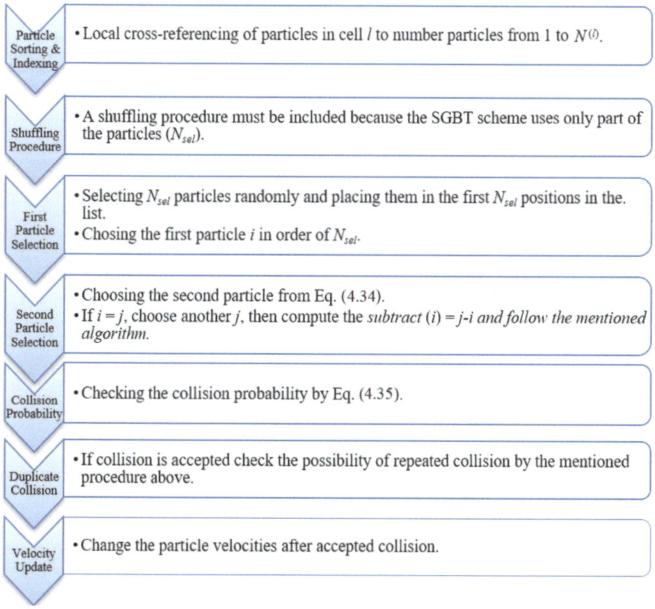

Fig. 4.24 The flowchart for the SGBT collision scheme

The nonequilibrium standing shock in argon gas is studied using SGBT with Bird's DSMC1S code. The SGBT algorithm replaces NTC for simulation, and results are compared with the NTC reference solution. The number of cells includes 3,000 for PPC 2, 6,000 for PPC = 1, and 12,000 for PPC = 0.5, all at Mach 2. Also, for Mach 30 based on 4000 particles, the number of cells is 800 and 2000 for PPC 5 and 2, respectively. Figure 4.25 compares dimensionless density and temperature profiles from SGBT and NTC for Mach 2 and 30, showing good agreement with NTC. Overall, SGBT maintains NTC accuracy when the number of particles in cell N is significant and N_{sel} is a fraction of N, while reducing computational cost since N_{sel} is chosen in pairs. In other words, the SGBT analyzes only N_{sel} particles, whereas SSBT examines all particles, resulting in a shorter convergence time for the SGBT.

The shock wave results obtained using the SGBT scheme showed that reducing the number of particles in the cell can bring the accuracy of the solution close to the reference value.

The temperature evolution at x = 0 was analyzed for SGBT and NTC at Mach 30 with PPC = 20, as shown in Fig. 4.26. SGBT behaves similarly to GBT for a larger number of particles and can achieve NTC accuracy with a properly chosen N_{sel}. SGBT converges faster due to its symmetric collision process, which reduces statistical dispersion—an additional advantage of this method.

The rarefied Fourier problem was selected to assess the convergence performance of the new algorithm in DSMC1.for code. Figure 4.27 compares SSBT, GBT, and SGBT with the reference solution. SSBT shows good accuracy, while SGBT

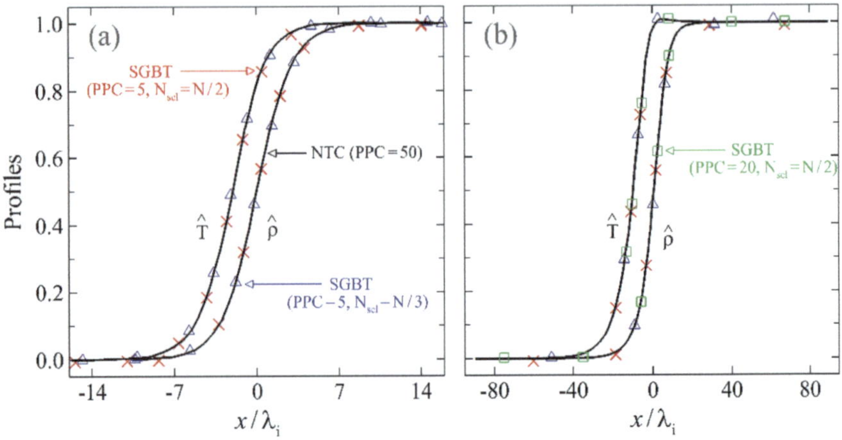

Fig. 4.25 Comparing the normalized density and temperature for a stationary normal shock using the NTC and SGBT schemes with the Mach numbers of **a** 2 and **b** 30. Reproduced from [26]

Fig. 4.26 Comparison of temperature changes over time at the point x = 0 at Mach = 30, using SGBT and NTC. Copied from [26] with permission

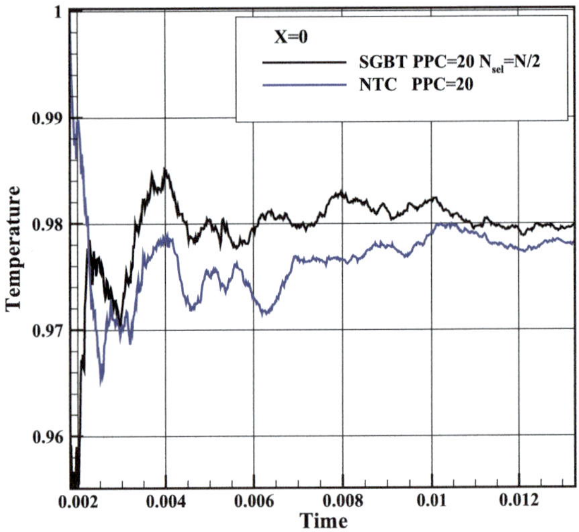

closely matches the reference heat flux (1512 W/m^2) with fewer samples than SSBT (Table 4.2). Increasing PPC improves convergence. All three methods (SGBT, SSBT, GBT) closely approximate the SBT reference data.

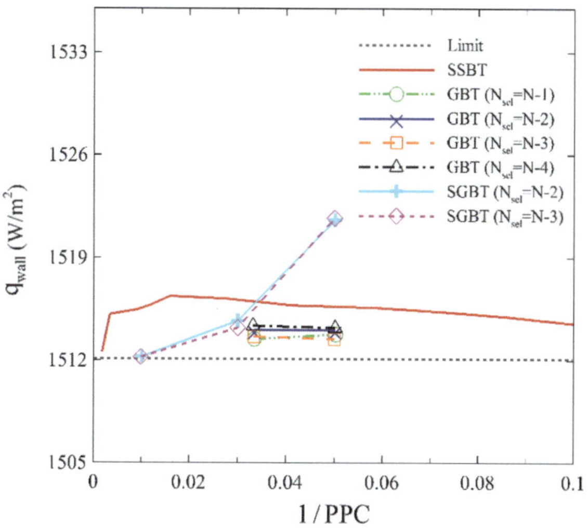

Fig. 4.27 Wall heat flux in terms of the reverse average number of particles per cell for all three collision schemes GBT, SSBT, and SGBT. Unpublished results from authors

Table 4.2 Wall heat flux with different cell numbers and constant total numbers of particles 24,000 for SGBT, N_{sel} = N–2, and N–3

N_{cell}	PPC	Δt	$\frac{dx}{dt}$	$Q_{SGBT(N-2)}\left(\frac{W}{m^2}\right)$	Samples
400	60	0.625 E-8	1.18	1512.1454	400 000
800	30	0.625 E-8	0.59	1514.5637	400 000
1200	20	0.2 E-8	1.23	1521.5206	400 000
				$Q_{SGBT(N-3)}\left(\frac{W}{m^2}\right)$	
400	60	0.625 E-8	1.18	1512.18384	400 000
800	30	0.625 E-8	0.59	1514.36834	400 000
1200	20	0.2 E-8	1.23	1521.63355	400 000

4.10 Note on the Accuracy of the Bernoulli Trials Family Schemes

As mentioned earlier, the NTC and NN schemes exhibit good performance at lower values of SOF (the ratio of the mean collision separation to the mean free path) [8]. So, it should be noted that low SOF does not always guarantee the correct solution if it is not accompanied by avoiding repeated collision between particles [27], a small time step compared to the mean collision time and mean transit time (around 0.2–0.3), and adequacy of the chosen flow field [8]. It was demonstrated that the MCS in the dominant gradient direction, i.e., normal to the surface in the boundary layer or normal to the shock in the shock interior, can be used as a metric for the DSMC simulation accuracy, rather than the global MCS [28].

On the other hand, for the BT family schemes, it is necessary to consider $0.5 < dt/dx < 1.5$, in addition to keeping both parameters of cell size (dx) and time step (dt) small [6, 17], where, dx is the normalized cell size by the gas mean free path, i.e. $\lambda = \frac{1}{(\sqrt{2}n\pi d^2)}$, and dt is the normalized time step by the mean collision time, i.e. $t_c = \frac{\lambda}{\sqrt{2k\frac{T}{m}}}$. In other words, the ratio dt/dx should be chosen in a way to fulfill the molecular statistically mild analog of the well-known Courant–Levy–Friedrich (CFL or simply Courant) condition, i.e., almost for any particle $c_i \frac{\Delta t}{\Delta x} < 1$, where c_i is the molecular velocity of any particle. It could be accepted that c_i is being replaced with the most probable speed, c_m [25].

4.11 Comparison Between the Collision Models

In this section, the comparison of the results obtained using the Bernoulli Trails family of collision pairs selection schemes, including the SBT, GBT, SSBT, and SGBT, is compared to that of the standard NTC model. The problems considered here are: collision frequency test, lid-driven micro cavity flow, and flow over circular cylinder.

4.11.1 *Micro Cavity Flow*

Performance comparison

Microcavity flow is simulated using different Bernoulli Trials collision selection schemes, including the SBT, GBT, SSBT, and SGBT, and their results are compared with those of Bird's NTC algorithm. The argon gas is considered with molecular mass and diameter of 6.64×10^{-26} kg and 4.092×10^{-10} m, respectively. The gas is confined in a lid-driven microcavity with a length of 1×10^{-6} m. Figure 4.28 indicates the geometry and flow condition of the considered cavity test case. The lid velocity and Kn are set to be equal to 100 m/s and 0.01, respectively. The walls of the cavity are at different temperatures. In this regard we set the temperature of the corners in such a way that both down corners of the cavity, i.e. A and D, are at the temperature of 350 K and the temperature of the corners B and C which are at the top of the cavity are set at 300 K. This cause the top and down wall of the cavity to be at uniform temperature of 300 and 350 K, respectively. However, the temperature of the left and right walls linearly changes from 300 K near the lid to 350 K near the bottom of the cavity.

To facilitate a detailed comparison between the collision schemes, a performance test is conducted to evaluate the computational efficiency of different collision partner selection schemes. The main focus of this section is on the SGBT algorithm. In this test, the collision algorithms simulate the cavity flow on a mesh with 200×200 grid

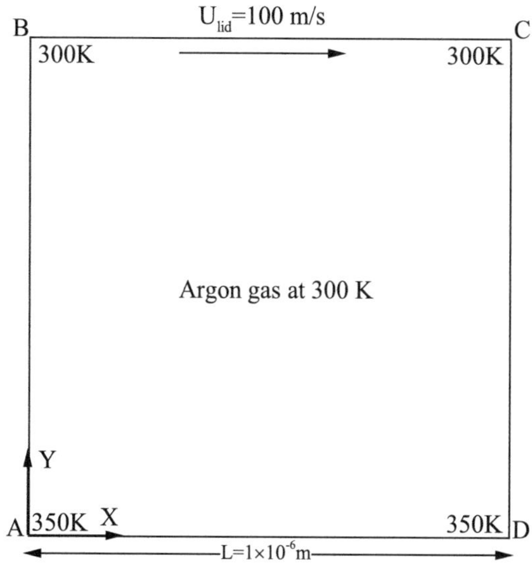

Fig. 4.28 Geometry and boundary condition of the cavity test case

cells and a particle per cell number of 10. The simulations start in an equilibrium state and continue until they reach a steady-state situation. The temperature of the gas near the cavity lid is considered the criterion for stopping the simulation. In this regard, a benchmark solution was performed, and the normalized difference between the gas temperature at the lid and the benchmark solution was calculated as the error in each cell on the lid. The stop point is devised in such a way that the summation of the error of all cells on the lid is less than 0.1, this reads as follows:

$$Error = \sum\nolimits_{lidcells} \frac{T_{lid} - T_{Benchmark}}{T_{Benchmark}} \leq 0.1 \tag{4.36}$$

The benchmark values are collected from a similar run with PPC = 50, which was performed for a sufficiently long time using the standard NTC scheme. Table 4.3 presents the simulation results of this performance test. In this table, to provide a rigorous comparison, some of the critical computational indexes of simulation from the SGBT scheme, as well as that of NTC, SBT, GBT, and SSBT schemes, are presented.

The CPU time spent to reach the stop point is presented in the first column of the table. The values are normalized based on the CPU time of the NTC scheme, except for N_{sel} = N–7 and N–8, the other values of N_{sel} of the SGBT scheme are faster than NTC. It is worth noting that to have physically correct results in BT schemes, the collisions that their collision probability exceeds the upper limit of 1 (W exceed ratio, W_E_R) to be bounded around 1%. For this reason, all runs in Table 4.3 have been performed using the same time step of 1.63×10^{-11} s, except for N–7 and N–8 whose time step is regulated in such a way that the W_E_R be bounded around 1%. This results in excessive computational time for these N_{sel} values in comparison to

Table 4.3 Corresponding results in the cavity flow, comparison of the NTC, SBT and two variants of the GBT scheme

Scheme	Normalized CPU time	W_E_R (%)	Col/Sel Ratio (%)	Normalized sample size
NTC	1	0	60.58	1.00
SBT	1.10	3.24	35.50	1.02
SSBT	0.68	0	31.56	0.64
GBT (N_{sel} = N–4)	0.62	4.99	52.71	0.64
SGBT (N_{sel} = N–2)	0.80	0.02	39.44	0.78
SGBT (N_{sel} = N–3)	0.61	0.05	45.03	0.62
SGBT (N_{sel} = N–4)	0.70	0.15	52.26	0.76
SGBT (N_{sel} = N–5)	0.68	1.28	61.68	0.75
SGBT (N_{sel} = N–6)	0.72	0.93	48.56	0.87
SGBT (N_{sel} = N–7)	1.22	0.91	42.45	1.60
SGBT (N_{sel} = N–8)	1.06	1.08	40.32	1.48

others. As the table shows, the SSBT scheme is faster than the SBT. However, the GBT, using N_{sel} = N–4, and the SGBT, using N_{sel} = N–3, are slightly faster than the SSBT. The GBT and SGBT reached the stop point in less time than the SBT and NTC, except for N–7 and N–8.

The coefficient of probability for GBT and SGBT collision schemes has been plotted in Fig. 4.29a. Additionally, the same coefficient for the SBT and the SSBT is shown in Fig. 4.29b. Both figures have been plotted for a cell with a PPC of 25 for which N_{sel} is supposed to be 20. As the probability coefficient for the GBT scheme is related to the order of the selected pair in the list of particles within the cell, we calculate the coefficient for the second particle in the list, i.e., $i = 2$. The values in Fig. 4.29b have been calculated for each particle in the cell, i.e. for each i. Both frames b and c show that, for the same conditions of time step and particle velocities, the SBT and GBT schemes have a greater probability coefficient than the SSBT and SGBT schemes, and thus, it is more probable to exceed the upper limit of the collision probability of 1. This is in accordance with the results presented in column 2 of Table 4.3 and frame c of Fig. 4.29, where we plotted the W_E_R values for GBT and SGBT after all runs were completed without decreasing the time step to bound the W_E_R around 1%. As this figure indicates, a greater probability coefficient for the GBT results in a higher probability of exceeding all values of N_{sel} compared to the SGBT.

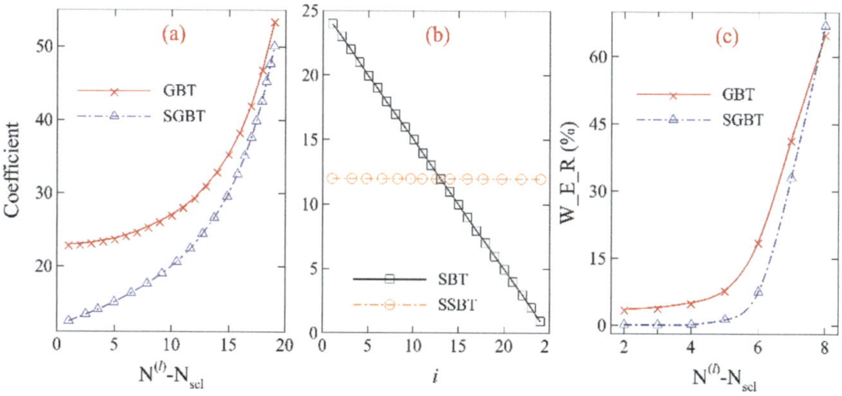

Fig. 4.29 Comparison of collision probability coefficient **a** GBT and SGBT, **b** SBT and SSBT, **c** comparison of W exceed ratio of GBT and SGBT schemes. Unpublished results from authors

The third column in Table 4.3 indicates the ratio of the number of collisions to the selection of pairs. For the SGBT algorithm, the decrease in N_{sel} from N–2 to N–8 results in an increase in the collision-to-selection ratio. To explain this behavior, we refer to the difference between the SSBT and SGBT schemes. Due to the small time step imposed by the condition that exceeding probabilities should be bounded, major parts of the selected pairs have little to no chance of colliding in the SSBT scheme. For this reason, like the GBT, the SGBT scheme was developed to decrease the number of selections, thereby increasing the probability of collision acceptance between selected pairs. However, in comparison to the GBT, this scheme has a uniform collision probability distribution leading to the lower W_E_R, as Fig. 4.29c shows. As mentioned above, the time step of every N_{sel} value in Table 4.3 is adjusted so that the W_E_R remains below 1%. For example, this is evident in the reduction of the collision to the selection ratio of cases N–6, N–7, and N–8 compared to N–5.

The sample size required to reach the stop point is reported in the last column of Table 4.3. The normalized values based on the sample size of the NTC scheme are presented in the table. This parameter provides a clearer understanding of the computational costs associated with each scheme in achieving the convergence condition. The SGBT has an optimum sample size for N_{sel} equal to N–3. Simulation with this N_{sel} decreased the sample size by up to 38 and 40% compared to the NTC and SBT, respectively. The sample size of the simulation using these N_{sel} values is approximately the same as the GBT with N-4 and SSBT.

Flow filed properties

To comprehensively evaluate the different collision partner selection schemes, the flow field properties are compared for various sets of Knudsen numbers. In this regard, we performed runs using various collision partner selection schemes, including NTC, SBT, GBT, SSBT, and SGBT. The Knudsen number is selected to encompass different rarefaction regimes, ranging from 0.01 at the slip flow regime to 10 as the upper limit of the transition regime. All runs reported here are performed using 200 × 200 grid cells and PPC is set to 10 at the initial condition. The GBT and SGBT simulations were performed using N_{sel} = N–4.

Figure 4.30 indicates the velocity magnitude contours obtained using different collision partner selection schemes at various Knudsen numbers. In each frame that presents the contour of velocity magnitude for a given Knudsen number, the flood on the background of the cavity indicates the velocity magnitude contours from the NTC scheme. To clearly demonstrate the results of the other collision schemes, the flow domain has been divided into columns, each representing a specific scheme. From left to right, the SBT (solid), the GBT (dashed), the SSBT (dash-dot), and the SGBT (dashed dot) are presented, respectively, in each frame on the NTC as the background. This allows for the comparison of different BT schemes with the NTC in part of the domain and with each other along column borders. From the figure,

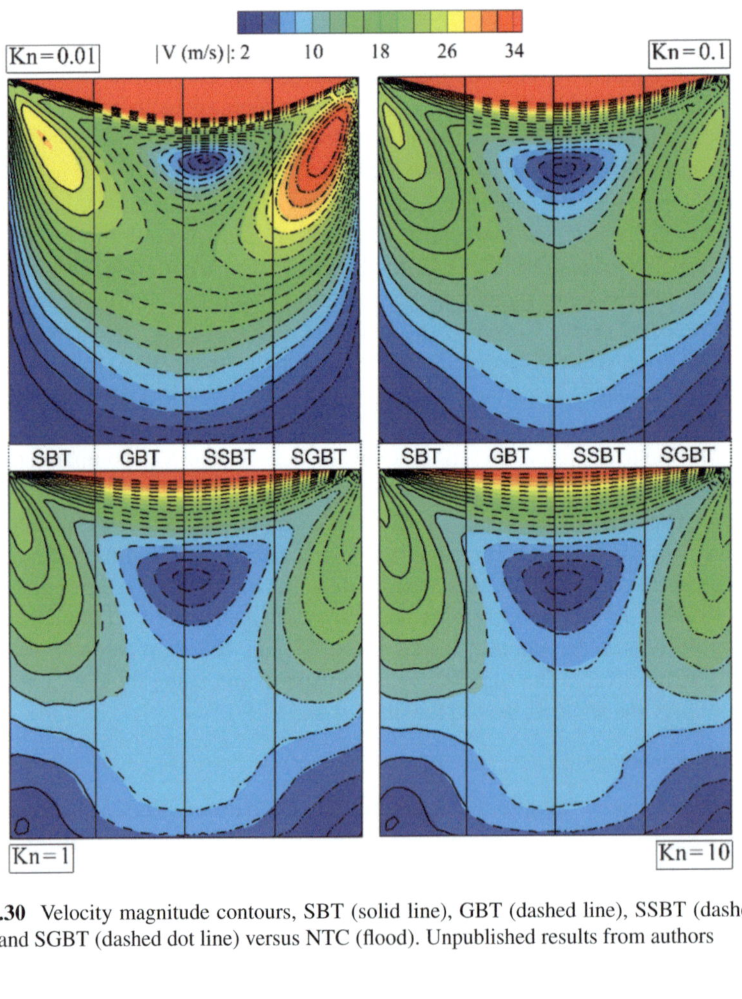

Fig. 4.30 Velocity magnitude contours, SBT (solid line), GBT (dashed line), SSBT (dashed dot line), and SGBT (dashed dot line) versus NTC (flood). Unpublished results from authors

a suitable agreement is observed among all schemes for simulating the velocity magnitude contours.

In order to investigate the particle pair selection schemes' capability to simulate the flow field properties, we present the temperature contours from the simulations at different rarefaction regimes in Fig. 4.31. The indication of the flood and lines is similar to that specified for the Fig. 4.30. An excellent agreement between the temperature fields calculated by all schemes is seen in Fig. 4.31.

Circulation and the location of the vortex center

In order to investigate the nonequilibrium effects on the vortex strength in the micro cavity, the circulation of the mean velocity field, Γ is defined as follows:

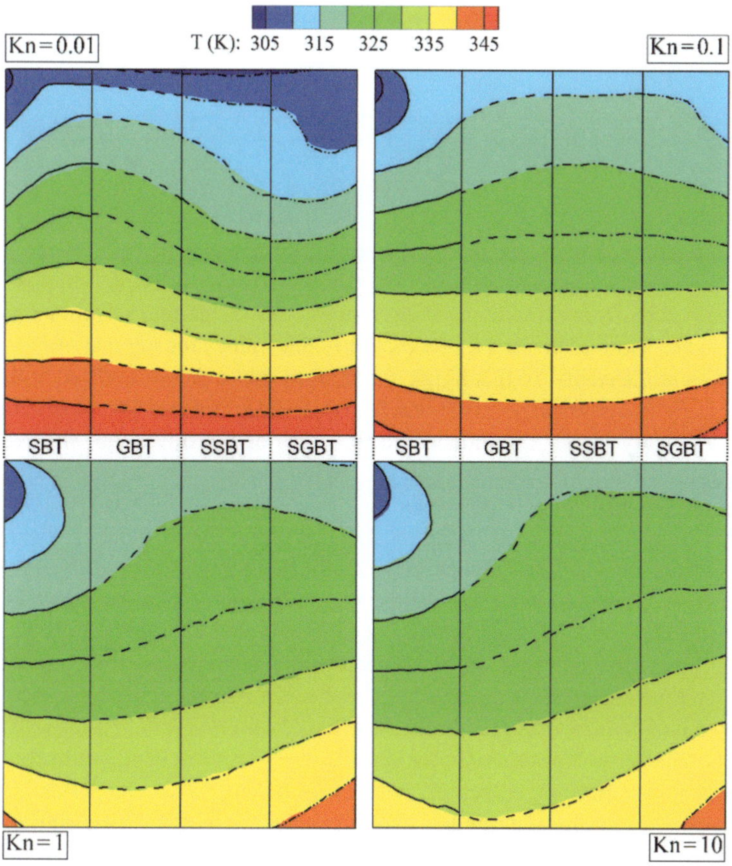

Fig. 4.31 Temperature contours, SBT (solid line), GBT (dashed line), SSBT (dashed dot line), and SGBT (dashed dot line) vs. NTC (flood). Unpublished results from authors

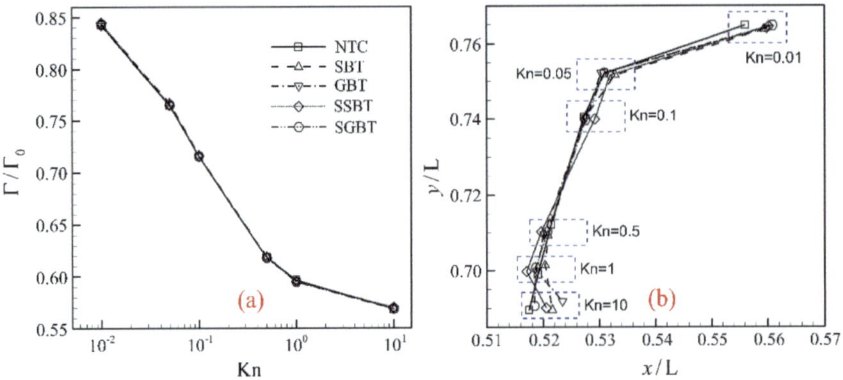

Fig. 4.32 **a** Circulation of the velocity vector and **b** location of the center of the vortex in the cavity. Unpublished results from authors

$$\Gamma = \oint V.ds = \int rot_n V.dA = \sum_{i,j}\left[\frac{(V_{yi+1} - V_{yi,j})}{\Delta x} - \frac{(V_{xj+1} - V_{xi,j})}{\Delta y}\right]\Delta x \Delta y \tag{4.37}$$

where the summation is taken over the entire computational cells of the cavity. Figure 4.32a shows the variation of flow circulation with the Knudsen number in the driven cavity from simulations using various collision partner selection schemes. As the figure shows, all schemes predict the same values for the magnitude of the circulation in the cavity. In this figure, circulation values are normalized with $\Gamma_0 = U_{wall} \times L$. From a physical point of view, it is observed that increasing nonequilibrium effects gradually hinder the induced shear forces of the driven lid, subsequently reducing the strength of the vortex in the cavity. Furthermore, the trend of such a reduction dramatically decreases as the Knudsen number exceeds 0.5.

Variation of the center of vortex location by the Knudsen number is shown in Fig. 4.32b. As the Knudsen number increases the Reynolds number decreases; therefore, viscous dissipation in the flow field enlarges. In other words, the induced kinetic energy from the top lid dissipates more notably in higher Knudsen numbers, which consequently shifts the center of the vortex toward the left to the geometric center of the cavity. Furthermore, it is observed that as the flow becomes rarefied, the center of the vortex moves away from the driven lid toward the bottom wall. It is also attributed to the reduced Reynolds number in the large Knudsen number, which shifts the center of the vortex toward the geometric center of the cavity. From a numerical point of view, it can be seen from the figure that the simulation of the flow using different collision partner selection schemes results in approximately the same coordinate for the center of the vortex.

Velocity and temperature profiles

Figure 4.33a, b show the normalized horizontal and vertical components of velocity vectors along the vertical and horizontal centerlines of the cavity, respectively. This

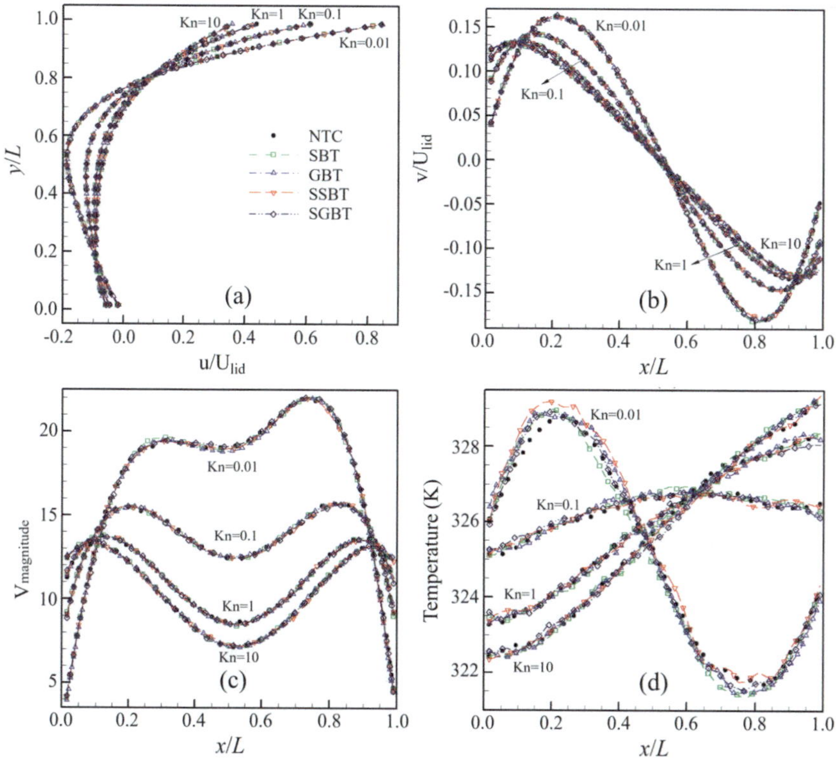

Fig. 4.33 Flow properties from different collision partner selection schemes along the centerline of the cavity; **a**, **b** *x* and *y* velocity components along the vertical and horizontal centerlines and **c**, **d** velocity magnitude and temperature along the horizontal centerline. Unpublished results from authors

figure illustrates that the reduced curvature of the horizontal velocity in the higher ranges of Knudsen number is the direct consequence of the diminished vortex strength in the micro cavity (see Fig. 4.32a). Frames c and d indicate the velocity magnitude and temperature along the horizontal centerline of the cavity. As the figure shows, all schemes perform the simulation with the same accuracy for all Knudsen numbers, and changing the rate of the rarefaction does not have any significant effect on the various scheme' ability to simulate the flow properties.

The dimensionless velocity slip and temperature jump on the driven lid surface are depicted in Fig. 4.34a, b. Both frames demonstrate excellent agreement between the NTC, SBT, GBT, SSBT, and SGBT results in predicting the velocity and temperature fields.

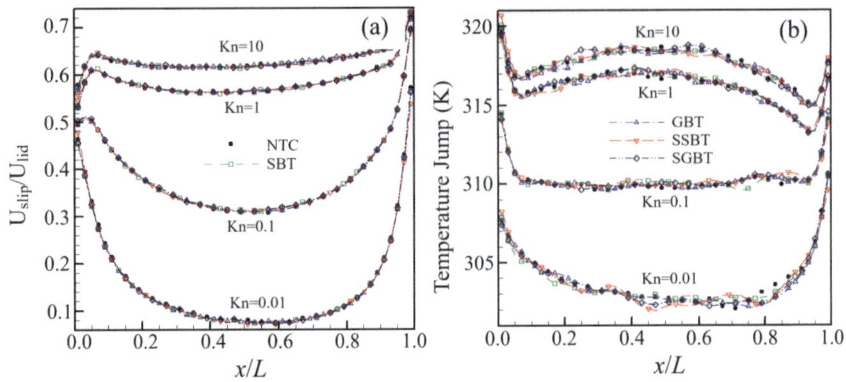

Fig. 4.34 **a** Normalized velocity slip and **b** temperature jump along the driven lid. Unpublished results from authors

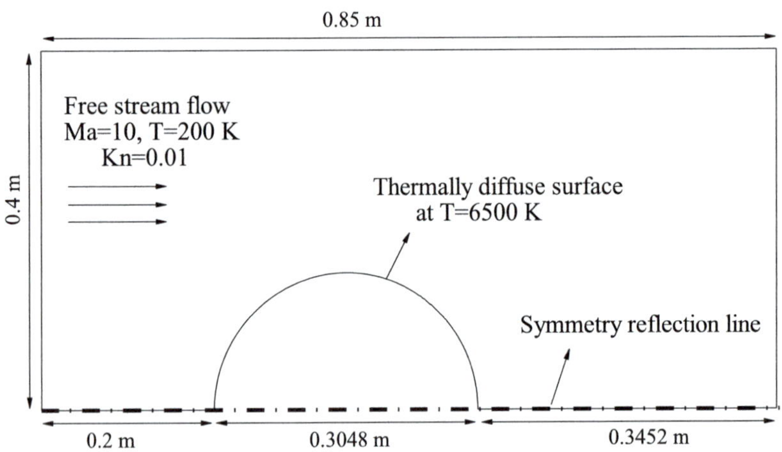

Fig. 4.35 Geometry and boundary conditions of the 2D hypersonic cylinder

4.11.2 Hypersonic Flow Over Circular Cylinder

To evaluate the performance of the different collision schemes, a benchmark problem of rarefied hypersonic flow over a cylinder was simulated. This benchmark problem illustrates a wide range of collision frequencies occurring in the computational domain, from the downstream back cylinder zone to the upstream stagnation point, and thus provides challenging conditions for evaluating any collision algorithm. A rarefied flow of argon gas is realized at Mach 10 ($U = 2634.1$ m/s) and a temperature of 200 K passes over a 12 inch circular cylinder with a fully diffusive surface at a surface temperature of 6500 K.

Figure 4.35 shows the computational domain and employed boundary conditions. Since the hypersonic flow past a cylinder is a problem where the collision frequency and mean free path experience significant gradients, it is necessary to use an adaptive subcell grid to perform the simulations [7]. For this, we implemented the GBT algorithm, accompanied by the TAS technique, in the DS2V code developed by G. Bird and described in his most recent monograph [11]. For all simulations carried out here, the reference solution is a run performed using the SBT-TAS algorithm with a sufficient number of PPC and fine grid size. The grid size and particles per cell independence test were performed for this case, as described in reference [22]. The suitable grid size and particle per cell (PPC) were determined to be 194 $\times$ 100 divisions and 11.5 particles per cell, respectively.

Flow field properties evaluation

The different algorithm's ability to simulate flow field properties is depicted in Fig. 4.36. Here, temperature contours from various collision pair selection schemes' results (solid lines) are compared with those of the reference solution of the SBT-TAS (flood background in each frame). The GBT and SGBT simulations have been performed using N_{sel} = N–4. The temperature contours show excellent agreement between the results of various schemes and those of the SBT-TAS solutions.

Another notable feature of flow in the hypersonic regime is the formation of a shock wave. Figure 4.37 illustrates the flow properties along the stagnation line ahead of the cylinder for different algorithms, compared with the SBT-TAS results as the reference. The Mach number and temperature profiles normalized by their free stream reference values are shown in Fig. 4.37a. Although results from SSBT, GBT, and SGBT schemes show suitable agreement with the SBT-TAS solution, the

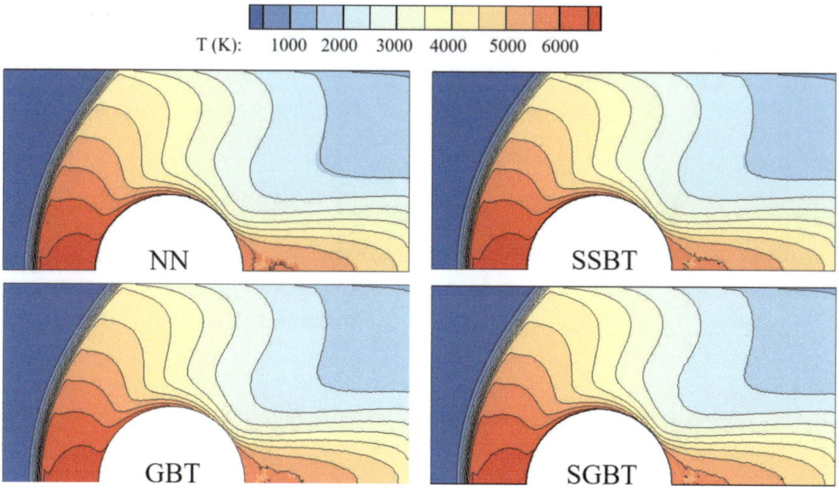

Fig. 4.36 Temperature contours: SBT-TAS (background contours) and other schemes (line contours). Unpublished results from authors

results of the NN for the Mach number profile indicate a slightly thicker shock wave. The density profile, normalized with respect to its free-stream solution, is depicted in Fig. 4.37b. In this figure, an excellent agreement is shown between all schemes and the reference solutions from SBT-TAS.

Surface properties evaluation

This subsection has reported the simulation results for velocity, temperature, heat flux, and pressure coefficient on the cylinder surface. The comparison of velocity and pressure coefficient distributions on the cylinder surface, which are shown in Fig. 4.38, indicates that the results of different collision pair selection algorithms agree with the results of the SBT-TAS scheme. All schemes, such as the SBT-TAS, have predicted the same maximum velocity. Moreover, the results also show the same behavior as the reference solution at the low back-pressure region of the cylinder, see Fig. 4.38a. The pressure coefficient distribution that is shown in Fig. 4.38b is in full agreement with the reference simulation results.

Figure 4.39a illustrates the temperature distribution normalized by the free stream temperature along the cylinder surface. The results of the NN, SSBT, GBT, and SGBT schemes with $N_{sel} = $ N–4 are generally in good agreement with the SBT-TAS solution. All schemes predicted the minimum temperature on the cylinder surface at approximately 138°. Some discrepancies are observed near the low-pressure region behind the cylinder. Another investigated parameter is the heat flux on the cylinder surface. It is shown in Fig. 4.39b. The results of the various schemes are in suitable agreement with the SBT-TAS results. Although the results coincide entirely at the rear region of the cylinder, small deviations are visible in front of the cylinder near the stagnation point region.

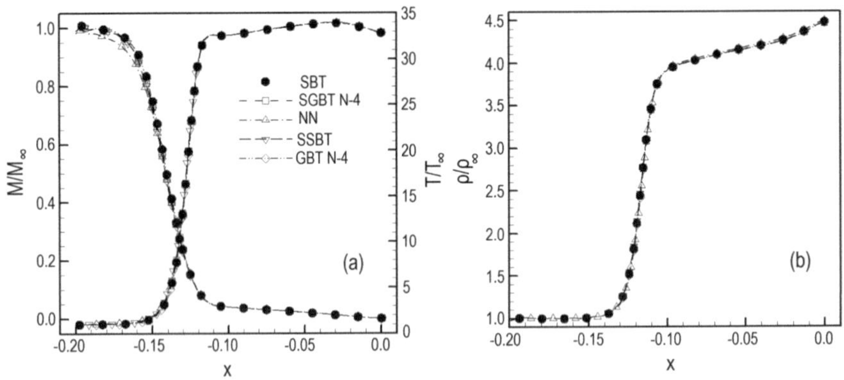

Fig. 4.37 **a** Normalized Mach number and temperature and **b** normalized density on the stagnation line. Unpublished results from authors

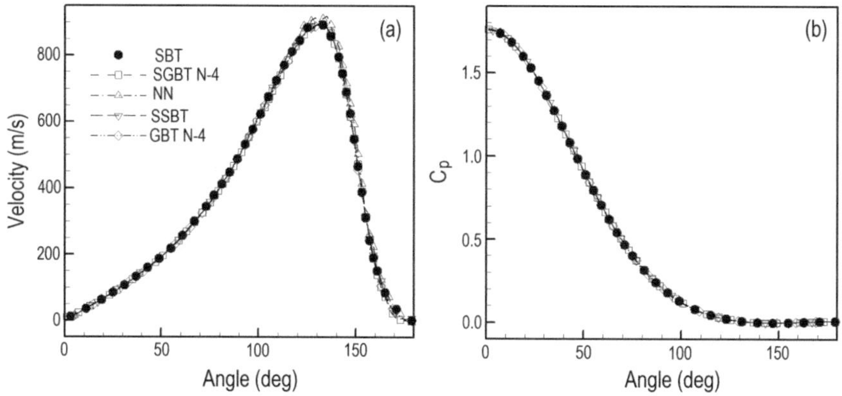

Fig. 4.38 **a** Velocity magnitude and **b** pressure coefficient distribution on the cylinder surface

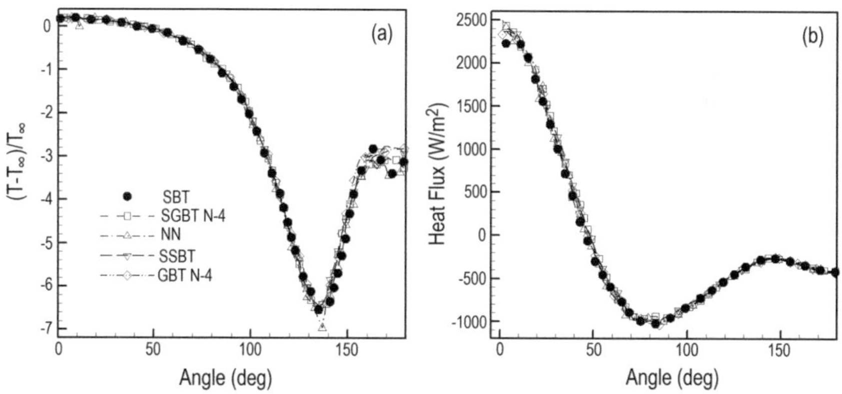

Fig. 4.39 **a** Normalized temperature and **b** heat flux on the cylinder surface

4.12 DSMC Collision Schemes for Small Deviation from Continuum Flow

At near-continuum conditions, the assumption of weak non-equilibrium effects enables the simplification and approximation of the Boltzmann equation, allowing for the use of a simpler form to derive new collision algorithms with improved computational properties. In addition to the new collision schemes considered in the previous sections, at low Knudsen numbers, there are new treatments related to the DSMC collision procedure that improve the collision performance, i.e., lead to a reduction in computational costs. In this section, the hybrid DSMC-Fokker–Planck method, the hybrid DSMC-LD scheme, the low-variance deviational simulation Monte Carlo (LVDSMC) technique, and collision limiter approaches will be introduced and discussed.

4.12.1 DSMC-Fokker-Plank Approach

A major challenge for the standard DSMC technique is its high computational cost when approaching the continuum regime. The stiffness of the collision operator, which results in massive computational costs in the near-continuum regimes for DSMC, can be treated by passing to the diffusion limit of the Boltzmann equation. In this context, the Fokker–Planck (FP) equation naturally arises as a diffusion approximation of the underlying Master equation [29]. The FP method enables computationally efficient simulations of rarefied gas flows in low and intermediate Knudsen regimes.

Consider the Boltzmann equation, as previously mentioned in Chap. 3 as follows:

$$\left(\frac{\delta F}{\delta t}\right)_{coll} = \frac{1}{m}\int_{\Re^3}\int_0^{4\pi}\left[F\left(v^*\right)F\left(v_1^*\right) - F(v)F(v_1)\right]g\sigma(\Omega.g)d\Omega dv_1 \tag{4.38}$$

where the mass distribution function of the velocity of the particle is given by F, velocities (v, v_1) and $\left(v^*, v_1^*\right)$ specify the pre-collision and the post-collision velocities of particles, respectively, the collision cross-section is given by σ, Ω is the angle between the velocities after and before the collision, and g is the relative velocity. If Kn is assumed to be small, the time derivative of F could be approximated by the FP equation as follows [30]:

$$\left(\frac{\delta F}{\delta t}\right)_{coll} \approx -\frac{\delta}{\delta v_i}(A_i F) + \frac{1}{2}\frac{\delta^2}{\delta v_i \delta v_j}\left(D^2 F\right) = S^{FP} \tag{4.39}$$

where, the $A_i\left(v', x, t\right)$ and $D(x, t)$ terms represent the drift and diffusion terms, respectively. Those are functions of the peculiar velocity $v' = \boldsymbol{v} - \boldsymbol{U}$, where the mean (or gas) velocity is defined as $\boldsymbol{U} = \int_{\Re^3}\boldsymbol{v}fd\boldsymbol{v}$. Assuming the Maxwellian molecular model, and considering the stress tensor and the thermal flux vector obtained from the Boltzmann equation [30]:

$$P_{ij} = \int_{\Re^3} v_i' v_j' S^{Boltz} d\boldsymbol{v} = -\frac{p}{\mu}\pi_{ij} \tag{4.40}$$

$$P_i = \frac{1}{2}\int_{\Re^3} v_i' v_j' S^{Boltz} d\boldsymbol{v} = -\frac{2}{3}\frac{p}{\mu}q_i \tag{4.41}$$

where the dynamic viscosity coefficient is specified by μ, p is the gas pressure, π_{ij} is the stress tensor, and q_i is the thermal flux. As a result, the Prandtl number, which is the ratio of the stress to the thermal flux, equals 2/3, the correct magnitude for monatomic gases. Applying the same approach to the S^{FP} operator and assuming the scalar diffusion coefficient, and that $\int_{\Re^3}\left(A_i v_j' + A_j v_i' + D\delta_{ij}\right)Fd\boldsymbol{v}$ is trace-free, it can be written as follows:

$$P_{ii} = \int_{\Re^3} \left(A_i v_j' + A_j v_i' + D\delta_{ij}\right) F d\boldsymbol{v} \quad (4.42)$$

$$P_i = \int_{\Re^3} \left(A_i v_j' v_j' + 2A_j v_i' v_j'\right) F d\boldsymbol{v} \quad (4.43)$$

where δ_{ij} is the Dirac delta. It is necessary to determine an accurate formula for the coefficients A and D. For the drift coefficient, A_i, a fourth-order polynomial relation with the molecular velocity is suggested [30], which ensures an accurate prediction for both the viscosity coefficient and the Prandtl number. The physical assumptions implied by the FP model are as [31]:

- The particle velocities change due to a permanent stochastic force, rather than discrete collisions.
- Particle interaction occurs due to coupling through the coefficients in the evolution equations, which are functions of stochastic moments of the local ensemble.
- Considering the extreme case of infinite Kn (i.e., no inter-molecular collisions occur), S^{FP} becomes zero, and thus the FP and DSMC models become identical.
- In the other extreme case of extremely small Kn, the Navier–Stokes equations are recovered by both the FP and DSMC models; the only concern here is the consistency of the macroscopic coefficients.

To start the simulations, a specific number of particles (with a weight of w_i) is generated based on a known distribution function $F(\boldsymbol{v}, x, t)$. It should be noted that the time step is determined according to the Courant number. As this algorithm is similar to DSMC, the molecular collision part of the DSMC method is substituted by the corresponding FP equation, (4.39). It was demonstrated that the error of the FP approach becomes more significant in zones with more intense flow gradients, i.e., recirculation zones [32]. The details for the FP approach considering diatomic molecules have been proposed in Refs. [33, 34, 35].

Combining DSMC and FP has gained popularity as an efficient particle method over the past few years. By coupling the DSMC with the Fokker–Planck scheme, the entire Knudsen number range can be covered by an accurate solution algorithm. Since the computational particles represent the mass density function in both DSMC and FP simulations, the coupling can be achieved without introducing interfaces [36]. Different criteria have been considered for switching between FP and DSMC as follows:

- Based on the number of collisions between gas molecules [36, 37, 38, 39]
- Based on the accuracy of each model (FP or DSMC) via the gradient-length Knudsen number (Kn_{GL}) [40].

Gorji and Jenny [36] suggested a switching criterion between DSMC and FP based on the ratio of the number of particles per cell to the number of collisions per cell, i.e., $\frac{N_c}{N_c^{coll}}$. If the number of collisions in a cell exceeds the number of particles, then FP will be performed; otherwise, DSMC will be used.

Figure 4.40 compares the viscosity and conductivity errors between the DSMC method and the hybrid FP-DSMC approach based on the number of collisions [36]. The results of the Couette flow problem have been reported. The different CPU times correspond to simulations with different numbers of particles per cell, i.e., 20, 40, 60, and 80. The CPU times are normalized by the CPU time of the hybrid FP–DSMC simulation corresponding to PPC = 20. It could be seen that the errors of both methods are in the same order for a given grid size, whereas the hybrid approach is more efficient.

Another hybrid FP-DSMC approach is based on the gradient-length Knudsen number ($\mathrm{Kn_{GL}}$) [40]. In this approach a Knudsen ratio is defined as:

$$\mathrm{KnR} = \frac{\mathrm{Kn_{GL}}}{\mathrm{Kn}} \tag{4.44}$$

is considered as the switching criterion. Actually, KnR is the ratio of the local non-equilibrium effects to the global non-equilibrium in the domain. The gradient-length Knudsen number is defined as (see Eq. (1.26) for more details):

$$\mathrm{Kn_{GL_\phi}} = \frac{\lambda}{\frac{\phi}{|\nabla\phi|}} = \frac{\lambda}{\phi}|\nabla\phi| \tag{4.45}$$

$$\mathrm{Kn_{GL}} = max\left\{\mathrm{Kn_{GL_U}}, \mathrm{Kn_{GL_T}}, \mathrm{Kn_{GL_P}}\right\} \tag{4.46}$$

For zones with a large KnR, the hybrid algorithm is effective in conjunction with DSMC due to its high accuracy. For other zones, FP estimates gas properties benefiting from low computational cost. In this approach, the algorithm starts with a DSMC calculation of all cells for an initial period of time and then calculates KnR for each cell using the local mean free path and local gradients. Then, corresponding

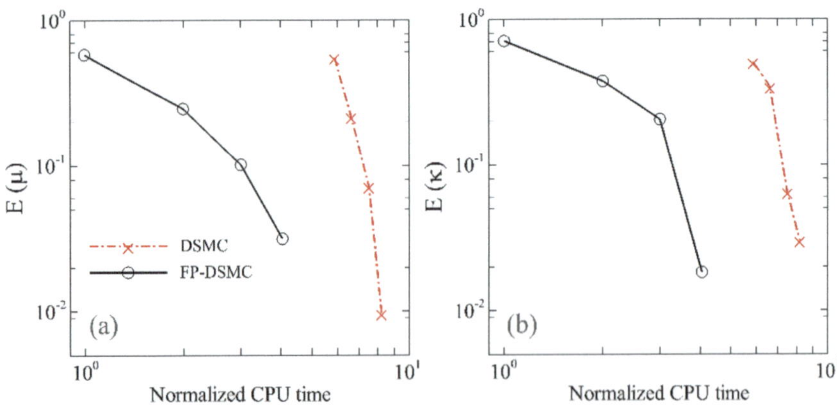

Fig. 4.40 The errors in the obtained viscosity (**a**) and heat conductivity (**b**) by DSMC and the hybrid FP–DSMC methods with respect to the normalized CPU time. Reproduced from [36]

Table 4.4 CPU time comparison between the hybrid algorithms with various $KnR_{threshold}$ in the Cavity flow [40]

Kn	Algorithm	$\frac{\left(\frac{\text{CPU time}}{PPC}\right)_{DSMC}}{\left(\frac{\text{CPU time}}{PPC}\right)_{algorithm}}$
0.05	Hybride ($KnR_{threshold} = 4$)	20
	Hybride ($KnR_{threshold} = 10$)	29.4
	FP	32
0.2	Hybride ($KnR_{threshold} = 4$)	19
	Hybride ($KnR_{threshold} = 10$)	22
	FP	25
0.5	Hybride ($KnR_{threshold} = 4$)	18
	Hybride ($KnR_{threshold} = 10$)	21
	FP	23
1	Hybride ($KnR_{threshold} = 4$)	10
	Hybride ($KnR_{threshold} = 10$)	11
	FP	12

to the KnR magnitude in each cell, it selects either the DSMC or FP algorithm to continue the simulations. Obviously, using a large KnR shifts the solver to use more FP in the domain, whereas a small KnR will set more DSMC calculations. Table 4.4 compares the required CPU times for the hybrid algorithm with those for the DSMC method in a simulation test case of Cavity flow. As expected, the hybrid approach becomes more efficient as Kn decreases. The required CPU time for the hybrid approach can be reduced by more than 20 times in the slip flow regime compared to the required CPU time for the DSMC method. Hybrid DSMC-FP solution agrees well with the pure DSMC results, see Ref. [40].

4.12.2 Low Diffusion (LD)-DSMC Approach

Burt and Boyd [41] proposed a low diffusion (LD) particle method for simulating compressible inviscid flows. It is considered an alternative method for simulating compressible, inviscid gas flows. The LD method is intended as an improvement over existing equilibrium particle methods, by significantly reducing both numerical diffusion effects and statistical scatter. The numerical procedures in the LD method are similar to those in the DSMC technique. The computational domain is divided into a large number of grid cells, and representative particles are tracked through these cells. During each time step, particles are moved through the grid according to assigned velocities, and bulk flow properties are computed by averaging quantities over all particles in each cell. The primary difference between this new method and DSMC lies in the procedures by which velocities used for particle movement are

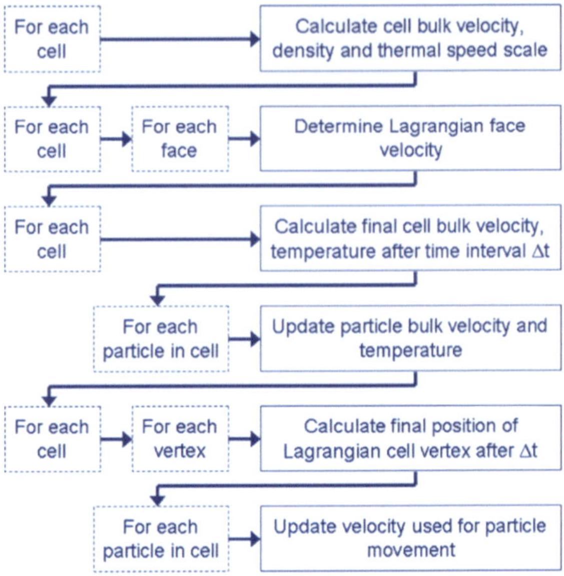

Fig. 4.41 Flow chart of simulation procedures performed in the LD method during each time step, replacing DSMC collision routines. Reproduced from [41]

updated during each time step [41]. In the LD method, standard DSMC collision routines are replaced with an alternative set of procedures, as shown in Fig. 4.41.

In the hybrid LD-DSMC algorithm, the simulated flow field is divided into continuum regions, which are assigned to the LD method, and nonequilibrium regions, which are assigned to the DSMC method. The information transfer procedures in the hybrid LD–DSMC scheme are somewhat more complicated due to differences between LD and DSMC particle properties, as well as the additional momentum and energy transport through Lagrangian cell faces in the LD method [42]. However, the use of similar particle methods in both rarefied and continuum regions enables relatively simple two-way information transfer, providing a basis for a strongly coupled hybrid algorithm suitable for application to either steady-state or unsteady-flow problems.

In the LD–DSMC hybrid simulation procedures, all cells in the fixed computational grid are periodically assigned one of four integer values, corresponding to the DSMC domain, the LD domain, or one of two buffer regions (Fig. 4.42). These buffer regions, designated as regions A and B, serve as a means of information exchange between the DSMC and LD methods. Each buffer region is two cells thick and is positioned along the boundary between DSMC and LD domains, such that a line drawn from a DSMC cell to a nearby LD cell will pass through at least two cells in buffer region A, then through at least two cells in region B. Standard DSMC collision calculations are performed in all cells within Region A, while all cells in Region B are included in LD method calculations [42].

The switching criterion between LD and DSMC has been suggested to be the Kn_{GL}. As recommended by Burt and Boyd [42], a cutoff value of 0.02 is suggested for Kn_{GL} as a switching criterion. Hence, for the cells with $Kn_{GL} < 0.02$, LD scheme

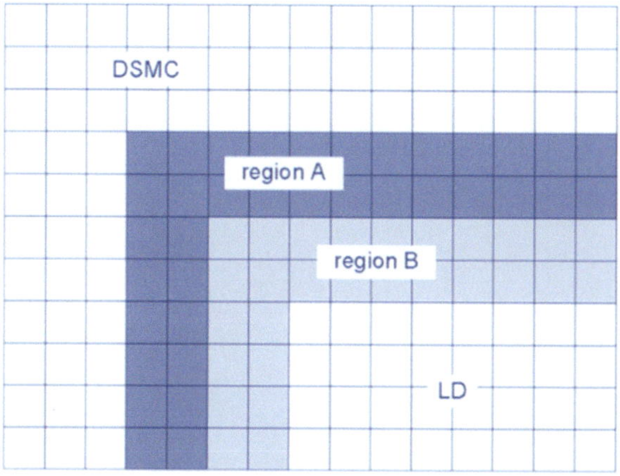

Fig. 4.42 Buffer regions in the hybrid LD-DSMC approach. Reproduced from Ref. [42]

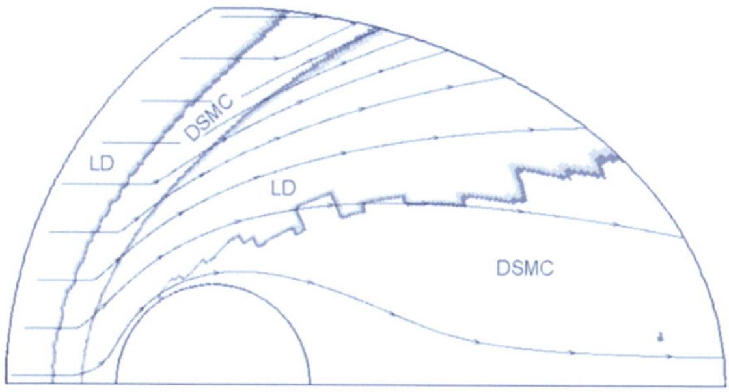

Fig. 4.43 Boundaries between LD and DSMC regions in the hybrid simulation. Reproduced from Ref. [42]

is implemented, and for the cells with $Kn_{GL} \geq 0.02$, DSMC method will be followed. Figure 4.43 shows the domain decomposition corresponding to different methods based on the continuum breakdown criteria. A detached bow shock is captured in the DSMC domain, and the low-density wake is also assigned to DSMC, while both freestream and high-density post-shock regions are assigned to the LD domain. Dark gray areas along the boundary between LD and DSMC domains denote cells within buffer region A, and neighboring light gray areas indicate cells within buffer region B.

Figure 4.44 compares the different flow distributions obtained by the DSMC method and hybrid LD-DSMC approach. The illustration shows profiles of density, translational temperature and rotational temperature along the stagnation streamline. There is relatively good overall agreement here for all three quantities, the disparities between hybrid LD-DSMC and DSMC results. All three hybrid simulation curves in Fig. 4.44 display a shock positioned slightly upstream of that in the DSMC simulation. A small (approximately 1.3%) unphysical increase in density is observed at

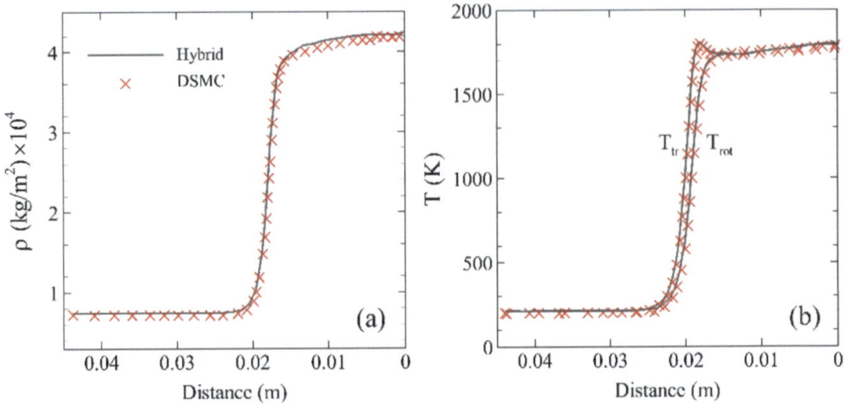

Fig. 4.44 Density, translational temperature and rotational temperature profiles along the stagnation streamline. Reproduced from Ref. [42]

the LD–DSMC interface along the downstream edge of the shock, accompanied by a corresponding decrease in translational temperature.

4.12.3 Low-Variance Deviational Simulation Monte Carlo (LVDSMC) Approach

Hadjiconstantinou and co-workers [43, 44, 45] proposed a variance reduction approach that addressed the limitations of Monte Carlo methods in the case of low-speed flows commonly found in small-scale devices. This approach focused on efficient methods for evaluating the collision integral. For a hard-sphere gas, the collision integral is as [43]:

$$\left[\frac{df}{dt}\right]_{\text{coll}}(\boldsymbol{r}, \boldsymbol{c}, t) = \aleph^2 \frac{\sqrt{\pi}}{4} \iiint \left(\delta_1' + \delta' - \delta_1 - \delta\right) \frac{ff_1}{\mathbb{N}^2} g\sigma d^2\Omega d^3\boldsymbol{v}_1 d^3\boldsymbol{v} \tag{4.47}$$

where $\aleph = \int f d^3\boldsymbol{v}$, $\delta' = \delta^3\left(\boldsymbol{v}' - \boldsymbol{c}\right)$ is the delta function, $\delta_1 = \delta^3(\boldsymbol{v}_1 - \boldsymbol{c})$, $\delta_1' = \delta^3\left(\boldsymbol{v}_1' - \boldsymbol{c}\right)$, and $\delta' = \delta^3\left(\boldsymbol{v}' - \boldsymbol{c}\right)$. Using importance sampling, the collision integral can be approximated by:

$$\left[\frac{df}{dt}\right]_{\text{coll}}(\boldsymbol{r}, \boldsymbol{c}, t) = \frac{\pi^{\frac{3}{2}}\mathbb{N}^2}{M} \sum_{i=1}^{M} \left(\delta_{1,i}' + \delta_i' - \delta_{1,i} - \delta_i\right) g_i \sigma_i \tag{4.48}$$

where M is the number of Monte Carlo samples. From the variance reduction point of view, an arbitrary Maxwell–Boltzmann distribution (f^{MB}) could be considered such that:

$$f = f^{\mathrm{MB}} + f^{d} \tag{4.49}$$

By substituting Eqs. (4.49) into (4.48), and considering zero value for $f^{\mathrm{MB}} f_1^{\mathrm{MB}}$ (f^{MB} is an equilibrium distribution), the collision integral could be obtained from the following single summation:

$$\left[\frac{df}{dt}\right]_{\mathrm{coll}} = \frac{\pi^{\frac{3}{2}} \aleph'}{M} \sum_{i=1}^{M} \left(\delta'_{1,i} + \delta_{i'} - \delta_{1,i} - \delta_i\right) \mathrm{sgn}\left(2f_i^{\mathrm{MB}} + f_i^{d}\right) \mathrm{sgn}\left(f_{1,i}^{d}\right) g_i \sigma_i \tag{4.50}$$

where $\aleph' = \int \left|2f^{MB}(\mathbf{v}) + f^{d}(\mathbf{v})\right| d^3\mathbf{v} \times \int \left|f^{d}(\mathbf{v})\right| d^3\mathbf{v}$ and sgn represents the sign function.

The variance reduction in cases where $f^{d} \ll f^{\mathrm{MB}}$ is considerable to the extent that at Mach of O (0.1) this approach provides considerable computational savings compared to direct methods which use (4.48). This is to be expected, since this approach allows one to avoid considering a large number of physically occurring collisions (the vast majority for $f^{d} \ll f^{\mathrm{MB}}$) with no net effect. Also, note that the above relations hold for any Maxwell–Boltzmann distribution. In other words, f^{MB} may be chosen to vary as a function of space and time so as to maximize computational efficiency by minimizing $f - f^{\mathrm{MB}}$. Applying this approach to a flow where the condition $f^{d} \ll f^{\mathrm{MB}}$ is not satisfied (e.g., shock wave) or choosing the "wrong" Maxwell–Boltzmann distribution should not affect the accuracy of the solution, only degrade its efficiency.

One particularly desirable feature of this approach is that the degree of variance reduction is larger for distributions that are closer to the equilibrium distribution (i.e., when f^{d} is smaller in magnitude), leading to a method that can practically capture very small deviations from equilibrium. This is in contrast to current methods (such as DSMC) whereas $f^{d} \to 0, f^{\mathrm{MB}}$ dominates the integrand landscape and thus leads to a statistical noise which is independent of f^{d} and a signal-to-noise ratio that decreases linearly with decreasing Mach number [43].

Figure 4.45 shows the results obtained by variance reduction approach for pressure driven flow in a two-dimensional channel and comparion with those obtained by linearized Boltzmann equation. The agreement is very good. It should be noted that the pressure gradient values are such that the nondimensional flow speed is of the order of 10^{-5}. Essentially noise-free calculations such as shown in Fig. 4.45 at flow speeds of this order would have been impossible to perform using DSMC with current current computational resources.

Following the variance reduction above mentioned concept, Hadjiconstantinou and co-workers [46–50], presented an efficient particle method for solving the Boltzmann equation named as low-variance deviational simulation Monte Carlo (LVDSMC) method. To do so, a new collision integral formulation was introduced which allows the method to retain the algorithmic structure of direct simulation Monte Carlo (DSMC). The variance reduction, achieved by simulating only the deviation

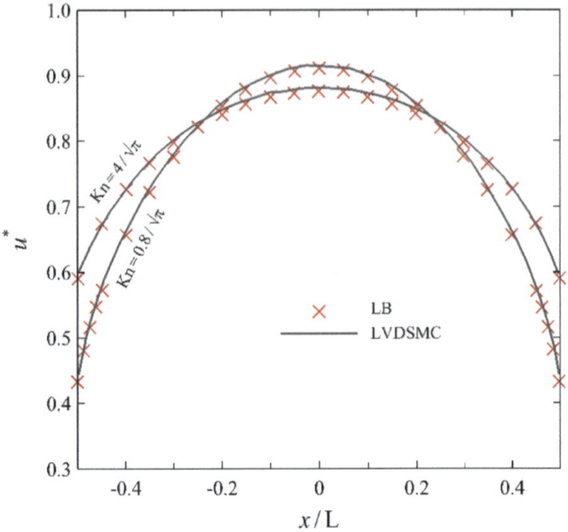

Fig. 4.45 Pressure-driven flow at $\mathrm{Kn} = \frac{0.8}{\sqrt{\pi}}$ and $\mathrm{Kn} = \frac{4}{\sqrt{\pi}}$. Comparison between variance reeduction approach and the numerical solutions of the linearized Boltzmann equation scirclesd. u^* is the nondimensional flow velocity normalized by the nondimensional logarithmic pressure gradient $\frac{\left(\frac{-L}{P}\right)}{\left(\frac{dP}{dx}\right)}$. Reproduced from Ref. [43]

from equilibrium, results in a significant computational efficiency advantage for low-speed flows compared to traditional particle methods such as DSMC. In particular, the resulting method can capture arbitrarily small deviations from equilibrium at a computational cost that is independent of the magnitude of this deviation. Consider an arbitrary Maxwell–Boltzmann distribution,

$$f^{\mathrm{MB}}(\boldsymbol{c}) = \frac{n_{\mathrm{MB}}}{\pi^{\frac{3}{2}} c_{\mathrm{MB}}^3} \exp\left[-\frac{(\boldsymbol{c} - \boldsymbol{u}_{\mathrm{MB}})^2}{c_{\mathrm{MB}}^2}\right] \tag{4.51}$$

In the work that follows this distribution will be identified with the local equilibrium distribution and thus $c_{\mathrm{MB}} = \sqrt{\frac{2kT_{\mathrm{MB}}}{m}}$ is the most probable speed based on the local equilibrium temperature T_{MB}, and $\mathbf{u}_{\mathrm{MB}}$ is the local equilibrium flow velocity. Upon substitution into Eq. (4.47), we obtain

$$\left[\frac{df}{dt}\right]_{\mathrm{coll}}(\boldsymbol{r}, \boldsymbol{c}, t) = \left[\frac{df^d}{dt}\right]_{\mathrm{coll}}(\boldsymbol{r}, \boldsymbol{c}, t) \tag{4.52}$$

$$\left[\frac{df^d}{dt}\right]_{\mathrm{coll}} = \frac{1}{2}\int\int\int \left(\delta_1' + \delta' - \delta_1 - \delta\right)\left(f^{\mathrm{MB}} f_1^d + f_1^{\mathrm{MB}} f^d + f^d f_1^d\right) \times g\sigma d^2\Omega d^3\boldsymbol{v} d^3\boldsymbol{v}_1 \tag{4.53}$$

$$\left[\frac{df^d}{dt}\right]_{\mathrm{coll}}(\boldsymbol{r}, \boldsymbol{c}, t) = \mathfrak{L}^{\mathrm{MB}}\left(f^d\right) + \mathfrak{J}\left(f^d, f^d\right) \tag{4.54}$$

since the integral involving $f^{\mathrm{MB}} f_1^{\mathrm{MB}}$ is identically zero. Equation (4.54) shows that the collision integral can be written as the sum of a linear $\{\mathfrak{L}^{\mathrm{MB}}(f^d)\}$ and a quadratic $\{\mathfrak{J}(f^d, f^d)\}$ term. Separating terms and using the fact that the linear terms are symmetric with respect to interchanging $\mathbf{v}$ and $\mathbf{v}_1$ gives

$$\left[\frac{df}{dt}\right]_{\mathrm{coll}}(\boldsymbol{r}, \boldsymbol{c}, t) = \frac{1}{2} \iiint (\delta_1' + \delta' - \delta_1 - \delta) f^d \left(2 f_1^{\mathrm{MB}} + f_1^d\right) g \sigma d^2 \Omega d^3 \boldsymbol{v} d^3 \boldsymbol{v}_1 \tag{4.55}$$

Hence, it is demanded to develop a particle formulation in which particles simulate the deviation from equilibrium f^{MB}. Particles may be positive or negative, depending on the sign of the deviation from equilibrium at the location in phase space where the particle resides [46]. According to Eq. (4.55), two types of collisions must be considered: those involving a deviational particle and the underlying Maxwell–Boltzmann distribution $f^{\mathrm{MB}} f_1^{\mathrm{MB}}$ and representing the linear term $\{\mathfrak{L}^{\mathrm{MB}}(f^d)\}$, and those between two deviational particles $f^d f_1^d$, representing the nonlinear term $\{\mathfrak{J}(f^d, f^d)\}$. Considering the first type of collision briefly, from (4.55) it could bee seen that the first particle drawn from f^d collides with second particle drawn from f^{MB} and leads to the creation of four particles: $\mathrm{sgn}(f^d)\delta_1'$, $\mathrm{sgn}(f^d)\delta_2'$, $-sgn(f^d)\delta_1$, $-sgn(f^d)\delta_2$. The existing particle $2[sgn(f^d)\delta_2]$ is cancelled by the new $-sgn(f^d)\delta_2$ particle, thus leading to a net creation of three particles. The $f^d f_1^d$ term can be interpreted and implemented analogously.

LVDSMC is a time-explicit particle method for solving the Boltzmann equation which retains the basic algorithmic structure of DSMC. As in DSMC, space is discretized in a number of cells in which the collision process is performed in a space-homogeneous fashion, while time integration is achieved by performing the desired number of time steps of duration Δt. Each time step consists of the following substeps [46]:

- Advection: Integrate $\frac{\partial f}{\partial t} + \boldsymbol{c} . \frac{\partial f}{\partial \boldsymbol{r}}$ for a time increment Δt. Apply boundary conditions to particles interacting with system boundaries.
- Collision: Integrate $\frac{\partial f}{\partial t} = \left[\frac{df}{dt}\right]_{\mathrm{coll}}$ for a time increment Δt.

Figure 4.46 show a comparison with DSMC results for the normalized flow velocity and shear stress at various time instants, including steady state. The same spatial and time discretization, as well as a comparable number of simulation particles, was used in the two methods. According to Fig. 4.46, the agreement between the results is excellent.

4.12.4 Collision Limiters and Near-Continuum Simulations

The DSMC method becomes numerically very expensive near or in the continuum regime when Knudsen numbers are low, due to the large number of collisions that

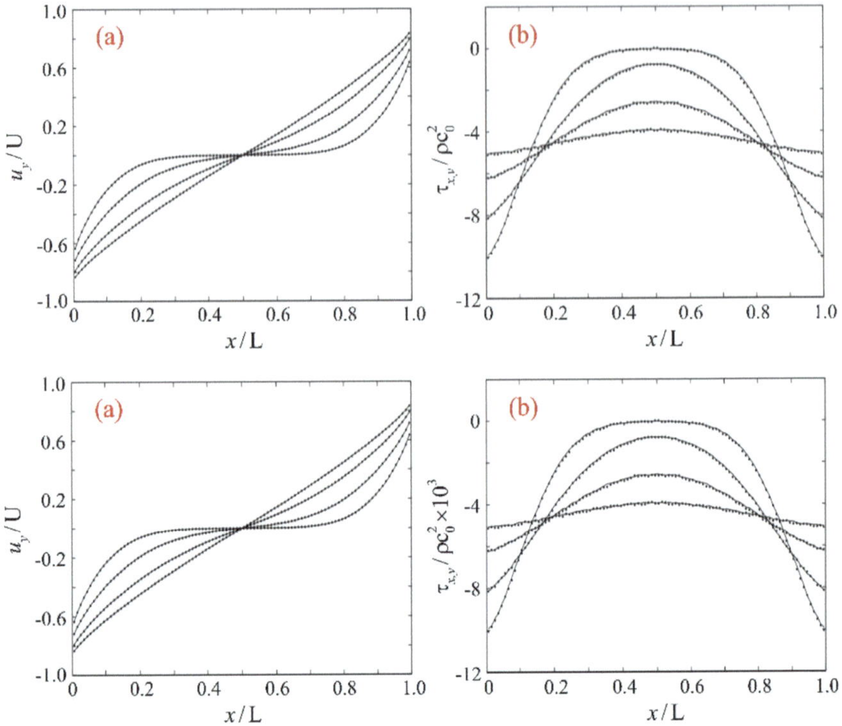

Fig. 4.46 Velocity profiles (**a**) and shear stress profiles (**b**) in impulsively started shear flow (Kn = 0.1, $U = 0.05c_0$) at various times. The continuous line represents the LVDSMC method, while the dots represent standard DSMC results. Reproduced from Ref. [46]

must be calculated. This high computational cost is mainly due to the collision procedure in the DSMC method. The collisional computing time can take up to 60% of the overall computing time, depending on the method used [51]. In order to reduce these costs, a common approach is to limit the number of possible collisions N_c in a cell to a constant value per time step as follows:

$$N_c = \eta N, \tag{4.56}$$

where, η is referred to as a limiter value and N is the number of particles per cell. The approach is valid in the continuum regime, where velocity relaxation to the equilibrium distribution is achieved after a certain number of collisions, and subsequent collisions do not alter the distribution further [52]. Based on the considered test case, different values have been reported in the range of $0.75 \leq \eta \leq 5$ [53, 54, 51, 55].

Another limiting approach proposed by Zhang et al. [56] is based on limiting the number of selections. The selection limiting procedure intervenes in the calculation of the selection number for the applied NTC scheme by multiplying the number of selected pairs (N_{sel}) by a function S dependent on the $\mathrm{Kn_{GL}}$. Here, the selection limiter

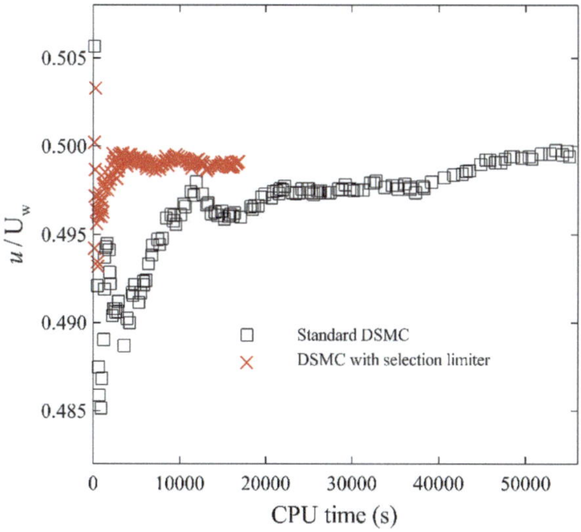

Fig. 4.47 Convergence time comparison between standard DSMC scheme and NTC scheme by selection limiter in the case of Couette flow. Reproduced from Ref. [56]

function S becomes unity in areas of large non-equilibrium, leaving N_{sel} unaltered but decreasing N_{sel} as the flow approaches the equilibrium state as a function of $\mathrm{Kn}_{\mathrm{GL}}$. Results for different test cases compared to the standard DSMC were in good agreement, whereas the evaluated collision limiter showed some deviations. Employing the limiter on the number of selections considerably reduces the required computational costs as illustrated in Fig. 4.47.

Following the abovementioned selection limiter method, Frieler and Groll [52] employed a constant selection limiter (CSL). In this approach, the number of selected pairs is set to a constant value per cell and time step, which is equal to twice the number of present test particles in the cell. If the number of selected pairs is lower than the limiter value, the collision calculation is unaltered.

References

1. Bird, G.A.: The perception of numerical methods in rarefied gas dynamics, In: Muntz, E.P., Weaver, D., Campbell, D. (eds.) Proceedings of 16th Int. Symposium on Rarefied Gas Dynamics, p. 211. AIAA, Washington, DC (1989)
2. Bird, G.A.: Shock wave structure in a rigid sphere gas. In: de Leeuw, J.H. (ed.) Proceedings of 4th International Symposium on Rarefied Gas Dynamic, p. 216. Academic, New York (1965)
3. Koura, K.: Transient Couette flow of rarefied binary gas mixtures. Phys. Fluids **13**(6), 1457–1466 (1970)
4. Ivanov, M.S., Rogasinskii, S.V.: Theoretical analysis of traditional and modern schemes of the DSMC method. In: Proceedings of 17th Symposiums on Rarefied Gas Dynamics, pp. 629–642. VCH, Weinheim (1990)
5. Yanitskiy, V.: Operator approach to direct simulation Monte Carlo theory in rarefied gas dynamics. In: Beylich, A. (ed.), Proceedings of 17th Symposiums on Rarefied Gas Dynamics, pp. 770–777. VCH, Weinheim (1990)

6. Koura, K.: Null-collision technique in the direct-simulation Monte Carlo method. Phys. Fluids **29**(11), 3509–3511 (1986)
7. LeBeau, G.J., Boyles, K.A., Lumpkin, F.E.: Virtual Sub-cells for the Direct Simulation Monte Carlo Method, pp. 2003–1031. AIAA, American Institute of Aeronautics and Astronautics (2003)
8. Bird, G.A.: The DSMC Method, CreateSpace Independent Publishing Platform (2013)
9. Bird, G.A., Gallis, M.A., Torczynski, J.R., Rader, D.J.: Accuracy and efficiency of the sophisticated direct simulation Monte Carlo algorithm for simulating noncontinuum gas flows. Phys. Fluids **21**(1) (2009)
10. Gallis, M.A., Torczynski, J.R.: Effect of collision-partner selection schemes on the accuracy and efficiency of the direct simulation Monte Carlo method. Int. J. Numer. Meth. Fluids **67**(8), 1057–1072 (2011)
11. Bird, G.A.: Sophisticated DSMC: Notes Prepared for a Short Course at the DSMC07 Meeting. Santa Fe, USA (2007)
12. Macrossan, M.N.: Searching for a near neighbor particle in DSMC cells using pseudo-subcells. J. Comput. Phys. **229**(17), 5857–5861 (2010)
13. Roohi, E., Stefanov, S., Shoja-Sani, A., Ejraei, H.: A generalized form of the Bernoulli Trial collision scheme in DSMC: derivation and evaluation. J. Comput. Phys. **354**, 476–492 (2018)
14. Stefanov, S.K.: On DSMC calculations of rarefied gas flows with small number of particles in cells. SIAM J. Sci. Comput. **33**(2), 677–702 (2011)
15. Stefanov, S.K. II.: Particle Monte Carlo algorithms with small number of particles in grid cells. In: Numerical Methods and Applications, pp. 110–117. Springer, Berlin (2011)
16. Amiri-Jaghargh, A., Roohi, E., Stefanov, S., Nami, H., Niazmand, H.: DSMC simulation of micro/nano flows using SBT–TAS technique. Comput. Fluids **102**, 266–276 (2014)
17. Taheri, E., Roohi, E., Stefanov, S.: On the convergence of the simplified Bernoulli trial collision scheme in rarefied Fourier flow. Phys. Fluids **29**(6) (2017)
18. Amiri-Jaghargh, A., Roohi, E., Niazmand, H., Stefanov, S.: DSMC simulation of low knudsen micro/nanoflows using small number of particles per cell. J. Heat Transfer **135**(10), 101008 (2013)
19. Su, C.C., Tseng, K.C., Cave, H.M., Wu, J.S., Lian, Y.Y., Kuo, T.C., Jermy, M.C.: Implementation of a transient adaptive sub-cell module for the parallel-DSMC code using unstructured grids. Comput. Fluids **39**(7), 1136–1145 (2010)
20. Goshayeshi, B., Roohi, E., Stefanov, S.: DSMC simulation of hypersonic flows using an improved SBT-TAS technique. J. Comput. Phys. **303**, 28–44 (2015)
21. Goshayeshi, B., Roohi, E., Stefanov, S. II.: A novel simplified Bernoulli trials collision scheme in the direct simulation Monte Carlo with intelligence over particle distances. Phys. Fluids **27**(10) (2015)
22. Stefanov, S., Roohi, E. and Shoja-Sani, A.: A novel transient-adaptive subcell algorithm with a hybrid application of different collision techniques in direct simulation Monte Carlo (DSMC). Phys. Fluids **34**(9) (2022)
23. Stefanov, S.K.: On the basic concepts of the direct simulation Monte Carlo method. Phys. Fluids **31**(6) (2019)
24. Shoja-Sani, A., Roohi, E., Stefanov, S.: Evaluation of the generalized Bernoulli trial-transient adaptive subcell (GBT-TAS) collision scheme in treating rarefied gas flows. Comput. Fluids **213**, 104740 (2020)
25. Taheri, E., Roohi, E., Stefanov, S.: A symmetrized and simplified Bernoulli trial collision scheme in direct simulation Monte Carlo. Phys. Fluids **34**(1) (2022)
26. Javani, M., Roohi, E., Stefanov, S.: Symmetrized generalized and simplified Bernoulli-trials collision schemes in DSMC. Comput. Fluids **272**, 106188 (2024)
27. Bird, G.A.: The DS2V/3V program suite for DSMC calculations. In AIP conference proceedings **762**(1), 541–546. American Institute of Physics (2005, May)
28. Josyula, E., Burt, J.: Review of rarefied gas effects in hypersonic applications. DTIC Document, Air Force Research Lab Wright-Patterson AFB OH (2011)

29. Gorji, M.H., Torrilhon, M.: Entropic Fokker-Planck kinetic model. J. Comput. Phys. **430**, 110034 (2021)
30. Gorji, M.H., Torrilhon, M., Jenny, P.: Fokker-Planck model for computational studies of monatomic rarefied gas flows. J. Fluid Mech. **680**, 574–601 (2011)
31. Bogomolov, S.V.: On Fokker-Planck model for the Boltzmann collision integral at the moderate Knudsen numbers. Math. Models Comput. Simul. **1**(6), 739–744 (2009)
32. RezapourJaghargh, V., Mahdavi, A., Roohi, E.: Shear-driven micro/nano flows simulation using Fokker Planck approach: investigating accuracy and efficiency. Vacuum **172**, 109065 (2020)
33. Gorji, M.H., Jenny, P.: A Fokker–Planck based kinetic model for diatomic rarefied gas flows. Phys. Fluids **25**(6) (2013)
34. Gorji, M.H., Jenny, P.: An efficient particle Fokker-Planck algorithm for rarefied gas flows. J. Comput. Phys. **262**, 325–343 (2014)
35. Sadr, M., Gorji, M.H.: A continuous stochastic model for non-equilibrium dense gases. Phys. Fluids **29**(12) (2017)
36. Gorji, M.H., Jenny, P.: Fokker–Planck–DSMC algorithm for simulations of rarefied gas flows. J. Comput. Phys. **287**, 110–129 (2015)
37. Jun, E.: Cubic Fokker-Planck-DSMC hybrid method for diatomic rarefied gas flow through a slit and an orifice. Vacuum **159**, 125–133 (2019)
38. Jun, E., Gorji, M.H., Grabe, M., Hannemann, K.: Assessment of the cubic Fokker–Planck–DSMC hybrid method for hypersonic rarefied flows past a cylinder. Comput. Fluids **168**, 1–13 (2018)
39. Küchlin, S., Jenny, P.: Parallel Fokker–Planck-DSMC algorithm for rarefied gas flow simulation in complex domains at all Knudsen numbers. J. Comput. Phys. **328**, 258–277 (2017)
40. Mahdavi, A., Roohi, E.: A novel hybrid DSMC-Fokker Planck algorithm implemented to rarefied gas flows. Vacuum **181**, 109736 (2020)
41. Burt, J.M., Boyd, I.D.: A low diffusion particle method for simulating compressible inviscid flows. J. Comput. Phys. **227**(9), 4653–4670 (2008)
42. Burt, J.M., Boyd, I.D.: A hybrid particle approach for continuum and rarefied flow simulation. J. Comput. Phys. **228**(2), 460–475 (2009)
43. Baker, L.L., Hadjiconstantinou, N.G.: Variance reduction for Monte Carlo solutions of the Boltzmann equation. Phys. Fluids **17**(5) (2005)
44. Hadjiconstantinou, N.G.: The limits of Navier-Stokes theory and kinetic extensions for describing small-scale gaseous hydrodynamics. Phys. Fluids **18**(11) (2006)
45. Hadjiconstantinou, N.G., Radtke, G.A., Baker, L.L.: On variance-reduced simulations of the boltzmann transport equation for small-scale heat transfer applications. J. Heat Transfer **132**(11) (2010)
46. Homolle, T.M., Hadjiconstantinou, N.G.: A low-variance deviational simulation Monte Carlo for the Boltzmann equation. J. Comput. Phys. **226**(2), 2341–2358 (2007)
47. Péraud, J.P.M., Hadjiconstantinou, N.G.: Efficient simulation of multidimensional phonon transport using energy-based variance-reduced Monte Carlo formulations. Phys. Rev. B—Condensed Matter Mater. Phys. **84**(20), 205331 (2011)
48. Péraud, J.P.M., Hadjiconstantinou, N.G.: An alternative approach to efficient simulation of micro/nanoscale phonon transport. Appl. Phys. Lett. **101**(15) (2012)
49. Radtke, G.A., Hadjiconstantinou, N.G.: Variance-reduced particle simulation of the Boltzmann transport equation in the relaxation-time approximation. Phys. Rev. E—Stat. Nonlinear, Soft Matter Phys. **79**(5), 056711 (2009)
50. Radtke, G.A., Hadjiconstantinou, N.G., Wagner, W.: Low-noise Monte Carlo simulation of the variable hard sphere gas. Phys. Fluids **23**(3) (2011)
51. Titov, E.V., Levin, D.A.: Extension of the DSMC method to high pressure flows. Int. J. Comput. Fluid Dyn. **21**(9–10), 351–368 (2007)
52. Frieler, T., Groll, R.: Micro-nozzle flow and thrust prediction with high-density ratio using DSMC selection limiter. Front. Space Technol. **4**, 1114188 (2023)
53. Bartel, T., Sterk, T., Payne, J., Preppernau, B.: DSMC simulation of nozzle expansion flow fields. In: 6th Joint Thermophysics and Heat Transfer Conference, p. 2047 (1994)

54. Sharma, A., Long, L.N.: Numerical simulation of the blast impact problem using the Direct Simulation Monte Carlo (DSMC) method. J. Comput. Phys. **200**(1), 211–237 (2004)
55. Titov, E., Zeifman, M., Levin, D.: Application of the kinetic and continuum techniques to the multi-scale flows in MEMS devices. In: 43rd AIAA Aerospace Sciences Meeting and Exhibit, p. 1399 (2005)
56. Zhang, R., Yao, W., Li, J.: A selection limiter of DSMC for near continuum flows. Commun. Nonlinear Sci. Numer. Simul.Numer. Simul. **13**(10), 2203–2212 (2008)
57. Radtke, G.A., Péraud, J.P.M., Hadjiconstantinou, N.G.: On efficient simulations of multiscale kinetic transport. Philos. Trans. R. Soc. A Math. Phys. Eng. Sci. **371**(1982), 20120182 (2013)

Chapter 5
Rarefied Pressure-Driven Flow

Abstract This chapter explores pressure-driven rarefied flows, focusing on Poiseuille flow in micro- and nano-channels. It presents analytical solutions over a broad spectrum of Knudsen numbers, assesses various velocity slip boundary conditions, and delves into heat transfer, wall interactions, and supporting experimental data. The chapter also covers step flow analysis and modeling of porous microchannels.

5.1 Pressure Driven Flow

5.1.1 Introduction and Governing Equations

This chapter aims to examine pressure-driven flows in micro- and nano-channels. The analytical solution of rarefied flow in micro- and nano-channels in a wide range of Knudsen numbers and the comparison of the accuracy of various slip boundary conditions for velocity profiles, pressure distribution, and flow rate are discussed [1]. Figure 5.1 illustrates a schematic of the microchannel and its boundary conditions.

The continuity equation is written as Eq. (5.1) in a microchannel with a high length-to-width ratio.

$$\frac{D\rho}{Dt}+\rho\nabla.V=0\rightarrow\frac{\partial\rho}{\partial t}+u\frac{\partial\rho}{\partial x}+v\frac{\partial\rho}{\partial y}+\rho\left(\frac{\partial u}{\partial x}+\frac{\partial v}{\partial y}\right)=0 \tag{5.1}$$

Considering steady and one-dimensional flow, Eqs. (5.2) and (5.3) are obtained:

$$u\frac{\partial\rho}{\partial x}+\rho\frac{\partial u}{\partial x}=0\rightarrow\frac{\partial}{\partial x}(\rho u)=0\rightarrow\rho u=cte \tag{5.2}$$

$$\rho_{in}u_{in}=\rho_{out}\ u_{out} \tag{5.3}$$

E. Roohi et al., *Advances in Direct Simulation Monte Carlo: From Micro-Scale to Rarefied Flow Phenomena*, https://doi.org/10.1007/978-981-96-8200-3_5

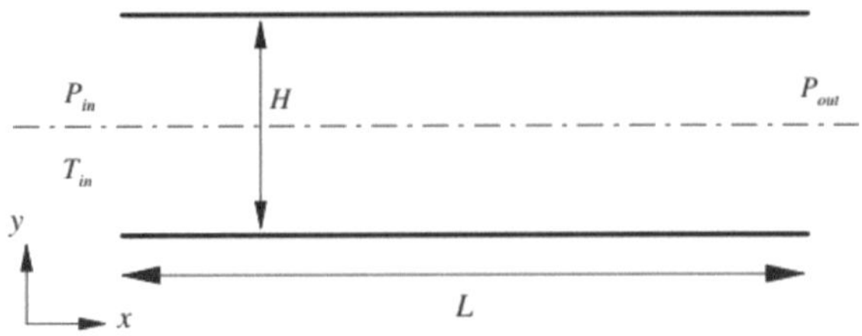

Fig. 5.1 Microchannel under pressure difference boundary condition

The momentum equation in the axial direction is written as follows:

$$\rho\left(\frac{\partial u}{\partial t}+u\frac{\partial u}{\partial x}+v\frac{\partial u}{\partial y}\right)$$
$$=\rho f_x-\frac{\partial P}{\partial x}+\frac{\partial}{\partial x}\left\{\mu\left[2\frac{\partial u}{\partial x}-\frac{2}{3}\left(\frac{\partial u}{\partial x}+\frac{\partial v}{\partial y}\right)\right]\right\}+\frac{\partial}{\partial y}\left\{\mu\left[\frac{\partial u}{\partial y}-\frac{\partial v}{\partial x})\right]\right\} \tag{5.4}$$

Given the absence of body forces, steady-state and one-dimensional flow (the vertical velocity component and velocity variations in the x-direction on the right side are negligible compared to variations in the vertical direction), Eq. (5.5) is obtained:

$$\rho\left(u\frac{\partial u}{\partial x}\right)=-\frac{\partial P}{\partial x}+\frac{\partial}{\partial y}\left\{\mu\frac{\partial u}{\partial y}\right\} \tag{5.5}$$

After double integrating over $dxdy$ on both sides of relation (5.5), Eq. (5.6) is obtained:

$$\int_0^h\int_0^L\rho\left(u\frac{\partial u}{\partial x}\right)dxdy=\int_0^h\int_0^L-\frac{\partial P}{\partial x}dxdy+\int_0^h\int_0^L\frac{\partial}{\partial y}\left\{\mu\frac{\partial u}{\partial y}\right\}dxdy \tag{5.6}$$

$$\rho\int_0^h(u(u_o-u_i))dy=\int_0^h-(P_o-P_i)dy+\int_0^L\left(\mu\left(\frac{\partial u}{\partial y}\right)_{(@y=h)}-\mu\left(\frac{\partial u}{\partial y}\right)_{(@y=0)}\right)dx \tag{5.7}$$

The integral over x is calculated for the first two terms of the above relation, and the integral over y is computed for the third term. On the other hand, the velocity gradient on the upper and lower walls is related as follows:

$$\left(\frac{\partial u}{\partial y}\right)@(y=h)=-\left(\frac{\partial u}{\partial y}\right)@(y=0)$$

Finally, the following equation is obtained:

$$(u_o - u_i)\int_0^h (\rho u)dy = (P_o - P_i)h - 2\left(\mu\left(\frac{\partial u}{\partial y}\right)@y = 0\right)L \tag{5.8}$$

Consequently, the momentum equation is obtained as (5.9)

$$\dot{m}(u_o - u_i) = (P_i - P_0)h - 2\tau L \tag{5.9}$$

Equation (5.10) is obtained by simplifying the above equation.

$$\frac{P_i - P_o}{P_o} - \frac{2\tau L}{P_o h} = \frac{\dot{m}(u_o - u_i)}{P_o h} \tag{5.10}$$

Assuming an ideal gas and also that the flow is isothermal, the right-hand side of the above equation simplifies to:

$$\begin{aligned}
&\dot{m} = \rho_o u_o h \Rightarrow \frac{\rho_o u_o h(u_o - u_i)}{P_o h} \\
&\frac{\rho_o u_o h(u_o - u_i)}{P_o h} \xrightarrow{continuity} \left(u_o = \frac{\rho_i u_i}{\rho_o}\right) \\
&\dashrightarrow \frac{\rho_o u_o u_i}{P_o}\left(\frac{\rho_i}{\rho_o}\right) - \frac{\rho_o u_o u_i}{P_o} \\
&\xRightarrow{ideal\ isothermal\ gas\ assumption} \left(\frac{\rho_i}{\rho_o} = \frac{p_i}{p_o}\right) \\
&\dashrightarrow \frac{\rho_o u_o u_i}{P_o}\left(\frac{p_i}{P_o}\right) - \frac{\rho_o u_o u_i}{P_o} \dashrightarrow\dashrightarrow \\
&\dashrightarrow \frac{\Delta P}{P_o} = \frac{2\tau L}{P_o h} + \frac{\rho_o u_o u_i}{P_o}\left(\frac{P_i}{P_o} - 1\right)
\end{aligned} \tag{5.11}$$

After some simplification, the following relation is obtained in terms of the inlet and outlet Mach numbers:

$$\begin{aligned}
&\frac{\Delta P}{P_o} = \frac{2\tau L}{P_o h} + \frac{\rho_o u_o u_i}{P_o}\left(\frac{P_i}{P_o} - 1\right) \to\to \frac{\Delta P}{P_o}\left(1 - \frac{\rho_o u_o u_i}{P_o}\right) = \frac{2\tau L}{P_o h} \to\to \\
&\frac{\Delta P}{P_o}\left(1 - \frac{\gamma \rho_o u_o u_i}{\gamma \rho_o R T_o}\right) = \frac{2\tau L}{P_o h} \to isothermalassumption \to T_i = T_o \\
&\to \frac{\Delta P}{P_o}\left(1 - \frac{\gamma u_o u_i}{\sqrt{\gamma R T_o} \times \sqrt{\gamma R T_i}}\right) = \frac{2\tau L}{P_o h} \to \frac{\Delta P}{P_o}(1 - \gamma M_o M_i) = \frac{2\tau L}{P_o h}
\end{aligned} \tag{5.12}$$

If the channel's inlet and outlet velocity (Mach number) are low, the convection term in Eq. (5.12) can be neglected.

Physically, the above equation means that the pressure difference is proportional to the shear stress. As the channel length increases, the shear stress within the channel rises, causing the pressure gradient to be used in overcoming this stress. Therefore, the flow passes through the channel at a lower speed. On the other hand, the ratio of the length to the height of the channel is large, so the right-hand side of the final relation (5.12) becomes large. Consequently, the pressure difference is used to overcome the right side, preventing an increase in the Mach number of the flow. Since the Mach number in microchannels is small, the effects of the inertia term are minimal. Hence, the equation that equates the frictional drag to the pressure force ($\frac{\Delta P}{P_o} = \frac{2\tau L}{P_o h}$), provides a reasonable solution. Note that these equations are not valid for micronozzles, as there is an area gradient, and the velocity and Mach number are high in them.

By simplifying Eqs. (5.12), (5.13) is obtained:

$$\begin{aligned}
&\mu \propto \rho c \lambda \\
&\acute{c} = \sqrt{\frac{8}{\pi} RT} \rightarrow \acute{c} \propto \sqrt{T} \\
&\mu \propto \frac{p}{T} \acute{c}\lambda \propto \frac{p}{(\acute{c})^2} \acute{c}\lambda \rightarrow \frac{p}{\acute{c}}\lambda \rightarrow \frac{\tau}{p_0} = \frac{p_0}{p_0 \acute{c}} \lambda \frac{u}{h} \\
&2\frac{L}{h}\frac{\tau}{p_0} = 2\frac{L}{h}\frac{\lambda}{\acute{c}}\frac{u}{h} \rightarrow 2\frac{L}{h} Kn_0 M_0 \\
&\frac{\Delta P}{P_o}(1 - \gamma M_o M_i) = \frac{2L}{h} Kn_0 M_0
\end{aligned} \tag{5.13}$$

The above relation indicates compressibility since $\frac{2L}{h} Kn_0 M_0$ is non-zero, then, $\frac{\Delta P}{P_o}$ is non-zero. The extent of this compressibility depends on the magnitude value of $Kn_0 M_0$.

5.1.2 Analytical Solution of the Velocity Field

By neglecting the convection terms, the differential form of the momentum equation converts to Eq. (5.14):

$$\frac{dp}{dx} = \mu \frac{\partial^2 u}{\partial y^2} \tag{5.14}$$

The flow is assumed to be fully developed, meaning that the pressure is considered a function of x only. The velocity and flow rate can be calculated by considering the dimensionless slip velocity relation.

$$U_s - U_{wall} = \frac{2-\sigma}{\sigma}\left[\mathrm{Kn}\left(\frac{\partial \mathrm{U}}{\partial \mathrm{n}}\right)_s + \frac{\mathrm{Kn}^2}{2}\left(\frac{\partial^2 \mathrm{U}}{\partial \mathrm{n}^2}\right)_s\right] \tag{5.15}$$

where n in the above equation is the direction perpendicular to the surface

$$\frac{dp}{dx} = \mu \frac{\partial^2 u}{\partial y^2} U = \mu \frac{u}{(-\frac{dp}{dx})h^2}, n = y/h \rightarrow \frac{\partial^2 U}{\partial n^2} = -1$$

The solution to the dimensionless equation is as follows:

$$\frac{\partial U}{\partial n} = -n + B_1 \rightarrow U = -\frac{1}{2}n^2 + B_1 n + B_2$$

$$n = 0 \rightarrow B_2 = \frac{2-\sigma}{\sigma}\left[\text{Kn}\left(\frac{\partial U}{\partial n}\right)_s + \frac{\text{Kn}^2}{2}\left(\frac{\partial^2 U}{\partial n^2}\right)_s\right]$$

$$n = 1 \rightarrow -\frac{1}{2} + B_1 + B_2 = \frac{2-\sigma}{\sigma}\left[\text{Kn}\left(\frac{\partial U}{\partial n}\right)_s + \frac{\text{Kn}^2}{2}\left(\frac{\partial^2 U}{\partial n^2}\right)_s\right] \rightarrow B_1 = \frac{1}{2}$$

$$\frac{\partial U}{\partial n} = -n + B_1 \rightarrow \left(\frac{\partial U}{\partial n}\right)_s = 0 + \frac{1}{2} = \frac{1}{2}$$

Note that $\left(\frac{\partial U}{\partial n}\right)_s(0) = \left(\frac{\partial U}{\partial n}\right)_s(1)$, because n is perpendicular to the surface. From the governing equation, we have:

$$\frac{\partial^2 U}{\partial n^2} = -1$$

Therefore:

$$U = -\frac{1}{2}n^2 + \frac{n}{2} + \frac{2-\sigma}{\sigma}[\text{Kn}\left(\frac{1}{2}\right) + \frac{\text{Kn}^2}{2}(-1)]$$

The dimensional velocity profile is expressed as follows:

$$\mu\frac{u}{\left(-\frac{dp}{dx}\right)h^2} = -\frac{1}{2}\left(\frac{y}{h}\right)^2 + \frac{\frac{y}{h}}{2} + \frac{2-\sigma}{\sigma}\left[Kn\left(\frac{1}{2}\right) + \frac{Kn^2}{2}(-1)\right]$$

$$u(y) = \frac{h^2}{2\mu}\frac{\partial P}{\partial x}\left(\frac{y^2}{h^2} - \frac{y}{h} + \frac{2-\sigma}{\sigma}(\text{Kn}^2 - \text{Kn})\right) \tag{5.16}$$

5.1.3 Analytical Solution of Flow Rate and Pressure Distribution

With the velocity profile obtained from Eq. (5.16), the mass flow rate can be calculated as follows:

$$\dot{m} = \int \rho u(y)dy = \rho\int_0^h \frac{h^2}{2\mu}\frac{\partial P}{\partial x}(\frac{y^2}{h^2} - \frac{y}{h} + \frac{2-\sigma}{\sigma}(\text{Kn}^2 - \text{Kn}))dy$$

$$\dot{m} = \rho \frac{h^2}{2\mu}\frac{dp}{dx}\left[\frac{h}{3} - \frac{h}{2} + \frac{2-\sigma}{\sigma}\left(\mathrm{Kn}^2 - \mathrm{Kn}\right)h\right], \rho = p/RT$$

$$\dot{m}dx = \frac{h^3}{2RT\mu}\left[-\frac{p}{6} + \frac{2-\sigma}{\sigma}(\mathrm{Kn}^2 p - \mathrm{Kn}p)\right]dp$$

Since, according to the definition of the molecular mean free path, the product of pressure and Knudsen number for isothermal flow is constant, the above equation is rewritten as follows:

$$\dot{m}dx = \frac{h^3}{2RT\mu}\left[-\frac{p}{6} + \frac{2-\sigma}{\sigma}((\mathrm{Kn_o}^2 p_o{}^2)/p - \mathrm{Kn_o}p_o)\right]dp$$

Now, the above integral is calculated along the length of the channel. Additionally, on the left side of the channel length appears, and on the right side, the integral is calculated from p_i to p_o.

$$\dot{m}l = \frac{h^3}{2RT\mu}\left[-\frac{p^2}{12} + \frac{2-\sigma}{\sigma}\left[(\mathrm{Kn_o}^2 p_o{}^2 \ln(p)) - \mathrm{Kn_o}p_o p\right]\right]$$

By substituting the integral limits into the above relation and defining $\Pi = p_i/p_o$, the above relation is rewritten as follows:

$$\dot{m} = \frac{h^3 p_0^2}{24\mu RTl}\left[\left(\Pi^2 - 1\right) + 12\frac{2-\sigma}{\sigma}\left(\mathrm{Kn_o}(\Pi - 1) - \mathrm{Kn_o^2}\ln(\Pi)\right)\right] \quad (5.17)$$

If the flow rate ratio in the slip condition to the flow rate without the slip boundary condition (ns) is calculated, the following relation is obtained:

$$\frac{\dot{m}}{\dot{m}_{ns}} = 1 + 12\frac{2-\sigma}{\sigma}\frac{\mathrm{Kn_o}}{\Pi + 1} - 12\frac{2-\sigma}{\sigma}\mathrm{Kn_o^2}\frac{\ln(\Pi)}{\Pi^2 - 1} \quad (5.18)$$

Now, we will derive the pressure distribution within the channel. Since $\frac{dP}{dx}$ is assumed to be constant in the momentum equation, this equation is not used. Instead, the pressure distribution can be obtained by equating the flow rates at two cross-sections.

$$\dot{m}dx = \frac{h^3}{2RT\mu}\left[-\frac{p}{6} + \frac{2-\sigma}{\sigma}\left(\frac{\mathrm{Kn_o^2}p_0^2}{P} - \mathrm{Kn_o}p_0\right)\right]dp$$

Moreover, the integration interval on the left side is from x to L, and on the right side, from p to p_0. Therefore, we have:

$$\dot{m}(L - x) = \frac{h^3}{2RT\mu}\left[-\frac{1}{6}\left(\frac{p^2}{2}\right) + \frac{(2-\sigma)\mathrm{Kn_o^2}p_0^2}{\sigma}\ln p - \frac{(2-\sigma)\mathrm{Kn_o}p_0}{\sigma}p\right]_p^{p_0}$$

$$\dot{m}(L-x) = \frac{h^3 p_0^2}{2RT\mu \times 12}\left(-\underbrace{\underbrace{\frac{p^2}{p_0^2}\Big|_p^{p_0}}_{\frac{1}{p_0^2}(p_0^2-p)}}_{1-\frac{p^2}{p_0^2}} + \frac{12(2-\sigma)\mathrm{Kn}_o^2}{\sigma}\ln\underbrace{\frac{p_0}{p}}_{-\ln\frac{p}{p_0}} - \frac{12(2-\sigma)\mathrm{Kn}_o}{\sigma}\underbrace{\frac{(p_0-p)}{p_0}}_{1-\frac{p}{p_0}} \right)$$

$$\xrightarrow{\frac{p}{p_0}=\tilde{P}} \frac{\dot{m}(L-x)24RT\mu}{h^3 p_0^2} = \left(\tilde{p}^2-1\right) + 12(\tilde{p}-1)\mathrm{Kn}_o\frac{2-\sigma}{\sigma} - 12\frac{2-\sigma}{\sigma}\mathrm{Kn}_o^2\ln\tilde{p}$$

The constant coefficients behind the x-dependent term on the right side are denoted by B, and we have:

$$B(L-x) = \tilde{p}^2 - 1 + 12\frac{2-\sigma}{\sigma}\mathrm{Kn}_o(\tilde{p}-1) - 12\frac{2-\sigma}{\sigma}\mathrm{Kn}_o^2 ln\tilde{p} \tag{5.19}$$

If Eq. (5.19) is solved, the pressure distribution in the microchannel is obtained. It should be noted that this equation must be solved implicitly. A sample Python code to solve this equation is provided in the appendix at the end of the chapter. This equation can be solved explicitly if the first-order boundary condition is used.

In Figs. 5.2 and 5.3, the dimensionless flow rate versus pressure ratio for different conditions and helium and air gases are plotted from experimental and numerical data with various boundary condition models.

At a constant pressure ratio, the dimensionless flow rate increases with the increase in the Knudsen number. This is because the average velocity profile increases with the Knudsen number, consequently leading to an increase in the flow rate.

According to Fig. 5.4, the pressure distribution and the pressure difference from the linear pressure are shown for two cases: with and without slip (Knudsen number equal to 0.3 and different pressure ratios). It can be seen that in the no-slip condition, the compressibility effect is dominant, and the pressure distribution becomes curved. In the slip flow case, the rarefaction effect partially neutralizes the compressibility effects, and the curvature of the pressure distribution curve is reduced.

It is important to note that the pressure difference from the linear pressure is very sensitive. A comparison of this parameter between the new model and the DSMC method is commonly used to evaluate new slip velocity models. Additionally, considering the constant product of pressure and Knudsen number for isothermal flow, the Knudsen number increases as the pressure drops along the flow path towards the end of the channel. This could even lead to a change in the flow regime.

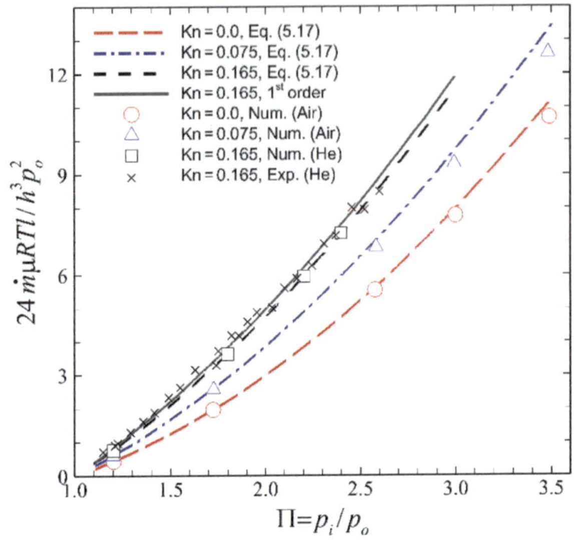

Fig. 5.2 Flow rate as a function of pressure ratio. Reproduced from Ref. [1]

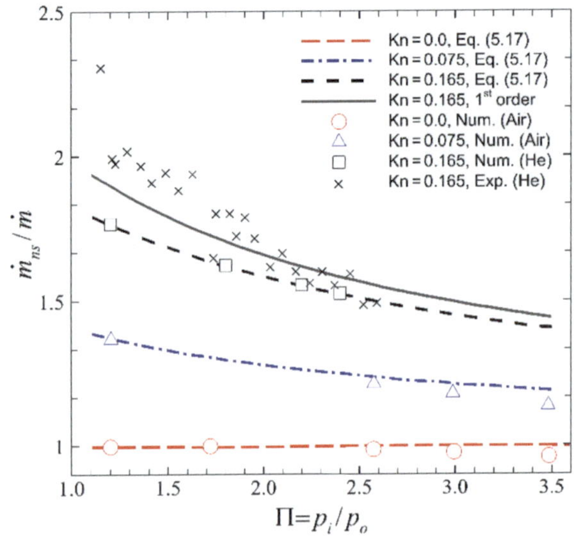

Fig. 5.3 Flow rate ratio versus pressure ratio. Reproduced from Ref. [1]

In Fig. 5.5, three channels with different heights and consequently different Knudsen numbers are examined, which are represented by ($h_1 > h_2 > h_3$), and ($Kn_3 > Kn_2 > Kn_1$), respectively. It is observed that in the channel with the smaller height, which has a higher Knudsen number, the slip velocity is also greater. Another point indicated in the figure is that as we get closer to the end of the channel, the slip velocity increases because along the channel, the pressure decreases, the Knudsen number increases, and therefore the slip increases.

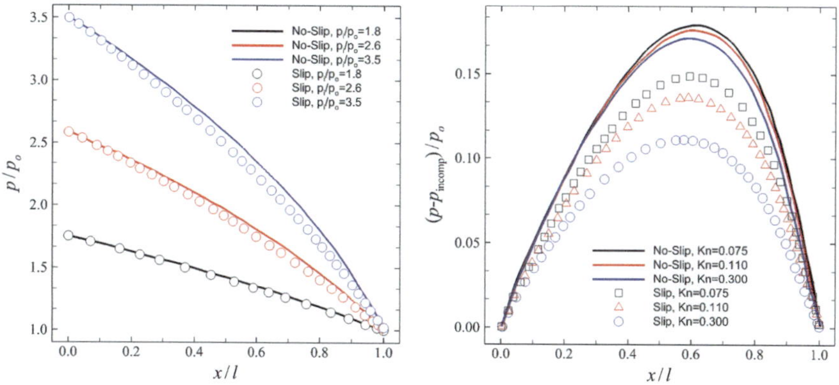

Fig. 5.4 Diagram of pressure ratio and the pressure deviation from the line pressure as a function of channel length. Reproduced from Ref. [1]

Fig. 5.5 Slip velocity as a function of channel length. Reproduced from Ref. [1]

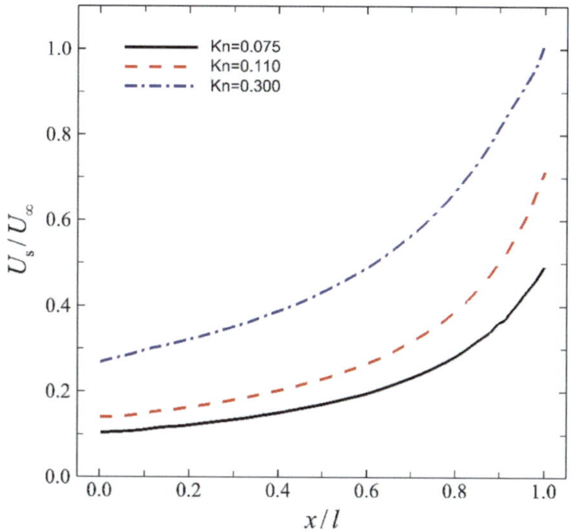

It should be noted that in the analytical solutions, the convective term of the momentum equation was neglected, while molecular solvers such as Direct Simulation Monte Carlo (DSMC) do not make this assumption. However, the solutions have little difference since the convective term is negligible in long channels. Also, the flow under study is not fully developed because the density (Knudsen number) changes at each cross-section. In other words, while the theory assumes a small velocity gradient, the velocity distribution changes from one section to another.

5.1.4 Discussion on the Second-Order Slip Velocity Boundary Condition

The following points are noteworthy regarding the slip boundary condition formula, velocity profile, and mass flow rate. With the second-order slip formula with both positive coefficients, i.e., the relation proposed in [1],

$$U_s - U_{wall} = \frac{2-\sigma}{\sigma}\left[\mathrm{Kn}\left(\frac{\partial \mathrm{U}}{\partial \mathrm{n}}\right)_s + \frac{\mathrm{Kn}^2}{2}\left(\frac{\partial^2 \mathrm{U}}{\partial \mathrm{n}^2}\right)_s\right] \tag{5.20}$$

The Knudsen number coefficient becomes negative in the velocity profile:

$$u(y) = \frac{h^2}{2\mu}\frac{\partial P}{\partial x}\left(\frac{y^2}{h^2} - \frac{y}{h} + \frac{2-\sigma}{\sigma}\left(\mathrm{Kn}^2 - \mathrm{Kn}\right)\right) \tag{5.21}$$

If we factor out a negative from the above relation, the coefficient of Kn^2 becomes negative. This same negative coefficient for Kn^2 appears in the mass flow rate formula as well:

$$\dot{m} = \frac{h^3 P_0^2}{24\mu RTl}\left[\left(\Pi^2 - 1\right) + 12\frac{2-\sigma}{\sigma}\left(\mathrm{Kn_o}(\Pi - 1) - \mathrm{Kn_o^2}\ln(\Pi)\right)\right] \tag{5.22}$$

The conclusion is that the velocity has reasonable accuracy because the effects of the second-order Knudsen term should be to reduce the over-predicted velocity by the first-order at high Knudsen numbers. In other words, the Knudsen number squared coefficient should be negative. However, the mass flow rate will be incorrect as in the dimensionless mass flow rate (S) derived from the above slip relation, and the Knudsen number squared coefficient also becomes negative.

$$\frac{\dot{m}}{\dot{m}_{ns}} = 1 + 12\frac{2-\sigma}{\sigma}\frac{\mathrm{Kn_o}}{\Pi + 1} - 12\frac{2-\sigma}{\sigma}\mathrm{Kn_o^2}\frac{\ln(\Pi)}{\Pi^2 - 1} \tag{5.23}$$

While processing the curve with experimental data shows that for accurate prediction of the flow rate, the Knudsen number squared coefficient must be a positive quantity ($S \approx 1 + 6c_1\mathrm{Kn} + 12c_2\mathrm{Kn}^2$), if the second-order Knudsen coefficient is considered negative in the slip boundary condition:

$$U_s - U_{wall} = \frac{2-\sigma}{\sigma}[C_1\mathrm{Kn}\left(\frac{\partial \mathrm{U}}{\partial \mathrm{n}}\right)_s - C_2\mathrm{Kn}^2\left(\frac{\partial^2 \mathrm{U}}{\partial \mathrm{n}^2}\right)_s] \tag{5.24}$$

The Knudsen number squared coefficient appears with a positive sign in the velocity profile and the dimensionless mass flow rate (S).

$$u(y) = -\frac{h^2}{2\mu}\frac{\partial P}{\partial x}\left(-\frac{y^2}{h^2} + \frac{y}{h} + \frac{2-\sigma}{\sigma}(C_1\mathrm{Kn} + 2C_2\mathrm{Kn}^2)\right) \tag{5.25}$$

$$S \approx 1 + 6c_1\mathrm{Kn} + 12c_2\mathrm{Kn}^2 \tag{5.26}$$

Therefore, the dimensionless flow rate in the transition regime will increase, which is consistent with experimental results.

5.2 The Unified Velocity Model

By presenting this unified model, Karniadakis and his colleagues: et al. [1] corrected the velocity distribution for all Knudsen numbers. Considering the boundary condition (5.27) for slip velocity, the momentum Eq. (5.28) is solved and the mass flow rate is calculated.

$$U_s - U_w = \frac{2-\sigma}{\sigma}\frac{\mathrm{Kn}}{1-b\mathrm{Kn}}\left(\frac{\partial U}{\partial n}\right) \tag{5.27}$$

$$\frac{dp}{dx} = \mu\frac{\partial^2 u}{\partial y^2} U = \mu\frac{u}{(-\frac{dp}{dx})h^2} \rightarrow \frac{\partial^2 U}{\partial n^2} = -1 \tag{5.28}$$

$$\frac{\partial U}{\partial n} = -n + B_1 \rightarrow U = -\frac{1}{2}n^2 + B_1 n + B_2 \tag{5.29}$$

$$n = 0 \rightarrow \text{Applying Boundary Conditions } B_2 = \frac{2-\sigma}{\sigma}\frac{\mathrm{Kn}}{1-b\mathrm{Kn}}\left(\frac{\partial U}{\partial n}(0)\right)$$

$$n = 1 \rightarrow \text{Applying Boundary Conditions } -\frac{1}{2} + B_1 + B_2$$

$$= \frac{2-\sigma}{\sigma}\frac{\mathrm{Kn}}{1-b\mathrm{Kn}}\left(\frac{\partial U}{\partial n}(1)\right), \frac{\partial U}{\partial n}(0) = \frac{\partial U}{\partial n}(1) \rightarrow B_1 = \frac{1}{2}$$

$$\frac{\partial U}{\partial n} = -n + B_1 \rightarrow \frac{\partial U}{\partial n}(0) = 0 + \frac{1}{2} = \frac{1}{2}$$

$$U = -\frac{1}{2}\left(\frac{y}{h}\right)^2 + \frac{1}{2}\frac{y}{h} + \frac{2-\sigma}{2\sigma}\frac{\mathrm{Kn}}{1-b\mathrm{Kn}} \tag{5.30}$$

$$\rightarrow u(y) = \frac{h^2}{2\mu}\frac{dp}{dx}\left[\left(\frac{y}{h}\right)^2 - \frac{y}{h} - \frac{2-\sigma}{\sigma}\frac{\mathrm{Kn}}{1-b\mathrm{Kn}}\right] \tag{5.31}$$

With the velocity distribution and by integrating it, the mass flow rate of the channel can be calculated as:

$$\dot{M} = \int \rho u(y)dy = \rho \int_0^h \frac{h^2}{2\mu}\frac{dp}{dx}\left[\left(\frac{y}{h}\right)^2 - \frac{y}{h} - \frac{2-\sigma}{\sigma}\frac{\text{Kn}}{1-b\text{Kn}}\right]dy \tag{5.32}$$

$$\dot{M} = \rho\frac{h^2}{2\mu}\frac{dp}{dx}\left[\frac{h}{3} - \frac{h}{2} - \frac{2-\sigma}{\sigma}\frac{\text{Kn}\,h}{1-b\text{Kn}}\right], \rho = p/RT \tag{5.33}$$

$$\dot{M}\,dx = \frac{h^3}{2RT\mu}\left[-\frac{p}{6} - \frac{2-\sigma}{\sigma}\frac{Kn_o p_o}{1-b\frac{Kn_o p_o}{p}}\right]dp \tag{5.34}$$

$$\int \frac{pdp}{p-b\text{Kn}_\text{o}p_o} = \int \frac{p-b\text{Kn}_\text{o}p_o + b\text{Kn}_\text{o}p_o}{p-b\text{Kn}_\text{o}p_o}dp = \int [1+\frac{b\text{Kn}_\text{o}p_o}{p-b\text{Kn}_\text{o}p_o}]dp$$

$$= p + b\text{Kn}_\text{o}p_o ln(p - b\text{Kn}_\text{o}p_o)$$

$$\dot{M}L = -\frac{h^3}{2RT\mu}\left(\frac{p^2}{12} + \frac{2-\sigma}{\sigma}Kn_o p_o(p + bKn_o p_o \ln(p - bKn_o p_o))\right) \tag{5.35}$$

$$\dot{M}L = -\frac{h^3}{2RT\mu}\left[\frac{p_0^2}{12} - \frac{p_i^2}{12} + \frac{2-\sigma}{\sigma}\text{Kn}_\text{o}p_o(p_o + b\text{Kn}_\text{o}p_o ln(p_o - b\text{Kn}_\text{o}p_o))\right.$$
$$\left. - \frac{2-\sigma}{\sigma}\text{Kn}_\text{o}p_o(p_i + b\text{Kn}_\text{o}p_o ln(p_i - b\text{Kn}_\text{o}p_o))\right]$$

$$\dot{M}L = -\frac{h^3}{2RT\mu}\left[\frac{p_0^2}{12} - \frac{p_i^2}{12} + \frac{2-\sigma}{\sigma}\text{Kn}_\text{o}p_o(p_o - p_i)\right.$$
$$\left. + \frac{2-\sigma}{\sigma}\text{Kn}_\text{o}p_o b\text{Kn}_\text{o}p_o ln\left(\frac{p_o - b\text{Kn}_\text{o}p_o}{p_i - b\text{Kn}_\text{o}p_o}\right)\right]$$

$$\dot{M}L = -\frac{h^3}{24RT\mu}\left[(p_o - p_i)(p_o + p_i) + 12\frac{2-\sigma}{\sigma}\text{Kn}_\text{o}p_o(p_o - p_i)\right.$$
$$\left. + 12\frac{p_o - p_i}{p_o - p_i}\frac{2-\sigma}{\sigma}\text{Kn}_\text{o}p_o b\text{Kn}_\text{o}p_o ln\left(\frac{p_o - b\text{Kn}_\text{o}p_o}{p_i - b\text{Kn}_\text{o}p_o}\right)\right]$$

$$\dot{M}L = +\frac{h^3\Delta p}{24RT\mu}\left[(p_o + p_i) + 12\frac{2-\sigma}{\sigma}\text{Kn}_\text{o}p_o\right.$$
$$\left. + 12\frac{2-\sigma}{\sigma}\text{Kn}_\text{o}p_o b\text{Kn}_\text{o}p_o\frac{1}{p_o - p_i}ln\left(\frac{p_o - b\text{Kn}_\text{o}p_o}{p_i - b\text{Kn}_\text{o}p_o}\right)\right]$$

$$\dot{M}L = +\frac{h^3\Delta p p_o}{24RT\mu}\left[(\Pi + 1) + 12\frac{2-\sigma}{\sigma}\text{Kn}_\text{o} + 12\frac{2-\sigma}{\sigma}\text{Kn}_\text{o}^2 bA\right] \tag{5.36}$$

$$A = \frac{p_o}{p_o - p_i}ln\left(\frac{p_o - b\text{Kn}_\text{o}p_o}{p_i - b\text{Kn}_\text{o}p_o}\right) = \frac{1}{\Pi - 1}ln\left(\frac{p_i - b\text{Kn}_\text{o}p_o}{p_0 - b\text{Kn}_\text{o}p_o}\right)$$

$$= \frac{1}{\Pi - 1} ln\left(\frac{\Pi - b\mathrm{Kn_o}}{1 - b\mathrm{Kn_o}}\right) \tag{5.37}$$

$$\dot{M} = \frac{h^3 \Delta p p_o}{24RT\mu L}\left[(\Pi + 1) + 12\frac{2-\sigma}{\sigma}\mathrm{Kn_o} + 12\frac{2-\sigma}{\sigma}\mathrm{Kn_o^2}\frac{b}{\Pi - 1} ln\left(\frac{\Pi - b\mathrm{Kn_o}}{1 - b\mathrm{Kn_o}}\right)\right] \tag{5.38}$$

As shown in Fig. 5.6, with the proper selection of coefficient *b*, the velocity profile predicted by the unified model provides good accuracy over a wide range of Knudsen numbers. The model's good accuracy at larger Knudsen numbers is related to the nature of the channel problem. At Knudsen numbers greater than 1, the entire space between the two channel walls falls within the Knudsen layer. Therefore, only collisions between molecules and the wall occur in the Knudsen layer. In other words, at higher Knudsen numbers, enough molecule-wall collisions occur that the flow behaves similarly to a continuum flow; only the nature of the collision has changed from intermolecular to wall-molecule collision. This feature allows a continuum model to provide reasonable results in the channel. Obviously, this model lacks such accuracy for flow over a single wall.

One of the issues with this model is that it performs very poorly in calculating the flow rate. Figure 5.7 shows that it performs worse than the Maxwell model ($b = 0$).

To correct this error, Karyadakis et al. [1] focused on improving the viscosity and proposed the following relation for the mass flow rate:

$$\frac{\dot{Q}}{W} = -\frac{h^3}{12\mu}\frac{dp}{dx}\left(1 + \frac{6\mathrm{Kn}}{1 - b\mathrm{Kn}}\right)C_r(\mathrm{Kn}) \tag{5.39}$$

On other hand, viscosity is defined as:

$$\mu = \frac{1}{2}\rho\lambda\bar{c} \tag{5.40}$$

Viscosity can be determined for two boundary values of low and high Knudsen numbers, and by bridging between these two solutions, a value for the entire range can be obtained. When the Knudsen number is small, the molecules are influenced by each other. Therefore, the characteristic length of the flow is equal to the molecular mean free path.

$$\mu_0 = \frac{1}{2}\rho\lambda\bar{c}$$

When the Knudsen number is large, the characteristic length of the flow is the channel height.

$$\mu = \frac{1}{2}\rho h\bar{c}$$

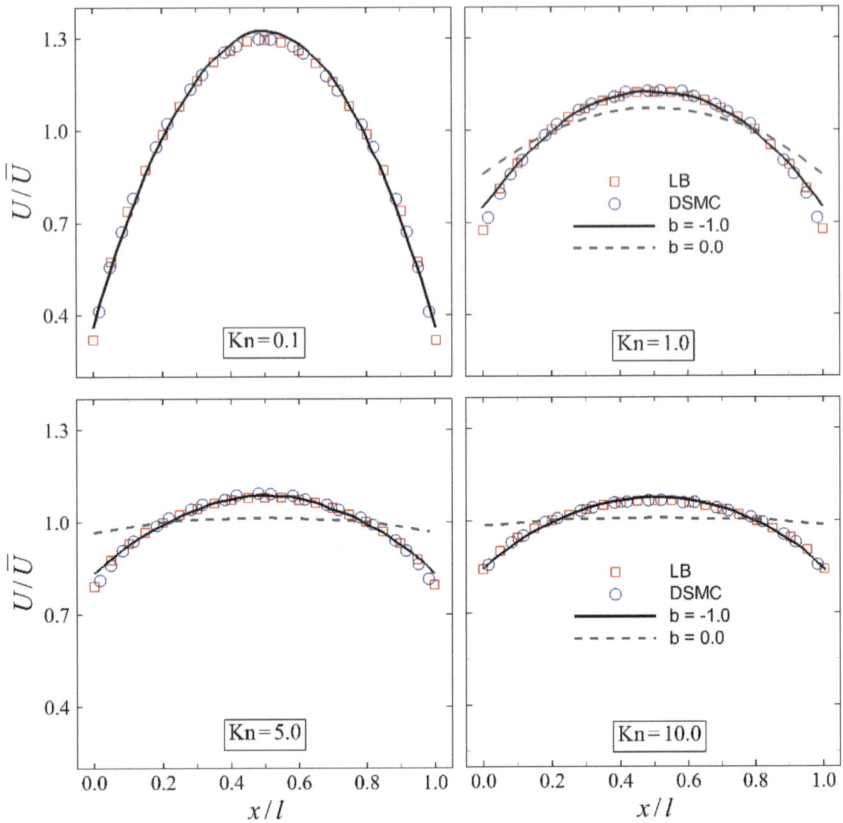

Fig. 5.6 Examination of the accuracy of the unified model in predicting the velocity profile compared to the linearized Boltzmann solution and DSMC at various Knudsen numbers. Reproduced from Ref. [1]

So, an equivalent characteristic length must be found that is accurate for the range between these two limits. Assuming a parallel circuit between these two limits, we have:

$$\frac{1}{\lambda_{eff}} = \frac{1}{\lambda} + \frac{1}{h} \tag{5.41}$$

$$\mu = \frac{\mu_0}{1 + \text{Kn}} \tag{5.42}$$

By using this equivalent characteristic length, the viscosity correction factor is written as Eq. (5.43):

$$C_r(Kn) = 1 + \alpha \text{Kn} \tag{5.43}$$

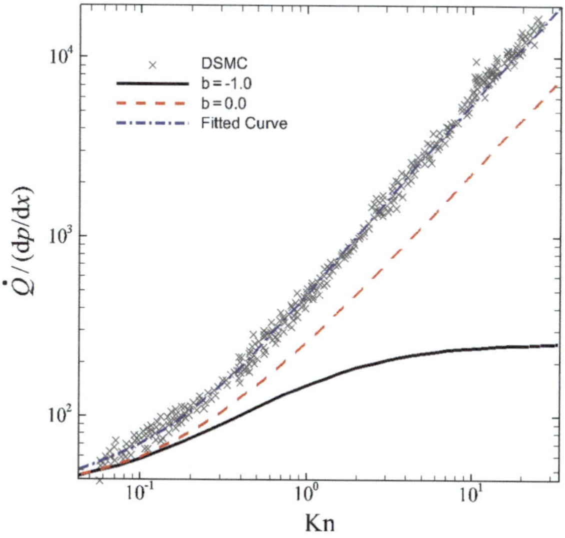

Fig. 5.7 Volumetric flow rate versus Knudsen number. Reproduced from Ref. [1]

The generalized velocity profile is obtained by considering the viscosity correction of Eq. (5.43).

$$u(y) = \frac{h^2}{2\mu}\frac{dP}{dx}\left[\left(\frac{y}{h}\right)^2 - \frac{y}{h} - \frac{2-\sigma}{\sigma}\frac{\text{Kn}}{1-b\text{Kn}}\right]$$

It is rewritten as follows:

$$u(y) = \frac{h^2}{2\mu_0}\frac{dP}{dx}\left[\left[\left(\frac{y}{h}\right)^2 - \frac{y}{h}\right](1+\alpha\text{Kn}) - \left(\left(\frac{2-\sigma}{\sigma}\right)\frac{\text{Kn}(1+\alpha\text{Kn})}{1-b\text{Kn}}\right)\right]$$

Based on the above equation, the mass flow rate is calculated as:

$$\dot{M} = \int \rho u(y)dy \rightarrow \rho\int_0^h \tfrac{h^2}{2\mu}\tfrac{dP}{dx}\left(\left(\left(\tfrac{y}{h}\right)^2 - \tfrac{y}{h}\right)(1+\alpha\text{Kn}) - \left(\tfrac{2-\sigma}{\sigma}\right)\tfrac{(\text{Kn})(1+\alpha\text{Kn})}{1-b\text{Kn}}\right)dy$$

$$\dot{M} = \frac{\rho h^2}{2\mu}\frac{dP}{dx}\left[\left(\frac{h}{3} - \frac{h}{2}\right)(1+\alpha\text{Kn}) - \left(\frac{2-\sigma}{\sigma}\right)\frac{\text{Kn}\,(1+\alpha\text{Kn})}{1-b\text{Kn}}h\right], \rho = P/RT$$

$$\dot{M}\,dx = \frac{h^3}{2RT\mu}\left[\left(-\frac{P}{6}\right)(1+\alpha\text{Kn}) - \underbrace{\left(\frac{2-\sigma}{\sigma}\right)\frac{\text{Kn}P(1+\alpha\text{Kn})}{1-b\text{Kn}}}_{A}\right]dP$$

$$\Rightarrow \int_0^L \dot{M}\,dx = \frac{h^3}{2RT\mu}\int_{P_i}^{P_o}\left[\left(-\frac{P}{6}\right)(1+\alpha\mathrm{Kn}) - \underbrace{\left(\frac{2-\sigma}{\sigma}\right)\frac{\mathrm{Kn}P(1+\alpha\mathrm{Kn})}{1-b\mathrm{Kn}}}_{A}\right]dP$$

$$A: \int\left(\frac{2-\sigma}{\sigma}\right)\left(\frac{(\mathrm{Kn}_0P_0)(1+\frac{\alpha\mathrm{Kn}_0P_0}{P})}{1-\frac{b\,\mathrm{Kn}_0P_0}{P}}\right)dP$$

$$= \int\left(\frac{2-\sigma}{\sigma}\right)(\mathrm{Kn}_0P_0)\left(\frac{1+\frac{\alpha\mathrm{Kn}_0P_0}{P}}{1-\frac{b\,\mathrm{Kn}_0P_0}{P}}\right)dP$$

$$\int\frac{1+\frac{\alpha Kn_0P_0}{P}}{1-\frac{bKn_0P_0}{P}} = -\int\frac{\frac{\alpha Kn_0P_0}{P}+1}{\frac{bKn_0P_0}{P}-1}$$

$$= -\int\frac{P+\alpha Kn_0P_0}{bKn_0P_0-P} = +\int\frac{P+\alpha Kn_0P_0}{P-bKn_0P_0} \overset{u=P-bKn_0P_0\to du=dP}{\rightarrow}$$

$$= \int\frac{u+bKn_0P_0+\alpha Kn_0P_0}{u} = \int\left(\frac{(bKn_0P_0+\alpha Kn_0P_0)}{u}+1\right)du$$

$$= (bKn_0P_0+aKn_0P_0)\int\frac{1}{u}du + \int 1du$$

$$= (bKn_0P_0+\alpha Kn_0P_0)lnu + u \overset{U=P-bKn_0P_0}{\rightarrow}$$

$$= (bKn_0P_0+\alpha Kn_0P_0)\ln(P-bKn_0P_0) + P - bKn_0P_0$$

$$\Rightarrow \int AdP = \left(\frac{2-\sigma}{\sigma}\right)(\mathrm{Kn}_0P_0)\{(b+\alpha)\mathrm{Kn}_0P_0(\ln(P-bKn_0P_0)) + P - bKn_0P_0\}$$

$$\dot{M}\,dx = \left[\int[-\frac{P}{6}-\left(\frac{P}{6}\alpha Kn\right)] - \int A\right]dP$$

$$\dot{M}L = \frac{h^3}{2RT\mu_o}\left[\int_{P_i}^{P_o}\{-\frac{P}{6}-\left(\frac{P}{6}\alpha Kn\right)\}dP\right]$$

$$-\left\{\left(\frac{2-\sigma}{\sigma}\right)Kn_oP_o[(b+\bar{\alpha})Kn_oP_o(\ln(P-bKn_oP_o)) + P - bKn_oP_o]\right\}_{P_i}^{P_o}$$

$$\rightarrow \dot{M} = \frac{h^3}{2RT\mu_oL}\left\{-\frac{P^2}{12}-\frac{1}{6}\alpha Kn_oP_oP - \left(\frac{2-\sigma}{\sigma}\right)Kn_oP_o\right.$$

$$\left.[(b+\bar{\alpha})Kn_oP_o(\ln(P-bKn_oP_o)) + P - bKn_oP_o]\right\}_{P_i}^{P_o}$$

$$\rightarrow \dot{M} = \frac{h^3}{2RT\mu_o L}\left\{-\frac{1}{12}\left(P_o^2 - P_i^2\right) - \frac{1}{6}\alpha Kn_o P_o (P_o - P_i) - \left(\frac{2-\sigma}{\sigma}\right)Kn_o P_o \right.$$
$$\left. [(b+\bar{\alpha})Kn_o P_o (\ln(P_o - bKn_o P_o) - \ln(P_i - bKn_o P_o)) + (P_o - P_i)]\right\}$$

$$\rightarrow \dot{M} = \frac{h^3}{2RT\mu_o L}\left\{-\frac{1}{12}(P_o - P_i)(P_o + P_i) - \frac{1}{6}\alpha Kn_o P_o (P_o - P_i) - \left(\frac{2-\sigma}{\sigma}\right)Kn_o P_o \right.$$
$$\left.\left[(b+\bar{\alpha})Kn_o P_o \ln\left[\frac{P_o(1 - bKn_o)}{P_o\left(\frac{P_i}{P_o} - bKn_o\right)}\right] + (P_o - P_i)\right]\right\}$$

$$\rightarrow \dot{M} = \frac{h^3(P_i - P_o)}{2RT\mu_o L}\left\{\frac{1}{12}((P_o + P_i) + 2\alpha Kn_o P_o) - \left(\frac{2-\sigma}{\sigma}\right)Kn_o P_o \right.$$
$$\left.\left[(b+\bar{\alpha})Kn_o \frac{P_o}{(P_i - P_o)}\ln\left(\frac{1 - bKn_o}{\frac{P_i}{P_o} - bKn_o}\right) - 1\right]\right\}$$

$$\rightarrow \dot{M} = \frac{h^3(P_i - P_o)P_o}{2RT\mu_o L}\left\{\frac{1}{12}(1 + \Pi + 2\alpha Kn_o) - \left(\frac{2-\sigma}{\sigma}\right)\right.$$
$$\left.\left[\frac{(b+\bar{\alpha})}{(\Pi - 1)}Kn_o^2 \ln\left(\frac{1 - bKn_o}{\Pi - bKn_o}\right) - Kn_o\right]\right\}$$

$$\rightarrow \dot{M} = \frac{h^3(P_i - P_o)P_o}{24RT\mu_o L}\left\{(1 + \Pi + 2\alpha Kn_o) + 12\left(\frac{2-\sigma}{\sigma}\right)\right.$$
$$\left.\left[\frac{(b+\bar{\alpha})}{(\Pi - 1)}Kn_o^2 \ln\left(\frac{\Pi - bKn_o}{1 - bKn_o}\right) + Kn_o\right]\right\}$$

$$\dot{M} = \frac{h^3 P_0 \Delta P}{24RT\mu_o L}\left[\Pi + 1 + \left(2\bar{\alpha} + 12\frac{2-\sigma}{\sigma}\right)Kn_0\right.$$
$$\left. + 12 \times \left(\frac{2-\sigma}{\sigma}\right)\frac{b+\bar{\alpha}}{\pi - 1}Kn_0^2 \ln\left(\frac{\Pi - Kn_0 b}{1 - Kn_0 b}\right)\right]$$

The dash-dot line in Fig. 5.7 shows the volumetric flow rate under the conditions of using the unified model and the viscosity correction (the above equation), which agrees well with the DSMC results.

By having the mass flow rate from the above equation and equating the flow rate at two arbitrary cross-sections, the following equation for the pressure distribution is obtained:

$$B(1 - x/L) = \tilde{P}^2 - 1 + 2(6 + \bar{\alpha})\frac{2-\sigma_v}{\sigma_v}Kn_0(\tilde{P} - 1)$$
$$+ 2(6b + \bar{\alpha})\frac{2-\sigma_v}{\sigma_v}K^2 n_0 \ln\left(\frac{\tilde{P} - bkn_0}{1 - bkn_0}\right) \quad (5.44)$$

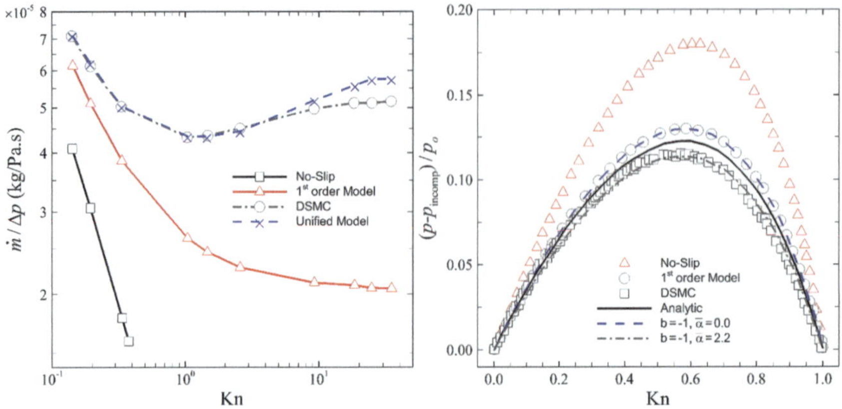

Fig. 5.8 The difference in pressure distribution from the linear distribution obtained from analytical and numerical solutions of the Navier–Stokes equation and the molecular solution (left) and the accuracy of different models (no-slip boundary condition, first-order boundary condition, and the unified model with viscosity correction) in predicting mass flow rate (right). Reproduced from Ref. [1]

As shown in Fig. 5.8, the pressure distribution is accurately calculated by applying appropriate coefficients for $\overline{\alpha}$ and b in Eq. (5.44), and since the pressure has been corrected, the mass flow rate is also calculated with acceptable accuracy over a wide range of Knudsen numbers. Note that the coefficients that improve the accuracy of the pressure difference do not necessarily have good accuracy in calculating the mass flow rate.

5.3 Accuracy Evaluation of Velocity Distribution

Figure 5.9 shows the velocity distribution from the molecular solution and analytical results for helium flow in a short channel at various positions along the channel. The second-order accurate relation by Karniadakis et al. [1] and their generalized dimensionless relation are presented in the following equations:

$$U^*(y, \mathrm{Kn}) = \frac{U(x, y)}{\overline{U}(x)} = \frac{-(\frac{y}{h})^2 + \frac{y}{h} + C_1\mathrm{Kn} + 2C_2\mathrm{Kn}^2}{\frac{1}{6} + C_1\mathrm{Kn} + 2C_2\mathrm{Kn}^2} \tag{5.45}$$

$$U^*(y, \mathrm{Kn}) = \left[\left(\frac{y}{h} - \left(\frac{y}{h}\right)^2 + \frac{\mathrm{Kn}}{1+\mathrm{Kn}}\Big/\left(\frac{1}{6} + \frac{\mathrm{Kn}}{1+\mathrm{Kn}}\right)\right)\right] \tag{5.46}$$

Equation (5.45) has first-order accuracy with the assumption of $C_2 = 0$ and second-order accuracy with the assumption of $C_2 = -0.5$. Equation (5.46), known as the unified equation, is valid for all Knudsen regimes in isothermal flows in a channel.

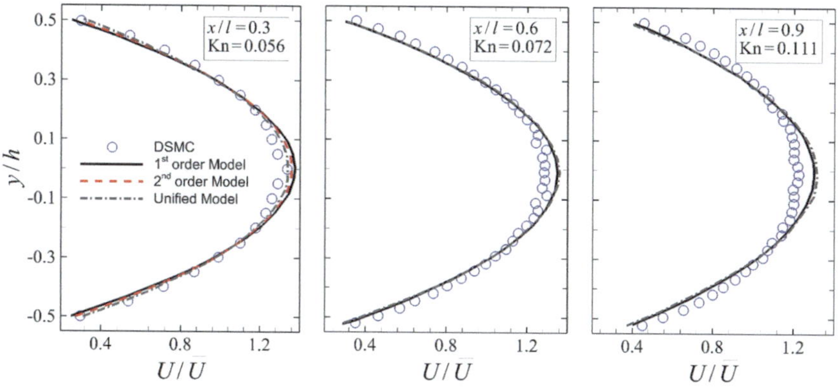

Fig. 5.9 Velocity distribution from molecular and analytical solutions in microchannel flow. Reproduced from Ref. [1]

To normalize the velocity in Fig. 5.9, the average velocity value corresponding to the numerical and analytical solutions is used. As the Knudsen number increases, the difference between the molecular and the analytical solutions increases. One reason for this is the approach to the channel exit and the effects of the outlet boundary condition applied in the molecular results. In Fig. 5.10, the velocity distribution from different analytical relations is compared with the results of molecular codes. Figure 5.10 shows the velocity profiles from the molecular method known as Information Preservation (IP) with two different numerical models, the DSMC method, the linearized Boltzmann method of Ohwada et al. [2], the second-order analytical formula $\left(u(x,y) = \frac{h^2}{2\mu}\frac{dp}{dx}\left[\left(\frac{y}{h}\right)^2 - \frac{y}{h} - \left(\frac{2-\sigma}{\sigma}\right)(\mathrm{Kn} - \mathrm{Kn}^2)\right]\right)$, and the unified relation for five different values of the Knudsen number.

The DSMC method is considered as the benchmark solution. The second-order formula provides good accuracy up to a Knudsen number of 0.451, but beyond that, its accuracy decreases significantly. Additionally, the unified formula predicts the maximum velocity with lower accuracy.

Another analytical relations for prediction of velocity and pressure distributions along the 2-D channel were proposed by Arkilic et al. [4] and evaluated by Stefanov et al. [5]. The two-dimensional rarefied gas flow in a long channel was analytically studied by Arkilic et al. [4]. They have obtained the asymptotic solution of the Navier–Stokes equations with slip-boundary conditions for very small channel aspect ratios ($\varepsilon \ll 1$). Normalized to the kinetic scales, the Arkilic solution for $p(x)$ and $U(x, y)$ takes the form

$$p(x) = -a_0 + \sqrt{a_0^2 + a_1\frac{x}{L} + a_2\left(1 - \frac{x}{L}\right)} \tag{5.47}$$

$$\frac{dp(x)}{dx} = \frac{(a_1 - a_2)}{2L\sqrt{a_0^2 + a_1\frac{x}{L} + a_2\left(1 - \frac{x}{L}\right)}} \tag{5.48}$$

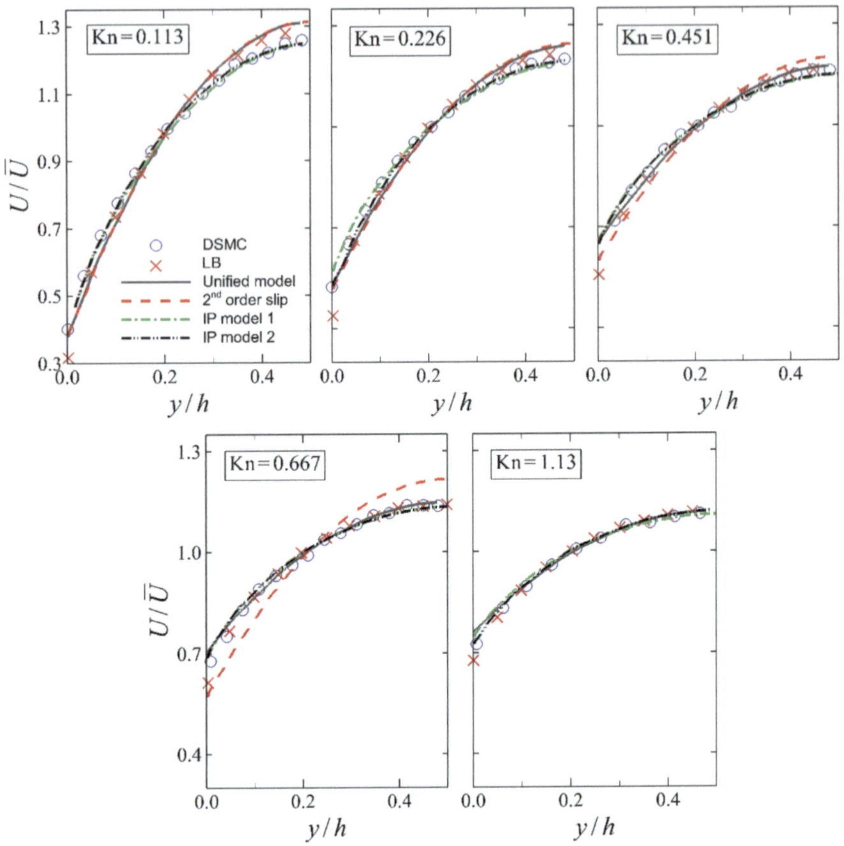

Fig. 5.10 Velocity distribution from various analytical models and the molecular solution. Reproduced from Ref. [3]

$$U(x, y) = \frac{\varepsilon L}{5\mathrm{Kn}\sqrt{\pi}} \frac{dp(x)}{dx} \left(1 - 4\left(\frac{y - 0.5}{h} \right)^2 + 4\beta \frac{\mathrm{Kn}}{p(x)} \right) \tag{5.49}$$

$$a_0 = 6\beta\mathrm{Kn},\ a_1 = 1 + 12\beta\mathrm{Kn},\ a_2 = P(P + 12\beta\mathrm{Kn}),$$
$$\beta = (2 - \alpha)/\alpha,\ P = p_{in}/p_{in}p_{out}$$

5.4 Velocity Variations with Knudsen Number

Figure 5.11a shows the variation of slip velocity and centerline velocity as a function of the Knudsen number from different numerical and analytical methods.

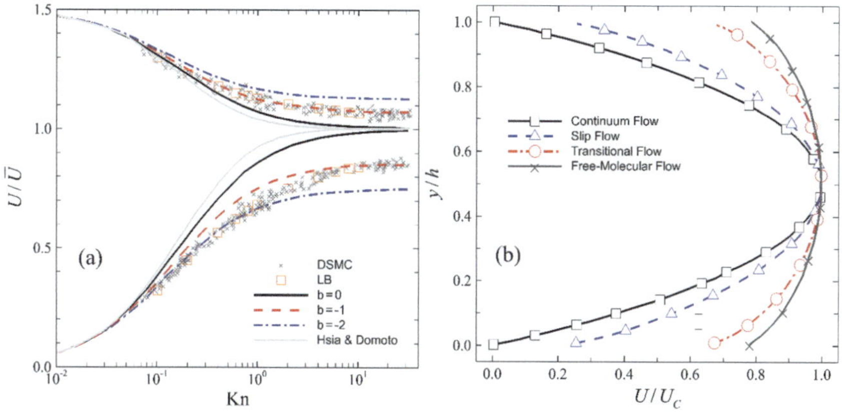

Fig. 5.11 Variations of the slip velocity and the centerline velocity to the Knudsen number (left) (Reproduced from Ref. [1]), and Velocity profile in Poiseuille flow for different Knudsen number regimes (right), where U_C is the centerline velocity. Reproduced from Ref. [6]

The linearized Boltzmann (LB) method and DSMC results are acceptable as benchmark results. The results of the unified model with three different values for b are shown in the figure. The condition $b = 0$ corresponds to the first-order Maxwell model, which loses accuracy with increasing Knudsen number. The results of the second-order model of Hsia and Domoto (with c_1=1, c_2=0.5) are also reported. Similar to the diagrams in Figs. 5.6 and 5.11b shows the variation of the dimensionless velocity distribution versus the Knudsen number in different flow regimes. As the Knudsen number increases, the velocity profile becomes flatter, and the velocity value on the wall approaches the maximum velocity.

5.5 Mass Flow Rate in Poiseuille Flow as a Function of Knudsen Number

Figure 5.12 shows the dimensionless mass flow rate of Poiseuille flow as a function of the Knudsen number ($k = \sqrt{\frac{\pi}{4}}\text{Kn}$) over a wide range, based on the hard sphere molecular model. The square symbols represent the numerical results of Ohwada et al. [2] and Kosugi et al. [7], the dashed-dotted line corresponds to the Sonne relation [8] for low Knudsen numbers, Eq. (5.50), the dashed line represents the Cherginani and Daneri relation [9], Eq. (5.51), and the solid line corresponds to the Takata and Funagan relation, Eq. (5.52). The latter two relations were derived for large Knudsen numbers. For large Knudsen numbers, the flow rate follows a logarithmic relation with the Knudsen number. Therefore, these two relations have a constant slope in Fig. 5.12 with the logarithmic horizontal axis.

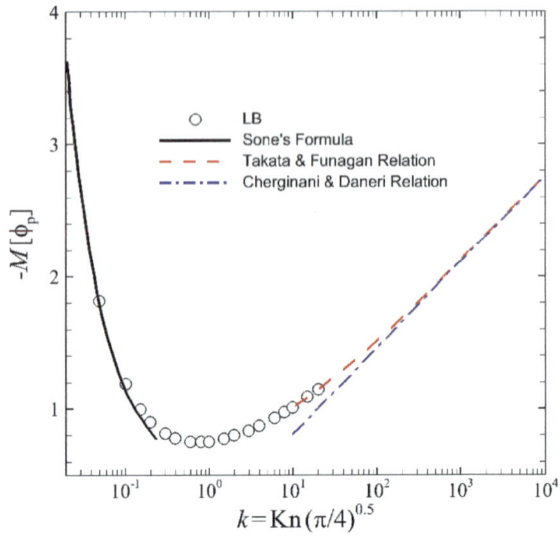

Fig. 5.12 Mass flow rate of Poiseuille flow as a function of the dimensionless parameter k. Reproduced from Ref. [10]

$$\dot{M} = -\frac{1}{\sqrt{\pi}\gamma_1 Kn}\left(\frac{1}{6} - k_0\sqrt{\frac{\pi}{4}}Kn + 2(a_1 - 2b_1)\left(\sqrt{\frac{\pi}{4}}Kn\right)^2\right) \quad (5.50)$$

$$\dot{M} \sim -\frac{1}{2\sqrt{\pi}}\left\{\left(\ln\left(\sqrt{\frac{\pi}{4}}Kn\right)\right)\right\} \quad (5.51)$$

$$\dot{M} = -\frac{1}{2\sqrt{\pi}}\left\{\left(\ln\left(\sqrt{\frac{\pi}{4}}Kn\right)\right) + \frac{3}{2}(1 - \gamma_e)\right\} + C_p(v) + O\left(\frac{(\ln Kn)^2}{Kn}\right) \quad (5.52)$$

In Eq. (5.50), γ_1 is the dimensionless viscosity and the coefficients k_0, a_1, and b_1 are the slip coefficients derived by Sone, whose values depend on the molecular model. In Eq. (5.51), $\gamma_e = 0.57721$ is the Euler's constant, and $C_p(v)$ is the function of the considered molecular model, for which values for several common molecular models are reported in Table 5.1.

Table 5.1 Values of $C_T(v)$ for several common molecular models [10]

Molecular models	Viscosity-temperature index	$\sqrt{\pi}C_p$
Hard Sphere	1/2	0.0310229
Variable Hard Sphere (Helium or Neon)	0.66	0.0237010
Variable Hard Sphere (Argon)	0.81	0.0147201
Variable Hard Sphere (Xenon)	0.85	0.0119477
Pseudo-Maxwellian	1	0

5.6 Pressure and Temperature in Compressible Poiseuille Flow

If the flow in the Poiseuille channel is not isothermal, the analytical relations presented in this chapter will not be applicable. Figure 5.13 shows the distribution of temperature, density, heat flux, and pressure profiles in a Poiseuille channel at a Knudsen number 0.1. The results of three methods, compressible Navier–Stokes equations, DSMC method, and NCCR[1] [1] method, are reported. Several interesting observations can be extracted from these two plots. The pressure distribution in the vertical direction is non-uniform. In other words, in microchannel flow, $dp/dy \neq 0$. Also, the temperature distribution in the center of the channel exhibits the phenomenon of minimum temperature. In other words, the temperature in the center of the channel reaches a minimum value. The Navier–Stokes equations, even with the slip boundary condition, cannot predict these two phenomena because these phenomena occur due to the violation of fundamental laws (stress–strain relations) and are unrelated to wall non-equilibrium [11]. The cause of the central minimum temperature phenomenon is the correlation of viscous shear stress and pressure gradient in the fundamental heat flux relation. Also, heat transfer from cold to hot is visible in the channel's dimensionless region between 0 and 0.3.

5.7 Correction of the Molecular Mean Free Path Near the Wall

One of the strategies to extend the application of Navier–Stokes equations in rarefied regimes is to modify the effective viscosity coefficient near the wall. In other words [12]:

$$\mu_{eff} = \frac{1}{\psi(z)}\mu_0 \tag{5.53}$$

$\psi(z)$ is a function that corrects the viscosity coefficient near the wall. In other words, considering the effects of rarefaction and molecule-wall collisions, the viscosity coefficient near the wall is assumed to be different from the bulk fluid viscosity coefficient. Considering the direct dependence of viscosity on the molecular mean free path, it can be written as:

$$\psi(z) = \frac{\lambda_0}{\lambda_{eff}(z)} \tag{5.54}$$

The value of λ_0 for the hard sphere molecule model is given by:

---

[1] Nonlinear coupled algebraic constitutive relations.

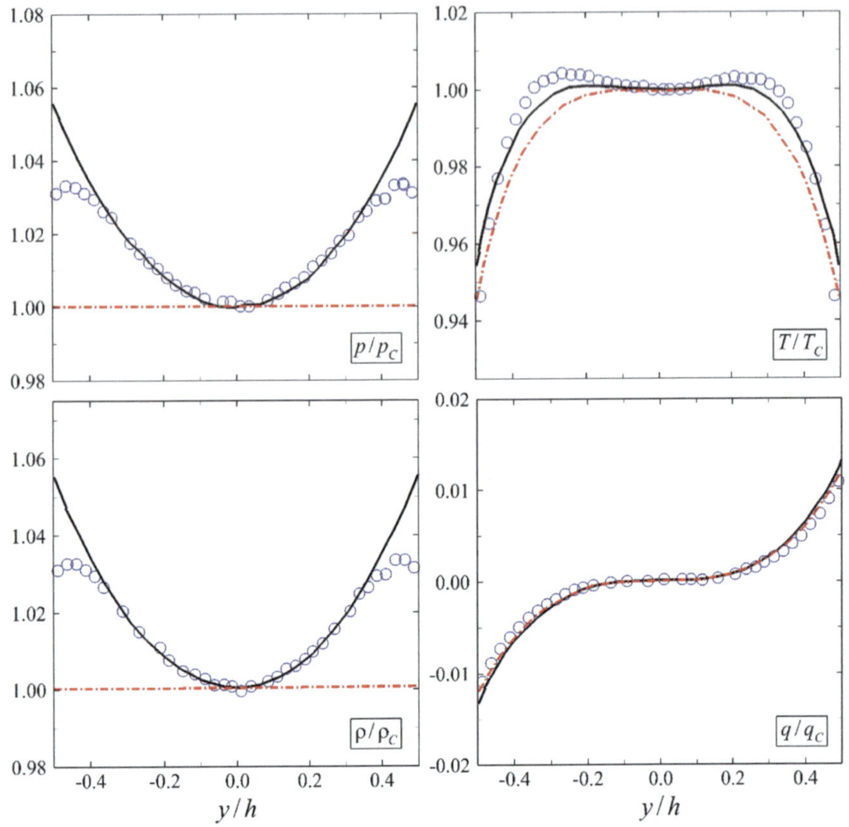

Fig. 5.13 Dimensionless pressure and temperature profiles in non-isothermal Poiseuille flow (the horizontal axis represents the dimensionless channel width). The solid line indicates the Navier–Stokes solution, the open circles represent the DSMC solution, and the filled circles correspond to the NCCR method. Reproduced from Ref. [11]

$$\lambda_0 = \frac{m}{\sqrt{2}\pi d^2 \rho} \tag{5.55}$$

For the first time, a researcher named Stops introduced the first position-dependent model for $\lambda_{eff}(z)$ in the form of $\psi(z) = \frac{e^{-z/\lambda}}{\lambda}$ [13]. The model introduced by Stops had some limitations. One of these limitations was that the probabilistic function Stops used to derive the corrected molecular free path length only provided accurate results under thermodynamic equilibrium conditions. However, in non-equilibrium systems, it failed to predict some essential characteristics of the field. A few decades later, Guo and his colleagues developed Stops' model and proposed the relation (5.56) for wall-bounded flows [14]:

$$\frac{\lambda_{eff}(z)}{\lambda_0} = 1 - \frac{1}{2}\left[y^2 E_1(z) + (1 - z)e^{-z}\right] \tag{5.56}$$

$$E_1(z) = \int_y^{\infty} \frac{e^{-z}}{z} dz$$

Recently, Abramov proposed the following relation for λ_{eff} [15]:

$$\frac{\lambda_{eff}(z)}{\lambda_0} = 1 - \frac{1}{2}\left[e^{-z} - yE_1(z)\right] \tag{5.57}$$

Shi et al. [12] compared the two recent relations with results from molecular dynamics and DSMC methods. The molecular mean free path was calculated as the average distance between two consecutive collisions (with either the wall or another molecule). As shown in Fig. 5.14, the measured molecular mean free path at different Knudsen numbers varies near the wall, and its value is approximately half of the mean free path size in the bulk flow. With the increasing Knudsen number, the region affected by the wall expands. The collisions between the wall and gas molecules are the main factor in reducing the mean free path near the wall.

In Fig. 5.15, the gas viscosity coefficient in Couette flow is plotted using the Eq. (5.53) and results of molecular dynamics (by calculating molecular stress and dividing it by the velocity gradient). It is observed that near the wall, the viscosity is reduced by approximately 50% of its value in the bulk flow. More precisely, at a Knudsen number of 0.05, the viscosity ratio is 49%; at a Knudsen number of 0.1, it is 48%; and at a Knudsen number of 0.2, it is 44%.

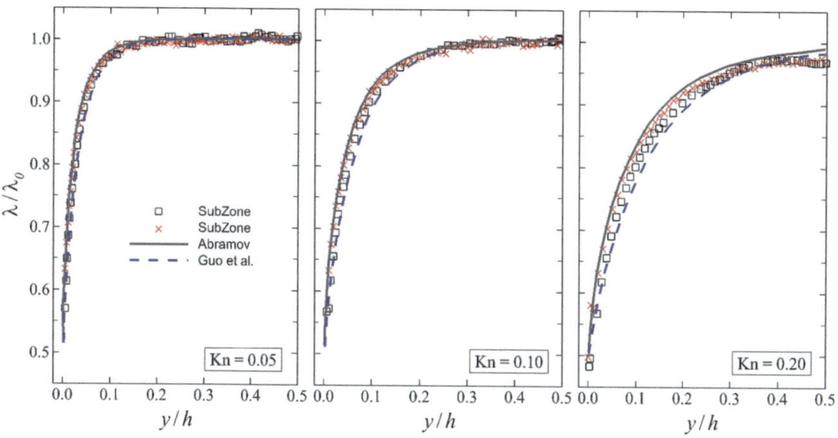

Fig. 5.14 Spatial variation of the molecular mean free path. Reproduced from Ref. [12]

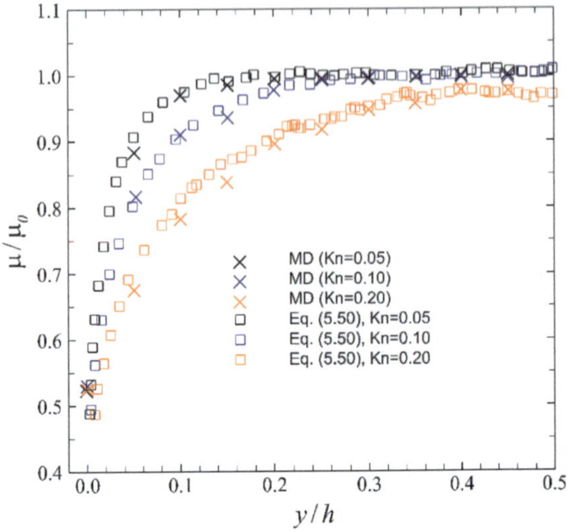

Fig. 5.15 The gas viscosity ratio distribution near the wall. Reproduced from Ref. [12]

5.8 Results Obtained from the Application of the Modified Molecular Mean Free Path

Dongari et al. [16] proposed the following power law (PL) distribution function to modify the molecular mean free path:

$$\Psi(y) = C(a + y)^n \tag{5.58}$$

where a and n are two constants. Given that $\lambda_{eff} = \beta_{PL}\lambda$, by performing some mathematical calculations, β is obtained from the following equation:

$$\beta_{PL} = 1 - \frac{1}{96}\left[\left(1 + \frac{H/2 - y}{a}\right)^{1-n} + \left(1 + \frac{H/2 + y}{a}\right)^{1-n}\right.$$
$$+4\sum_{i=1}^{8}\left(1 + \frac{H/2 - y}{\operatorname{acos}[(2i-1)\pi/32]}\right)^{1-n} 4\sum_{i=1}^{8}\left(\left(1 + \frac{H/2 + y}{\operatorname{acos}[(2i-1)\pi/32]}\right)\right)^{1-n}$$
$$\left. +2\sum_{i=1}^{7}\left(1 + \frac{H/2 - y}{\operatorname{acos}[(i\pi)/16]}\right)^{1-n} + 2\sum_{i=1}^{7}\left(1 + \frac{H/2 + y}{\operatorname{acos}[(i\pi)/16]}\right)^{1-n}\right] \tag{5.59}$$

In Fig. 5.16, the distribution of the function β from Stops relation, the power-law relation, Eq. (5.56), the Arlemark et al., and Molecular Dynamics solutions at different Knudsen numbers are shown. For Knudsen numbers greater than 0.2, this function is less than 1 in almost the entire channel. In the middle regions of the flow, the power-law model predicts a value for β either slightly less than or greater than the

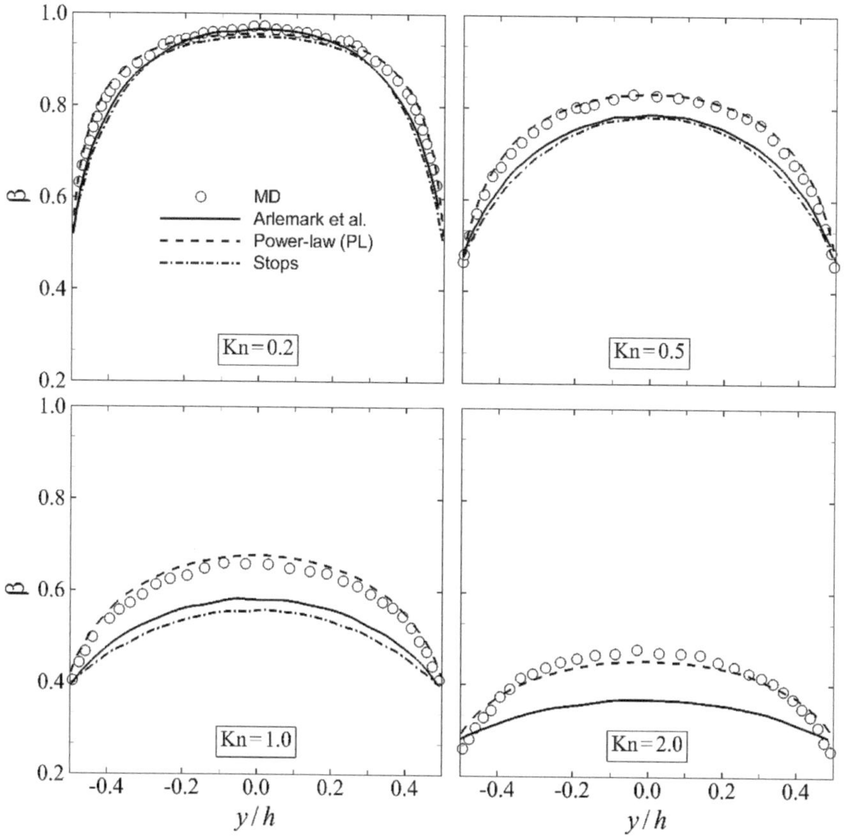

Fig. 5.16 Variation of the dimensionless mean free path through the channel width at various Knudsen numbers. Reproduced from Ref. [16]

MD results. However, this model shows better accuracy overall than other available models. Figure 5.17 compares the velocity profile from the power-law model applied in the Navier–Stokes equations with the results of the linearized Boltzmann equation solution, the power-law model, the R26 Equations (26 regularized equations), and the Navier–Stokes equations with first- and second-order boundary conditions. At a Knudsen number of 0.113, the power-law model has a desirable accuracy compared to the linearized Boltzmann solution. Still, the standard Navier–Stokes solution with first or second-order slip boundary conditions has a slight error in the central regions of the flow. As the Knudsen number increases to less than 1, the power-law model maintains accuracy except near the wall. However, as the Knudsen number exceeds 1, the model's accuracy decreases throughout the field. It is worth noting that combining this model with more complex slip boundary conditions, such as the unified model, will likely maintain its accuracy over a wider range of Knudsen numbers. Figure 5.18 shows the dimensionless mass flow rate graph versus the inverse of the Knudsen

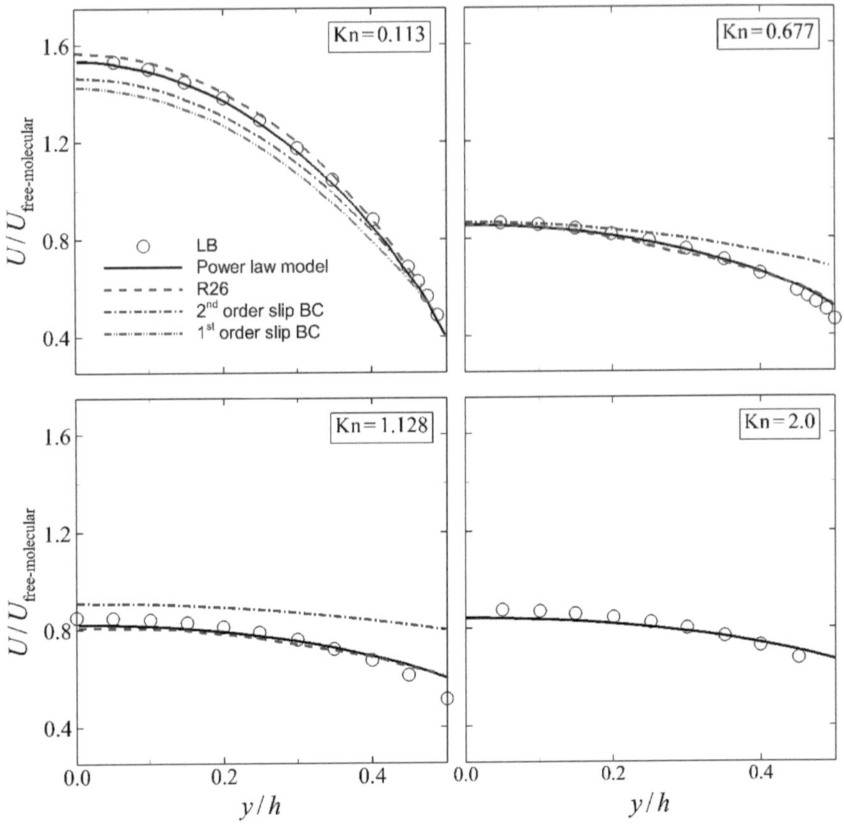

Fig. 5.17 Comparison of the dimensionless velocity in a channel at various Knudsen numbers, using the linearized Boltzmann solution, the power-law model, the R26 equations, and the Navier–Stokes equations with first- and second-order boundary conditions. Reproduced from Ref. [16]

number for the aforementioned models. It is observed that the first-order slip models fail to predict the Knudsen number minimum phenomenon.

Furthermore, second-order slip models report the Knudsen minimum value slightly lower than the solution of the Boltzmann equation. In other words, second-order models show errors in mass flow rate predictions for Knudsen numbers above 0.5. The power-law model and second-order boundary conditions predict desirable solutions for the mass flow rate up to Knudsen number 10.

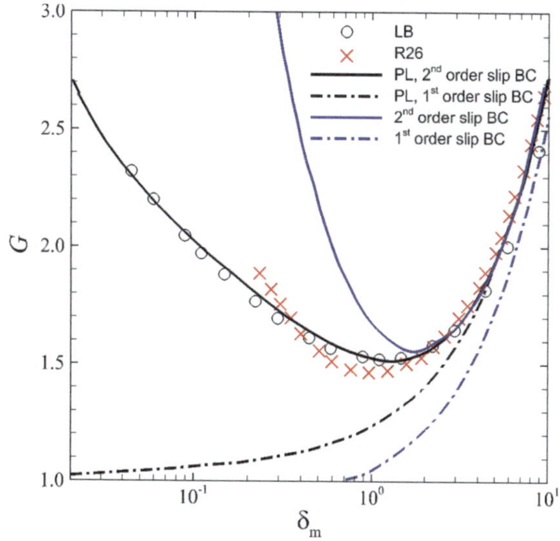

Fig. 5.18 Dimensionless mass flow rate plotted against the reciprocal of the Knudsen number. Reproduced from Ref. [16]

5.9 Heat Transfer in Poiseuille Flow

Consider Poiseuille flow in the slip regime, as Fig. 5.19a shows. The pressure increases towards the right, and the temperature becomes uniform. As a result, the density decreases to the right. The velocity profile is as shown in Fig. 5.19a. Imagine an observer on the channel's centerline moving with the macroscopic flow velocity. So, the velocity profile that the observer perceives is shown in Fig. 5.19b, where the velocity at the center point is zero and increases as it moves away from the center point.

Suppose molecules are approaching point O from the left and right. Assume that the flow field is symmetrical about the centerline passing through point O. The molecules approaching point O from the left are represented by molecules from point A. This point is located at a 45° angle above the centerline and a distance of

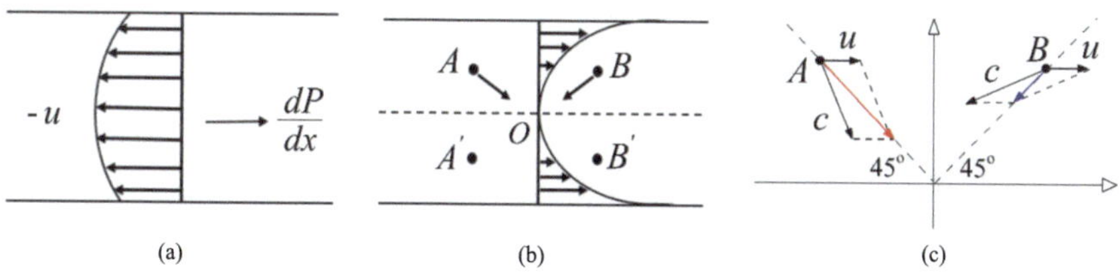

Fig. 5.19 **a** Velocity profile in Poiseuille flow; **b** relative velocity profile for an observer at the origin; **c** comparison of momentum and energy transferred from points A and B to the origin

the mean free path length from the center. Due to the mean free path distance from point O, the molecules transfer the gas state at the point where the last collision (i.e., point A) occurred. Similarly, the molecules approaching point O from the right are represented by molecules coming from point B.

As shown in Fig. 5.19c, the velocity of the molecules coming from A (red arrow) is the sum of a horizontal vector with magnitude u and a downward vector c, where c is the average thermal velocity of the molecules. Similarly, the velocity of the molecules coming from B is the sum of a horizontal vector with magnitude u and a downward vector c. It is obvious from the figure that the red vector is larger than the blue vector, and this means that, on average the molecules approaching the origin O from the left move faster than the molecules approaching from the right.

Note that the density at point A is less than B because the density decreases along the channel. In other words, fewer molecules with high velocity approach point O from the left while more molecules with low velocity approach point O from the right. Due to this, there is no particle flux from left to right or vice versa, which means that the relative macroscopic velocity of the flow at the origin is zero (Fig. 5.19b). However, when comparing the energy transferred by the molecules approaching the origin from A and B, it is observed that the molecules approaching point A have more transferred energy. This is because energy transfer, compared to mass transfer, is obtained from a higher-order moment of the velocity distribution function (proportional to the square of the velocity). This means that energy flows from left to right. Since the relative macroscopic velocity at point O is zero, the transferred energy is purely thermal transfer. Therefore, there is a heat flux in the direction of the pressure gradient. As the momentum equations for Poiseuille flow show, the pressure gradient is related to the second derivative of the velocity field (according to Eq. (5.14) and Stokes flow), which is the reason for the curvature of the velocity profile.

Therefore, in rarefied flows, besides Fourier heat transfer (proportional to the temperature gradient), heat transfer in the direction of the pressure gradient (the second derivative of velocity) is also possible. It should be noted that this heat transfer is of the second order to the Knudsen number. So, if the temperature gradient is weak, heat transfer can occur in the direction of the pressure gradient. This heat transfer effect is known as the Sone equation, discussed in Chap. 1 with examples of cold-to-hot heat transfer. It is important to note that this term does not appear in the governing equation for the temperature field because the flow is in equilibrium (its value becomes zero over a closed control volume). Thus, it does not induce a temperature field.

5.10 Gas Behavior Near the Wall

Barisik and Beskok [17] investigated the behavior of argon gas flow near the wall using the molecular dynamics method. In the molecular dynamics method, in addition to the wall rarefaction effects, the effects of surface adsorption and wall surface

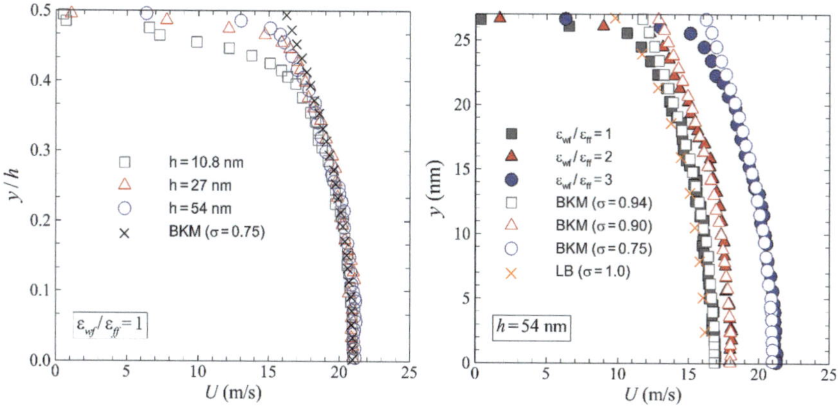

Fig. 5.20 Velocity profile at Knudsen number $\sqrt{\frac{\pi}{4}}$ in channels with different Tangential Momentum Accommodation Coefficients (TMAC) and heights. Reproduced from Ref. [17]

forces are also taken into account. Therefore, the velocity profile near the wall is different from the velocity profile predicted from the unified equation of Beskok and Karniadakis (BKM) and the linearized Boltzmann (LB) solution of Ohwada et al. (see Fig. 5.20). The smaller the channel dimensions, the more area the wall affects. In the Lennard–Jones molecular model, the parameter ε in the figure represents the ratio of the wall molecule potential to the gas molecule potential. The result of this discussion is that the velocity profiles extracted from the Navier–Stokes equation with slip boundary conditions and even the Boltzmann equation are inaccurate near the walls. In other words, if the channel thickness is reduced to below 100 nm, accurate results near the wall will be achievable using the molecular dynamics method.

5.11 Experimental Results for Velocity Profiles in the Slip Regime

For the first time, Frantantonio et al. investigated the velocity profile in microchannels using Molecular Tagging Velocimetry (MTV) [18]. Figure 5.21 shows the velocity profile in a helium-acetone mixture at a Mach number of 0.7 and a Knudsen number of 0.008 in four consecutive time intervals. The slip velocity at the walls was calculated to be approximately 2 m/s. Due to the turbulence in the molecular data, a comparison between the experimental and analytical results is impossible.

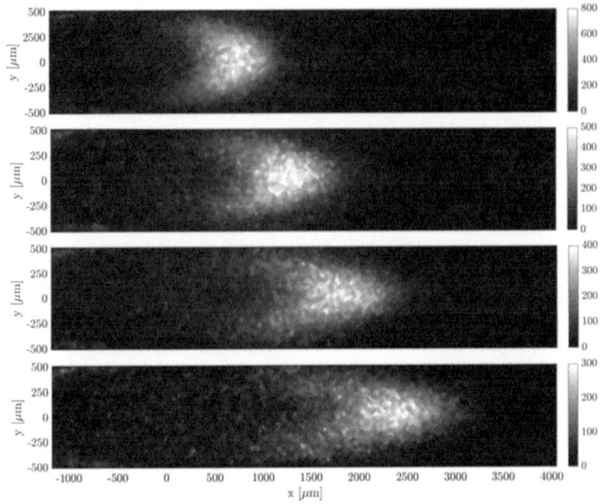

Fig. 5.21 Velocity profiles at Mach number 0.07 and Knudsen number 0.008 over four consecutive time intervals. Copied from open access Ref. [18]

5.12 Step Flow

Micro-steps, formed by connecting nanochannels of different diameters, exhibit flow and heat transfer behaviors that continuum models cannot deal with due to the significant molecular mean free path relative to characteristic lengths. These flows must be treated as rarefied gas flows and analyzed using kinetic theory principles. This section evaluates the cold-to-hot heat transfer phenomena in nano-step geometries and the Knudsen number effects on the micro-step flow behavior. Also, an analytical solution for a two-dimensional micro-step flow is investigated.

5.12.1 Cold to Hot Heat Transfer

Mahdavi and Roohi [19] investigated heat transfer and hydro/thermal behaviors in nano-step geometries using the Direct Simulation Monte Carlo (DSMC) method. Their study highlights the effects of hydrodynamic separation on the heat lines and shows that a "cold-to-hot" heat transfer occurs when the wall and inlet temperatures are close.

A backward-facing step through which nitrogen gas flows is considered (Fig. 5.22). The setup includes four walls: walls 2 and 3 are isothermal at 500 K, and walls 1 and 4 are set at 300 K. The Knudsen number (Kn) is 0.01, the inlet temperature is 300 K, and the pressure ratio is 2. Figure 5.22 shows a hydrodynamic separation zone (S_H-hydrodynamic separation length) where the gas circulates near wall 3. Dashed lines determine the heat lines, and heat transfer behavior in the geometry shows two distinct directions: (1) heat lines move from the hot wall (wall 3) to the cold wall (wall 4), similar to straight channels, and (2) some heat lines deviate from their established path and move back toward the inlet near the separation zone.

This deviation region is called the thermal separation zone, with its extent denoted as S_T.

Figure 5.23 presents velocity contours with streamlines and temperature contours with heat lines for three different wall temperatures. The results indicate that variations in wall temperature influence the thermal separation zone. Specifically, increasing the wall temperature leads to a reduction in both the hydrodynamic separation length and the thermal separation length. When $T_{wall2,3} = 360$ K, an unusual "cold-to-hot" heat transfer phenomenon is observed, where heat lines move from a low-temperature region back toward a hotter region near the inlet (as shown in Figs. 5.23a and 5.24d). This occurs due to a sudden temperature drop at $X/L = 0$ caused by flow expansion, creating a cold region. Heat lines from walls 2 and 3 enter this region but are deflected toward the hotter region near the inlet.

Fourier's law cannot explain this phenomenon, as it is insufficient for rarefied flow conditions. Instead, the weakly nonlinear form of the Boltzmann equation, derived by

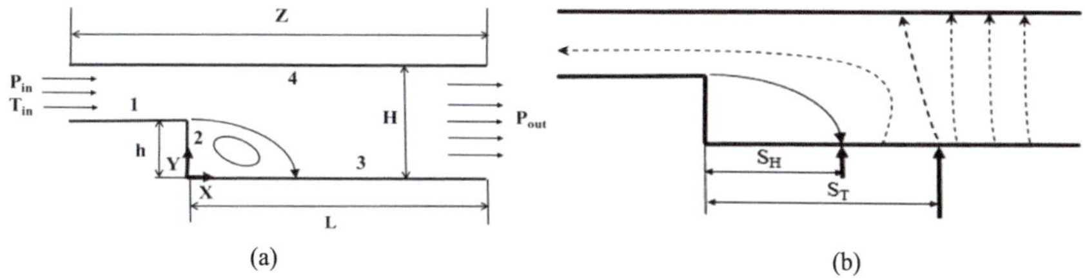

Fig. 5.22 Schematic of the backward-facing step flow: **a** overall geometry illustrating the hydrodynamic separation zone, **b** thermal separation zone (S_T) and hydrodynamic separation zone (S_H)

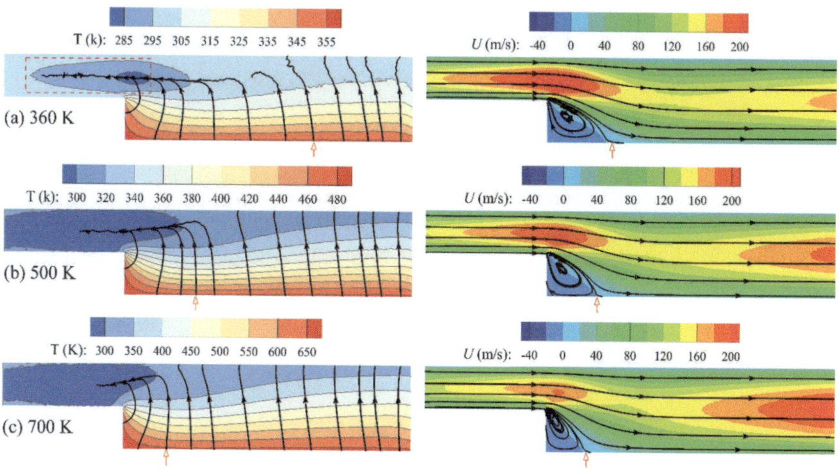

Fig. 5.23 Velocity contours with streamlines and temperature contours with heat lines at various $T_{wall2,3}$, Kn = 0.01. Reproduced from Ref. [19]

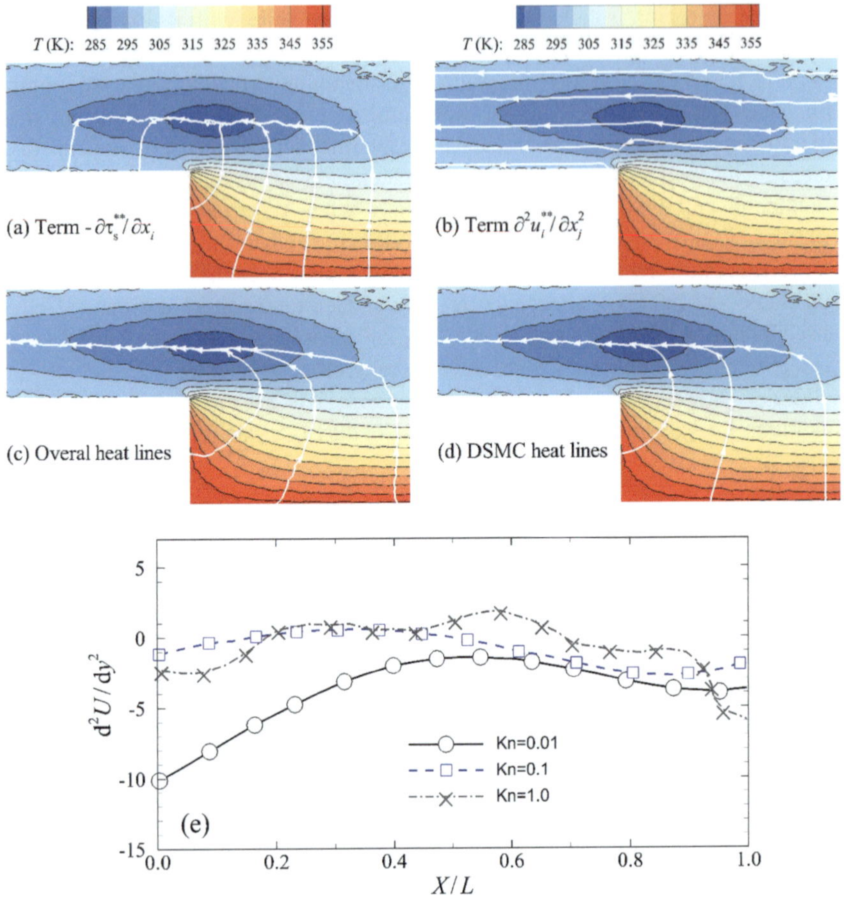

Fig. 5.24 **a–d**: Heat lines plotted in the zoomed view of the step geometry, **e**: second derivative of velocity at $Y/L = 0.2$. Reproduced from Ref. [19]

Sone, is employed. These equations account for the temperature dependence of the gas's thermal conductivity. Since the variations in Knudsen number and temperature are small, using Sone's constitutive laws is appropriate. The heat flow vector in these equations incorporates higher-order effects necessary to describe this behavior, which is given as follows [19]:

$$Q = Kn^{**}Q_1 + Kn^{**2}Q_2 + Kn^{**3}Q_3 \tag{5.60}$$

$$Q_1 = 0 \tag{5.61}$$

$$Q_2 = -\frac{5}{4}\gamma_2 \frac{\partial \tau^{**}}{\partial x_i} \tag{5.62}$$

$$Q_3 = -\frac{5}{4}\gamma_2\frac{\partial\tau^{**}}{\partial x_i} - \frac{5}{4}\gamma_5\tau^{**}\frac{\partial\tau^{**}}{\partial x_i} + \frac{1}{2}\gamma_3\frac{\partial^2 u^{**}}{\partial {x_j}^2} \tag{5.63}$$

The employed parameters are defined as follows:

$$Kn^{**} = Kn\frac{\sqrt{\pi}}{2} \tag{5.64}$$

$$\tau^{**} = \frac{T - T_0}{T_0},\ u^{**} = \frac{u}{\sqrt{2RT}},\ x = \frac{x}{L} \tag{5.65}$$

Numerical results reveal that the term $\frac{\partial^2 u^{**}}{\partial {x_j}^2}$ has a greater magnitude than other parameters, indicating that the behavior of heat lines is determined by the interaction between the Fourier term $-\frac{\partial\tau^{**}}{\partial x_i}$ and $\frac{\partial^2 u^{**}}{\partial {x_j}^2}$. Figure 5.24a shows that the Fourier term moves from the hot to cold region, while Fig. 5.24b illustrates that isolines of the second derivative of velocity are directed toward the inlet. Figure 5.24c demonstrates the overall heat lines aligning well with DSMC results (Fig. 5.24d). It also shows that the Fourier term is dominant near wall 3 (lower half), while $\frac{\partial^2 u^{**}}{\partial {x_j}^2}$ dominates in the upper half, near the velocity expansion region.

When the inlet and wall temperatures differ moderately, the temperature gradient is small, limiting the Fourier term's effect. In such cases, the hydrodynamic separation zone creates a significant velocity gradient, making $\frac{\partial^2 u^{**}}{\partial {x_j}^2}$ dominant and enabling "cold-to-hot" heat transfer, as observed at $T_{wall2,3} = 360$ K. However, as wall temperature increases (e.g., $T_{wall2,3} = 500$ K or 700 K), the temperature gradient becomes more significant, and the hydrodynamic separation zone weakens, reducing the influence of $\frac{\partial^2 u^{**}}{\partial {x_j}^2}$. Consequently, the cold-to-hot transfer vanishes as they move back toward the inlet. The reduction in differences between the hydrodynamic separation length (S_H) and thermal separation length (S_T) with increasing wall temperature is attributed to the balance between the Fourier term and $\frac{\partial^2 u^{**}}{\partial {x_j}^2}$.

Figure 5.24e illustrates the distribution of $\frac{\partial^2 u^{**}}{\partial y^2}$ along wall-3 for three Knudsen numbers (Kn = 0.01,0.1, 1). At Kn = 0.01, a large negative magnitude of $\frac{\partial^2 u^{**}}{\partial y^2}$ is observed, indicating a flux from the outlet toward the inlet. For higher Knudsen numbers (Kn = 0.1 and Kn = 1), variations in $\frac{\partial^2 u^{**}}{\partial y^2}$ along the channel are minimal, limiting its contribution and resulting in a smaller thermal separation zone.

5.12.2 Effects of Knudsen Number

The standard DSMC method faces significant computational challenges near the continuum regime due to the stiffness of the collision operator, especially in low-Knudsen regions. In contrast, the Fokker–Planck (FP) method enables computationally efficient simulations for rarefied gas flows in low and intermediate Kn regimes by approximating the diffusion limit of the Boltzmann equation. Combining DSMC and FP has become popular in balancing accuracy and efficiency. Mahdavi and Roohi [20] proposed a hybrid DSMC-FP method to investigate Kn and Re's effects on the micro-step flow behaviour that switches between the two models based on their respective accuracies. This hybrid algorithm uses DSMC in regions with intense gradients, such as shock waves and wake regions, to leverage its high accuracy while employing FP in low-Kn low-Mach-number regions to reduce computational costs effectively.

One of the typical phenomena in backward-facing step flows is flow separation, which is significant for many engineering applications. Vortex flow forming in the separation zone may affect some properties, such as forces and mass flow. The effect of Knudsen number on flow behavior in step geometry is investigated in Fig. 5.25. As the flow enters the larger section, momentum carries it downstream, creating a low-pressure region at the step corner. This low-pressure region causes part of the flow to turn around and fill the area, initiating a negative pressure zone. Consequently, the flow separates at the bottom wall, forming a concave vortex that reattaches further downstream. For Kn $\leq$ 0.001, a secondary separation zone with a convex vortex forms on the top wall, as shown in Fig. 5.25.

The separation zone in microscale flows is associated with a positive pressure gradient, similar to the continuum regime. Figure 5.26a, b highlight that the positive

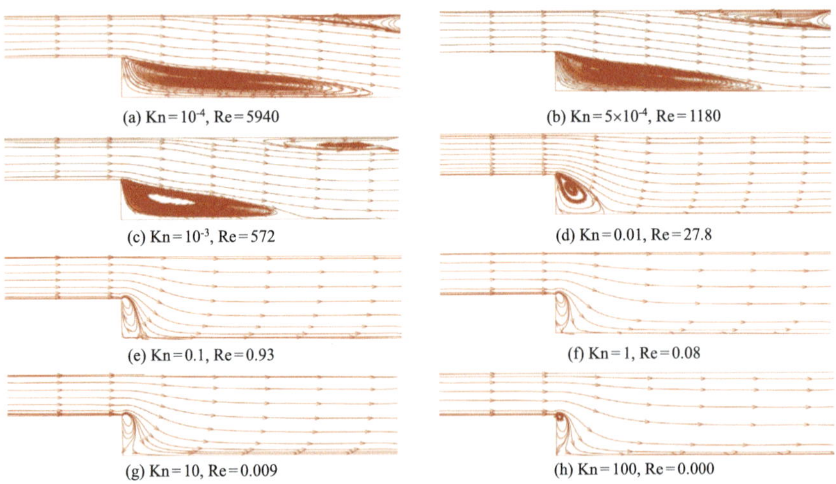

Fig. 5.25 Streamlines in the micro-step geometry. Reproduced from Ref. [20]

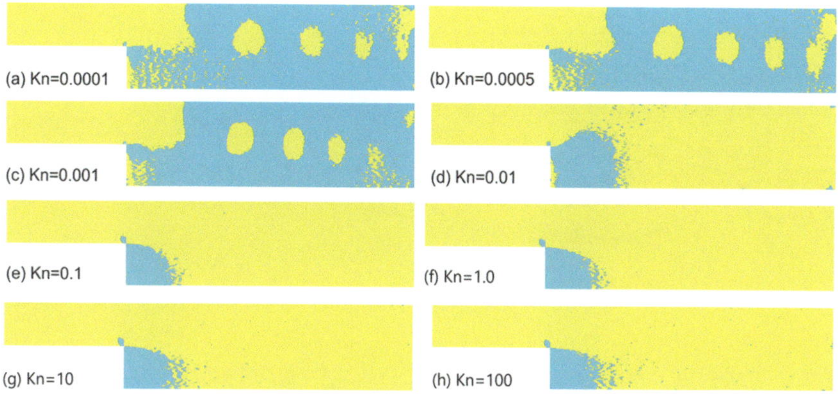

Fig. 5.26 Contours of dP/dX, where Blue: dP/dX > 0 and Yellow: dP/dX < 0. Reproduced from Ref. [20]

pressure gradient (dP/dX > 0) is more prominent in lower-Kn cases. A large separation zone at the step corner is caused by an extended positive pressure gradient, particularly for Kn = 0.001, 0.0005, and 0.0001, where most channels consist of such regions, leading to a second recirculation zone on the top wall. For Kn = 10 and 100, as shown in Figs. 5.25g, h, 5.26g, h, the separation zone and pressure gradient variations diminish.

Figure 5.27a illustrates the pressure gradient over wall 3. The region with $\left(\frac{dP}{dX}\right)^{*}>0$ is more dominant and extensive in lower-Kn cases (e.g., Kn = 0.0001), resulting in a stronger and more extended circulation region. As Kn decreases, the peak of $\left(\frac{dP}{dX}\right)^{*}$ shifts from wall 2 toward the exit, causing the $\left(\frac{dP}{dX}\right)^{*}>0$ zone to cover the entire channel. This behavior explains the formation of the second recirculation zone on the top wall for Kn = 0.001, 0.0005, and 0.0001, as seen in Fig. 5.26a, b. Figure 5.27b shows that the length of the separation zone over wall 3 (X_s/L) decreases as Kn increases. Figure 5.27c illustrates that, for Kn < 0.001, the variation in the mass flow rate becomes negligible. Figure 5.27d illustrates the average inlet velocity as a function of the Knudsen number. The inlet velocity remains nearly constant for Kn < 0.0005 but sharply decreases as Kn increases up to 0.1. Beyond this point, changes in velocity are minimal, with the minimum velocity observed at Kn = 1, indicating a Knudsen minimum.

While vortex formation in continuum flows is influenced by Re, micro gas flows differ due to low Re and rarefied effects. In the laminar regime, the reattachment length of the primary vortex increases nonlinearly with Re. However, as shown in Fig. 5.27b, the separation length decreases with Kn and increases nonlinearly with Re at high Kn and low Re. As Kn increases, the convex vortex on the top wall disappears, and the concave vortex on wall 3 shrinks, becoming confined to the vertical wall for Kn ≥ 10.

The separation zone in rarefied gas flows is influenced by momentum similar to continuum flows and velocity slip, which is unique to rarefied regimes. Higher

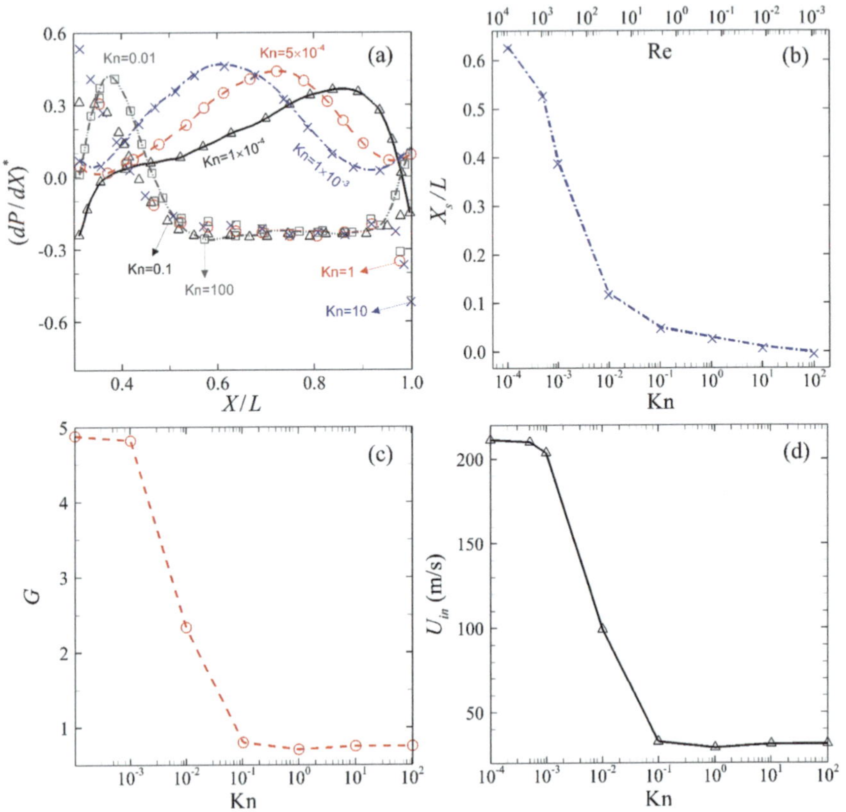

Fig. 5.27 Various flow properties at the investigated Knudsen numbers. Reproduced from Ref. [20]

momentum extends the separation region in the continuum regime, while rarefaction affects flow via wall slip and diffusion. Wall slip enhances mass flow rate and momentum near walls, preventing flow separation by energizing the flow over walls, while diffusion suppresses the formation of the separation zone. According to Fig. 5.28a, b, the slip velocity peaks at the junction, especially at the start of wall2 (s = 0), where the flow moves downwards for all Kn values. Slip velocity on wall-2 approaches zero as it moves toward wall-3, indicating the separation zone—negative slip velocity at lower Kn values and positive at higher Kn $\geq$ 0.01. Surprisingly, beyond Kn = 10, slip velocity stabilizes. The separation zone extends significantly at very low Kn (0.0001–0.001) due to increased momentum in near-continuum conditions, leading to sudden pressure drops. A high slip velocity is observed over wall-2 in the near-continuum regime. An unusual behavior is observed over wall-4, as shown in Fig. 5.28c, d. In the near-continuum regime (Kn = 0.0001–0.001), slip velocity drops to almost zero for $X/L > 0.6$ but reaches a peak of 58 m/s around $X/L \approx 0.4$. This anomaly is attributed to this regime's intense adverse pressure drop, leading to reduced rarefaction effects.

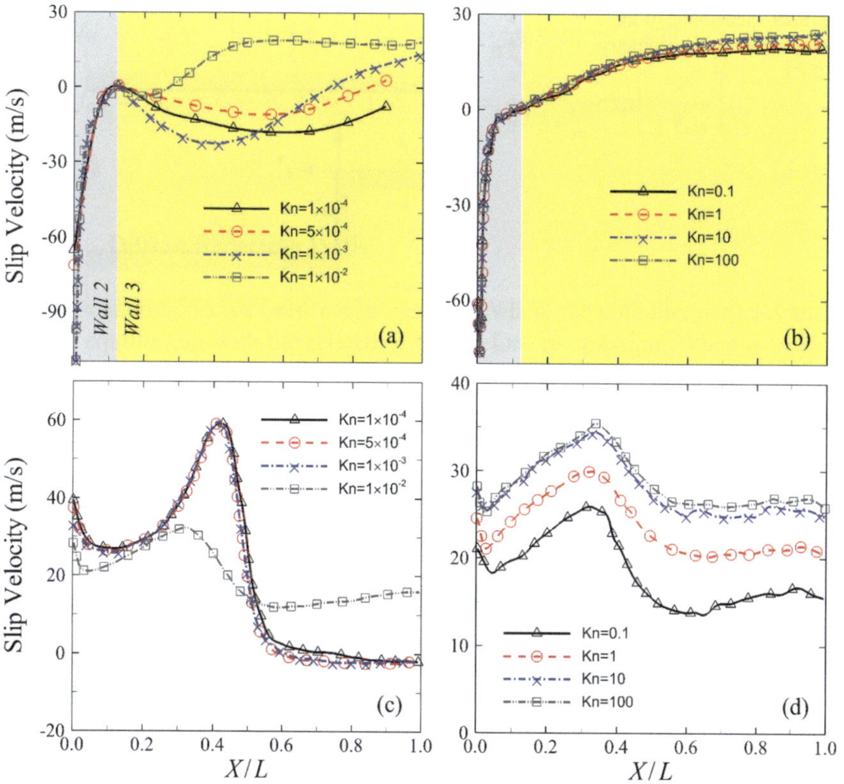

Fig. 5.28 Velocity slips for various Knudsen numbers; **a**, **b** Over walls 2 and 3, and **c**, **d** over wall-4. Reproduced from Ref. [20]

5.12.3 Analytical Solution

Manela and Gibelli [21] considered a two-dimensional micro-step flow of a rarefied gas over a backward facing step in the limit of large Knudsen numbers. The free-molecular problem is solved analytically for both diffuse and specular-reflecting channel boundaries, and the solutions are validated through comparison with direct simulation Monte Carlo calculations. Also, the results for the density- and temperature-drop-driven flows are analyzed and contrasted. The study considers a perfect monatomic gas flowing through a two-dimensional channel with a step at $x^* = 0$ (Fig. 5.29). The inlet and outlet sections are connected to equilibrium reservoirs, where the gas maintains equilibrium densities and temperatures.

Gas particles entering these sections follow Maxwellian equilibrium distributions, which are strictly valid for free-molecular conditions, but end-effect corrections must be considered at finite Knudsen numbers. However, for simplicity and given the study's focus on collisionless and near-collisionless regimes, it is also applied to finite Knudsen numbers. Gas-surface interactions are modeled using the Maxwell

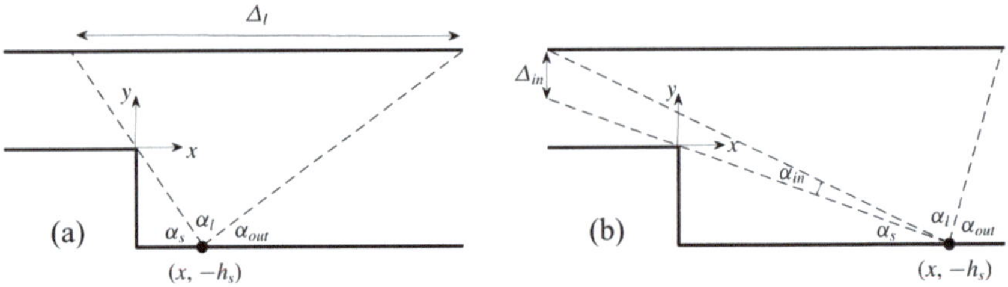

Fig. 5.30 Geometrical construction of the impermeability condition at a point $(x, -h_s)$ along the downstream wall l_d. **a** A point location where the inlet section and the left part $-l_u \leq x \leq 1+l_d-\Delta_l$ of the upper boundary are obscured. **b** A point where the lower part $0 \leq y \leq 1 - \Delta_{\text{in}}$ of the inlet section is obscured. The angles α_s, α_l, α_{out} and α_{in} denote the section angles associated with each boundary. Copied from open access Ref. [21]

these lines confine the c_x/c_y the ratio of in-plane molecular velocity components pertaining to particle emittance from each surface. Two cases are distinguished:

1. For $0 \leq x \leq h_s l_u$ (Fig. 5.31(a)): Particles arrive from the upper boundary within $-x\big/h_s \leq x \leq l_d$ (denoted as Δ_l) but not from the inlet.
2. For $h_s l_u < x \leq l_d$ (Fig. 5.31b): Particles come from the entire upper boundary and the inlet section within $1 - h_s - h_s l_u/x \leq y \leq 1$ (denoted as Δ_{in}).

The contributions of incoming particles from each surface to the impermeability condition (5.67) are analyzed through particle kinematics. For particles reflected from the outlet ($(l_d - x)/(1 + h_s) \leq c_x/c_y < \infty$ with both $c_x, c_y \leq 0$). Using Eq. (5.68), the integral expression for their contribution is derived:

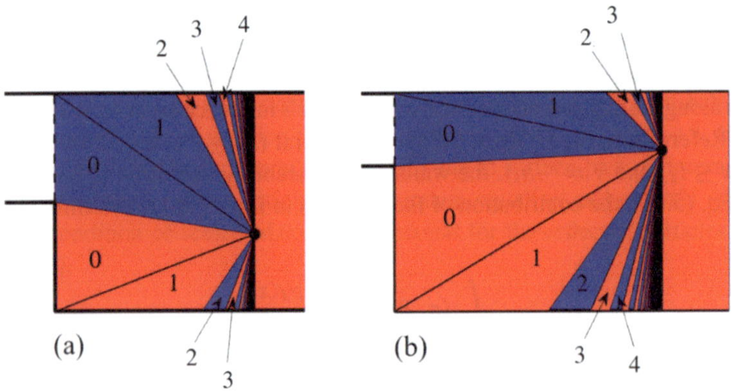

Fig. 5.31 Division of the $x > 0$ downstream part of a specularly reflecting channel into sections of particles originating from the inlet (blue) and outlet (red) at a given location (x_p, y_p). Channels with step sizes **a** $h_s = 1$ and **b** $h_s = 2$ are presented. Copied from open access Ref. [21]

$$\int\limits_{c\cdot\hat{n}<0} (c\cdot\hat{n}) f_{\text{out}}\, \mathrm{d}c = \int\limits_{-\infty}^{0} \int\limits_{-\infty}^{((x-l_d)/(1+h_s))|c_y|} \int\limits_{-\infty}^{\infty} c_y f_{\text{out}}\, \mathrm{d}c_z\, \mathrm{d}c_x\, \mathrm{d}c_y = \frac{\rho_{\text{out}}\sqrt{T_{\text{out}}}}{4\sqrt{\pi}}\left(\frac{l_d - x}{\sqrt{(1+h_s)^2 + (l_d - x)^2}} - 1\right) \tag{5.72}$$

For particles arriving from the step wall, ($x/h_s \le |c_x/c_y| < \infty$ with $c_x \le 0$ and $c_y \le 0$) using Eq. (5.69), the integral expression for their contribution is derived, representing the effect of step-reflected particles on the impermeability condition. The argument in $\rho_s = \rho_s(y)$ establishes a relation between each x-location along l_d and a corresponding y-position along the step for a particle arriving with a velocity ratio c_x/c_y. Applying an appropriate change of variables and integrating results:

$$\int\limits_{c\cdot\hat{n}<0} (c\cdot\hat{n}) f_s \mathrm{d}c = -\frac{x}{4\sqrt{\pi}} \int\limits_{-h_s}^{0} \frac{(p+h_s)\rho_s(p)}{\left[(p+h_s)^2 + x^2\right]^{3/2}} \mathrm{d}p \tag{5.73}$$

Considering to the upper l boundary ($(x - l_d)/(1 + h_s) \le c_x/|c_y| < x/h_s$, with $-\infty < c_x < \infty$ and $-\infty < c_y \le 0$):

$$\int\limits_{c\cdot\hat{n}<0} (c\cdot\hat{n}) f_l \mathrm{d}c = -\frac{(1+h_s)^2}{4\sqrt{\pi}} \int\limits_{-x/h_s}^{l_d} \frac{\rho_l(p)}{\left[(1+h_s)^2 + (x-p)^2\right]^{3/2}} \mathrm{d}p \tag{5.74}$$

For the case in Fig. 5.30b, the same expression applies to the l-surface, but with the lower integration limit $-x/h_s$ replaced by $-l_u$. Using Eq. (5.68), the additional contribution of the inlet section for $h_s l_u < x \le l_d$ is derived, accounting for the effect of particles entering the inlet on the impermeability condition.

$$\int\limits_{c\cdot\hat{n}<0} (c\cdot\hat{n}) f_{in} \mathrm{d}c = \int\limits_{-\infty}^{0} \int\limits_{\left(\frac{(x+l_u)}{(1+h_s)}\right)|c_y|}^{\left(\frac{x}{h_s}\right)|c_y|} \int\limits_{-\infty}^{\infty} c_y f_{in}\, \mathrm{d}c_z\, \mathrm{d}c_x\, \mathrm{d}c_y = \frac{1}{4\sqrt{\pi}}\left(\frac{l_u + x}{\sqrt{(1+h_s)^2 + (l_u + x)^2}} - \frac{x}{\sqrt{h_s^2 + x^2}}\right) \tag{5.75}$$

Summing Eqs. (5.71)–(5.74) in (5.70), the impermeability condition along $0 \le x \le h_s l_u$ is obtained by incorporating the contributions of particles from the outlet, step, and inlet sections.

$$2\rho_d(x) - x\int_{-h_s}^{0} \frac{(p+h_s)\rho_s(p)\mathrm{d}p}{\left[(p+h_s)^2+x^2\right]^{3/2}} - (1+h_s)^2 \int_{-x/h_s}^{l_d} \frac{\rho_l(p)\mathrm{d}p}{\left[(1+h_s)^2+(x-p)^2\right]^{3/2}} \tag{5.76}$$

$$= \rho_{\text{out}}\sqrt{T_{\text{out}}}\left(1 - \frac{l_d - x}{\sqrt{(1+h_s)^2 + (l_d - x)^2}}\right)$$

whereas for $h_s l_u \le x \le l_d$

$$\frac{x}{\sqrt{{h_s}^2 + x^2}} - \frac{l_u + x}{\sqrt{(1+h_s)^2 + (l_u + x)^2}} + \rho_{\text{out}}\sqrt{T_{\text{out}}}\left(1 - \frac{l_d - x}{\sqrt{(1+h_s)^2 + (l_d - x)^2}}\right) \tag{5.77}$$

Using similar arguments, the impermeability conditions for all solid walls are derived. For brevity, the equations for the lower upstream wall ($-l_u \le x \le 0,\ \mathrm{y} = 0$) and the step wall ($-h_s \le y \le 0,\ \mathrm{x} = 0$) are obtained respectively:

$$2\rho_u(x) - \int_{-l_u}^{l_d} \frac{\rho_l(s)\mathrm{d}s}{\left[(x-\mathrm{s})^2+1\right]^{\frac{3}{2}}} = 1 - \frac{l_u+x}{\sqrt{1+(l_u+x)^2}} + \rho_{\text{out}}\sqrt{T_{\text{out}}}\left(1 - \frac{l_d-x}{\sqrt{1+(l_d-x)^2}}\right) \tag{5.78}$$

and

$$2\rho_s(\mathrm{y}) - (\mathrm{y}+h_s)\int_0^{l_d} \frac{s\rho_s(s)\mathrm{d}s}{\left[\mathrm{s}^2 + (\mathrm{y}+h_s)^2\right]^{\frac{3}{2}}} + (\mathrm{y}-1)\int_0^{l_d} \frac{s\rho_l(s)\mathrm{d}s}{\left[\mathrm{s}^2 + (1-\mathrm{y})^2\right]^{\frac{3}{2}}}$$
$$= \rho_{\text{out}}\sqrt{T_{\text{out}}}\left(\frac{\mathrm{y}+h_s}{\sqrt{l_d^2 + (\mathrm{y}+h_s)^2}} + \frac{1-\mathrm{y}}{\sqrt{l_d^2 + (1-\mathrm{y})^2}}\right) \tag{5.79}$$

Along the upper ($\mathrm{y} = 1$) wall, for the case $-l_u \le x \le -\frac{l_d}{h_s}$ (no particles arrive from either the step or l_d walls) the impermeability conditions, respectively, are:

$$2\rho_s(\mathrm{y}) - (\mathrm{y}+h_s)\int_0^{l_d} \frac{s\rho_s(s)\mathrm{d}s}{\left[\mathrm{s}^2 + (\mathrm{y}+h_s)^2\right]^{\frac{3}{2}}} + (\mathrm{y}-1)\int_{l_d}^{l_d} \frac{s\rho_l(s)\mathrm{d}s}{\left[\mathrm{s}^2 + (1-\mathrm{y})^2\right]^{\frac{3}{2}}}$$
$$= \rho_{\text{out}}\sqrt{T_{\text{out}}}\left(\frac{\mathrm{y}+h_s}{\sqrt{l_d^2 + (\mathrm{y}+h_s)^2}} + \frac{1-\mathrm{y}}{\sqrt{l_d^2 + (1-\mathrm{y})^2}}\right) \tag{5.80}$$

and

$$= -1 - \frac{l_u + x}{\sqrt{1 + (l_u + x)^2}} - \rho_{\text{out}}\sqrt{T_{\text{out}}}\left(1 - \frac{x}{\sqrt{1+x^2}}\right) \tag{5.81}$$

and for the cases $-\frac{l_d}{h_s} \leq x \leq 0$ (where the step surface is obscured) and $0 \leq x \leq l_d$ (particles arrive from all walls) the impermeability conditions are:

$$= -1 - \frac{l_u + x}{\sqrt{1 + (l_u + x)^2}} - \rho_{\text{out}} \sqrt{T_{\text{out}}} \left(1 - \frac{l_d - x}{\sqrt{(1 + h_s)^2 + (l_d - x)^2}}\right) \quad (5.82)$$

By combining Eqs. (5.76)–(5.82), a system of coupled integral equations is formed that calculate how a substance's density changes at the system's boundaries ($\rho_u(x)$, $\rho_s(y)$, $\rho_d(x)$ and $\rho_l(x)$). When the inlet and outlet reservoirs are at the same conditions ($\rho_{\text{out}} = T_{\text{out}} = 1$), a uniform solution ($\rho_u(x) = \rho_s(y) = \rho_d(x) = \rho_l(x) = 1$) holds. For non-trivial cases of ρ_{out} and T_{out}, the equations are solved numerically by discretizing the boundary fluxes. By discretizing the fluxes and using Simpson's rule along the boundaries to approximate the integral terms, the impermeability conditions were solved numerically for any valid combination of $0 \leq \rho_{\text{out}} \leq 1$ and $0 < T_{\text{out}} \leq 1$. This produced non-homogeneous, equations that were solved using MATLAB code, requiring only a minor computational cost compared to the DSMC solution.

The probability density function in Eq. (5.66) can be determined after the fluxes have been calculated, and the hydrodynamic fields can then be computed using the proper quadratures over the velocity space. In particular, the density, pressure, as well as velocity components, can be calculated as:

$$\begin{cases} \rho(x, y) = \int\limits_{-\infty}^{\infty} f \, \mathrm{dc}, \; u_x(x, y) = \dfrac{1}{\rho(x, y)} \int\limits_{-\infty}^{\infty} c_x f \, \mathrm{dc}, \text{ and } u_y(x, y) = \dfrac{1}{\rho(x, y)} \int_{-\infty}^{\infty} c_y f \, dc \\ P(x, y) = \dfrac{2}{3} \int\limits_{-\infty}^{\infty} \left[(c_x - u_x)^2 + \left(c_y - u_y\right)^2 + c_z^2 \right] f dc \end{cases} \quad (5.83)$$

Also, the temperature is calculated by the scaled state equation of ideal gas $T = p/\rho$. At each (x, y) location, the integrations in (5.83) must average contributions from particles arriving from different boundaries based on channel geometry. Similar to boundary flux calculations, some boundaries may be obscured at certain points.

5.12.3.3 Specular Reflecting Walls

Specular interactions occur when the incident molecules collide with a solid surface and rebound elastically as if hitting a perfectly smooth wall. For a specular-reflecting wall ($\alpha = 0$), the macroscopic impermeability conditions are automatically satisfied since particles undergo mirror-like reflections without altering the probability density function. Inside the channel, particles can follow either the inlet or outlet Maxwellian distribution (see Eq. 5.68). The solution thus reduces to distinguishing at each (x, y) location between the distributions based on the direction of the in-plane velocity components (c_x, c_y).Using Eq. (5.83) and the equation of state, the hydrodynamic fields in the upstream part of the channel $(x \leq 0)$ are found to be uniform values [21]:

$$\begin{cases} \rho(x \leq 0, y) = \frac{1}{2}(1 + \rho_{out}) \\ u_x(x \leq 0, y) = \frac{1}{\sqrt{\pi}(1+\rho_{out})}\left(1 - \rho_{out}\sqrt{T_{out}}\right), u_y(x \leq 0, y) = 0 \\ p(x \leq 0, y) = \frac{1}{2} - \frac{2u_x}{3\sqrt{\pi}} + \frac{u_x}{3} + \rho_{out}\left(\frac{T_{out}}{2} + \frac{2u_x\sqrt{T_{out}}}{3\sqrt{\pi}} + \frac{u_x{}^2}{3}\right) \end{cases} \tag{5.84}$$

In the downstream region $(x > 0)$, particle sorting and hydrodynamic field calculations become more complex. In the downstream region, particles with $c_x < 0$ originate solely from the outlet reservoir, while those with $c_x > 0$ may come from either the inlet or outlet. The step wall acts as a reflecting source, reversing the x-velocity component of particles from the outlet. This kinematic constraint leads to non-uniform and two-dimensional flow behavior.

By tracking the origin of particles at all desired (x, y) locations in the downstream part, the distribution function f (x, y, c) in the five-dimensional phase space is determined. The hydrodynamic fields are then computed using Eq. (5.83). Figure 5.31 illustrates this process by dividing the $x > 0$ region based on particle origins (inlet or outlet) for step sizes hs = 1 (Fig. 5.31a) and hs = 2 (Fig. 5.31b). The numbers in the figure represent N_{coll}, the number of sidewall collisions before reaching a given point.

5.12.3.4 Results for Free-Molecular Regime

According to Fig. 5.32, the ballistic analytical solution is validated by comparing it with collisionless DSMC results under free-molecular conditions. The agreement is very good in all cases, confirming both methods' accuracy and the analytical solution's minimal computational effort. In a diffuse-reflecting channel, articles reflected from the step wall cannot reach the upstream region (x < 0) directly and can only reach after an additional collision with the downstream wall (l_d) or the upper boundary in the x > 0 region. In a specular-reflecting channel, step-emitted particles are entirely prevented from traveling upstream at any stage. In both cases, the "invisibility" of the step to the upstream section leads to sharp variations in the flow field at x = 0. These gradients decrease with a reduction in the Knudsen number since molecular collisions facilitate indirect momentum transfer from the step to the upstream section.

According to the specular wall analysis, hydrodynamic fields remain constant for $x < 0$ as predicted by Eq. (5.84), while x and y variations are observed in diffuse-reflecting channels. A key observation is that the density field remains constant in the temperature-driven specular case even for x > 0 (Fig. 5.32c). This constancy also holds in the y-direction, making the flow incompressible regardless of the imposed temperature drop. Additionally, x-velocity amplitudes are higher in the density-driven case than the temperature-driven case despite being driven by the same imposed pressure drop ($p_{out} = T_{out} = 0.1$). Furthermore, at fixed outlet conditions, velocity amplitudes are generally larger in the specular-reflecting setup compared to the diffuse case.

The streamline contours in Fig. 5.33 show that the flow remains fully attached to the step in all cases, with notably low flow speeds near the step. The speed jump at x = 0 (corresponding to the sharp gradients in density and x-velocity from Fig. 5.32) diminishes as the distance y from the step edge singularity increases.

To investigate the effects of problem parameters on the gas mass flow rate, considering the specular-reflecting case and Eq. (5.81), we have:

$$\dot{m}_{\text{specular}} = \frac{1}{2\sqrt{\pi}}\left(1 - \rho_{\text{out}}\sqrt{T_{\text{out}}}\right) \tag{5.85}$$

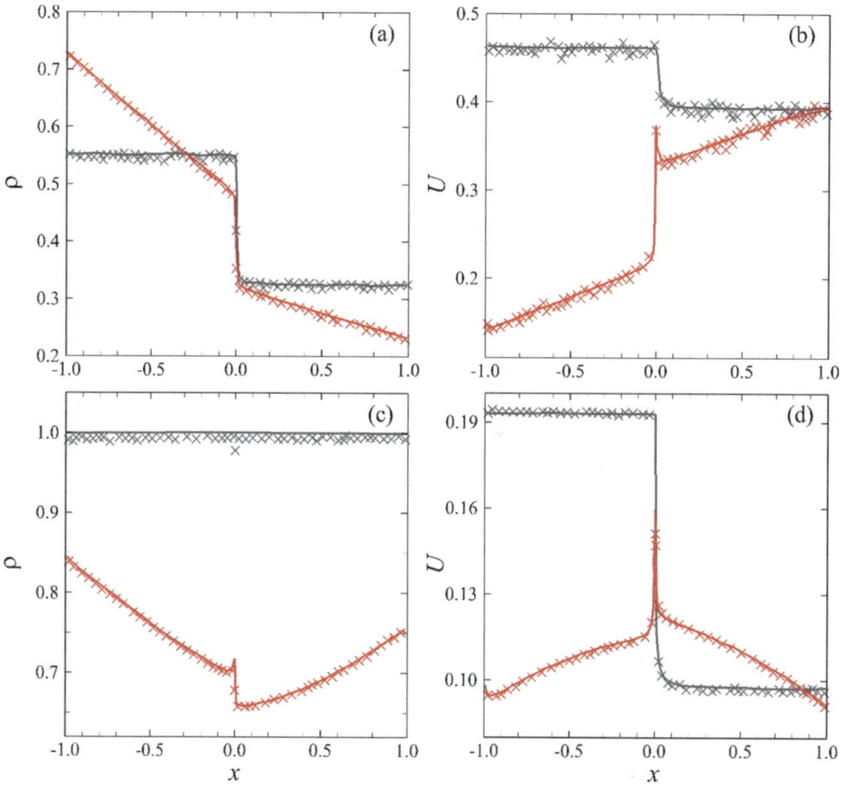

Fig. 5.32 Comparison between the free-molecular analytic (lines) and DSMC (symbols) solutions for (**a**, **b**) density-driven and (**c**, **d**) temperature-driven flow setups. The black solid lines and crosses denote results for a channel with diffuse walls, and the red dashed curves and circles mark data for a channel with specular walls. All comparisons are made at $y = 0.002$ along $-1 \leq x \leq 1$. Reproduced from Ref. [21]

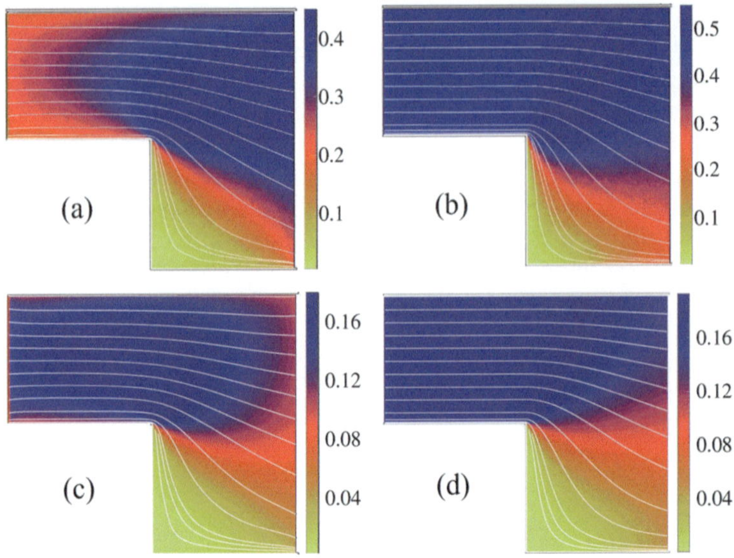

Fig. 5.33 The free-molecular flow field in (**a**, **b**) density-driven and (**c**, **d**) temperature-driven flow set-ups: (**a**, **c**) and (**b**, **d**) show results for channels with diffuse and specular walls, respectively. The contours and colourmaps show the flow streamlines and gas speeds, respectively. Reproduced from Ref. [21]

The linear dependence of $\dot{m}$ and ρ_{out} and the square-root dependence on T_{out} correspond to the higher x-velocities observed in the density-driven case, as shown in Fig. 5.32 (dashed lines and circles in Fig. 5.32b, d). In the specular-reflecting case, channel geometry does not affect the mass flow rate ($\dot{m}$), whereas in diffuse-scattering walls, a particle entering from the inlet may not exit through the outlet under diffuse conditions, which leads to reduced flow speed (as shown in Figs. 5.32 and 5.33) and a lower overall mass flow rate. This indicates the decreased permeability of diffuse channels to free-molecular gas transfer.

The effect of step size is highlighted in Fig. 5.34 for two cases, density-driven and temperature-driven flow. Unlike the specular case, where $\dot{m}$ remains unaffected by h_s, in the diffuse case, $\dot{m}$ increases monotonically with h_s. Although the step is not directly visible to the inlet section (since no particles are emitted from the step directly toward the inlet), it affects axial flow velocity and mass flow rate through back-reflected particles from the outlet and influences density fluxes at the upper ($\rho_l(x)$) and downstream ($\rho_d(x)$.) walls.

Each panel in Fig. 4.35 presents flow streamlines and gas speed (color maps) for different combinations of T_{out} and h_s in the downstream region $x \in [0,1]$, $y \in [-h_s, 0]$ for $\rho_{\text{out}} = 1$. The results are based on the analytical solution, as the extremely low flow velocities ($|u| \ll 0.01$) near the step in these smaller thermodynamic non-uniformities make DSMC simulations inefficient due to a significant decrease in the

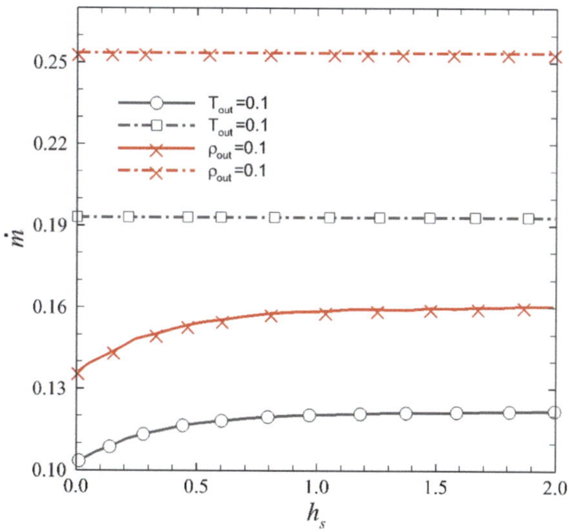

Fig. 5.34 Effects of the step heighth_s, on the mass flow rate $\dot{m}$ in the free-molecular regime. Solid and dashed curves show results for channels with diffuse and specular walls, respectively. $\rho_{\text{out}} = 0.1$ and $T_{\text{out}} = 0.1$ notations pertain to density-driven (with $T_{\text{out}} = 1$) and temperature-driven (with $\rho_{\text{out}} = 1$) flow setups, respectively. Reproduced from Ref. [21]

simulation signal-to-noise ratio. The analytical solution becomes particularly useful for studying the free-molecular flow field in this limit.

Unlike the flow pattern in Fig. 5.33, the results in Fig. 5.35 reveal a detached zone near the step surface, where streamlines originate and end at the step edges (0, 0) and (0, –h_s). These streamlines, marked by dashed curves, expand with increasing $T_{\text{out}} \rightarrow 1$ and h_s. This expansion is accompanied by decreased flow velocity amplitude, as shown in the speed colourmaps. This detached region differs from recirculation patterns at lower Knudsen numbers, where closed streamline contours typically form (see Fig. 5.37d–f). However, this type of flow detachment is significant, as the gas in this zone does not reach the channel outlet and remains isolated from the bulk flow. A similar behavior (not shown here) was also observed in density-driven flows ($\rho_{\text{out}} \rightarrow 1$, $T_{\text{out}} = 1$) and specular-reflecting channels. In these configurations, where higher velocities occur (see Figs. 5.33b and 5.34a, b), relatively larger $\rho_{\text{out}} \rightarrow 1$ values are needed to produce detached gas layers of similar size to those in Fig. 5.29.

5.12.3.5 Effect of Molecular Collisions

This section investigates the validity range of the free-molecular description in the backward step flow at high but finite Knudsen numbers and the qualitative impact of infrequent molecular collisions on the gas flow field. Figure 5.36 explores the breakdown of the free-molecular regime as the Knudsen number decreases. In the density-driven case (Fig. 5.36a, b), the collisionless description remains valid up to $\text{Kn} \approx 1$, where molecular collisions slightly smooth out the sharp jump at $x = 0$. However, for $\text{Kn} \lesssim 1$, molecular collisions become significant, leading to large

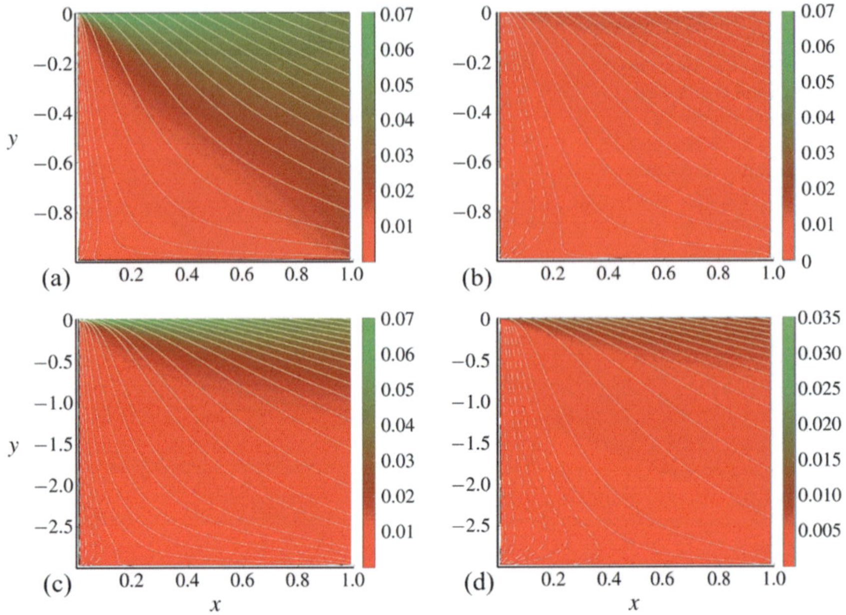

Fig. 5.35 Effects of the step size $((a, b)h_s = 1; (c, d)h_s = 3)$ and outlet temperature $((a, c)T_{\text{out}} = 0.4; (b, d)T_{\text{out}} = 0.8)$ on the flow streamlines (curves) and gas speed (colourmaps) in the vicinity of the step wall (zone III) at free-molecular conditions. In all cases, $\rho_{\text{out}} = 1$, and the channel walls are diffusely reflecting. The dashed curves depict streamlines originating and ending at the $(0, 0)$ and $(0, -h_s)$ edges of the step surface, respectively. Reproduced from Ref. [21]

deviations from the collisionless results, as seen in the Kn $=$ 0.1 curve (marked by triangles). A similar breakdown occurs in the temperature-driven case (Fig. 5.36c, d), but an important observation is that noticeable differences (up to 4%) appear even atKn $=$ 100. This indicates that the free-molecular regime collapses at much higher Knudsen numbers in low-speed flow configurations (see Fig. 5.33a, d). Even minor molecular collisions in these cases cause a visible departure from the incompressible free-molecular state.

In Fig. 5.37, for the diffuse-reflecting case, the flow remains almost unchanged between Kn $=$ 5 and Kn $=$ 1 (Fig. 5.37a, b), closely resembling the collisionless-flow map in Fig. 5.33a. This aligns with Fig. 5.36a, b, confirming that the free-molecular description remains valid up to Kn $\approx$ 1. However, Fig. 5.37c shows increased gas speed and irregularities in the flow streamlines near the step. At slightly lower Knudsen numbers, these irregularities evolve into a distinct recirculation zone near the step. Thus, the transition between attached and detached flow profiles, where flow speeds increase in bulk and recirculating velocities first emerge at this step, marking the breakdown of the free-molecular description.

In the specular-reflecting case (Fig. 5.37d–f), flow detachment occurs at much higher Knudsen numbers than diffuse-reflecting walls. As seen in Fig. 5.15c, d, the breakdown of the free-molecular regime happens at Kn $\approx$ 100. With decreasing Kn,

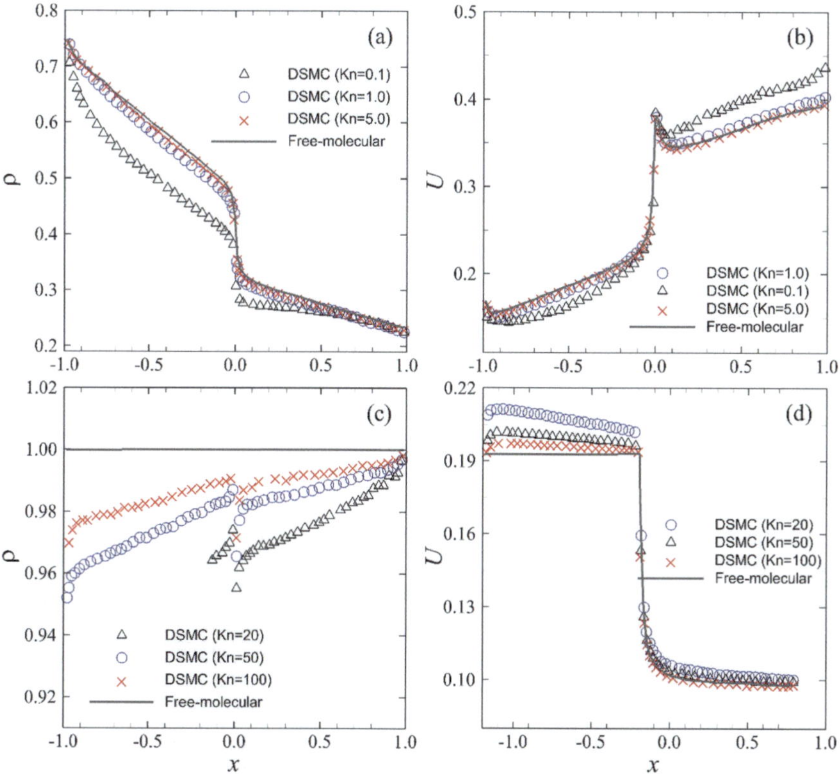

Fig. 5.36 Breakdown of the free-molecular regime with decreasing Knudsen number: x-variations at $y = 0.01$ of the (a, c) density and $(b, d)x$-velocity fields for $(a, b)\rho_{out} = 0.1$, $T_{\text{out}} = 1$ and diffuse-reflecting boundaries; $(c, d)\rho_{\text{out}} = 1$, $T_{\text{out}} = 0.1$ and specular-reflecting boundaries. The solid line shows the free-molecular solution and the symbols mark DSMC data at the indicated values of Kn. Reproduced from Ref. [21]

step detachment appears at Kn and intensifies as Kn decreases further, increasing velocity amplitude and recirculation zone size. The temperature-driven flow, which has lower speeds than the density-driven case, is more prone to detachment at higher rarefaction levels.

5.13 Porous Microchannels

Unconventional gas resources, characterized by low-permeability reservoirs, have become critical hydrocarbon sources. The prediction of production from unconventional reservoirs remains a contentious and complex issue, primarily due to its intricate dependence on pore structures and the flow properties of porous media. Such reservoirs, modeled as low-permeability porous media, consist of numerous

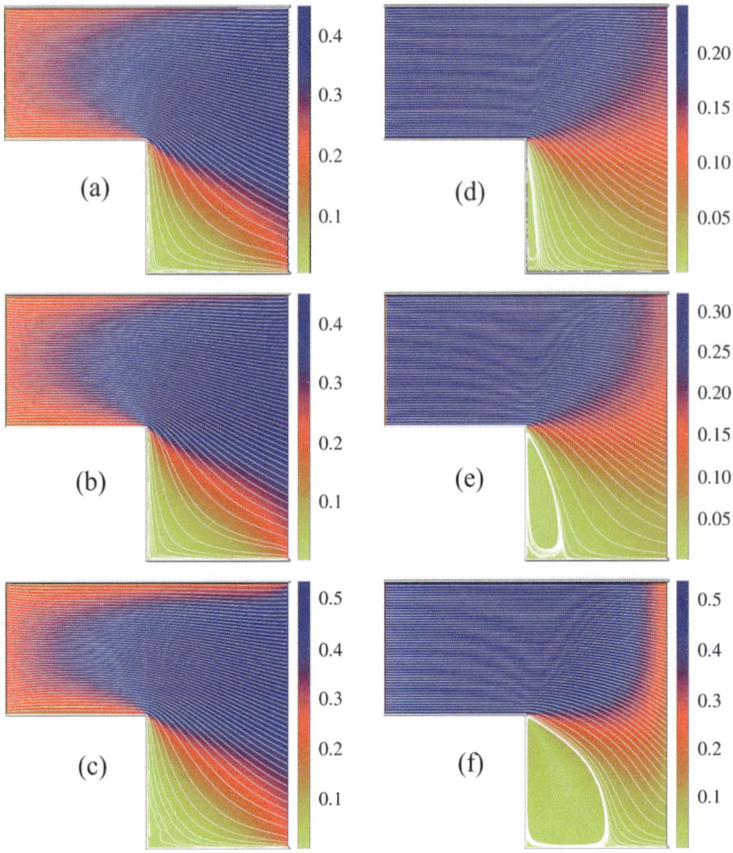

Fig. 5.37 Effects of the Knudsen number ((a, d)Kn $= 5$; (b, e)Kn $= 1$; (c, f)Kn $= 0.1$) and outlet reservoir conditions ($(a-c)\rho_{out} = 0.1$, $T_{\text{out}} = 1$ and diffuse-reflecting walls; $(d-f)\rho_{\text{out}} = 1$, $T_{\text{out}} = 0.1$ and specular-reflecting walls) on the flow streamlines (curves) and gas speed (colourmaps). Reproduced from Ref. [21]

shapeless particles dispersed in space, and fluids flow through interstitial volume within this media, i.e., pores. Evaluating essential properties like permeability and porosity requires understanding their dependence on pore structure characteristics. While conventional Darcy's law traditionally estimates permeability, these reservoirs experience significant rarefaction effects as pore sizes approach the mean free path of gas molecules, necessitating advanced modeling approaches.

Shariati et al. [22] simulated the flow properties and permeability enhancement due to rarefaction effects in porous microchannels across a wide range of Knudsen numbers using the DSMC method. A Python code was developed to randomly distribute solid circular particles with a specified radius within the channel with no interpenetration between the obstacles, and snappyHexMesh was utilized for mesh generation in OpenFOAM. The DSMC solver, dsmcFoamPlus, was employed to simulate rarefied gas flow. Key parameters such as porosity were analyzed for their impact on flow behavior, including velocity profile, apparent gas permeability and mass flow rate. To investigate how decreasing the porosity affects the morphology of the porous structure and the rarefaction in microscale media, simulations were performed for gas flow in a porous microchannel with porosities ranging from 0.4 to 1. They show that the method can simulate microporous media with up to 40% porosity while providing the accuracy required to analyze the transport mechanisms in such media.

5.13.1 Porosities

Porosity, defined as the fraction of the void region in a domain, influences the structure of porous media and the induction of rarefaction. Figure 5.38 shows the streamwise velocity distribution of porous microchannels at different porosities. As solid particles are added, reducing porosity, the flow path becomes increasingly tortuous, decreasing flow velocity. In Fig. 5.1a, the flow path is completely straight, but each additional solid particle acts as an obstacle, causing the flow to fluctuate. By Fig. 5.1h, where the domain is nearly filled with solid particles, the flow velocity throughout the domain significantly decreases to its minimum.

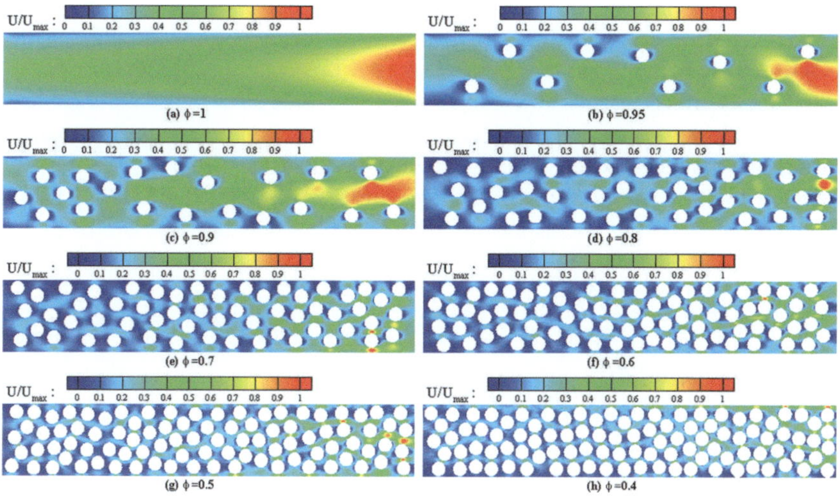

Fig. 5.38 Streamwise velocity distribution of porous microchannel at different porosities. In all the cases $\mathrm{Kn}_{in} = 0.1$, $P_{in}/P_{out} = 3$. Copied from open access Ref. [22]

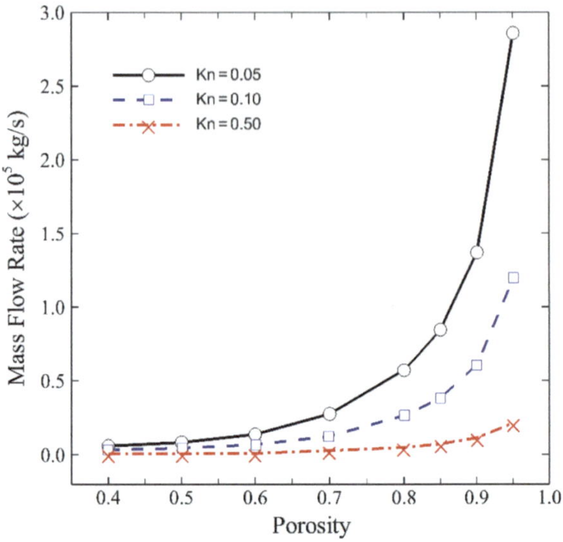

Fig. 5.39 Variation of mass flow rate with porosity at different Knudsen numbers. Reproduced from Ref. [22]

5.13.2 Mass Flow Rate

Figure 5.39 presents the variation of the mass flow rate at different porosities. As shown, reducing porosity significantly influences the mass flow rate, which is more pronounced at higher porosities. In particular, at Kn $=$ 0.05, both slip velocity and viscous flow are the most dominant transport mechanisms. Reducing porosity considerably affects viscous flow at this Knudsen number since the pressure gradient drives the flow. However, the dominant transport mechanism is mostly Knudsen diffusion at higher Knudsen numbers. Hence, the molecules are not derived by the pressure gradient but by their interaction with the solid boundaries. As a result, inducing rarefaction in porous media makes the mass flow rate less altered by the structure.

5.13.3 Permeability

Permeability is a critical property of porous media, reflecting their capacity to transmit fluid flows. In microporous channels, the wide range of pore sizes results in fluid flow extending beyond the slip flow regime (Kn $>$ 0.1), where the continuum assumption breaks down. Consequently, the intrinsic permeability, typically a characteristic of the porous structure, no longer holds its conventional value. Instead, it is redefined as the apparent gas permeability (K_{app}), which accounts for the deviations introduced by rarefaction effects in such flow regimes.

Fig. 5.40 Permeability of porous media normalized by the diameter of solid particles ($d_p = 1000nm$) at different inlet Kn. Reproduced from Ref. [22]

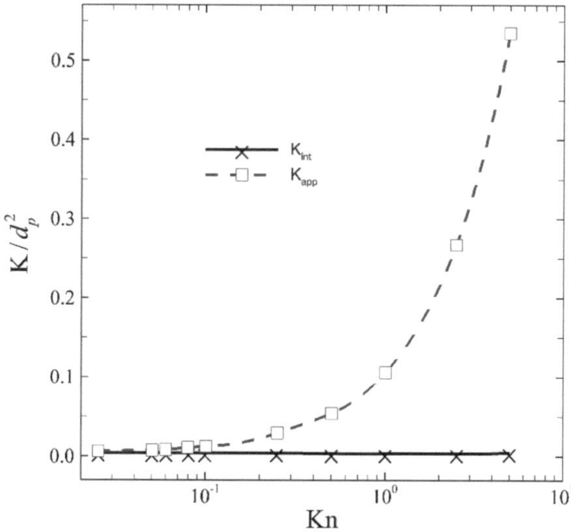

Figure 5.40 shows the permeability variation at different Knudsen numbers, where the inlet pressure is reduced while maintaining a pressure ratio of 3. The intrinsic permeability (K_{int}) remains independent of the Knudsen number, whereas the apparent permeability (K_{app}) increases significantly as the inlet pressure decreases. The figure also demonstrates that K_{app} converges to K_{int} in the continuum regime, confirming that apparent permeability deviates from intrinsic permeability only when rarefaction effects are present in the flow.

Over the past decades, researchers have developed apparent permeability models incorporating morphological parameters and various transport mechanisms to predict porous media's permeability accurately. According to Fig. 5.41, At low Knudsen numbers within the slip flow regime, the simulation aligns with the Klinkenberg and Beskok-Karniadakis (B-K) models, which account for viscous flow and slippage at solid boundaries. Although models by Sakhaee-Pour, Bryant, Mohammadmoradi, Kantzas, and Zhao use the Klinkenberg first-order equation with varying slippage factors, they are less accurate in predicting permeability. At higher Knudsen numbers, the results deviate and follow Kawagoe's model, which incorporates viscous flow, slip velocity, Knudsen diffusion, and the effect of tortuosity, providing a more accurate representation of flow mechanisms and porous media morphology.

Also, Shariati et al. [23] used the DSMC method and dsmcFoam + to analyze flow characteristics on permeability in microporous media with complex obstacles across various Knudsen numbers in rarefied and transitional regimes. They utilize the PoreSpy package to generate and analyze the morphological features of porous media. The generator module within PoreSpy provides functions, such as Blobs, which create porous structures by generating an image of random noise, applying a Gaussian blur, and normalizing values for the image's direct thresholding to obtain binary images, i.e., solid and void. PoreSpy facilitates data extraction from 3D

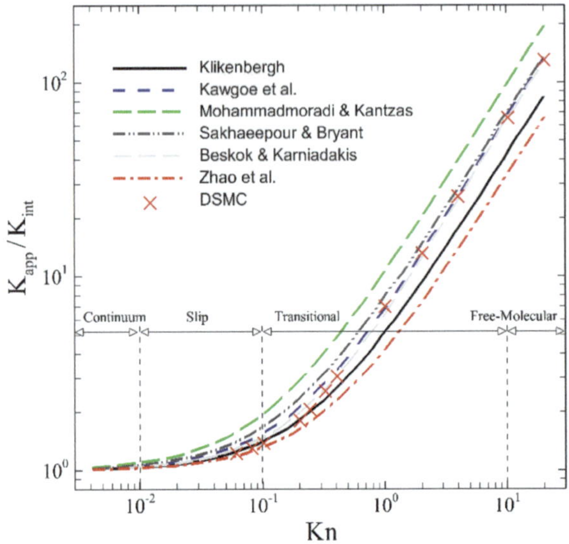

Fig. 5.41 Comparison of different permeability models with the current simulation result. Reproduced from Ref. [22]

images of porous materials, commonly obtained via X-ray tomography. Figure 5.42 illustrates the porous structure and meshed area generated using SnappyHexMesh [23].

Figure 5.43 compares the DSMC results with various apparent permeability models for a microporous channel. DSMC results align with the Klinkenberg and Beskok models at low Knudsen numbers, while at higher Knudsen numbers, the results shift between the Zhao and Lu models. The Zhao model predicts a lower permeability ratio due to its slippage factor of 0.8 instead of 1. In contrast, the Lu model overestimates it, as their study did not consider porous particles. At high Knudsen numbers, DSMC results at T = 300 K agree with Kawagoe's model, which uses spherical particles with DSMC. Additionally, increasing the heat flux ratio from 0 to –0.5 enhances the permeability ratio, with results at $q_W = -0.3$ matching Yang and Weigand's DSMC-based models for porous media with square and circular barriers.

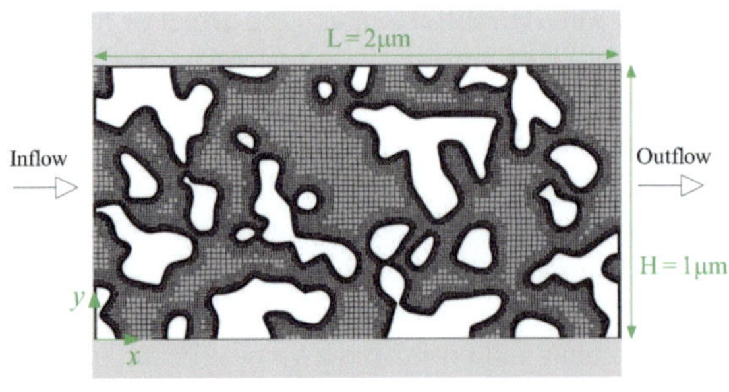

Fig. 5.42 Schematic of a generated porous structure with poreSpy package and meshed region. Reproduced from Ref. [23]

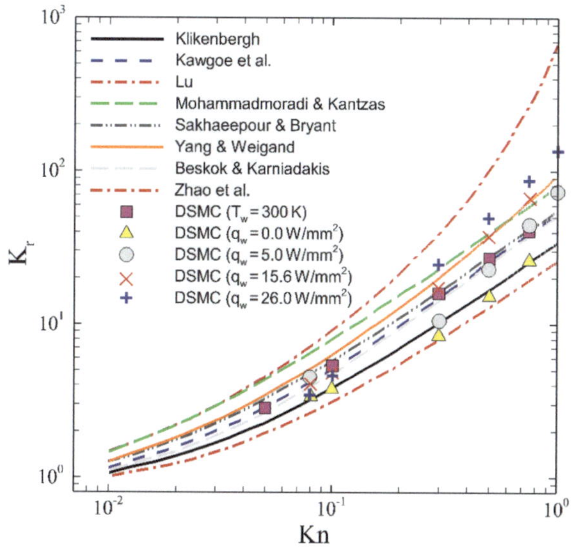

Fig. 5.43 Comparison of the permeability ratio (k_r) model with the current simulation result for the porosity of 0.8. Reproduced from Ref. [23]

Figure 5.44 illustrates permeability as a function of porosity and flow regime in porous media. Intrinsic permeability (K_{int}) decreases with reduced porosity due to its dependence on the morphology of porous media. Apparent permeability (K_{app}) follows a similar trend but varies more intensely at higher porosities, indicating that the impact of porous structures on apparent permeability lessens as porosity decreases. The permeability ratio (K_{app}/K_{int}) highlights the dominant transport mechanism: at high porosities, the equivalent pore size is considerable compared to the mean free path of the gas molecules, and viscous flow dominates due to higher molecular collisions than the collision between the molecules and the solid walls. While at lower porosities, Knudsen diffusion becomes significant as collisions between gas molecules and solid walls increase.

5.13.4 *Velocity Distribution*

Figure 5.45, the simulations of argon gas flow with porosities ranging from 0.5 to 0.9, reveal that solid particles disrupt velocity distribution within the microporous channels. As porosity increases, the flow path becomes less convoluted, leading to higher flow velocities [23].

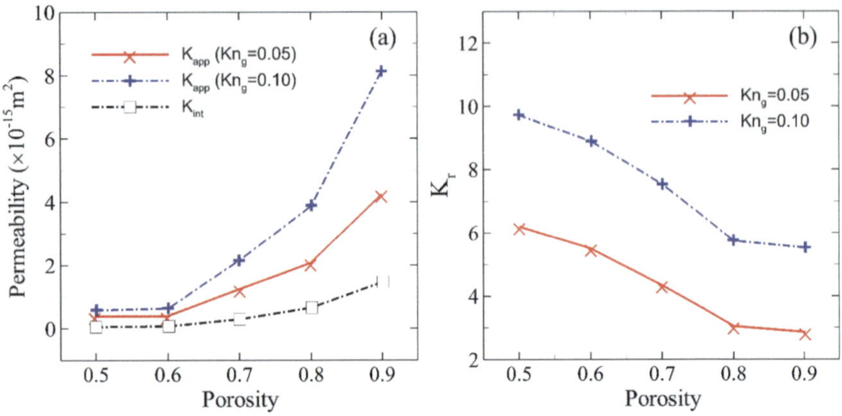

Fig. 5.44 Apparent (K_{app}), intrinsic (K_{int}) permeabilities and their ratio at various porosities. Reproduced from Ref. [23]

5.13.5 *Tortuosity*

Tortuosity is a key parameter in porous media, measuring the ratio of the average microscopic flow path length to the length of the medium in the streamwise direction. Figure 5.46 illustrates that the tortuosity factor increases as porosity decreases, indicating a more convoluted flow path in lower porosities. However, the slippage effect smooths the flow channels with higher Kn, reducing the hydraulic tortuosity factor despite the complex structure. As the Knudsen number increases, wall boundaries cease to act as friction surfaces for gas particles. Instead, the slip velocity facilitates the flow of gas particles through the pores, reducing the tortuosity of the flow path.

5.14 Summary

In this chapter, pressure-driven flows were comprehensively examined. Analytical solutions of the governing equations for the flow, along with the slip velocity boundary conditions, were presented. Analytical relations for velocity profiles, flow rate, and pressure distribution with different orders of accuracy were derived as a function of the Knudsen number, and the accuracy of these relations was evaluated. Modifying the molecular mean free path length to enhance the accuracy of analytical relations for a wider range of Knudsen numbers was discussed. Heat transfer and temperature distribution were investigated, and the latest experimental achievements in rarefied pressure-driven flow were mentioned.

On behalf of micro-step flow, the cold-to-hot transfer phenomenon results from the interaction between a higher-order term in the heat flux formula, dependent on the second derivative of velocity, and the Fourier term, observed when there was a

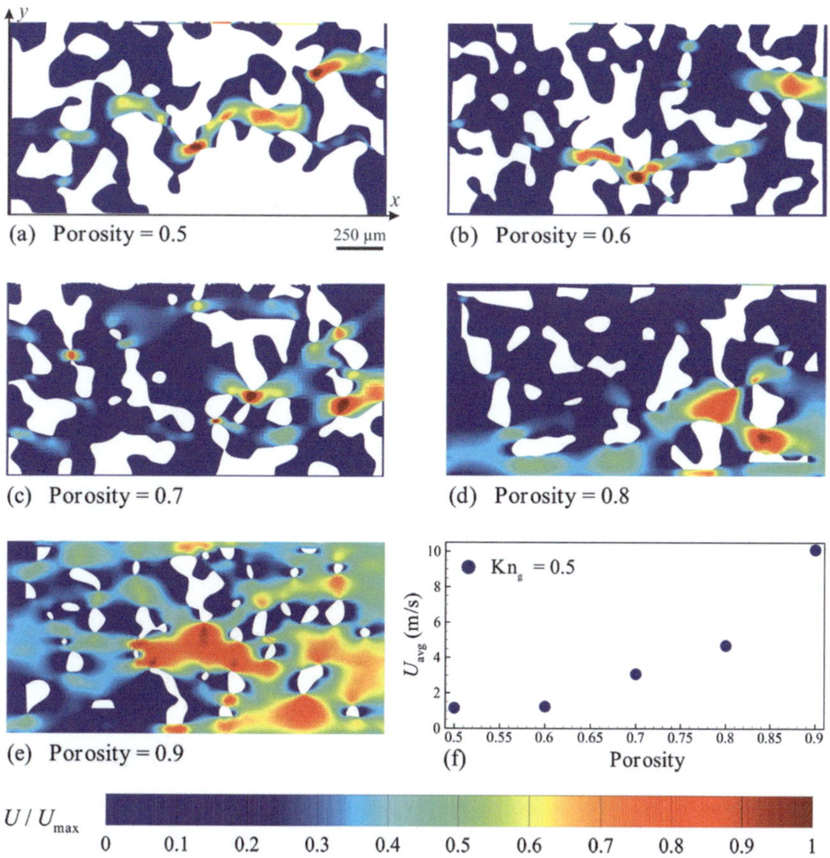

Fig. 5.45 Influence of various porosity on velocity profile in the microporous channel at $Kn_{inlet} = 0.5$, $P_{in}/P_{out} = 2$. Copied from open access Ref. [23]

large separation zone and low-temperature gradient. The cold-to-hot transfer effect diminishes as wall temperature or Knudsen number increases. A second separation region is identified on the top wall of the step for Re > 400. Increasing Kn at a constant pressure ratio caused this secondary recirculation zone to disappear and reduced the separation lengths of the primary vortex. For low Kn conditions (0.0001 < Kn < 0.001), a vortex was detected on the top wall of the micro-step geometry, driven by an extended positive pressure gradient. For Kn > 1, the recirculation zone was confined to the vertical wall. Additionally, for Kn < 0.001, the flow experiences velocity slip and temperature jump, proving that the classical slip regime range is not valid for step geometries. At finite Knudsen numbers, the collisionless-flow regime breaks down at higher Knudsen numbers for lower gas speed flows, leading to step flow separation and recirculation. It is shown that conventional flow detachment, as a recirculating flow reattaching along the downstream channel wall, occurs at high but finite Knudsen conditions, which are not captured by the collisionless description.

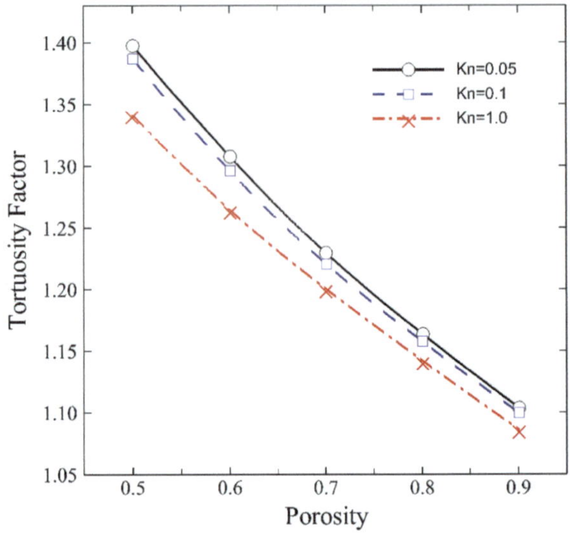

Fig. 5.46 Tortuosity (τ_h) versus porosity at various Kn. Reproduced from Ref. [23]

Unlike in a diffuse-reflecting channel, the step is completely obscured from the upstream part of a specular channel, resulting in uniform flow along its upstream section. This leads to higher mass flow rates in the specular-reflecting configuration, which remain unaffected by the step geometry (unlike the diffuse-reflecting setup). The study rationalizes the higher flow rates observed in the latter by comparing temperature-drop-driven and density-drop-driven flows. The linear and square-root dependencies of the specular channel flow rate on density and temperature drop to the lower gas velocities in the temperature-driven case, making it more susceptible to flow separation.

At porous microchannel flows, the hydraulic tortuosity factor and permeability ratio increase as porosity decreases. With higher Knudsen numbers, tortuosity decreases due to slip flow at solid boundaries, causing the permeability ratio to rise. The comparison of different apparent permeability models reveals that the B-K (Beskok and Karniadakis) and Klinkenberg models are only valid up to the transition regime (Kn < 0.1 and $\frac{q_w}{q_b} < -0.1$). At higher Knudsen numbers, where Knudsen diffusion and tortuosity become significant, the results align with Kawagoe's model, which accounts for these effects. This result demonstrates that Knudsen diffusion, usually neglected at slip flow regime, is crucial at higher Knudsen numbers. Additionally, for higher heat flux ratios ($\frac{q_w}{q_b} > -0.1$), the results match second-order models like Yang, highlighting the importance of Knudsen diffusion and heat flux effects in influencing apparent permeability in the transition regime.

Problems

1. Explain why, although the resultant tangential velocity of the reflected molecules on the diffuse wall is zero, the velocity slip is non-zero?

2. Prove that on a specular wall, thermal creep, shear stress, and heat flux do not occur are zero.
3. If the unsimplified equation of Poiseuille flow in a channel with length L and height h is as follows:

$$\rho u \frac{du}{dx} = -\frac{dP}{dx} + \mu \frac{\partial^2 u}{\partial y^2}$$

Show that in Poiseuille flow with low Reynolds numbers and a high aspect ratio (L/h), we have:

$$\frac{d^2 P}{dx^2} = 0$$

4. Using the first-order slip boundary condition, derive the velocity profile, flow rate, and pressure distribution.

$$U_s - U_w = \frac{2-\sigma}{\sigma} Kn \left(\frac{\partial U}{\partial n} \right)$$

5. A researcher considered the Poiseuille flow equation in dimensional form and used the following dimensional boundary condition to solve the equation. Derive the velocity profile, flow rate, and pressure distribution.

$$u_z = \sigma_p \lambda \frac{du}{dy}$$

6. The temperature jump boundary condition, which includes the vertical velocity gradient term, is as follows:

$$T + \frac{2}{5R}\sqrt{\frac{2}{R}} a_{TII} \frac{k_w}{\rho\sqrt{T_w}} \nabla_n T = T_w + \frac{1}{R} a_{VII} \frac{\mu_w}{\rho} ((\nabla u \cdot n) \cdot n)$$

   a. Calculate the second term on the right-hand side for a horizontal flat plate and a vertical flat plate
   b. Calculate the second term on the right-hand side for planar Couette and cylindrical Couette.

7. This question aims to solve the problem of the flow around a sphere in isothermal flow with the environment and the very small flow velocity compared to the average molecular velocity. Under these conditions, the energy equation is simplified as follows:

$$\nabla^2 \tau = 0$$

where $\tau = \frac{T-T_0}{T_0}$ is the dimensionless temperature and T_0 is the temperature of the environment and the surface of the sphere. The boundary condition given in question 6 is simplified in this case as follows:

$$\tau = -\frac{3}{2}\left[\frac{|U_i|}{(2RT_0)^{\frac{1}{2}}}\right]d_4{}^{*}cos\theta,\ \mathrm{at} r = 1$$

$$\tau \to 0 \text{ as } r \to \infty$$

$$d_4{}^{*} = -0.446749$$

Solve the Laplace equation governing the dimensionless temperature in spherical coordinates with the given boundary condition and show that the solution is as follows:

$$\tau = -\frac{3}{2}\frac{|U_i|}{(2RT_0)^{\frac{1}{2}}}d_4{}^{*}\frac{1}{r^2}cos\theta$$

Plot the temperature distribution around the sphere from the above equation on the surface *($r = 1$)*. r is the radial component of the flow dimensionless to the sphere's radius.

8. Show that the Poiseuille number in pressure-driven flow with first and second-order boundary conditions is as follows:

$$C_f Re = \frac{24}{(1 + 6C_1 Kn)}$$

$$C_f Re = \frac{24}{(1 + 6C_1 Kn + 12C_2 Kn^2)}$$

Appendix

The Python code written for solving the nonlinear second-order pressure distribution Eq. (5.44) is provided below. This code assumes that $\overline{\alpha} = 2.2$, $b = -1$, the pressure ratio PR $= 2.5$, the exit pressure is 1 atmosphere, and the exit Knudsen number is 0.055.

```
import numpy as np
import matplotlib.pyplot as plt
from scipy.optimize import brentq
import pandas as pd

# Constants
ALFA= 2.2
bb = -1
SIGMA= 1
KN_OUT= 0.055
COF= (2 - SIGMA) / SIGMA
COF2= 0
POUT= 1
PR= 2.5
PIN= POUT* PR

# Pre-calculate constant part of B
PR1 = PR - 1
B = PR ** 2 - 1 + 2 * (6 + ALFA)* COF* KN_OUT * PR1

# Arrays for result
num_points = 300 # Reduce for faster execution; increase for full resolution
Q_vals = np.linspace(0, 1, num_points)
DeltaP_vals = np.zeros(num_points)

# Root-finding loop
for i, Z in enumerate(Q_vals):
  def f_root(P):
    if P <= bb * KN_OUT: # Avoid invalid log
      return 1e6
    term_log = np.log((P - bb * KN_OUT) / (1 - bb * KN_OUT))
    return P ** 2 - 1 + 2 * (6 + ALFA)* COF* KN_OUT * (P - 1) + \
        2 * (6 * bb + ALFA)* COF2* KN_OUT ** 2 * term_log - B * (1 - Z)

  try:
    P_sol = brentq(f_root, 0.1, 10.0, maxiter=100)
  except ValueError:
    P_sol = np.nan

  PL= POUT+ (PIN- POUT)* (1 - Z)
  DeltaP_vals[i]= (P_sol - PL)/ POUT

# Plot results
plt.plot(Q_vals, DeltaP_vals, label="ΔP/POUT")
plt.xlabel("Q")
```

```
plt.ylabel("ΔP (normalized)")
plt.grid(True)
plt.title("Normalized Pressure Difference vs Q")
plt.legend()
plt.tight_layout()
plt.savefig("delta_pressure_fast.png", dpi=300)
plt.show()

# Save results
df = pd.DataFrame({"Q":Q_vals, "DeltaP": DeltaP_vals})
df.to_csv("delta_pressure_fast.csv", index=False)
```

References

1. Karniadakis, G., Beskok, A., Aluru, N.: Microflows and Nanoflows: Fundamentals and Simulation, vol. 29. Springer Science & Business Media (2006)
2. Ohwada, T., Sone, Y., Aoki, K.: Numerical analysis of the Poiseuille and thermal transpiration flows between two parallel plates on the basis of the Boltzmann equation for hard-sphere molecules. Phys. Fluids A **1**(12), 2042–2049 (1989)
3. Roohi, E., Darbandi, M.: Extending the Navier-Stokes solutions to transition regime in two-dimensional micro- and nano-channel flows using information preservation scheme. Phys. Fluids **21**, 82001 (2009)
4. Arkilic, E.B., Schmidt, M.A., Breuer, K.S.: Gaseous slip flow in long microchannels. J. Microelectromech. Syst. **6**, 167–178 (1997)
5. Stefanov, S., Roohi, E., Shoja-Sani, A.: A novel transient-adaptive subcell algorithm with a hybrid application of different collision techniques in direct simulation Monte Carlo (DSMC). Phys. Fluids **34**(9) (2022)
6. Agrawal, A., Kushwaha, H.M., Jadhav, R.S.: Microscale Flow and Heat Transfer. Springer International Publishing, Cham (2020)
7. Kosuge, S., Sato, K., Takata, S., Aoki, K.: Flows of a binary mixture of rarefied gases between two parallel plates. In: Capitelli, M. (ed.) Rarefied Gas Dynamics, pp. 150–155 (2005)
8. Sone, Y.: Molecular Gas Dynamics: Theory, Techniques, and Applications. Birkhäuser, Boston (2007)
9. Cercignani, C., Daneri, A.: Flow of a rarefied gas between two parallel plates. J. Appl. Phys. **34**, 3509–3513 (1963)
10. Takata, S., Funagane, H.: Poiseuille and thermal transpiration flows of a highly rarefied gas: over-concentration in the velocity distribution function. J. Fluid Mech. **669**, 242–259 (2011)
11. Myong, R.S.: A full analytical solution for the force-driven compressible Poiseuille gas flow based on a nonlinear coupled constitutive relation. Phys. Fluids **23**(1), 012002 (2011)
12. Xie, J., Borg, M.K., Gibelli, L., Henrich, O., Lockerby, D.A., Reese, J.M.: Effective mean free path and viscosity of confined gases. Phys. Fluids **31**(7), 072002 (2019)
13. Stops, D.W.: The mean free path of gas molecules in the transition regime. J. Phys. D Appl. Phys. **3**(685), 696 (1970)
14. Guo, Z.L., Shi, B.C., Zheng, C.G.: An extended Navier-Stokes formulation for gas flows in the knudsen layer near a wall. Europhys. Lett. **80**, 24001 (2007)
15. Abramov, R.V.: Gas near a wall: Shortened mean free path, reduced viscosity, and the manifestation of the knudsen layer in the Navier-Stokes solution of a shear flow. J. Nonlinear Sci. **28**, 833–845 (2018)
16. Dongari, N., Zhang, Y., Reese, J.M.: Modeling of Knudsen layer effects in micro/nanoscale gas flows. J. Fluids Eng. **133**(7) (2011)

17. Barisik, M., Beskok, A.: Molecular free paths in nanoscale gas flows. Microfluidics Nanofluidics **18**(5–6), 1365–1371 (2015)
18. Fratantonio, D., Rojas-Cárdenas, M., Barrot, C., Baldas, L., Colin, S.: Velocity measurements in channel gas flows in the slip regime by means of molecular tagging velocimetry. Micromachines **11**(4), 374 (2020)
19. Mahdavi, A.-M., Roohi, E.: Investigation of cold-to-hot transfer and thermal separation zone through nano step geometries. Phys. Fluids **27**(7) (2015)
20. Mahdavi, A., Roohi, E.: A study on micro-step flow using a hybrid direct simulation Monte Carlo–Fokker–Planck approach. Phys. Fluids **34**(6) (2022)
21. Manela, A., Gibelli, L.: Free-molecular and near-free-molecular gas flows over backward facing steps. J. Fluid Mech. **889**, A22 (2020)
22. Shariati, V., Ahmadian, M.H., Roohi, E.: Direct Simulation Monte Carlo investigation of fluid characteristics and gas transport in porous microchannels. Sci. Rep. **9**(1), 17183 (2019)
23. Shariati, V., Roohi, E., Ebrahimi, A.: Numerical study of gas flow in super nanoporous materials using the direct simulation Monte-Carlo method. Micromachines **14**(1), 139 (2023)

Chapter 6
Thermally Driven Micro- and Nano-Flows

Abstract This chapter aims to study thermal flows at the micro- and nano-scales. If a mass flow is generated in a rarefied fluid due to a temperature difference between points in a body or the distribution of an applied temperature on the boundaries of the rarefied fluid, the resulting flow is called a thermal flow. The types of flows resulting from temperature differences in rarefied gases are as follows [1]:

- Thermal creep flow
- Reverse thermal creep flow
- Thermal stress slip flow
- Nonlinear thermal stress flow
- Flow generated at the corner of a plate with a temperature different from its surroundings (thermal edge flow)
- Flow caused by a temperature discontinuity on either side of a plate (radiometric flow)

In the following, these flows will be introduced.

6.1 Thermal Creep Flow

When two reservoirs, initially at equal pressure but maintained at different temperatures, are connected by a small-diameter tube as shown in Fig. 6.1, thermal creep flow is induced between them [2].

$$T_1 > T_2, \quad P_2 = P_1 \tag{6.1}$$

The mass flow rates of reservoirs 1 and 2 can be derived from Eqs. (6.2)–(6.5), respectively. The mass flow rate ratio from reservoir 1 to reservoir 2 is then calculated. Using the proportionality relations $T \propto \overline{c}^2$ and $\rho \propto n$, substitutions are made. In the next step, by simplifying the written ratio using the equation of state, we arrive at Eq. (6.5):

E. Roohi et al., *Advances in Direct Simulation Monte Carlo: From Micro-Scale to Rarefied Flow Phenomena*, https://doi.org/10.1007/978-981-96-8200-3_6

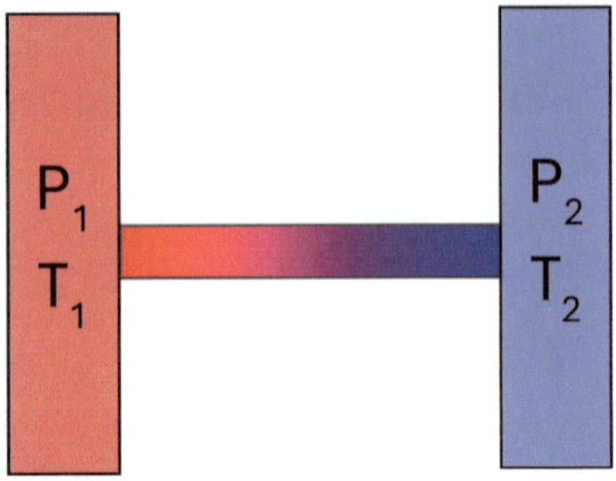

Fig. 6.1 Two reservoirs with equal pressure and different temperatures

$$\dot{m}_1 \propto mn_1\bar{c}_1 \quad (6.2)$$

$$\dot{m}_2 \propto mn_2\bar{c}_2 \quad (6.3)$$

$$P = \rho RT, \quad \frac{P_1}{P_2} = 1 \quad (6.4)$$

$$\frac{mn_1\bar{c}_1}{mn_2\bar{c}_2} \cong \frac{\rho_1}{\rho_2}\left(\frac{T_1}{T_2}\right)^{0.5} = \frac{P_1}{P_2}\left(\frac{T_2}{T_1}\right)^{0.5} = \left(\frac{T_2}{T_1}\right)^{0.5} \leq 1 \quad (6.5)$$

Since the temperature of the first reservoir is higher, the mass flow rate from the second reservoir exceeds that of the first. This difference generates a flow from the colder reservoir (reservoir 2) to the warmer reservoir (reservoir 1). Hence, thermal creep flow is created from cold to hot sides.

After a certain period, as the flow from the cold reservoir to the warm reservoir continues, the pressure in the warm reservoir increases. This pressure difference generates another flow (Poiseuille flow) from the warm reservoir back to the cold reservoir. Eventually, these two flows balance in the steady-state condition, resulting in equal mass flow rates in both directions, and the net flow stops.

If the steady-state pressure ratio of these reservoirs is expressed as a function of temperature, we arrive at Eq. (6.6). The parameter γ in this equation depends on the gas type, reactivity, surface dimensions, and most importantly, the Knudsen number (Fig. 6.2).

$$\frac{P_H}{P_C} = \left(\frac{T_H}{T_C}\right)^{\gamma} \quad (6.6)$$

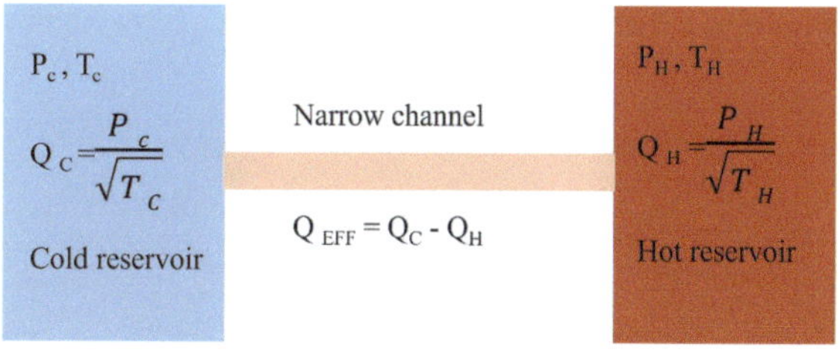

Fig. 6.2 Two reservoirs with thermal creep flow in steady state

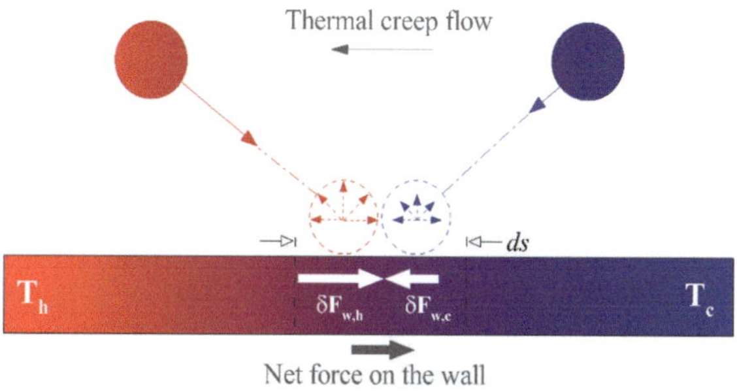

Fig. 6.3 Mechanism of thermal creep flow generation (Copied with permission from Ref. [34])

But in reality, what occurs as thermal creep flow is the flow over a plate where one side is cold, and the other side is hot. The explanation for the formation of the flow is depicted in Fig. 6.3. As seen in the figure, the particles that collide with the surface element (ds) have less momentum if they come from the cold region. Therefore, the particles in the region with a higher temperature also transfer more momentum (more force) to the surface.

$$\text{momentum flux} \propto mnc^2, \quad c \propto \sqrt{T} \tag{6.7}$$

Since the position of the surface is fixed, a force is exerted on the molecules in the opposite direction from the wall, causing the flow from the cold side to the hot side.

Another explanation for the cause of this phenomenon is illustrated in Fig. 6.4. Molecules from the hot region have a higher velocity and thus transfer more momentum. The resultant velocity of the molecules diffusely reflected from the wall is perpendicular to the wall. Therefore, the hot molecules have a greater resultant velocity, and since the wall is stationary, the flow moves from cold to hot.

The slip velocity has two components: the velocity caused by pressure-driven flow, and the other is caused by creep flow [2] (Fig. 6.5).

$$U_s = \frac{1}{2}[(1-\sigma)U_\lambda + \sigma U_\omega] + U_c, \quad U_c = \frac{3}{4}\frac{\mu R}{P}\frac{\partial T}{\partial x} \tag{6.8}$$

With the velocity profile, the flow rate can be calculated, which consists of one component due to the pressure difference and another due to the creep flow [2]

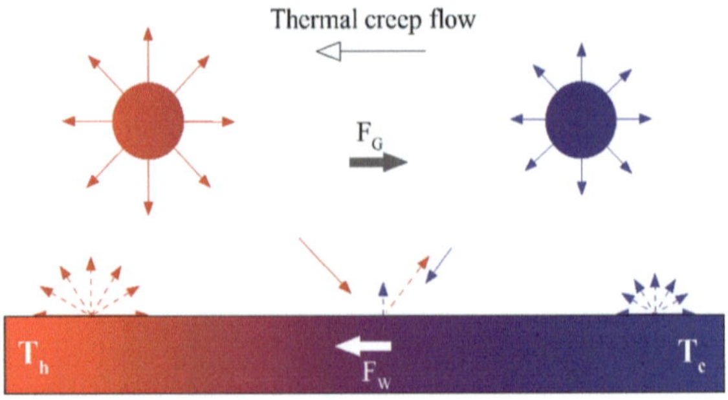

Fig. 6.4 The method of generating thermal creep flow on a diffuse wall (Reproduced from Ref. [3])

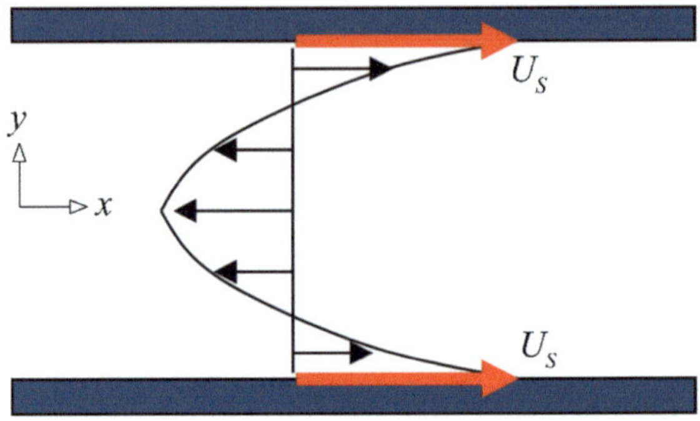

Fig. 6.5 Slip velocity in a channel

$$\dot{M} = -\frac{h^3 P}{12\mu RT}\frac{dP}{dx}\left[1 + 6\frac{2-\sigma_v}{\sigma_v}\left(Kn - Kn^2\right)\right] + \frac{3}{4}\frac{\mu h}{T}\frac{dT}{dx} \tag{6.9}$$

Since thermal creeping flow moves in the direction of increasing pressure, there is also a flow due to the pressure difference in the opposite direction of the creeping flow. In the steady-state, the flow rates of these two flows are equal, and the total flow rate will be zero. If we set the flow rate equal to zero in the equation, $\frac{dp}{dx}$ is derived under equilibrium conditions [2].

$$\frac{dP}{dx} = \left(\frac{9\mu^2 R}{h^2 P}\right)\frac{\frac{dT}{dx}}{1 + 6\frac{2-\sigma_v}{\sigma_v}\left(Kn - Kn^2\right)} \tag{6.10}$$

σ_v is the momentum accommodation coefficient. Then, the resulting equation is integrated, where the average values of $\overline{p}$, $\overline{\mu}$, and $\overline{Kn}$ are used during the integration [2].

$$P_1 - P_2 \approx \left(\frac{9\overline{\mu}^2 R}{h^2 \overline{P}}\right)\frac{(T_1 - T_2)}{1 + 6\frac{2-\sigma_v}{\sigma_v}\left(\overline{Kn} - \overline{Kn}^2\right)} \tag{6.11}$$

From the above equation, it can be concluded that by decreasing h or $\overline{p}$, the pressure difference between the sides increases. Also, decreasing h is equivalent to increasing the Knudsen number. In addition, it is understandable that decreasing $\overline{p}$ is equivalent to increasing the Knudsen number.

The following examines the equations for velocity and pressure from Sone's perspective [1]. Sone used the linearized Boltzmann equations and arrived at Eq. (6.12) for the slip velocity related to the thermal creep flow [1]:

$$U_c = -\left(\frac{\pi RT}{2}\right)^{1/2} \frac{\lambda}{T} \frac{dT}{dx} \left[K_1 + \frac{1}{2} Y_1 \left(\frac{2y}{\sqrt{\pi}\lambda}\right)\right], \tag{6.12}$$

where λ is the gas mean free path. The term $\frac{1}{2} Y_1 \left(\frac{2y}{\sqrt{\pi}\lambda}\right)$ is "Knudsen layer effects," which were not considered in the equation presented in [2]. The coefficient K_1 for the hard sphere model is -0.6463, and for the BGK model is -0.38316. This coefficient differs from the one presented by Maxwell.

6.1.1 Physical Description of Thermal Creep Flow

The slip velocity relation due to the temperature gradient is valid from the solution of the linearized Boltzmann equation for flow with a low-temperature gradient on the wall. Ohwada et al. extracted the velocity profile of thermal creep flow from the solution of the linearized Boltzmann equation in various rarefaction regimes [4]. Niimi studied the thermal creep flow of a rarefied gas between two parallel plates with a linear temperature gradient applied to the walls and observed that the velocity profile becomes flatter with decreasing Knudsen number [5]. In Fig. 6.6, the velocity profile is plotted against the Knudsen number (indicated on the graph) from Niimi's paper, where the horizontal axis represents the dimensionless velocity, with k is a constant and $h = (2RT_0)^{-1}$.

The coordinate system is centered in the middle of the channel. The velocity profile is a logarithmic function of the Knudsen number at high Knudsen numbers. This figure shows that the velocity profile becomes flatter as the Knudsen number decreases in thermal creep flow. Additionally, the variation of mass flow rate with the modified Knudsen number ($k = \sqrt{\frac{\pi}{4}}$ Kn) for thermal creep flow is shown in Fig. 6.7.

In this figure, the dimensionless thermal mass flow rate is reported:

$$M_T = \frac{\dot{m}_{th}}{\rho_{in} \beta H \sqrt{2RT_{in}}}, \tag{6.13}$$

where $\dot{m}_{th}$ is the thermal mass flow rate, ρ_{in} is the inlet density of the channel, and β is the product of the dimensionless temperature ratio and the channel aspect ratio $\left(\frac{H}{L}\right)$, which is defined as follows:

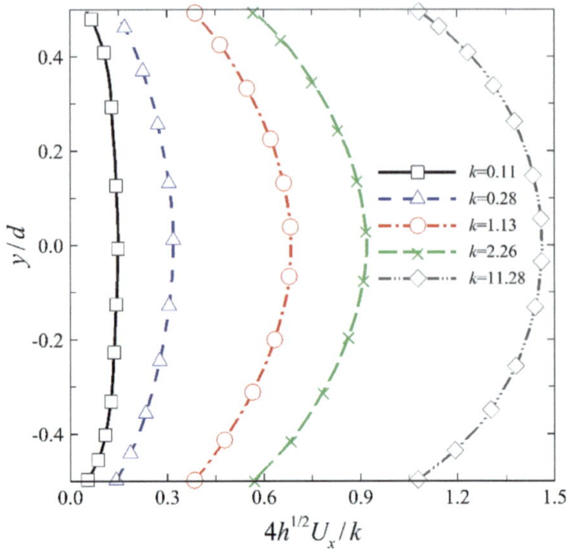

Fig. 6.6 Velocity profile in a channel at different Knudsen numbers (shown in the figure) (Reproduced from Ref. [5])

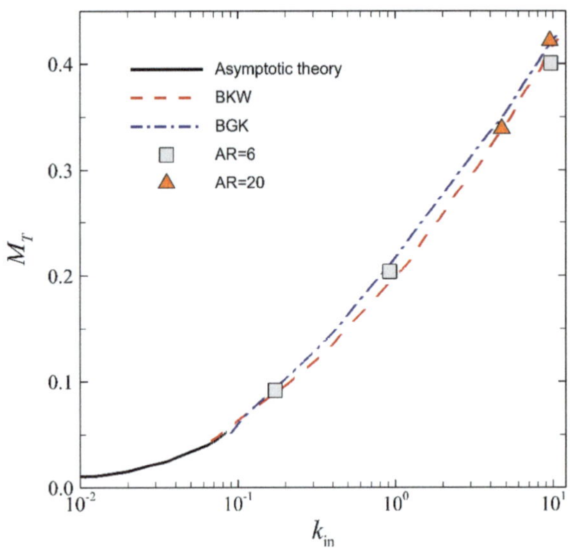

Fig. 6.7 Variation of mass flow rate with the inlet Knudsen number; symbols are corresponding to DSMC data for straight duct with $\Delta T_w/T_{in} = 0.02$ (Reproduced from Ref. [6])

$$\beta = \left(\frac{\Delta T}{T_{in}}\right)\left(\frac{H}{L}\right)$$

$$\Delta T = T_{out} - T_{in}$$

In Fig. 6.7, the simulation results for a two-dimensional microchannel with a simple cross-section are compared with aspect ratios of 6 and 20 and a wall temperature gradient ($\Delta T_{wall}/T_{in}$) of 0.02, over a wide range of Knudsen numbers, along with the numerical solutions of BGK, BKW, and Sone results [1]. It is observed that under thermal creep conditions, the Knudsen minimum phenomenon does not exist, and the increase in flow rate is logarithmic.

For an accurate description of the thermal creeping phenomenon, diffuse wall reflection conditions are considered [7]. Note that thermal creep flow does not occur for a wall with specular reflection because this wall is practically insulated, and heat and momentum transfer from the wall to the gas molecules do not occur.

If it is assumed that initially, there is no flow in a channel where the wall temperature increases along its surface (the constant $C > 0$), according to Fig. 6.8, the wall is warmer for larger x_2. In this case, the mass flow rate of molecules moving with positive velocity is in equilibrium with the mass flow rate of molecules moving with negative velocity. Consider an arbitrary position p on the wall. Molecules coming from warmer regions bring negative momentum in the horizontal direction to position p, while molecules coming from cooler regions bring positive momentum in the horizontal direction to position p because the velocity of molecules in warmer regions is, on average, higher than the velocity of molecules in colder regions. The molecules entering the wall at position p have brought negative tangential momentum in the resultant state. On the other hand, both groups of molecules are reflected diffusely, i.e., isotropically, and therefore, on average, have no horizontal momentum after reflection. This means that momentum transfer from the gas to the wall occurs in the negative x_2 direction. Due to the fixed position of the wall, a force in the x_2 direction acts on the gas through Newton's third law, and the gas starts to flow in the x_2 direction. When the flow accelerates, the generated flow from left to right increases the positive momentum transfer from the colder region (and decreases the negative momentum transfer from the warmer region) to position p. Finally, when the momentum transfer from the cold and hot molecules to the wall becomes equal, the acceleration of velocity stops. In this condition, the shear stress on the wall also becomes zero. This behavior is in stark contrast to the behavior of Poiseuille flow, in which non-zero shear stress occurs to balance the applied pressure gradient.

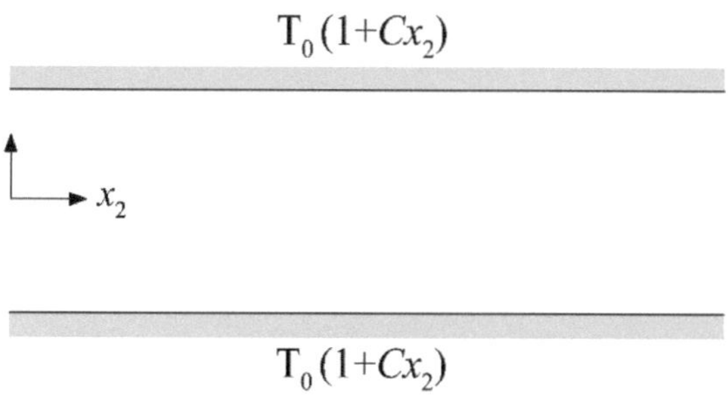

Fig. 6.8 Boundary conditions in the channel

6.1.2 Thermal Creep Flow at Zero Knudsen Number

The condition for thermal creep flow is that particles approaching a wall from a distance of one molecular mean free path (λ) but from different directions must come from points with a temperature difference. When the Knudsen number is very small, the molecular mean free path is much smaller than the characteristic length of the flow properties and is practically negligible. Therefore, practically within the molecular mean free path range, temperature variations is insignificant. For example, suppose in standard atmospheric conditions, where the mean free path is about 100 nm, the temperature changes by 100 K over a 10-cm wall length. That means the rate of temperature change per 1 cm is equal to 10 K or 0.0001 K per molecular mean free path. Therefore, molecules approaching the element dS from positions λ on the right or λ on the left effectively experience no temperature difference to impart momentum difference to the wall. Hence, the amount of thermal creep at zero Knudsen number is zero.

Note that the magnitude of thermal creep is a function of the product of the Knudsen number and the temperature gradient, see Eq. (6.12). In other words, thermal creep will exist if the temperature gradient is strongly amplified with the decrease in the Knudsen number. This issue is visible at the tip of the thin vanes of a radiometer (which will be introduced later). In other words, even at very low Knudsen numbers, the radiometer has a weak rotation because of the very small thickness at the top of the vane tip, which causes the temperature to change. Therefore, the temperature gradient is very strong; in other words, the local Knudsen number based on the radius of curvature of the vane tip is significant and not zero.

6.1.3 Thermal Creep Flow at Infinite Knudsen Number

Note that according to the definition of thermal creep, molecules from the hot and cold sides enter the wall with unequal momentum, and the reaction of this momentum from the wall to the molecules is applied as a force from cold to hot. At very large Knudsen numbers, the force applied to the wall is non-zero and exists, but the number of molecules and the intermolecular collisions to transfer the wall force to the fluid are very low. In other words, the number of molecules that collide with the wall and give different momentum is small. The molecules reflected from the wall mostly collide with other walls rather than with other molecules. Therefore, the momentum difference on the wall cannot be transferred to the fluid. As a result, the effects of thermal creep are almost limited at large Knudsen numbers.

6.1.4 Analysis of Thermal Creep Flow Equations

The simplified form of the momentum equation for a channel is as follows:

$$\frac{dp}{dx} = \mu \frac{\partial^2 u}{\partial^2 y} \tag{6.14}$$

Assuming zero pressure difference, the momentum equation for thermal creeping flow simplifies as follows:

$$\mu \frac{\partial^2 u}{\partial^2 y} = 0 \tag{6.15}$$

Given that the above equation is derived under the assumption of continuum flow, it is not applicable for Knudsen numbers greater than 0.001. It must be coupled with appropriate slip boundary conditions for small Kn. By integrating the above momentum equation twice, we have:

$$\begin{aligned} \frac{\partial u}{\partial y} &= \frac{\tau}{\mu} = A_1 \\ u &= A_1 y + A_2 \\ y &= 0 \Rightarrow u_s = A_2 \end{aligned} \tag{6.16}$$

So, in other words, the first coefficient is a function of the amount of wall shear stress, and the second coefficient is a function of the slip value on the wall. The first-order boundary condition for thermal creeping flow can be expressed as follows:

$$u_s = c_1 \lambda \frac{\partial u}{\partial y} + c_2 \lambda \frac{\partial T}{\partial x} \tag{6.17}$$

In the previous section, we saw that in thermal creep flow, the shear stress on the wall in the steady state equals zero. So, we will have

$$\frac{\partial u}{\partial y} = \frac{\tau}{\mu} = A_1 = 0 \tag{6.18}$$

Assuming a linear wall temperature distribution, the temperature field gradient is constant, and therefore, the slip velocity is also constant and a function of the Knudsen number. As a result, the velocity profile at very low Knudsen numbers is a vertical line (see Fig. 6.6). Sone [1], with asymptotic theoretical analysis, showed that at low Knudsen numbers, the velocity profile is proportional to the Knudsen number and functions of Knudsen layer effects. Under these conditions, Takata and Funagan [7] showed that the stress tensor is diagonal and isotropic ($P_{\text{ij}} = 0$). Moreover, the velocity profile near the wall follows the following equation:

$$a + b(1/2 - x_1)\ln(1/2 - x_1) + c(1/2 - x_1) \tag{6.19}$$

Similar to the discussion presented in previous chapters, the velocity profile follows the function *xlnx*, and its gradient diverges at the boundary. According to Sone's analysis [1], at low Knudsen numbers, the mass flow rate can be expressed as follows:

$$|\dot{M}| = \sqrt{\frac{\pi}{4}}\left(K + 2b\sqrt{\frac{\pi}{4}}Kn\right)Kn \tag{6.20}$$

The constants b and K are the slip coefficients and functions of the considered molecular model. As the Knudsen number increases, the nonlinear behavior of the profile is visible in the Knudsen layer region adjacent to the wall. Then, as the Knudsen layer grows, this nonlinear, region extends throughout the entire flow field. Niime [5] and Takata and Funagan [7] showed that at very large Knudsen numbers, the mass flow rate of thermal creeping flow is proportional to:

$$\dot{M} = \frac{1}{4\sqrt{\pi}}(\ln\left(\sqrt{\frac{\pi}{4}}Kn - \frac{3}{2}\gamma_e + \frac{1}{2}\right) + C_T(v) + O\left(\frac{(\ln Kn)^2}{Kn}\right), \tag{6.21}$$

where $\gamma_e = 0.57721$ is the Euler constant and the function $C_T(v)$ is a function of the considered molecular model, for which the values for several common molecular models are reported in Table 6.1.

The mass flow rate in a channel with thermal creep flow (which is called thermal transpiration flow [1]) is plotted in Fig. 6.9 for a wide range of rarefaction parameters. Notably, there is an important relation in thermodynamics called Onsager's reciprocal relation, which states that the heat flux of Poiseuille flow is equal to the mass flow rate of thermal creep flow and vice versa. The proof of this relation and further explanations are provided in Ref. [7].

Also, Su et al. [8] showed that at low Knudsen numbers, the velocity in the Knudsen layer of thermal creep flow in a channel follows the following relation:

$$\frac{U_d(x_2)}{U_d(0)} = \sum_{n=0}^{2}\sum_{m=0}^{2} c_{nm},\, x_2^n (x_2 \ln x_2)^m \tag{6.22}$$

Table 6.1 The value of $C_T(v)$ for several common molecular models

Molecular model	Viscosity-temperature index	$\sqrt{\pi}C_T$
Hard sphere	1/2	0.120791
Variable hard sphere (He or Ne)	0.66	0.0832157
Variable hard sphere (Ar)	0.81	0.0471379
Variable hard sphere (Xe)	0.85	0.0373583
Pseudo-Maxwellian	1	0

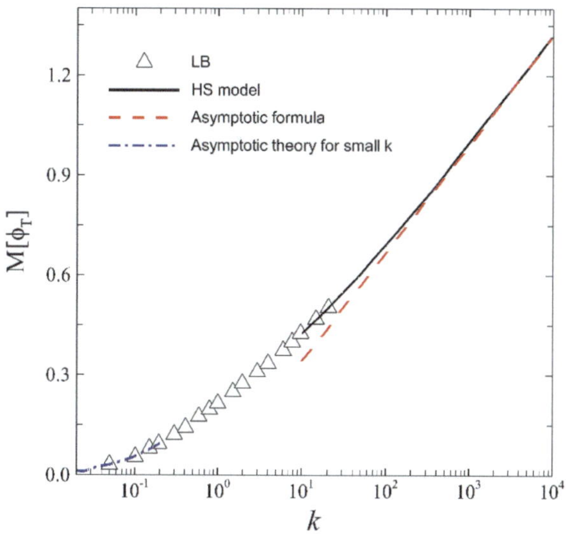

Fig. 6.9 Variation of the mass flow rate of thermal creep flow in a channel to parameter k. The solid lines indicate the data taken from the database based on HS model [7], the dashed lines the asymptotic formula [7], the dash-dotted lines the asymptotic theory for small k [1], and the symbols the numerical data linearized Boltzmann [4] (Reproduced from Ref. [7])

6.2 Inverse Thermal Creep

According to Mohammadzadeh and Struchtrup [3], inverse creep flow occurs when the majority of gas molecules approaching the plate at a higher speed from the hotter region of the flow are reflected in a "specular" (mirror-like) manner from the wall due to their high velocity. Therefore, these molecules do not exchange any tangential momentum with the surface. At the same time, most slower gas molecules approaching the plate from the colder region of the flow are reflected diffusely from the plate and will exchange tangential momentum with the surface of the wall. This type of molecule reflection is called velocity-dependent Maxwell reflection [1]. As a reaction to these molecular collisions, a force is applied to the wall towards the hotter region, or if the wall is kept stationary, the rarefied gas is driven from the hot side to the cold side, called inverse creep flow. This mechanism is shown in Fig. 6.10. Although it is unlikely that non-uniform behavior in the reflection of cold and hot molecules will actually be seen, this flow can be theoretically analyzed.

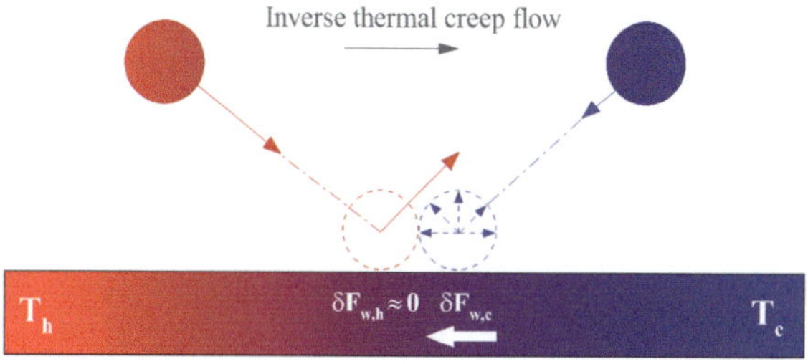

Fig. 6.10 Reflection of molecules from the boundary, velocity function, and inverse thermal creep flow (Copied with permission from Ref. [34])

6.3 Thermal Stress Slip Flow

When the Boltzmann equation is expanded asymptotically in terms of the Knudsen number, the term representing the velocity slip appears to be of the order Kn^2. In the boundary condition, the second-order velocity expansion in terms of the Knudsen number results in this term. This expression is proportional to the vertical variations of the horizontal temperature gradient. Essentially, if the vertical variations of the horizontal gradient at the boundary are non-zero, this type of flow will occur [1].

$$\frac{\partial}{\partial y}\left(\frac{\partial T}{\partial x}\right) \neq 0 \tag{6.23}$$

When the temperature gradient normal to the boundary is not uniform along the boundary, or when the isothermal surfaces are not parallel to the boundary, the flow occurs in the direction where the isothermal lines converge or diverge, depending on whether the boundary temperature is lower or higher than the gas temperature. Fig. 6.11 shows an example of a case where this condition is satisfied.

Note that satisfying the condition $\frac{\partial}{\partial y}\left(\frac{\partial T}{\partial x}\right) \neq 0$ inside the field does not generate the flow, but this condition must be satisfied at the geometry boundaries for thermal stress slip flow to occur. According to Fig. 6.11, the thermal stress slip flow direction is from the hot side to the cold side. If there are two non-concentric cylinders, according to Fig. 6.12a, with different uniform temperatures T_1 and T_2, a slip flow due to thermal stress will occur because the gradients $\frac{\partial T}{\partial r}$ depends on the angular coordinate Ø. So, due to the existence of $\frac{\partial}{\partial Ø}\left(\frac{\partial T}{\partial r}\right)$, thermal stress slip flow is created.

Lockerby et al. [9], using the thermal stress from the Burnett equation (the first term of Eq. 6.31) in the slip velocity equation (known as the Maxwell-Burnett slip equation), analyzed the streamlines in the problem of flow between two eccentric cylinders with the Navier–Stokes equations and reported results with acceptable accuracy (Fig. 6.12b). The direction of the thermal stress slip is from the hot to cold region.

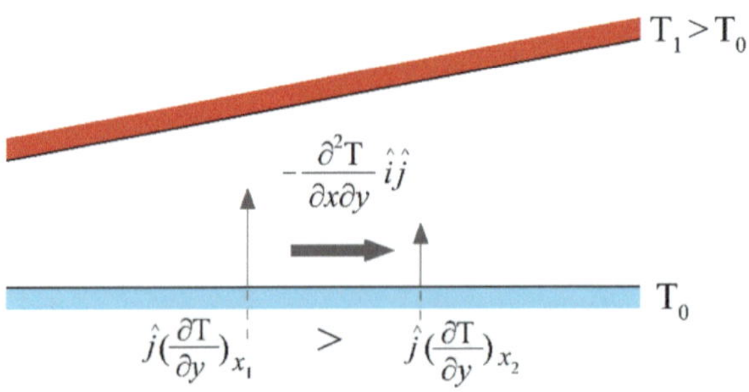

Fig. 6.11 Thermal stress slip flow between two isothermal walls (the upper wall is inclined, and the lower wall is flat) (Reproduced from Ref. [1])

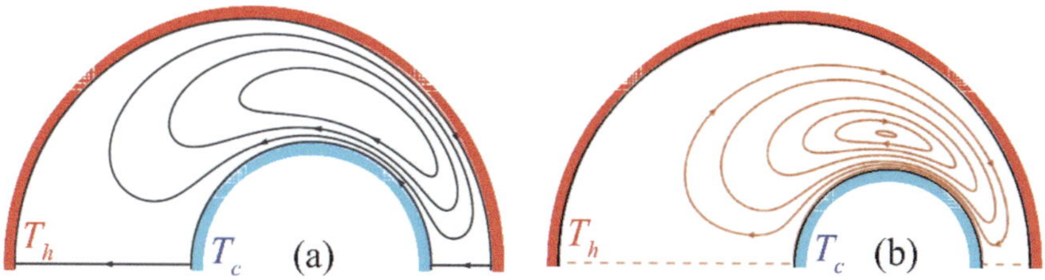

Fig. 6.12 Thermal slip flow between two non-concentric cylinders: **a** solution of the linearized Boltzmann equation and **b** solution of the Navier–Stokes equation with slip boundary conditions from the Burnett equation (Reproduced from Ref. [9])

6.3.1 Mechanism of Thermal Stress Slip Flow

Like in Fig. 6.13, consider a small element *ds* on the isothermal surface T_0 and two oblique isothermal lines T_1 above and T_{-1} below the element, four points (molecules) A, B, C, and D are drawn at an equal distance λ from *ds* [1].

We have: $(T_C > T_D, T_A > T_B)$. In this section, the goal is to prove the following expression:

$$stress \approx T_A - T_B + T_C - T_D \approx \lambda^2 \frac{\partial^2 T}{\partial x \partial y} \tag{6.24}$$

The stress equals the momentum difference of the molecules that collide with *ds* from points *A* and *B*.

Fig. 6.13 Cause of thermal stress flow generation (Reproduced from Ref. [1])

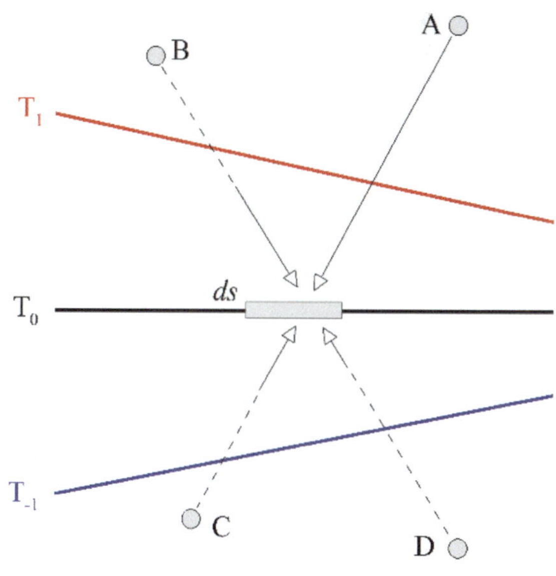

$$stress \approx ((\rho c)c_A - (\rho c)c_B) \approx c_A^2 - c_B^2 \approx T_A - T_B \tag{6.25}$$

T_A and T_B are expressed based on the second-order Taylor expansion of the mean free path length.

$$\begin{aligned} T_A = \mathrm{T}(\lambda, \lambda) &= \mathrm{T}(0,0) + \lambda \frac{\partial T}{\partial x}(0,0) + \lambda \frac{\partial T}{\partial y}(0,0) \\ &+ \frac{\lambda^2}{2}\left[\frac{\partial^2 T}{\partial x^2}(0,0) + 2\frac{\partial^2 T}{\partial x \partial y}(0,0) + \frac{\partial^2 T}{\partial y^2}(0,0)\right] \end{aligned} \tag{6.26}$$

$$\begin{aligned} T_B = \mathrm{T}(-\lambda, \lambda) &= \mathrm{T}(0,0) - \lambda \frac{\partial T}{\partial x}(0,0) + \lambda \frac{\partial T}{\partial y}(0,0) \\ &+ \frac{\lambda^2}{2}\left[\frac{\partial^2 T}{\partial x^2}(0,0) - 2\frac{\partial^2 T}{\partial x \partial y}(0,0) + \frac{\partial^2 T}{\partial y^2}(0,0)\right] \end{aligned} \tag{6.27}$$

$$T_A - T_B = 2\lambda^2 \frac{\partial^2 T}{\partial x \partial y}(0,0) + 2\lambda \frac{\partial T}{\partial x}(0,0) \tag{6.28}$$

Therefore:

$$T_A - T_B \approx \lambda^2 \frac{\partial^2 T}{\partial x \partial y} \tag{6.29}$$

In examining the control volume of a closed boundary ABCD, if C-D is not a solid surface, thermal stress slip flow does not occur because if there is no wall (Fig. 6.14), the input and output mass fluxes across the boundaries are equal. The momentum (stress) changes in the closed boundary ABCD becomes zero. In kinetic theory, when a virtual boundary is considered, the mass flux is equal to ρc; but if this boundary is solid (for example, a wall), the flow flux is reduced to $\frac{1}{2}\rho c$.

If $\frac{\partial^2 T}{\partial x \partial y}$ exists, but there is no solid wall in a closed boundary; it is concluded that thermal stress slip flow does not occur because the net momentum on all boundaries will be zero. According to Eq. (6.29), the thermal stress slip is of the order of Kn^2 and is negligible at low Knudsen numbers.

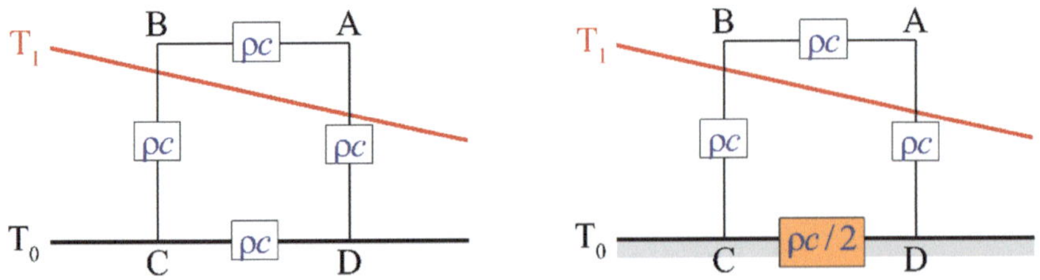

Fig. 6.14 Analysis of thermal stress slip flow in a closed control volume (Reproduced from Ref. [1])

6.3.2 Thermal Stress from Kogan's Perspective

Kogan et al. [10] added the stress term from the Burnett equation ($P_{ij}^{(2)}$) to the Navier–Stokes equations. This term is divided into two components: thermal stresses $P_{ij}^{(T)}$ and non-thermal stresses $P_{ij}^{(*)}$.

$$P_{ij} = P\delta_{ij} + P_{ij}^{(1)} + P_{ij}^{(2)} + \ldots, \quad P_{ij}^{(2)} = P_{ij}^{(T)} + P_{ij}^{(*)} \tag{6.30}$$

The term $P_{ij}^{(T)}$ consists of the thermal terms of the Burnett stress, as it is only derived from the temperature derivatives, which represent thermal stress flow, and consists of two terms:

$$P_{ij}^{(T)} = K_2 \frac{\mu^2}{\rho T}\left[\frac{\partial^2 T}{\partial x_i \partial x_j}\right] + K_3 \frac{\mu^2}{\rho T^2}\left[\frac{\partial T}{\partial x_i}\frac{\partial T}{\partial x_j}\right] \tag{6.31}$$

It is worth mentioning that a similar approach is applied to heat flux:

$$q_{ij} = 0 + q_i^{(1)} + q_i^{(2)} + \ldots, \quad q_i^{(2)} = \alpha_1 \frac{\mu^2}{\rho T}\frac{\partial u_j}{\partial x_j}\frac{\partial T}{\partial x_i} + \ldots \tag{6.32}$$

Assuming $\mu_* \approx \rho_* a_* \lambda_*$ and $\theta = (\Delta T)_* / T_*$, the order of magnitude of the following terms is expressed as stated in Eq. (6.33):

$$\begin{aligned} &P_{ij}^{(1)} \approx Kn\rho_* a_* u_*, \quad P_{ij}^{(k\geq 3)} \leq Kn^3 \rho_* a_* u_* \theta \\ &P_{ij}^{(*)} \approx Kn^2 \rho_* u_*^2, \quad P_{ij}^{(T)} \approx Kn^2 \rho_* a_*^2 \theta \\ &q^{(1)} \approx Kn\rho_* a_*^3, \quad q^{(2)} \approx Kn^2 \rho_* a_*^2 u_* \end{aligned} \tag{6.33}$$

For example, to prove the expression $P_{ij}^{(T)}$, the following reasoning is used:

$$\frac{\mu^2}{\rho T}\left[\frac{\partial^2 T}{\partial x_i \partial x_j}\right] \rightarrow P_{ij}^{(T)} \approx Kn^2 \rho_* a_*^2 \theta \tag{6.34}$$

$$\begin{cases} \frac{\partial^2 T}{\partial x_i \partial x_j} = \frac{\Delta T}{L_*^2} \\ \mu_* \approx \rho_* a_* \lambda_* \end{cases} \tag{6.35}$$

$$\frac{(\rho_* a_* \lambda_*)^2}{\rho_* T_*}\frac{\Delta T}{L_*^2} = \frac{\lambda_*^2}{L_*^2}\frac{\Delta T}{T_*}\rho_* a_*^2 = kn^2 \theta \rho_* a_*^2 \tag{6.36}$$

6.4 Nonlinear Thermal Stress Flow

Sone [1], by examining the linearized Boltzmann equations, proved that there are terms of the order of the Knudsen number that do not appear in the Navier–Stokes equations, which are of the order of the Knudsen number accuracy. For example, the following expression does not exist in the Navier–Stokes equations and is the cause of nonlinear, thermal stress flow:

$$\frac{\gamma_7}{T_{H0}}\frac{\partial T_{H0}}{\partial x_i}\frac{\partial T_{H0}}{\partial x_j}\left(\frac{u_{jH1}}{\gamma_2\sqrt{T_{H0}}}-\frac{1}{4}\frac{\partial T_{H0}}{\partial x_j}\right), \tag{6.37}$$

where $\gamma_2 = 1.922284066$ and $\gamma_7 = 1.758705$, and the subscripts H and the corresponding number indicate the order of the Grad-Hilbert expansion parameter. Unlike the two previous flows that were caused by the temperature distribution and gradient at the boundaries, this flow is generated due to the presence of the terms of Eq. (6.37) in the constitutive laws of rarefied flows, and its direction is also from hot to cold [1]. In other words, if the boundaries have a constant temperature in rarefied flow, a flow will be induced if the terms of Eq. (6.37) are non-zero.

Unlike thermal creep flow, thermal stress slip flows and nonlinear, thermal stress flows occur in regions where the isotherms are not parallel, and the boundary's temperature distribution is uniform. One example of nonlinear, thermal stress flow is two concentric elliptical cylinders with constant temperature and a ratio of the outer to inner surface temperature of 5 ($T_1/T_0 = 5$) at low Knudsen numbers, as shown in Fig. 6.15 (from the author's results). The Knudsen number is defined as λ_{ave}/L, where L is the smaller radius of the outer ellipse, and λ_{ave} is the molecular mean free path at the initial average temperature of the field. As seen in the figure, at a Knudsen number of 0.005, there is only one counter-clockwise vortex in the field, which is created due to nonlinear, thermal stress because the isotherms in this problem are not parallel, and the temperature difference is also significant. By increasing the Knudsen number to 0.02, two vortices are seen in the field; the larger vortex originates from the nonlinear, thermal stress, while the smaller, clockwise-rotating vortex is caused by a phenomenon like the thermal edge, which is discussed in the next section.

As the Knudsen number increases, the inner vortex grows, the larger vortex shrinks, and the flow forms in the opposite direction of the nonlinear, thermal stress. The reason for the formation of the inner vortex is as follows: the temperature of the inner wall is constant, but the first-order Knudsen thermal slip phenomenon occurs on it. Due to the presence of thermal slip and the non-uniform temperature distribution on this wall, which is especially stronger in its upper regions due to the high curvature, the second-order Knudsen thermal creep flow forms on this wall. This flow, which moves towards the top (curved tip) of the lower wall, is like the thermal edge flow on a wall colder than the environment (the flow on the thermal edge on a wall warmer than the environment is explained in the next section). After hitting the symmetry boundary, it reverses and forms a vortex due to the equivalence of its magnitude with the nonlinear, thermal stress flow. Since the term behind this

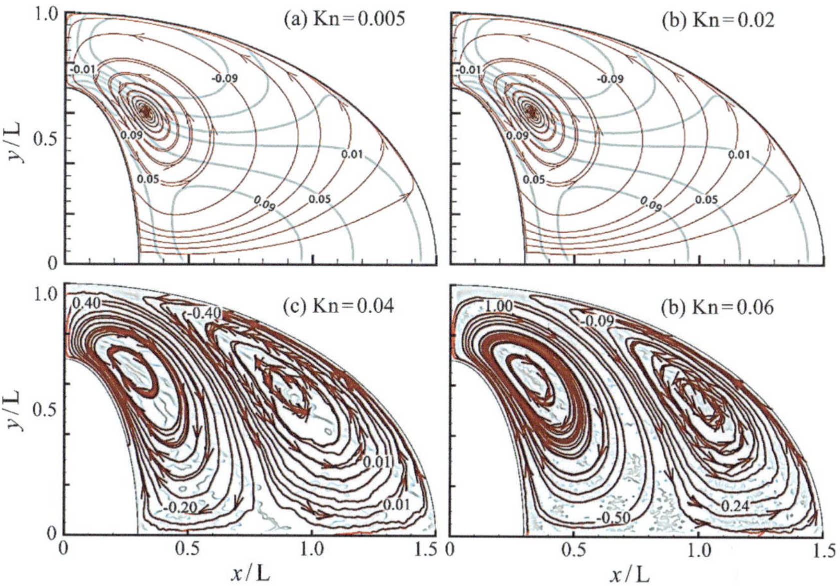

Fig. 6.15 Thermal flow between two concentric elliptical cylinders with constant temperature at different Knudsen numbers. The temperature of the outer cylinder is T_1, and the temperature of the inner cylinder is T_0. ($T_1/T_0 = 5$). The numbers indicate the dimensionless velocity values (Unpublished results from authors)

sentence is of the second-order Knudsen number in the asymptotic expansion, which is larger than the term behind the nonlinear, thermal stress term, as the Knudsen number increases, the strength of this vortex dominates the outer vortex.

6.5 Thermal Edge Flow

If a uniformly heated plate in rarefied flow has a sharp edge, a flow called thermal edge flow is created. This flow was first studied by Aoki et al. [11] and then by Sone and Yoshimoto [12], and its mechanism is like thermal creep flow.

According to Fig. 6.16, a flat plate with a uniform temperature T_1 is placed in an environment with a temperature T_0, where $T_1 > T_0$. According to the heat conduction equation, temperature variations exist in the air surrounding the plate. Due to the uniform temperature of the plate, the isothermal lines far from the corners of the plate are parallel to the plate. However, near the corners of the plate, the isothermal lines curve sharply, see Fig. 6.16. It is important to note that if $\mathrm{Kn}_{local} = \frac{\lambda}{r}$ is defined, since at the tip of the plate r is very small, the local Knudsen number at the corners becomes very large, and at the corners, the flow deviates from the continuum conditions.

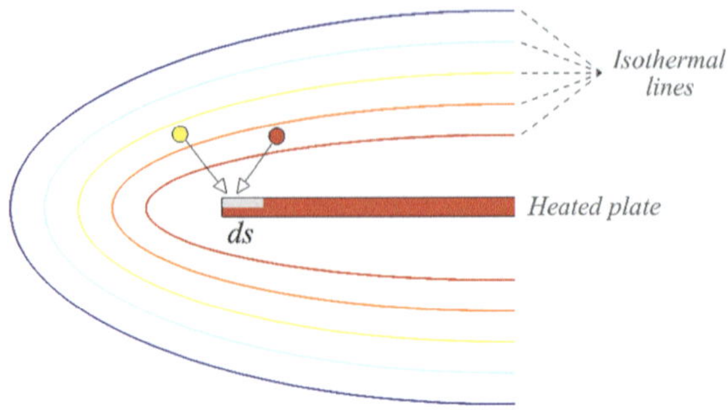

Fig. 6.16 Thermal edge flow over the edge of a flat plate

In kinetic theory, no exact solution has been presented for the temperature field at the corner of a plate. Assuming the validity of the solution of the continuum heat conduction equation around a plate with a temperature different from the environment, the following temperature distribution is obtained in the polar coordinate system [1]:

$$T(r,\theta) = T_1\left[1 - ar^{\frac{1}{2}}\sin\left(\frac{\theta}{2}\right) + \ldots\right] a > 0 \tag{6.38}$$

Assume that the molecules that collide with the surface dS near the edge of the plate are composed of two groups: molecules from region A and molecules from region B. If Eq. (6.38) is written for the two regions A and B, each located at a radial distance λ from the corner of the plate, as shown in Fig. 6.16, we have:($\alpha_0 > \alpha_1 > 0$)

$$T_A = T_1\left[1 - (\alpha_0 + \alpha_1)\lambda^{1/2}\right], \quad T_B = T_1\left[1 - (\alpha_0 - \alpha_1)\lambda^{1/2}\right] \tag{6.39}$$

Since the distance of A and B to the plate is equal to λ, a molecule coming from these points does not collide with any other molecule before it reaches the surface. Suppose a molecule from A and B collide with the surface dS. In that case, a momentum difference is transferred to the plate, which is proportional to the temperature of the particles. In other words:

$$\frac{T_A - T_B}{\Delta x} = \frac{T_1\left[1 - k_1\lambda^{1/2}\right] - T_1\left[1 - k_2\lambda^{1/2}\right]}{\sqrt{2}\lambda} = \frac{T_1}{\lambda^{1/2}} \propto \frac{1}{\sqrt{Kn}} \tag{6.40}$$

Since point B is closer to the plate ($T_B > T_A$), a thermal creep flow is generated from A (cold) to B (hot) because $0 \neq \frac{\partial T}{\partial r}$. Based on Eq. (6.40), the magnitude of the induced velocity by the thermal edge flow can be calculated:

$$\Delta T = \lambda\frac{dT}{dx} \& \frac{dT}{dx} \propto \lambda^{-0.5} \tag{6.41}$$

Therefore, $\Delta T \propto \lambda^{0.5}$. Given that the nature of thermal edge flow is similar to thermal creep flow, the induced velocity of the thermal edge flow can be calculated as follows:

$$U_{s,creep} \propto Kn\frac{dT}{dx} \rightarrow U_s = Kn \times Kn^{-0.5} = Kn^{0.5} \tag{6.42}$$

Therefore, the strength of the thermal edge flow is of the order of $Kn^{0.5}$ and is weaker than the thermal creep flow.

6.5.1 *Thermal Edge Compressor*

One of the applications of thermal edge flow is creating a type of compressor. Many flat plates that generate heat are stacked on each other, as shown in Fig. 6.17. According to the discussed flow mechanisms, a significant flow occurs around the plates, which increases as the passing gas becomes rarefied and can act as a compressor. Figure 6.18 shows the flow generation mechanism in this compressor.

6.6 Radiometric Flow

Another type of thermal flow is the flow around a plate where one side is hot, and the other is cold and placed in a rarefied environment (Fig. 6.19). Due to the temperature difference between the two edges, a flow is generated, known as radiometric flow [14]. Sir William Crookes first observed the radiometric flow inside a bulb containing rarefied air, and the vanes were painted white on one side and black on the other. The Crookes radiometer is also called a light mill. When these vanes were exposed to heat or light, they began to rotate.

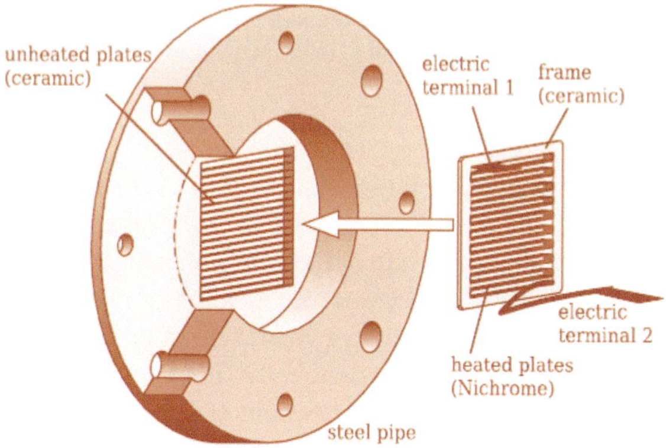

Fig. 6.17 A unit of the thermal edge compressor. The unheated plate is ceramic, and the heated plates are arranged in Nichrome heating strips. The smaller holes are used as channels for cooling water, while the larger holes are for bolt connections (Reproduced from Ref. [13])

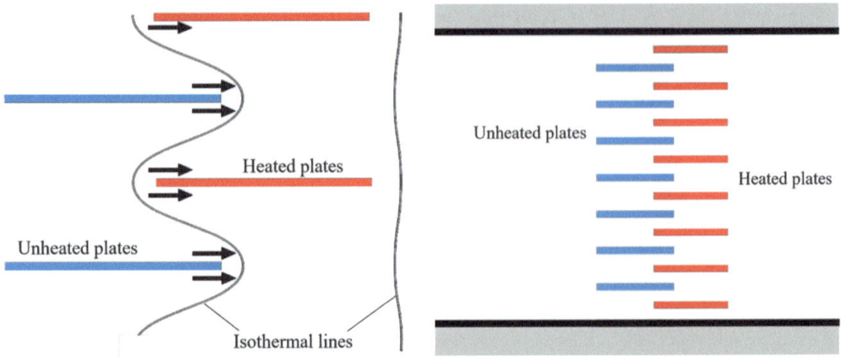

Fig. 6.18 Thermal edge flow: **a** Arrangement of heated and unheated plates; **b** Isothermal lines and thermal edge flow generated by the plates in the chamber (Reproduced from Ref. [13])

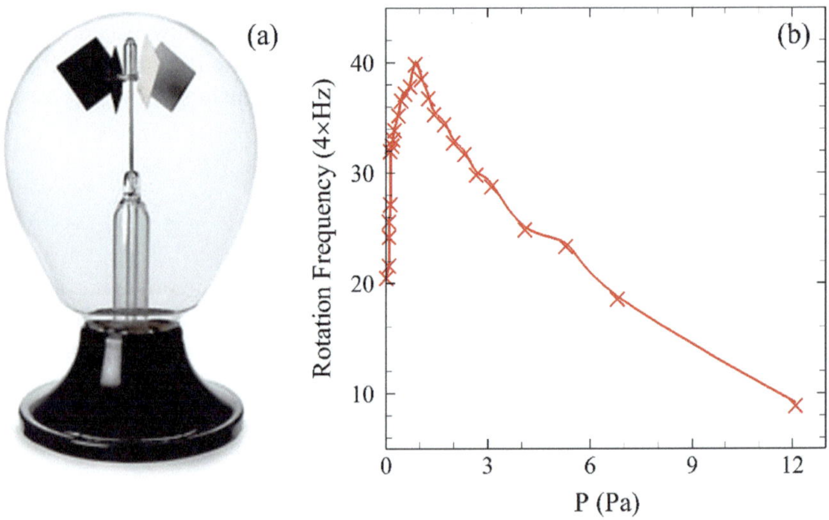

Fig. 6.19 **a** Crookes radiometer and **b** its measured rotational speed (data extracted from experimental results provided in this video: https://www.youtube.com/watch?v=r7NEI_C9Yh0)

It demonstrates energy transfer from light or heat into motion and is often used to demonstrate basic physical principles. At first, Crookes used a light to shine on it, and it started moving clockwise, and the brighter the light, the quicker it spins. When light heats the black side of the vanes, the air on that side moves faster, causing the vanes to rotate. The black surface absorbs more energy than the white surface, which reflects the light, leading to a greater effect on the black face. When one puts canned air duster on the Crookes radiometer, it moves counterclockwise. If there was no gas in the radiometer it would not spin at all. There must be some ideal pressure that makes it work well and it cannot be zero or ambient pressure, see Fig. 6.19b. Note

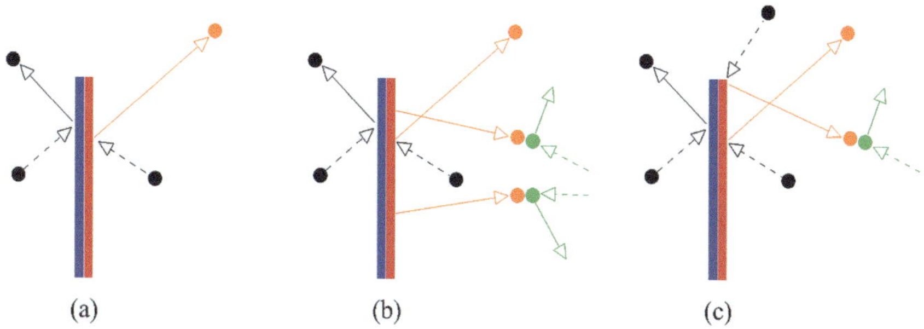

Fig. 6.20 Cause of radiometric force: **a** High-speed molecules reflected from the hot surface, **b** high-speed molecules upon collision with surrounding molecules, and **c** near the edge (Reproduced from Ref. [15])

that this graph is specific to the studied radiometer. In this case, the mean free path of the gas molecules is 1 cm.

To better understand the physics of radiometric flow, consider Fig. 6.20, where the temperature on the right side of the vane is higher than the left side, and the left side's temperature is equal to the ambient temperature [15]. The molecules that collide with the left side of the vane return at a speed close to their pre-collision speed. In contrast, molecules colliding with the vane's right side are reflected at higher speeds because the molecular speed is proportional to the surface temperature. This means that the resultant molecules receive momentum towards the right, and as a reaction, the vane receives momentum to the left. Therefore, force is applied from the hot side to the cold side of the plate. In other words, at large Knudsen numbers, all molecules reflected from the middle of the plate contribute to generating the radiometric force.

In conditions where the Knudsen number is small, the molecules reflected from the wall effectively prevent new molecules from colliding with the hot plate, and as a result, the flux of colliding molecules to the surface decreases (Fig. 6.20b). This phenomenon does not occur near the edges of the plate. Consequently, at low Knudsen numbers, the molecules near the edge significantly contribute to generating the radiometric force (Fig. 6.20c).

6.6.1 Radiometric Force

The researchers' primary goal was to establish a relation between the force applied to the vanes, the pressure inside the bubble and the mechanism of the vanes' rotation. West investigated this phenomenon and reached the following conclusions [16]:

- At very low pressures, the force is proportional to $\sqrt{T}$.
- As the pressure increases, the phenomenon causing the rotation of the vanes will be the thermal creep flow at the edges of the vanes.
- The temperature gradient at the edges is much stronger than along the vanes.

Finally, West proposed the relation between force and pressure, where (a and b) are constant.

$$F \propto \frac{1}{\frac{a}{P} + \frac{P}{b}} \tag{6.43}$$

The following relation results from a circuit containing two parallel resistors, which is inversely proportional to pressure at high Knudsen numbers and proportional to pressure at low Knudsen numbers. So, if the Knudsen number tends towards infinity, the force approaches zero because the pressure becomes zero. In this condition, no molecules transfer momentum and induce motion. If the Knudsen number tends towards zero, the force also tends towards zero (see the explanation in Sect. 6.1.3). The maximum force will be at an intermediate Knudsen number (pressure). Relation (6.43) is obtained based on experimental observations.

During this period, Einstein also presented a theory that described the state of the radiometric force under two different conditions [17]:

1. If a vane, with one side cold (C) and the other side hot (H), is placed in a gas with a uniform temperature T, the following relations hold:

$$p_C \propto n_C \sqrt{T_C} \qquad p_H \propto n_H \sqrt{T_H} \tag{6.44}$$

$$\Delta p = p_H - p_C \propto n_{ave}\left(\sqrt{T_H} - \sqrt{T_C}\right) = \frac{p_{ave}}{\sqrt{T_{ave}}}\left(\sqrt{T_H} - \sqrt{T_C}\right) \tag{6.45}$$

In Eq. (6.45), it is assumed that $n_{cold} = n_{hot} \sim n_{ave}$. The equality of the numerical flux of particles on the sides of the vane under free molecular conditions is proven analytically in Chap. 2 of Sone's book. It is not necessarily correct under other conditions. It should be noted that the pressure difference is equivalent to the momentum flux difference, and momentum is proportional to the molecular velocity, which in turn is proportional to the square root of the temperature. Therefore, the square root of temperature appears in the above relations. If the temperature difference between the two sides of the vane is small, it is substituted in the above relation [18].

$$T_H = T_{ave} + \frac{\Delta T}{2} \tag{6.46}$$

$$T_C = T_{ave} - \frac{\Delta T}{2} \tag{6.47}$$

$$\Delta p = p_H - p_C = \frac{p_{ave}}{\sqrt{T_{ave}}}\left(\sqrt{T_{ave} + \frac{\Delta T}{2}} - \sqrt{T_{ave} - \frac{\Delta T}{2}}\right) \tag{6.48}$$

$$\Delta p = p_H - p_C = \frac{p}{\sqrt{T}}\sqrt{T}\left(\sqrt{1 + \frac{\Delta T}{2T}} - \sqrt{1 - \frac{\Delta T}{2T}}\right) \tag{6.49}$$

It is assumed that T_{avg} is the ambient temperature ($T_{avg} = T$). The expansion of Eq. (6.48) is used to simplify the pressure difference expression:

$$(1+\in)^{1/2} = 1 + \frac{1}{2}\in \tag{6.50}$$

$$\Delta p = p_H - p_C = p\left(1 + \frac{\Delta T}{4T} - 1 + \frac{\Delta T}{4T}\right) = p\frac{\Delta T}{2T} \tag{6.51}$$

According to Einstein's theory, at low Knudsen numbers, particles in a region at a distance of the mean free path of molecules (λ) around the tip of the radiometer have the greatest effect on exerting the radiometric force. As a result, the force applied to the vane (F), assuming the force's effective area is $A = \lambda L$ and L is the length of the vane's side, will be as follows:

$$\Delta p = \frac{F}{A}, A = \lambda L \rightarrow F/L = p\lambda\frac{\Delta T}{2T} \tag{6.52}$$

2. If the vane has a uniform temperature and is placed inside a gas with a significant temperature gradient (in other words, the phenomenon of thermophoresis occurs)

By substituting $\Delta T = \lambda\frac{dT}{dx}$ at small Knudsen numbers, Eq. (6.52) is obtained.

$$F/L = -\frac{1}{2}P\frac{\lambda^2}{T}\frac{dT}{dx} \tag{6.53}$$

According to Fig. 6.20, at high Knudsen numbers, the force is due to the surface-force and is proportional to pressure:

$$F \propto p \propto \frac{1}{Kn}$$

At low Knudsen numbers

$$F \propto P\frac{\lambda^2}{T}\frac{dT}{dx} \propto n\lambda^2\frac{dT}{dx} \tag{6.54}$$

According to Eq. (6.41), ΔT is proportional to $\lambda^{0.5}$. Using this assumption, the force at low Knudsen numbers is proportional to:

$$F \propto n\lambda^2\frac{dT}{dx} = \lambda\lambda^{-0.5} = \lambda^{0.5} = \sqrt{Kn} \tag{6.55}$$

Thus, the radiometric force is proportional to $\sqrt{Kn}$ at low Knudsen numbers and proportional to $\frac{1}{Kn}$ at high Knudsen numbers. Sexl proposed the following relation for analyzing the radiometric force [16]:

$$F = -\frac{14.72}{n+5}\frac{p\lambda^2}{T}\Delta T, \tag{6.56}$$

where n is the degree of freedom of the molecules, the theories of Einstein [1] and Epstein [2] can be expressed as follows:

$$F = -cLp\frac{\lambda^2}{T}\frac{dT}{dx}, \begin{cases} c = \pi & \text{Einstein} \\ c = 8.95 & \text{Epstein} \end{cases} \tag{6.57}$$

Seldon also proposed the following relation for the force using Einstein's relation (6.57):

$$F/L = \frac{1}{2}p\frac{\lambda^2}{T}\frac{dT}{dx} \tag{6.58}$$

$$F = \frac{1}{2}pA\frac{\lambda}{T}\frac{dT}{dx} \tag{6.59}$$

$$F = \frac{1}{2}pA\left(1 + \frac{\lambda}{2T}\frac{dT}{dx} - 1 + \frac{\lambda}{2T}\frac{dT}{dx}\right) = \frac{1}{2}pA\left(1 + \frac{\lambda}{2T}\frac{dT}{dx} - \left(1 - \frac{\lambda}{2T}\frac{dT}{dx}\right)\right) \tag{6.60}$$

Assuming $\frac{\lambda}{T}\frac{dT}{dx} \ll 1$ and using the expansion of relation (6.61), we will arrive at relations (6.62) and (6.63).

$$1 + \frac{1}{2}\in = (1 + \in)^{1/2} \tag{6.61}$$

$$F = \frac{1}{2}pA\left(\sqrt{\frac{T + \lambda\frac{dT}{dx}}{T}} - \sqrt{\frac{T - \lambda\frac{dT}{dx}}{T}}\right) \tag{6.62}$$

$$F = \frac{1}{2}pA\left(\sqrt{\frac{T_H}{T_g}} - \sqrt{\frac{T_C}{T_g}}\right) \tag{6.63}$$

It is assumed that the temperature difference between the cold and hot sides of the vane is small so that the temperatures on both sides can be extracted using a Taylor expansion. The force considered as the main force causing the rotation of the vanes is the force created near the edges of the vane. Thus, the edge effect is dominant at low Knudsen numbers, but the side effects also become noticeable at high Knudsen numbers. In Fig. 6.21, the results of Einstein's relation (Eq. 6.57 with coefficient c $= \pi$), Epstein's relation (Eq. 6.57 with coefficient c $= 8.95$), the relations related to free molecular flow [19], and the experimental results by Brough and Litvin [20] are compared. It is observed that Einstein's theory at low Knudsen numbers and the

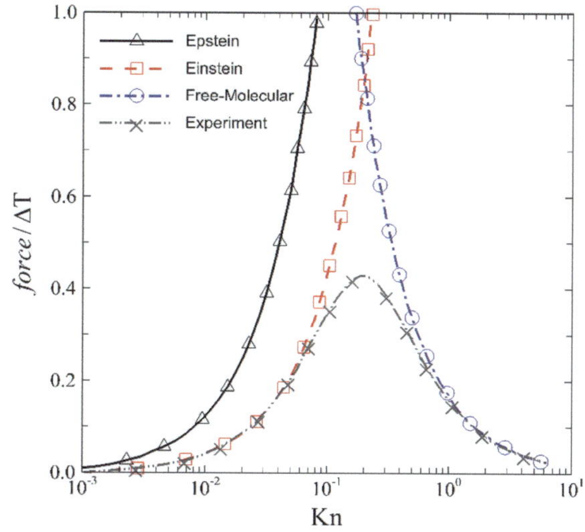

Fig. 6.21 Comparison of the theoretical results of Einstein, Epstein, and free-molecular theories with the experimental results of Broch and Litwin [20] for force per unit temperature gradient as a function of the Knudsen number (Reproduced from Ref. [16])

free molecular theory at high Knudsen numbers have the most agreement with the experimental results.

6.6.2 *Calculation of the Force on Radiometric Vanes Under Different Conditions*

This section calculates the force applied to a vane in radiometric flow under different conditions. The vane is modeled as a two-dimensional plate with negligible thickness, as shown in Fig. 6.22. It is assumed that the temperature of the vane is constant, the distance between the bulb walls and the vane is L, the mean free path is λ, and the characteristic velocity of molecules colliding from the right and left with the vane is $\overline{c}_{\pm\lambda} = \sqrt{2kT_{\pm\lambda}/m}$ which are different due to the temperature difference between the right and left sides of the vane. Similarly, the molecular flux entering the vane from each section is equal to ${}^{nc}/_{2}$ [18].

It is assumed that according to the continuity law, the collisional flux from the sides of the vane to the vane is equal to each other, then the difference in the collisional momentum from the sides is calculated as follows:

$$\frac{1}{2}n_{+\lambda}\overline{c}_{+\lambda} = \frac{1}{2}n_{-\lambda}\overline{c}_{-\lambda} \tag{6.64}$$

$$\text{molecular momentum flux} = \left(\frac{1}{2}nc\right)(mc) \tag{6.65}$$

$$\text{molecular momentum difference } = \frac{1}{2}nm\overline{c}(\overline{c}_{-} - \overline{c}_{+}) \tag{6.66}$$

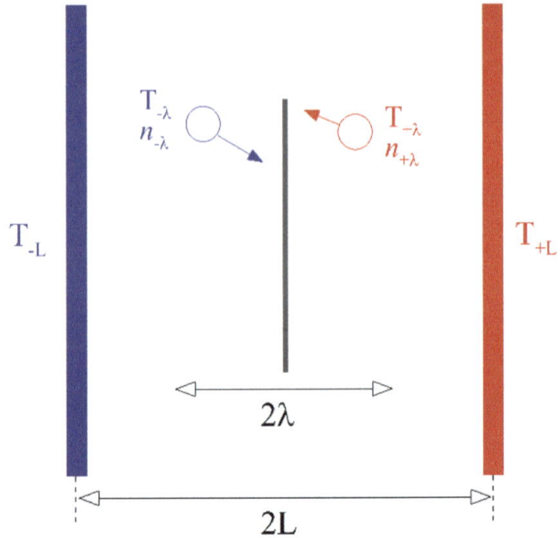

Fig. 6.22 Simulation of a radiometric fin

$$\frac{\overline{c}_{-} - \overline{c}_{+}}{2\lambda} = \frac{\partial \overline{c}}{\partial x} \tag{6.67}$$

The velocity $\overline{c}_{\pm\lambda} = \sqrt{2kT_{\pm\lambda}/m}$ is squared, and the derivative is taken from both sides to obtain the relation (6.68).

$$\overline{c}\frac{\partial \overline{c}}{\partial x} = \frac{k}{m}\frac{\partial T}{\partial x} \tag{6.68}$$

If Eqs. (6.67) and (6.68) are substituted into the equation for the molecular momentum difference, we arrive at Eq. (6.69):

$$\text{Molecular momentum flux difference } \approx -nk\lambda\partial_x T \tag{6.69}$$

Then, using the equation of state $p = nkT$, the molecular momentum difference is also expressed as follows:

$$\text{Molecular momentum difference} \approx -p\lambda\partial_x T/T \tag{6.70}$$

What has been derived here is the momentum of the colliding molecules. But regarding the reflected molecules, if the wall is diffuse since it is assumed that the air surrounding the wall has different temperatures on either side and the plate's temperature is constant, the effect of the bulk of the reflected molecules from the right and left ultimately cancels each other out, as they return with speeds proportional to the wall's temperature. Furthermore, if the wall is specular, the components of each molecule cancel each other out in pairs.

The above analysis, which was used to calculate the force (momentum difference), is valid only for the distance λ from the vane. Three different cases are considered if we want to calculate the total force acting on the plate.

6.6.2.1 First Case $o(\lambda) \approx o(L)$

In this case, the mean free path (λ) is approximately equal to the distance from the wall to the vane. Therefore, the control volume (the color area in Fig. 6.23) fully encompasses the inside of the bulb, and the estimate obtained for the force is valid for the total force exerted on the plate. In this case, molecules are affected by the last wall they collide with (either vane or bulb walls) because they do not collide with other molecules. As a result, the force is influenced by the temperature difference between the walls.

$$\lambda \frac{\partial T}{\partial x} = \Delta T \tag{6.71}$$

$$\frac{F}{A} = nk\lambda \partial_x T = nk\Delta T, \quad A = d^2 \tag{6.72}$$

$$F = nk\Delta T d^2 \tag{6.73}$$

So, the force is proportional to n, and as n decreases, meaning the number of molecules available for collision decreases, the force also decreases. On the other hand:

$$\text{Kn} = \frac{1}{\sqrt{2}\pi d^2 Ln} \quad kn \propto \frac{1}{n} \quad F \propto \frac{1}{kn} \tag{6.74}$$

Here, d is the particle diameter, in (6.73) its the length of the fin. Therefore, with high Knudsen numbers, small force is obtained.

6.6.2.2 Second Case $d < \lambda < L$

In this case, where the schematic of the problem is shown in Fig. 6.24, the governing equation of the field (except in the region of λ around the blade) is the heat conduction equation. This means the control volume encompasses the entire field, and the presented analysis applies to the whole plate.

$$F = nk\lambda \partial_x T d^2$$

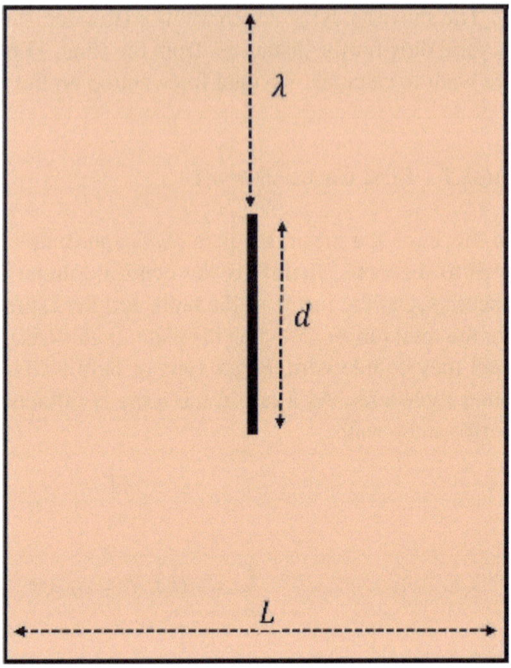

Fig. 6.23 Control volume of a radiometric fin at high Knudsen number

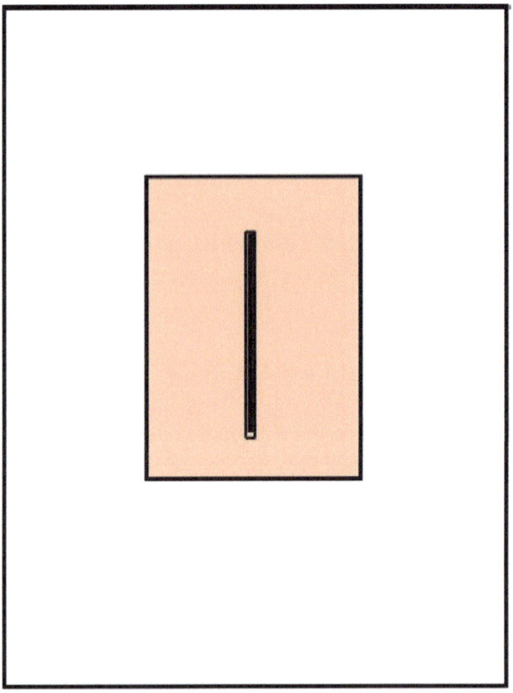

Fig. 6.24 Control volume of a radiometric fin at moderate Knudsen numbers

Fig. 6.25 Control volume of a radiometric fin—low Knudsen numbers

6.6.2.3 Third Case ($\lambda < L, \lambda < d$)

In these conditions, whose schematic is shown in Fig. 6.25, and the Knudsen number is very small, the relation obtained for the force in Sect. 6.6.2 is incorrect for the whole chamber. Therefore, the force applied to this surface originates from the molecules located within a distance of λ from it, and the distant molecules have little effect on the plate. In this case, the effective region of the blade is $d\lambda$.

$$\frac{F}{A} = kn\lambda\partial_x T, \quad A = d\lambda \tag{6.75}$$

$$F = kn\lambda^2\partial_x Td \tag{6.76}$$

6.6.3 Micro-Radiometric Pump

6.6.3.1 Introduction

Recently, Shahabi et al. [21] showed that the micropump with sawtooth teeth, designed in 2011 by Donkov et al. [22], operates based on the principles of radiometric flow. Figure 6.26 shows the geometry of the pump proposed by Shahabi et al. [21], where the temperature of the upper wall is higher than the lower wall ($T_1 < T_2$). A series of sawtooth teeth are placed on the lower wall, where the vertical surface of these teeth has diffusive reflection (solid lines), and their inclined surface has specular reflection (dashed lines). A strong radiometric flow is generated above the sawtooth, and the flow starts to move from right to left in the microchannel (the red arrow indicates the direction of the radiometric flow).

If the lower wall does not have teeth, only conductive heat transfer occurs between the upper and lower walls. Since a temperature gradient in the *horizontal direction* is required to create thermal creep flow, a height difference can be created instead of creating a temperature difference (creating teeth). If the inclined side of these teeth is specular, the pump's efficiency increases. The advantages of the radiometric pump are as follows:

1. It has no moving parts.
2. No oil or liquid is used in it.
3. It can utilize the wasted energy from other devices.
4. Like radiometric flow, it can work at different pressures.

Initially, both sides of the tooth were considered to have diffuse reflection, and the temperature of the upper and lower walls was supposed to be the same. However, the resulting velocities were very low and chaotic, and no flow was generated. When the inclined side is considered specular, flow is created. To generate thermal creep flow at the tip of the tooth, not only is an asymmetric geometry required, but there must also be a temperature difference on the sides of the tooth tip. Note that the temperature value is not applied on specular surfaces. Still, these surfaces act as insulating surfaces (in practice, the boundary condition of zero heat flux and stress is applied on these surfaces, and the temperature is determined during the numerical solution). Due to the presence of the diffuse side with a lower temperature on the right

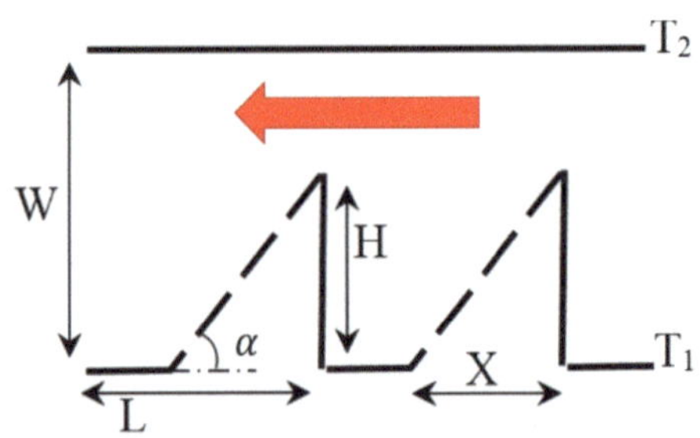

Fig. 6.26 Gas micropump ($T_1 < T_2$)

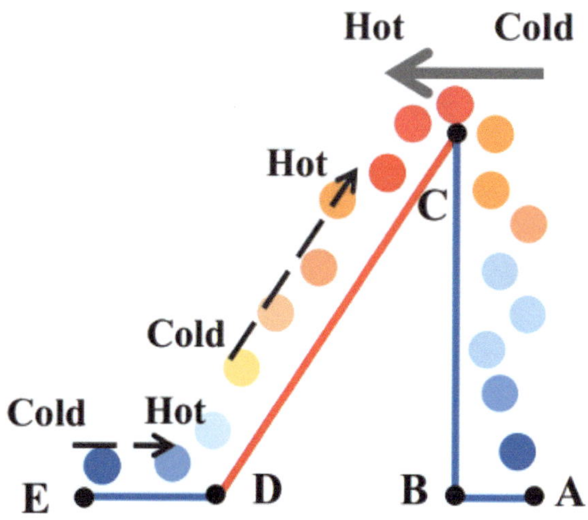

Fig. 6.27 Mechanism of flow generation around a ratchet (Copied from open access Ref. [21])

side of the tooth tip, when the molecule collides with this wall and exchanges energy, the molecule has less momentum than the molecule colliding with the specular side; because the temperature near the specular wall is higher. Finally, at the tip of the tooth, thermal creep flow is generated from right to left, see Fig. 6.27.

6.6.3.2 Streamlines at Different Knudsen Numbers

The streamlines, along with the velocity magnitude contours ($V = \sqrt{u^2 + v^2}$) are plotted in Fig. 6.28. The characteristic length of the geometry is assumed to be $-\frac{H}{2}$, and the Knudsen number is defined as $\mathrm{Kn} = \frac{\lambda}{(W-\frac{H}{2})}$ (see Fig. 6.26). The flow is generated in the negative x-direction, and at low Knudsen numbers, the highest velocity occurs above the tip of the tooth. With increasing Knudsen number to 0.1, the strength of the radiometric force increases, and the vortex formed adjacent to the lower surface becomes larger. The strength of the vortex formed adjacent to the lower surface increases with decreasing Reynolds number. Also, with increasing Knudsen number, the flow expansion after passing the tooth tip covers the entire field above the tooth tip, and as a result, the flow velocity changes in the horizontal layers above the tip, with the highest velocity occurring farther from the tip. Also, the effect of the vertical diffuse wall decreases because the large vortex formed covers the entire space between the wall with specular and diffuse reflection.

In Fig. 6.28, thermal stress does not play a role in forming the vortex adjacent to the inclined wall. This type of vortex is known as the Moffatt vortex [1], which occurs in Stokes flow, i.e., flow with low Reynolds number. According to Moffatt's analytical solution [23], any corner with an angle less than 146.3° in low Reynolds number flow causes the formation of a Moffatt vortex. The flow above the specular

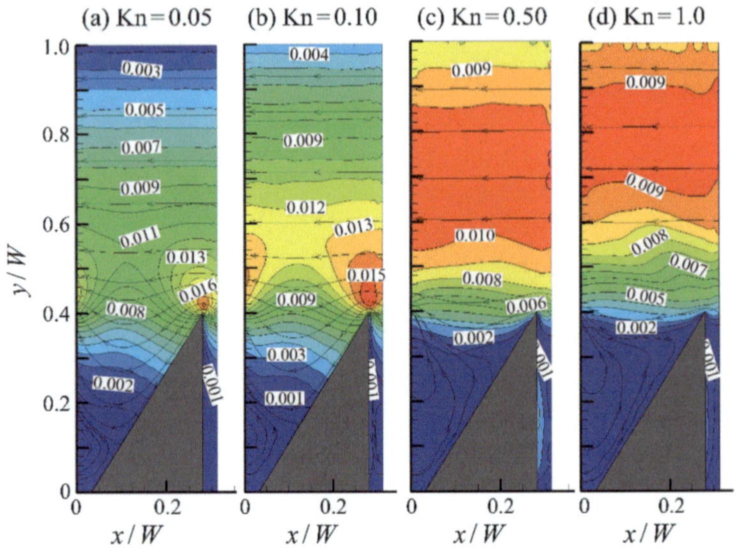

Fig. 6.28 Velocity contours along with streamlines in a radiometric micropump (Copied from open access Ref. [21])

surface acts like the moving wall of a cavity in which a vortex is formed. Further explanations about Moffatt vortices are provided in Sect. 6.9.

The variation of the mean velocity on the vertical line connecting the channel's lower surface to the channel's upper surface is plotted against the Knudsen number for a vane with a height of 400 nm in Fig. 6.29. The condition $x/\mathrm{H} = 2.812$ corresponds to the line that connects the tip of the tooth to the upper surface. As the Knudsen number approaches zero, the flow velocity becomes zero. Also, the flow velocity decreases with the increase in the Knudsen number in the transition regime. For low Knudsen numbers, it can be assumed that the velocity is proportional to the relation $u \sim F_x H/\mu$, where F_x is the axial force per unit area, H is the tooth height, and μ is the gas viscosity. DSMC results show that at low Knudsen numbers, the velocity is proportional to the square root of the Knudsen number, and since there is no flow at large Knudsen numbers, it can be assumed that in the free molecular regime, we have $\mathrm{u} \sim 1/\mathrm{Kn}$. Given that the velocity value in the two ranges of low and high Knudsen numbers has been estimated according to the mentioned relations, for the entire range of Knudsen numbers, the following function is proposed for the variation of velocity with the Knudsen number: $\mathrm{u} = \frac{1}{\mathrm{aKn}^{0.5} + \frac{\mathrm{b}}{\mathrm{Kn}}}$. The coefficients of this approximation are shown in Fig. 6.29.

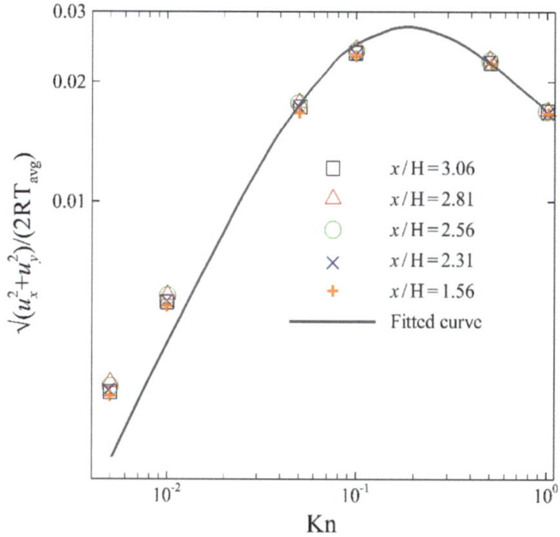

Fig. 6.29 Variations in velocity magnitude as a function of the Knudsen number at different positions for a ratchet with a height of 400 nm (Reproduced from Ref. [21])

6.6.4 Horizontal Forces

Westphal [24], as well as Broch and Litvin [20], found from experimental results that the ratio of axial force to the temperature gradient along the horizontal axis follows the relation:

$$\frac{F_x}{\Delta T} \propto \frac{1}{\frac{a}{Kn} + bKn} \tag{6.77}$$

where *a* and *b* are constants that depend on the type of gas and the geometry of the radiometer. Based on Einstein's theory at low Knudsen numbers, particles in a region extending a distance of the order of the molecular mean free path around the tip of the radiometer have the greatest effect in applying the radiometric force. The force exerted on the tooth is calculated as follows according to relation (6.78):

$$F_x = pH\lambda^2\nabla T/T \Rightarrow \frac{F_x}{\nabla T} \propto n\lambda^2 \propto \frac{1}{p} \propto Kn \tag{6.78}$$

Considering the discussion presented regarding relation (6.54), the radiometric force on the vane at low Knudsen numbers can be estimated with the function $F_x \propto Kn^{0.5}$. At high Knudsen numbers, ΔT between the two channel walls should be replaced with $\lambda\nabla T$ because the molecules have few intermolecular collisions within the distance of collision with the upper and lower surfaces. Therefore, the collision space is no longer limited to a region of the order of λ. The value of λ should be replaced with *H*. As a result, the axial force will be given by:

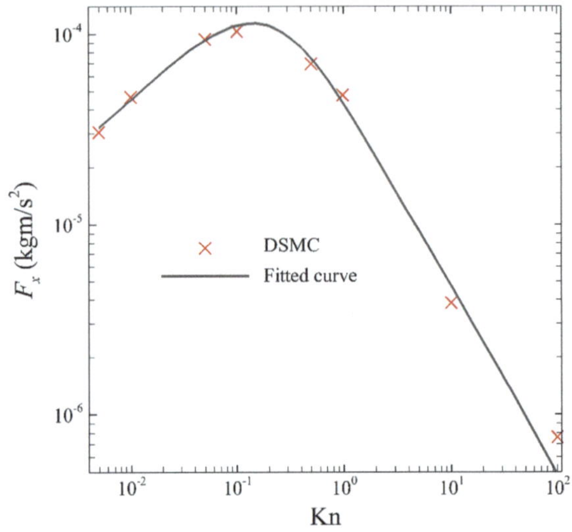

Fig. 6.30 Variations in axial force as a function of the Knudsen number (Reproduced from Ref. [21])

$$F_x = pH^2\Delta T/T \Rightarrow \frac{F_x}{\Delta T} \propto n \propto p \propto \frac{1}{Kn} \tag{6.79}$$

The results obtained from the DSMC solution at high Knudsen numbers show that the scale of the axial force F_x in the present study follows the order of $Kn^{-0.5}$. Figure 6.30 shows the axial force applied to the tooth over a wide range of Knudsen numbers. Additionally, the curve fitted to the DSMC data follows the equation in the form:

$$F_x = \frac{1}{\frac{a_1}{Kn^{0.5}} + b_1 Kn} \tag{6.80}$$

where the values of the constants a_1 and b_1 are shown in Fig. 6.30, as shown, the maximum value of the radiometric force occurs near the upper boundary of the slip regime, which is consistent with the experimental predictions of Brook and Litvin [20].

6.6.5 *Effect of Tooth Height*

The effect of changing the tooth height on the flow field pattern at a Knudsen number of 0.1 and a temperature difference of 300 K between the two walls is shown in Fig. 6.31. The velocity contours and streamlines are plotted for 6 different tooth heights, including $\hat{H} = H/W = 0, 0.0125, 0.1, 0.2, 0.3, 0.7$. Figure “a” corresponds to zero tooth height, meaning a channel with the lower wall parallel to the upper wall, except that the lower wall consists of two diffuse regions and one specular region

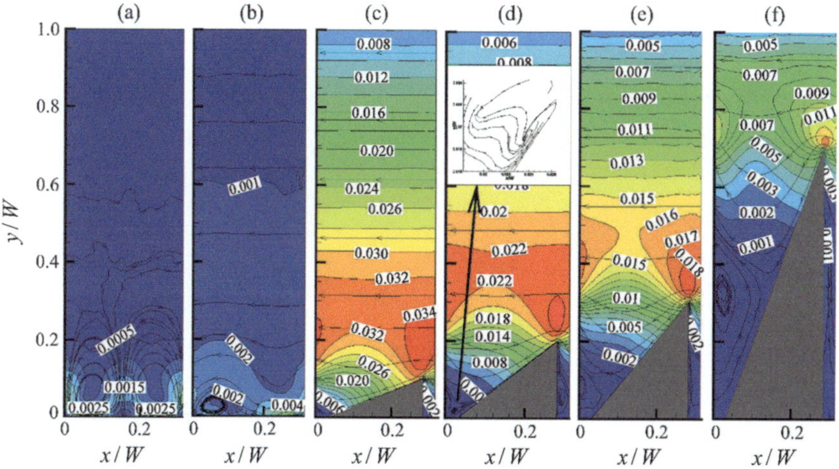

Fig. 6.31 Effect of changing tooth height on the velocity behavior and streamlines in the solution field at Knudsen number 0.1 and a temperature difference of 300 K; **a** H/W = 0.0, **b** H/W = 0.0125, **c** H/W = 0.1, **d** H/W = 0.2, **e** H/W = 0.3, and **f** H/W = 0.7 (Copied from open access Ref. [21])

(equal to the length of the inclined side of the tooth in the previous conditions). At $\hat{H}$ = H/W = 0, two vortices are formed on the lower wall.

The reason for the formation of these two vortices is similar to thermal edge flow. The flow temperature adjacent to the specular surface is higher than the two diffuse surfaces. Therefore, thermal creep flow approaches from both directions, from the diffuse wall (colder) to the specular wall (hotter) and collides with each other. Since these two flows have the same strength, they cannot neutralize each other, forming two vortices.

When the tooth height increases in Figure "b" ($\hat{H}$ = 0.0125), a thermal creep is created on the small vertical wall of the tooth, pushing the flow upward and then onto the inclined wall (similar to the mechanism of flow movement from A to B and then B to C in Fig. 6.27). After the flow reaches the end of the inclined wall, it encounters the thermal edge flow moving from the diffuse wall to the right. This encounter of two flows with almost equal strength creates a vortex on the diffuse wall. When the height increases to $\hat{H}$ = 0.1 (Figure c), the mechanism of flow movement on the tooth is similar to the previous case. Still, the flow generated on the inclined surface of the tooth is stronger than the thermal edge flow (left to right); therefore, the flow on the diffuse wall also moves to the left. A small Moffatt-type vortex is also formed near the vertical wall because the flow cannot be completely attached at a 90-degree angle change (Figure d). With a further increase in height, the flow mechanism near the vertical wall does not change, but another vortex is formed near the inclined wall due to the increase in angle. Further increasing the tooth height increases the strength of this vortex (Figures e and f). The axial force is shown in Fig. 6.32 at Knudsen number 0.1 and height $\hat{H}$ = 0.1 is the maximum. Thus, these conditions represent the optimal operating conditions for the pump investigated.

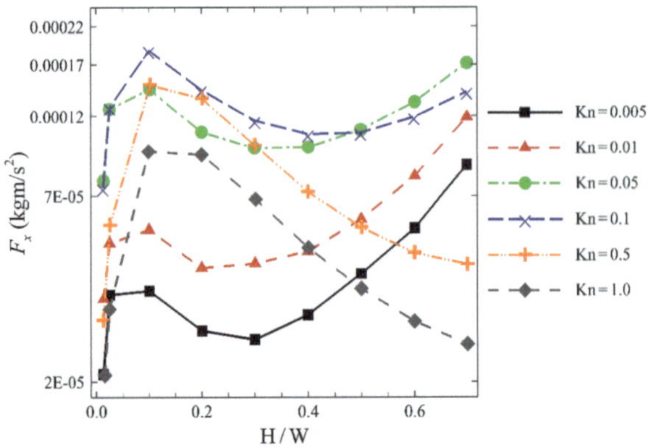

Fig. 6.32 Variations in horizontal force with fin height at different Knudsen numbers (Reproduced from Ref. [21])

6.6.6 Novel Geometries for Radiometric Micropumps

Lotfian and Roohi [25] proposed numerous geometries for radiometric pumps. In Fig. 6.33, seven proposed geometries are shown in three groups. Blue and red lines correspond to cold and hot walls with temperatures of T_C and T_H, respectively. Also, the dashed lines indicate the specular surfaces. The first group proposes three types of pumps designed based on the Crookes radiometer. Geometry G1 presents a channel with a Crookes radiometer vane on the lower wall. Geometries G2 and G3 are radiometric pumps in which a right triangle or an isosceles triangle replaces the vane. The second group includes radiometric pumps where the left side of their vane has specular reflection properties. Figure 6.33 shows specular surfaces with dashed lines. The third group proposes radiometric pumps that include dual triangular vanes. For example, in geometry 6, each triangular vane has half the width of the triangle used in G3. Geometry G7 also has the same dual triangular pattern on the upper wall; therefore, the geometry of Fig. 6.7 is called a zigzag channel. Geometry G7 is equivalent to using sawtooth teeth in geometry 6 in half of the solution domain. Periodic (repeating) boundary conditions are considered on both sides of the simulation domain. Therefore, there is no pressure difference across the channel. The vertical vane of geometry 1 or the triangular vanes (isosceles or right triangle) have a periodic length *L* with height *H* and width *B*.

For all geometries, the ratios $L/W = 5/16$ and $H/W = 0.1$ are considered. The ratio of $H/W = 0.1$ is proposed as the optimal tooth height for the maximum induced velocity and radiometric force [21]. Shahabi et al. [21] proposed a dimensionless periodic ratio equal to $L/W = 5/16$ for geometry 4. In addition, the results presented in the work of Shahabi et al. [21] showed that variations in the dimensionless periodic ratio have very little effect on the performance of geometry 4. The temperature of the horizontal surfaces on the lower wall is set to T_C, and a temperature ratio equal to $T_H/T_C = 2$ is considered between the hot walls (red lines) and the cold walls (blue lines). The temperature difference is perpendicular to the channel axis since the temperature cannot be applied to the specular wall. As a result, the upper boundary

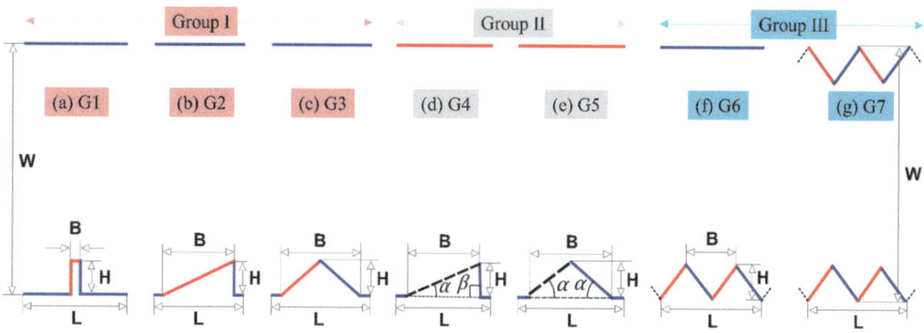

Fig. 6.33 Schematic of seven designed geometries for radiometric pumps categorized into three groups (Reproduced from Ref. [25])

Table 6.2 Geometric specifications and operating conditions of proposed radiometric pumps ($Kn = \lambda / W_{eff} = 0.1$)

Geometry	B/W	Side angles	$Kn = \lambda/H$
G1	$0.0095/4$	$\alpha = 90°$	1
G2, G4	$1/4$	$\alpha = 21.8°, \beta = 90°$	0.96
G3, G5	$1/4$	$\alpha = 38.65°$	0.96
G6, G7	$0.625/4$	$\alpha = 52°$	0.95, 0.90

is considered the hot wall in geometries 4 and 5. In other geometries, the left side of the vane/sawtooth is set to the hot temperature, while the upper wall temperature is assumed to be T_C. Other geometric specifications of the pumps, including the slope angles of the left and right sides of the vane, which are indicated by α and β, respectively, are reported in Table 6.2.

If the Knudsen number (Kn) is defined as the ratio of the molecular mean free path of the gas to the characteristic length of the pump geometry, and the scale of the characteristic length is the effective height of the channel, Kn is varied below 0.1. In this geometry, the effective channel height is defined as $W_{eff} = W - A/L$, where A/L represents the equivalent height of the vane or triangles. In geometries 1 and 6, the vane area is $A = HB$; in geometries 2 to 5, the vane area is $A = HB/2$, and in geometry 7, the vane area is $A = 2HB$. This Knudsen number reports the fluid conditions far from the vanes in the open channel above them. If Kn is defined based on the height of the vane or sawtooth (H), the simulated geometries will have a Knudsen number ranging from 0.9 to 1 (see the last column of Table 6.2). This Knudsen number reports the gas conditions near the vanes. Figure 6.34 shows the regions affected by both Knudsen numbers defined above.

Figure 6.35 shows the variation of mass flux with Knudsen number for geometries G1 and G7. As the Knudsen number increases from 0.1 to 0.5, the mass flux in geometries 1 and 7 increases by factors of 4.5 and 10, respectively. Moreover, the mass flux increases in both geometries as the Knudsen number increases to 0.5, and when the Knudsen number in the transition regime decreases, the mass flux. However, the maximum mass flux in both geometries occurs in the $0.1 < Kn < 0.5$.

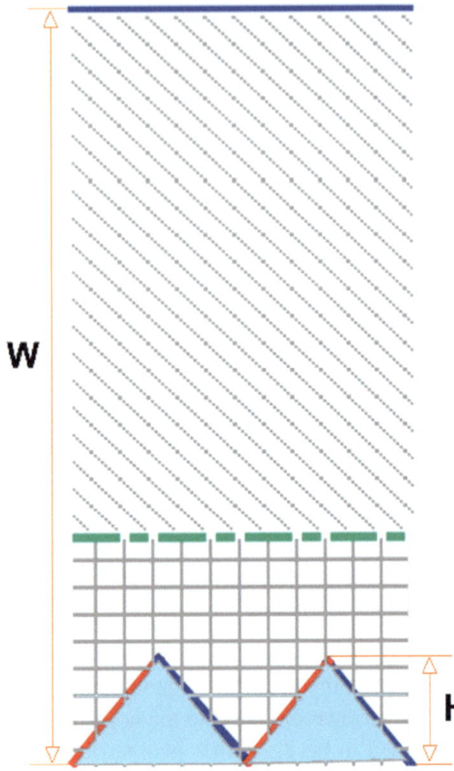

Fig. 6.34 Effective regions for two defined Knudsen numbers: one based on the characteristic length of the effective channel height and the other based on the fin/sawtooth height (Reproduced from Ref. [25])

The difference in mass flux at Kn = 0.5 and Kn = 0.9 for G7 compared to G1 is relatively small. The rate of decrease in mass flux for geometry 7 is also larger at higher Knudsen numbers.

6.6.6.1 Flow Mechanism

The fluid flow mechanism in all the configurations considered in Fig. 6.33 is almost identical. Due to the temperature difference, a radiometric force is generated in the positive axial direction on the vane near the tips of the vane/sawtooth. The reaction of this force is transmitted to the gas in the vicinity of the vane edge through momentum exchange and in the opposite direction. The momentum exchange within the gas occurs due to molecular collisions between the molecules reflected from the walls and the molecules near the wall. In the investigated geometries that are in the range of Kn $\sim$ 1, this momentum exchange occurs within a range of one mean free path above the vane tip. Therefore, the fluid flow develops in this range and even beyond it (depending on the shape of the vane). The temperature gradient around the vane tip and on the lower left horizontal surface of the vane or the cold inclined side in

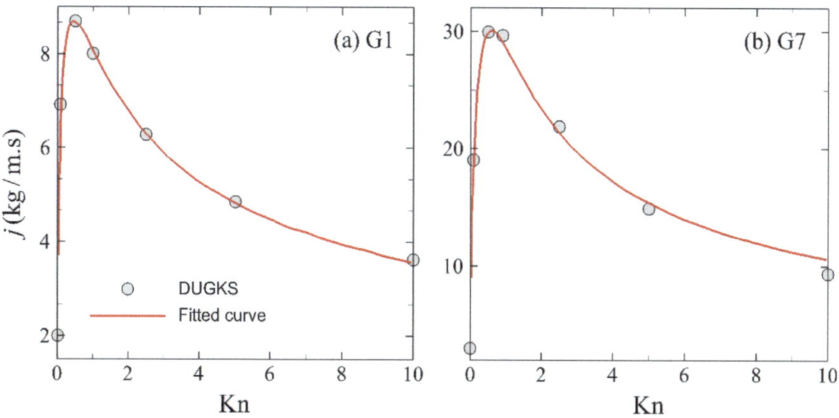

Fig. 6.35 Variations in mass flux with the Knudsen number for G1 (**a**) and G7 (**b**) (Reproduced from Ref. [25])

G6 and G7 drives the flow in two opposite directions. Although the temperature of the left wall (G1 to G5) or the cold inclined surfaces (G6 and G7) is constant in all geometries, as expected in rarefied flow, due to the occurrence of a temperature jump in the fluid adjacent to the boundary, a temperature gradient is created near the boundaries. As a result of this flow, a net rightward thermal flow is created on the left horizontal wall or the vane surfaces. The nature of this flow is similar to thermal creep flow, and therefore, here it is called "rightward flow generated by the same mechanism as thermal creep flow" or, for simplicity, "rightward flow. A schematic of the radiometric flow mechanism and net rightward flow is shown in Fig. 6.36, where the rightward flow flows from a point near the relatively colder point (the middle of edge CD in the right part of Fig. 6.36) towards the relatively hotter point (point B in the right part of Fig. 6.36). It seems that the rightward flow is weaker than the radiometric flow at the tip, and this flow is blocked by the frictional force of the left vertical wall (in G1 to G5) and the inclined wall (G6 and G7)[1] [1]. In the investigated Knudsen number, the resultant flow moves to the left due to sudden temperature changes around the vane(s) and thus in opposition to the rightward flow generated on the left horizontal wall of the vane/sawtooth [1].

6.7 Comparison of the Strength of Different Thermal Flows

Table 6.3 compares the order of magnitude of different thermal flows.

[1] Note that the strength of the rightward flow relative to the radiometric flow varies from one geometry to another, and its strength changes at low Knudsen numbers.

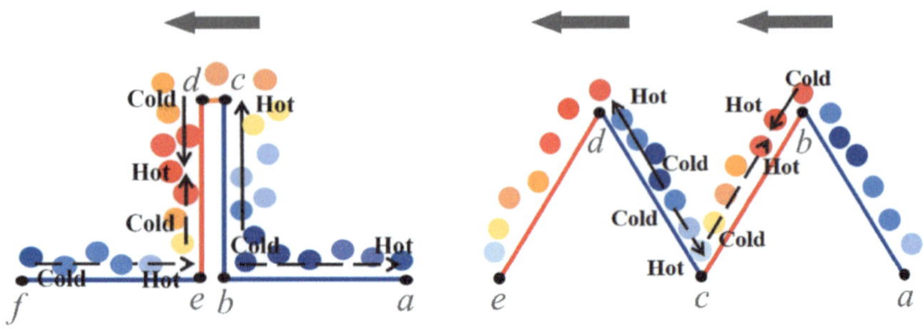

Fig. 6.36 Radiometric flow mechanism (solid line arrow) and flow similar to thermal creep flow (dashed line arrow) around an isosceles triangular tooth and G1. The blue and red circles represent cold and hot molecules (Reproduced from Ref. [26])

Table 6.3 Comparison of the order of magnitude of different thermal flows

Flow Order	Thermal flows
$O(\mathrm{Kn})$	Thermal creep flow
$O(\mathrm{Kn}^2)$	Thermal stress slip flow
$O(\mathrm{Kn})$	Nonlinear, thermal stress flow
$O(\mathrm{Kn}^{0.5})$	Thermal edge flow (low Knudsen numbers)
For high Knudsen numbers $O(\mathrm{Kn}^{-1})$ For low Knudsen numbers $O(\mathrm{Kn}^{0.5})$	Radiometric flow

6.8 Gas Mixture Separation with Radiometric Pumps

In addition to transferring flow, radiometric pumps can separate binary mixtures of gases [26]. The primary reason for this is that the thermal velocity of molecules is inversely related to the molecular mass, i.e. $v \propto \sqrt{kT/m}$. Therefore, a molecule that has a lower weight gains a higher thermal velocity compared to heavier molecules. On the other hand, the mean free path of gases is a function of the inverse of the molecular diameter, and the larger the molecular diameter, the smaller the mean free path of the gas. In high Knudsen number regimes, where the radiometric force is proportional to the mean free path, less force is applied to the molecular species with a larger diameter.

The more precise reason for gas mixture separation in radiometric pumps is as follows: the mass flux of each molecular species is a function of the pressure gradient, temperature gradient, and concentration gradient, meaning:

$$J^A = M_T^A \cdot \frac{\nabla T}{T} + M_P^A \cdot \frac{\nabla P}{P} + M_c^A \cdot \frac{\nabla c}{c} \tag{6.81}$$

$$J^B = M_T^B \cdot \frac{\nabla T}{T} + M_P^B \cdot \frac{\nabla P}{P} + M_c^B \cdot \frac{\nabla c}{c} \tag{6.82}$$

where the coefficients M_T, M_P, and M_c are the coefficients of dependence of the species' mass flux on the temperature gradient, pressure gradient, and concentration gradient, respectively. A is assumed to be the lighter species, and B is the heavier species. Although only the temperature difference is applied to the radiometric pump problem, pressure and concentration distributions are created in the field, which induce mass flux. The coefficient M_P has very little dependence on the molecular mass, but the coefficient M_T is strongly dependent on the molecular mass. Therefore, the mass flow rate and velocity of the lighter species can be greater than the mass flow rate of the heavier species. The mixture velocity is defined as follows:

$$\vec{v} = \frac{m_A \overrightarrow{v_A} + m_B \overrightarrow{v_B}}{m_A + m_B} \tag{6.83}$$

Therefore, the velocity and diffusion flux ($\overrightarrow{J_D^B} = n_i(\overrightarrow{v^B} - \vec{v})$) of the heavier species will be in the opposite direction of the velocity and diffusion flux of the lighter species, and gas mixture separation occurs.

Figure 6.37 shows the molar ratio of the mixture with initial conditions of 50% helium–50% xenon at a Knudsen number of around 1 in geometries 1, 6, and 7 and the streamlines of the diffusion velocity. The molar ratio of helium on the right side of the vane decreases from the initial value but increases on the left side of the vane, as helium is the lighter gas. On the other hand, the heavier gas xenon rises on the right side of the vane.

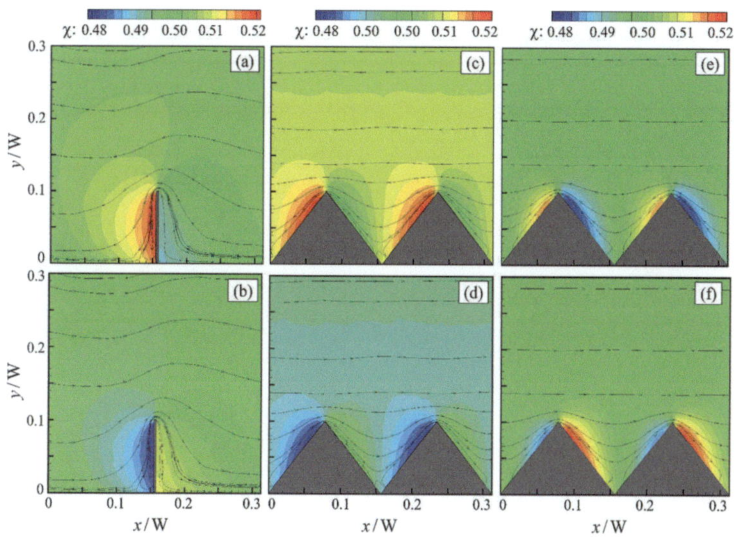

Fig. 6.37 Contours of Helium molar fraction (upper) and Xenon molar fraction (lower) in three different geometries; G1 (**a**, **b**), G6 (**c**, **d**), and G7 (**e**, **f**) (Reproduced from Ref. [26])

6.9 Stokes Flow

Since the flow velocity in microflows is usually negligible, and given that vortices are formed in the geometry of radiometric micropumps, which have a viscous nature, this section examines the Stokes flow equations and examples [27]. When the Reynolds number is very small, meaning the density is low, the fluid's viscosity is high, and the velocity is low, Stokes flow is observed. This type of flow is commonly encountered in chemical engineering or materials science, where slow and viscous non-Newtonian fluids are dealt with. However, in microflows, where the viscosity is usually low, the Reynolds number is small due to the small characteristic length, and Stokes flow is observable.

To derive the Stokes flow equation, we first normalize the momentum equation with the following definitions:

$$x_i^* = \frac{x_i}{h}, \quad v_i^* = \frac{v_i}{U}, \quad P^{**} = \frac{P - P_0}{\frac{\mu U}{h}} = \frac{P - P_0}{\frac{1}{2}\rho U^2}\left(\frac{Re}{2}\right) \tag{6.84}$$

Regarding the normalization of pressure, since the velocity in Stokes flows is very low ($P^{**} = \frac{p - p_0}{\frac{\rho u^2}{2}}$), the pressure is non-dimensionalized using the stress.

The dimensional momentum equation is expressed as follows:

$$\frac{\partial}{\partial t}(\rho V) + \nabla.(\rho VV) = -\nabla P + \nabla.\tau \tag{6.85}$$

The normalized momentum equation is:

$$Re\frac{DV^*}{Dt^*} = -\nabla^* P^{**} + \nabla^{*2} V^* \tag{6.86}$$

Given that $Re \to 0$, the following equations are obtained:

$$\nabla^* P^{**} = \nabla^{*2} V^* \tag{6.87}$$

$$\nabla P = \nabla.\tau \tag{6.88}$$

$$\tau = \mu \nabla V \tag{6.89}$$

6.9.1 Other Expressions of the Stokes Equation

By using the relations $\tau = \mu \nabla V$ and $\nabla P = \nabla.\tau$, we have:

$$\nabla P = \mu \nabla^2 V \tag{6.90}$$

Based on the stress tensor relation, which includes pressure and viscous stress, another expression of the Stokes equation, Eq. (6.91), is extracted:

$$T_{ij} = -p\delta_{ij} + \tau_{ij} \tag{6.91}$$

$$\nabla.T = 0 \tag{6.92}$$

Additionally, by using the mathematical relation (6.93) and assuming incompressible flow, we arrive at relation (6.96) where ω is vorticity:

$$\nabla \times (\nabla \times A) = \nabla(\nabla \cdot A) - \nabla^2 A \tag{6.93}$$

$$\nabla \times (\nabla \times V) = \nabla(\nabla \cdot V) - \nabla^2 V \tag{6.94}$$

$$\nabla \times (\omega) = 0 - \nabla^2 V \tag{6.95}$$

$$\nabla P = -\mu \nabla \times \omega \tag{6.96}$$

If in Stokes flow, the density is very low ($\rho \to 0$), the same relations are obtained again. Therefore, the results of Stokes flow are independent of density. In the non-dimensional form, we have:

$$\nabla^{*2} P^{**} = 0 \tag{6.97}$$

In potential flow, $\nabla^2 \varphi = 0$, where φ is the potential function. However, in Stokes flow, pressure satisfies the Laplace equation. Therefore, the pressure equation can be easily solved with given boundary conditions.

Assuming $\omega^* = \frac{\omega}{\frac{U}{L}}$, the vorticity equation in non-dimensional form is as follows:

$$Re\frac{D\omega^*}{Dt^*} = Re\omega^*.\nabla V^* + \nabla^{*2}\omega^* \tag{6.98}$$

$$Re \to 0 \quad \nabla^{*2}\omega^* = 0 \tag{6.99}$$

In the derived equations, there is no dependence on time. Thus, Stokes flow is quasi-steady, meaning it becomes steady very quickly. The validity of this conclusion is based on the use of the term $\frac{t}{L/U}$ for non-dimensionalizing time, and since $u \to 0$, there is no time dependency.

Another advantage of Stokes flows is that all the governing equations are linear, allowing them to be solved using superposition. An additional advantage is the direct consequence of its linearity. If the mirror image of all properties is substituted into the equations, the equations remain unchanged. Therefore, Stoke's flow from right to left is no different from the flow from left to right. If the geometry is symmetric, the velocity field resulting from Stokes flow is also symmetric, and the upstream and downstream flows are completely identical. Figure 6.38a shows an experimental example of symmetric Stokes flow around a square.

CFD simulations were also performed to compare with experimental data. The flow domain consists of a square block region, with a width, denoted as L, of 1 cm. The dimensions of the flow field extend to $4L$ in height and $10L$ in length. The fluid enters from the left edge and exits from the right edge, while the remaining edges

(a) Experimental observation at very low Re (Reproduced from Ref. [27]).

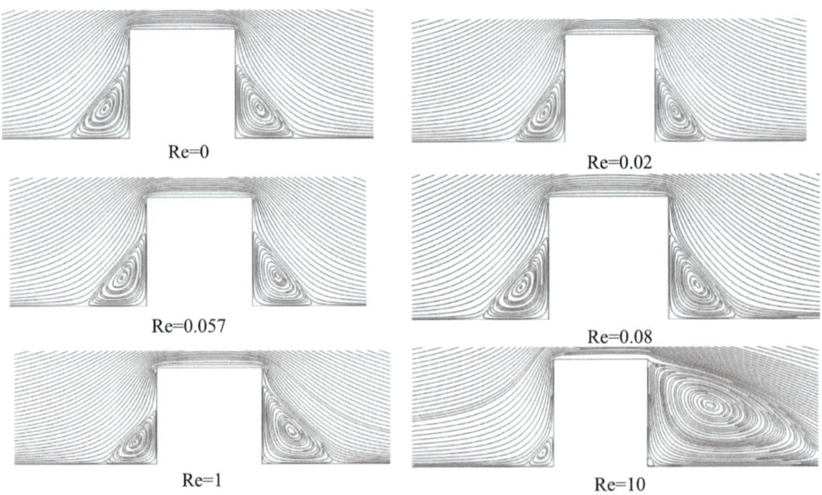

(b) CFD simulations using COMSOL

Fig. 6.38 Low Reynolds flow around a square obstacle; **a** experimental picture and **b** CFD results (Yulong Liang, graduate student in Johns Hopkins University) are designated as walls, thereby simulating the flow characteristics around a block. For this case, water at a temperature of 20 °C is employed as the working fluid.

As depicted in Fig. 6.38 Re $= 0$, in the regime where $Re \ll 1$, the streamlines adjacent to the block region exhibit high symmetry. Moreover, wake length is $L_w \approx 0.56L$. In cases where Re is not significantly smaller than 1, noticeable asymmetry becomes apparent in the flow field. Additionally, it is observed that the L_w increases with higher velocities. Employing a binary search approach, one can determine the critical Reynolds number, at which the wake behind and after the block becomes asymmetric, to be $Re = 0.057344$. This can be considered as the limit of failure of Stokes flow assumption.

6.9.2 *Derivation of the Equation* $\nabla^4 \psi = 0$

A vector potential function B is first defined to derive the streamline equation.

$$B = \psi \nabla g \tag{6.100}$$

$$V = \nabla \times B \tag{6.101}$$

$$\omega = \nabla^2 B = -\nabla \times \nabla \times B \tag{6.102}$$

Using the two-dimensional and compressibility assumptions:

$$g = z \quad B = i_z \psi \quad V = \nabla \times B \tag{6.103}$$

$$V = (\nabla \times \psi) i_z \tag{6.104}$$

$$\nabla \times V = \nabla \times (\nabla \times \psi) i_z \tag{6.105}$$

$$\omega_z = \frac{\partial v}{\partial x} - \frac{\partial u}{\partial y} = \frac{\partial}{\partial x}\left(-\frac{\partial \psi}{\partial x}\right) - \frac{\partial}{\partial y}\left(\frac{\partial \psi}{\partial y}\right) \tag{6.106}$$

$$-\omega_z = \nabla^2 \psi \tag{6.107}$$

$$\nabla^2 \omega_z = 0 \tag{6.108}$$

By using Eqs. (6.107) and (6.108), Eq. (6.109) is derived:

$$\nabla^4 \psi = 0 \tag{6.109}$$

$$\nabla p = -\mu \nabla \times \omega_z \tag{6.110}$$

This means:

$$\frac{\partial \mathrm{p}}{\partial \mathrm{x}} i + \frac{\partial p}{\partial y} j = -\mu \left(\frac{\partial \omega_z}{\partial y} i - \frac{\partial \omega_z}{\partial x} j \right) \tag{6.111}$$

$$\frac{\partial \mathrm{p}}{\partial \mathrm{x}} = -\mu \frac{\partial \omega_z}{\partial y} = \mu \frac{\partial (\nabla^2 \psi)}{\partial y} \tag{6.112}$$

$$\frac{\partial \mathrm{p}}{\partial \mathrm{y}} = \mu \frac{\partial \omega_z}{\partial x} = -\mu \frac{\partial (\nabla^2 \psi)}{\partial x} \tag{6.113}$$

6.9.3 Moffatt Vorticity

This section describes fluid flow analysis around a corner with angle α (like Fig. 6.39) to introduce the Moffatt vorticity. It explains that by defining the Reynolds number locally, due to the no-slip condition in a continuum flow, the Reynolds number on the wall is zero and very small near it. This indicates a Stokes flow near the wall, where the bottom wall moves with velocity *u*.

The far-field boundary conditions influence the local Stokes flow, and Moffatt vorticity may be generated in local corner flows where part of the flow has a large overall Reynolds number.

The solution to the equation $\nabla^4 \psi = 0$ using the method of separation of variables for corner flows in polar coordinates results in the following expression:

$$\psi(r, \theta) = r^{n+1}[A \cos(n+1)\theta + B \sin(n+1)\theta$$

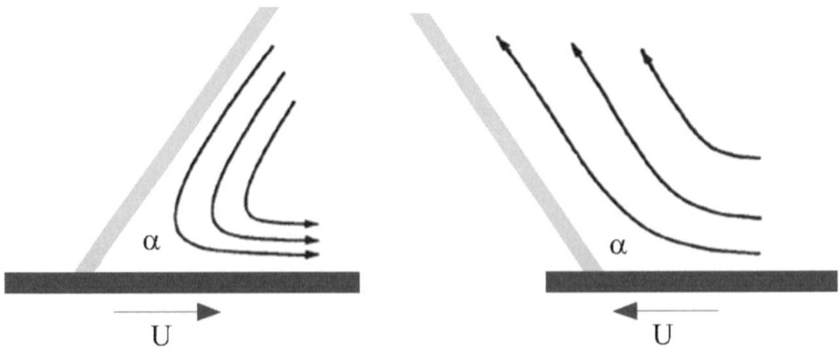

Fig. 6.39 Flow in a corner with a moving bottom wall (Reproduced from Ref. [27])

$$+C\cos(n-1)\theta + D\sin(n-1)\theta] \tag{6.114}$$

The value of n is determined according to the corner angle.

$$\psi(r,\theta) = A\cos 2\theta + B\sin 2\theta + C\theta + D \quad (n=-1) \tag{6.115}$$

$$\psi(r,\theta) = r(A\cos\theta + B\sin\theta + C\theta\cos\theta + D\theta\sin\theta) \quad (n=0) \tag{6.116}$$

$$\psi(r,\theta) = r^2(A\cos 2\theta + B\sin 2\theta + C\theta + D) \quad (n=1) \tag{6.117}$$

To obtain the coefficients A, B, C, and D, the following boundary conditions are used:

$$\psi(r,\theta=0) = 0 \quad \psi(r,\theta=\alpha) = 0 \tag{6.118}$$

$$v_r(r,\theta=0) = U = \frac{1}{r}\frac{\partial\psi}{\partial\theta} \quad v_r(r,\theta=\alpha) = U = \frac{1}{r}\frac{\partial\psi}{\partial\theta} \tag{6.119}$$

To obtain the appropriate solution, n must satisfy the above boundary conditions; in this case, n = 0 is the most desirable solution.

$$\psi(r,\theta) = \frac{-Ur}{\sin^2\alpha - \alpha^2}\left[\alpha^2\sin\theta - \left(\sin^2\alpha\right)\theta\cos\theta + (\sin\alpha\cos\alpha - \alpha)\theta\sin\theta\right] \tag{6.120}$$

The potential flow predicts solutions that are mostly free of vortices; however, vortices may form even in a potential flow. In the previous case (Fig. 6.39), the bottom wall was moving, but the issue commonly occurs in reality is the flow over a flat plate where separation occurs at an angle α (Fig. 6.40). In this case, there is no actual inclined wall; instead, a virtual wall (separation boundary $\psi = 0$) divides the flow into two parts. Therefore, there are two sets of solutions.

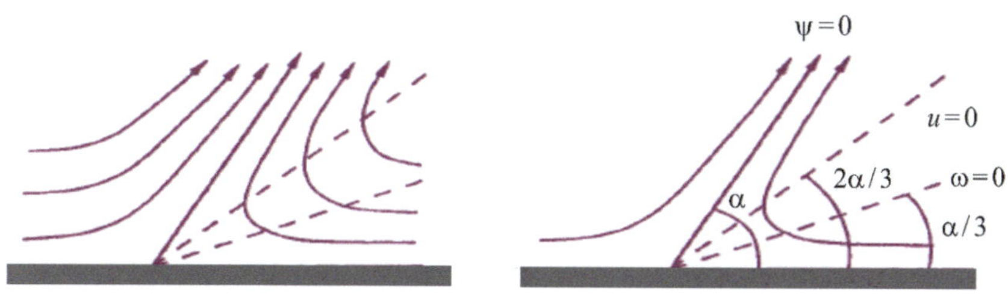

Fig. 6.40 Separation of flow over a plate, left: real flow; right: modeled flow (Reproduced from Ref. [27])

To obtain the desired n, if $n \leq 0$, infinite force is generated at $r = 0$. If $n = 1$ and the no-slip velocity condition is satisfied, we have:

$$\psi \approx Ar^2 \cos 2\theta \tag{6.121}$$

In this solution, no separation occurs. Therefore, $n = 2$ is the best answer.

$$\psi_2(r, \theta) = r^2[A(\cos 3\theta - \cos \theta) + B(\sin 3\theta - 3 \sin \theta)] \tag{6.122}$$

If $\psi = 0$ is used, the separation angle, a function of A and B, is determined. However, the exact separation angle is determined by considering flow properties outside the Stokes region. In the analysis of flow around a sharp edge, the most suitable value of n is $n = {}^1\!/_2$. However, even these solutions do not capture the actual separation that occurs in reality.

$$\psi_{1/2}(r, \theta) = r^{3/2}\left[A_{1/2}\left(\cos \frac{3}{2}\theta - \cos \frac{1}{2}\theta\right) - B_{1/2}\left(\sin \frac{3}{2}\theta - 3 \sin \frac{1}{2}\theta\right)\right] \tag{6.123}$$

If B = 0, the flow follows the pattern shown in Fig. 6.41a, and if A = 0, the flow follows the pattern shown in Fig. 6.41b. Depending on the conditions and the physics of the problem, A and B should be chosen appropriately. If $A, B \neq 0$, separation occurs.

In examining the flow around a wedge, there are two cases: symmetric, as shown in Fig. 6.42a, and asymmetric, as shown in Fig. 6.42b. However, the above answers are only valid for the wedge if the condition $\beta > 146.3$ is met. Now, if $\beta < 146.3$, as in Fig. 6.43, a set of vortices, rotating in opposite directions, forms near the wall or corners.

Moffatt obtained an analytical solution for these conditions in 1964. Consider a V-shaped corner with a rotating cylinder above, as shown in Fig. 6.44. The rotation of the cylinder induces the first vortex, and the first vortex, in turn, induces the second vortex; similarly, other vortices rotate one another.

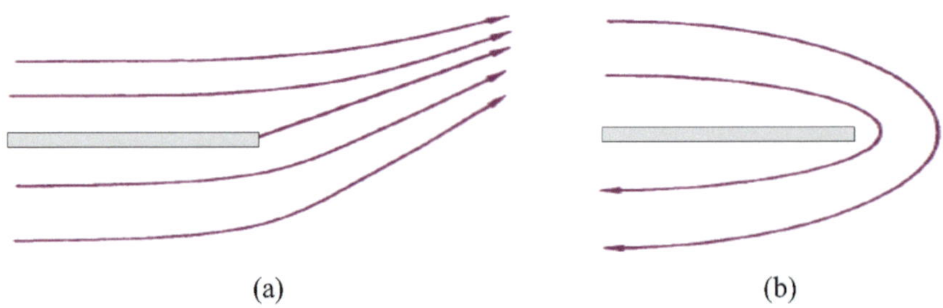

Fig. 6.41 Flow around a flat plate (Reproduced from Ref. [27])

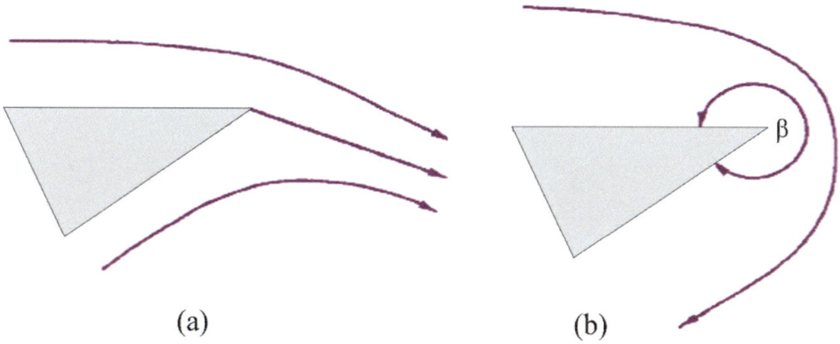

Fig. 6.42 Flow around a wedge (Reproduced from Ref. [27])

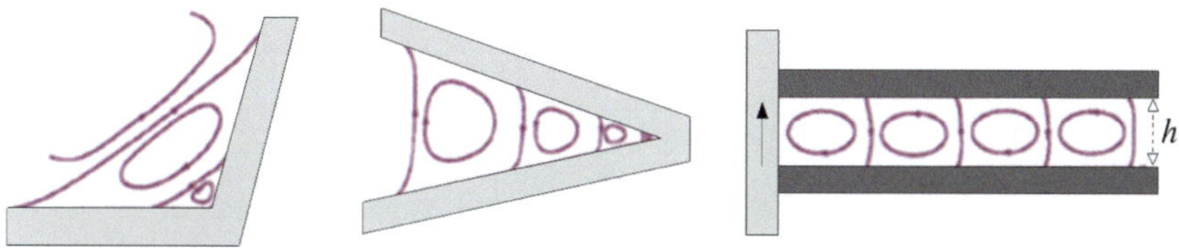

Fig. 6.43 Moffatt vorticity in a corner and cavity (Reproduced from Ref. [27])

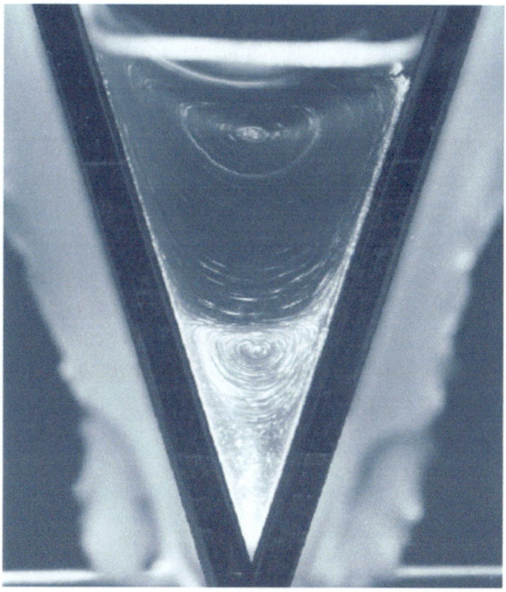

Fig. 6.44 Formation of Moffatt vortices in a triangular corner (Reproduced from Ref. [27])

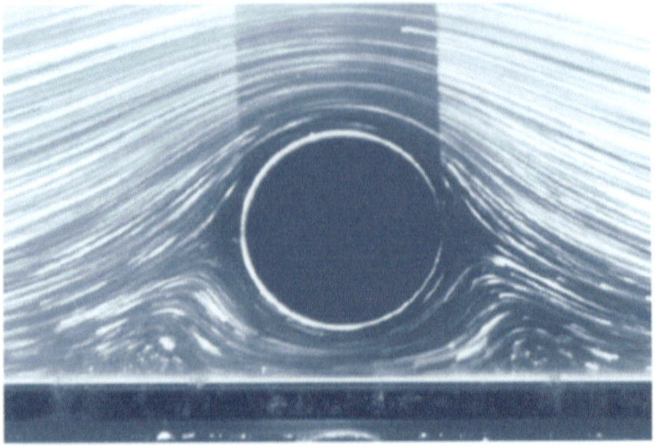

Fig. 6.45 Stokes flow around a cylinder near a wall (Reproduced from Ref. [27])

Fig. 6.46 Flow inside a channel with a fluid wall (Reproduced from Ref. [27])

The flow, passing over a cylinder near a wall, creates two vortices, as shown in Fig. 6.45. If the distance between the wall and the cylinder is small, the strength of the vortices increases, and with further reduction of this distance, the number of vortices also increases.

Moffatt vortices are seen in gaps, cavities, and holes. Figure 6.46 shows that barriers (plates) of height h and a distance of L from each other are placed inside a channel on the channel walls. Therefore, a fluid wall is created instead of a solid wall, and the fluid interacts with fluid rather than with a solid wall. Consequently, there is less drag. As a result, a channel with a diameter of d is formed, which has less pressure drop compared to a channel with the same diameter but with solid walls.

Moffatt's vortices near the corner, between two solid boundaries, or between a solid boundary and a free surface, are observed in Stokes flow. Moffatt demonstrated through similarity analysis that a sequence of vortices is formed, with their size and intensity gradually decreasing. An example of Moffatt's vortices in a triangular corner is shown in Fig. 6.47.

6.9.4 Vortex Formation in a Radiometric Pump

The effect of the micro-pump vane height in the radiometric pump was discussed in Sect. 6.6.5. Based on the previous discussion on Moffatt's vortices, the aforementioned analysis is reported here. Section 6.6.5, H = 0 was initially considered (Fig. 6.48a). Since the dashed line surface represents a specular surface, the thermal

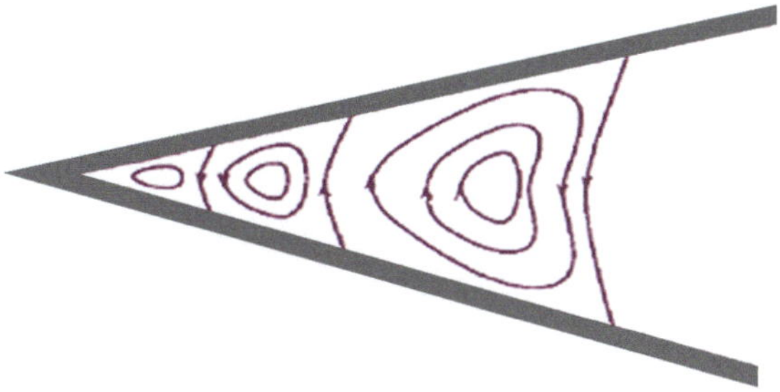

Fig. 6.47 Moffatt vortices in a triangular corner (Reproduced from Ref. [27])

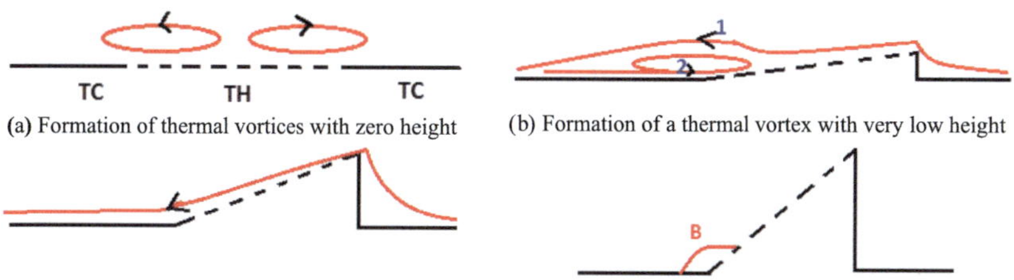

(a) Formation of thermal vortices with zero height
(b) Formation of a thermal vortex with very low height
(c) Flow over a ratchet at the optimal height
(d) Display of the optimal angle β

Fig. 6.48 Schematic of vortex formation and flow formation over the ratchet at various heights

creep phenomenon does not occur on this wall because no momentum from the molecules colliding from the cold and hot regions of the flow is transferred to the specular wall.

Under these conditions, two vortices are formed because the temperature in the specular region is higher during the numerical solution than in the diffusive region. Note that the boundary condition of the specular wall is the zero heat flux, and the condition of the diffusive wall is the temperature jump condition. Therefore, according to the thermal creep flow laws, the flow moves from T_c to T_H. As a result, two vortices meet in opposite directions in the center of the specular wall, no flow is created, and the pump does not transfer any fluid.

In the next step, in Sect. 6.6.5, the height is increased by a very small amount (H = 0.005), and as soon as the height increases, the radiometric flow is generated at the tip of the formed triangle (Flow 1 in Fig. 6.48b). A rightward flow (Flow 2) originates from thermal creep and is also generated on the left horizontal side. A vortex is formed from the interaction between Flow 1 and Flow 2.

The height is increased further in the third case (Fig. 6.48c). Then, the radiometric flow overcomes the thermal creep, and the flow moves completely near the wall; therefore, a vortex is not formed next to the specular wall.

According to Fig. 6.48d, as the height increases further because $\beta < 146.3$, a Moffatt vortex resulting from Stokes flow will be created. As H increases more, the strength of the vortex increases, and the pump performance decreases. Between these two states, there is $\beta = 146.3$, which shows the optimal state. In this case, no vortex is formed, and the optimal H of the pump is determined using it.

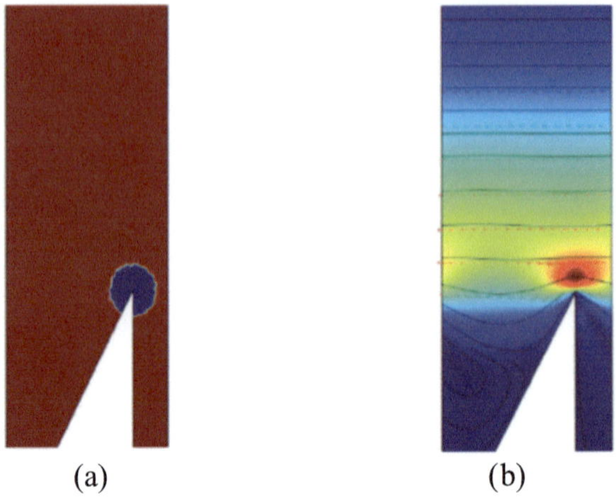

Fig. 6.49 Concentrated force around the ratchet (**a**) and Stokes flow in a homo-thermal geometry with fixed force density (**b**) (Unpublished results from the authors)

In summary, at small blade vanes, vortices resulting from thermal phenomena are formed, while at higher heights, Moffatt vortices appear. Between these two conditions, there is an optimal state at $\beta = 146.3$. The following analysis is provided to understand the cause of vortex formation better.

At low Knudsen numbers, viscous and thermal stress terms play a role in vortex formation. To eliminate the thermal stress term, force-driven Stokes flow at the tip of the blade is simulated in a similar geometry under isothermal conditions. Figure 6.49 shows a schematic of the geometry corresponding to isothermal flow in the pump, created under the influence of a constant and concentrated force at the tip of the tooth.

As the numerical results in this figure show, the flow is directed along the negative x-axis, and a counter-clockwise vortex forms on the inclined surface with a specular mirror property. Therefore, thermal stress does not play a role in forming the vortex adjacent to the inclined wall. This type of vortex is known as a Moffatt vortex, which was discussed earlier. According to Moffatt's analytical solution [23], any corner with an angle less than 146.3 in a low Reynolds number flow forms a Moffatt vortex. The flow over the specular surface acts like a moving wall of a cavity, inside which a vortex is formed. Therefore, although thermal stress effects can influence the strength and shape of the vortex, the primary cause of vortex formation is the viscous effects.

6.10 Thermophoresis

Thermophoresis is the phenomenon in which small particles suspended in a fluid move due to a temperature gradient. This motion arises because the warmer side of the particle experiences higher thermal energy, leading to an imbalance in the momentum transfer from gas molecules striking its surface. As a result, the particle is pushed toward the cooler region. Thermophoresis plays a crucial role in various scientific and industrial applications, such as aerosol technology, microfluidics, and environmental science. It is significant in processes like particle deposition in exhaust systems, air filtration, and the formation of dust rings in astrophysical environments. The efficiency of thermophoretic transport depends on factors like particle size, fluid viscosity, temperature gradient, and the properties of the surrounding medium. This phenomenon is considered in problem 6 at the end of this chapter.

6.11 Species Separation Using a Difference in Accommodation Coefficients (Thermal Edge Flow)

Recently, it is shown that a simple parallel plate configuration, with only a difference in surface properties (diffuse-specular) along its bottom plate and a temperature difference between the diffuse upper and lower plates, can induce species separation. Lotfian and Roohi [28] examined the species separation in slip and transitional regimes in this geometry.

Figure 6.50 schematically illustrates a conceptual geometry for gas mixture separation in a configuration comprised of two parallel plates, labeled as Plate A and Plate B. Plate A is at the top with a cooler temperature T_C, while Plate B is at the bottom with a warmer temperature T_H ($T_c = 300$ K and $T_h = 900$ K). A symmetry boundary condition is applied along the left and right vertical boundaries. At Plate A and the right half of Plate B, a diffusive boundary condition is applied with $\alpha = 1$ and $\alpha_{R,B} = 1$, respectively. In contrast, the left half of Plate B has a specular boundary condition with $\alpha_{L,B} = 0$. The ratio of the length to the height of the configuration (L/D) is assumed to be one. This geometry follows the suggestion of Ref. [28], which suggested it for flow induction.

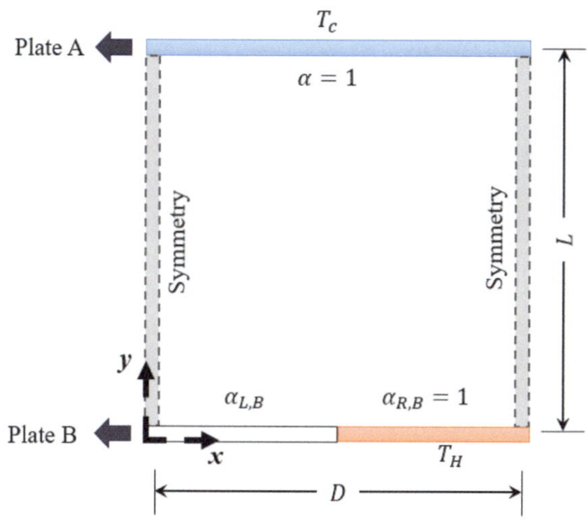

Fig. 6.50 Schematic of the considered geometry (reproduced from Ref. [28])

The average mean free path for each species is given by the following expression [30]:

$$\lambda_i = \left[\frac{p_0}{k_B T_0} \pi \sum_{j=1}^{2} \chi_{0_j} \left(r_i + r_j\right)^2 \left(1 + \frac{m_i}{m_j}\right)^{-\frac{1}{2}} \right]^{-1}, \tag{6.124}$$

where χ_{0j} and r respectively represent the initial mole fraction of species j and the radius of the species at the beginning of the simulation, where $r = d/2$. Here, the parameters k_B, p_0 and T_0 respectively denote the Boltzmann constant, the initial pressure, and the initial temperature. The parameter L remains constant across all studied configurations, although λ varies. Defining Kn as λ_{mix}/L, where λ_{mix} is calculated using:

$$\lambda_{mix} = \sum_{i=1}^{2} \chi_{0_i} \lambda_i, \tag{6.125}$$

where χ_{0i} is the initial mole fraction of species, while L represents the distance between plates A and B.

The simulation employs a mixture of helium (He) and xenon (Xe), with each gas comprising 50% of the mixture. The simulated cases result in Knudsen numbers (Kn_{mix}) of 0.05, 0.1, 0.5, and 1. Helium has a molecular diameter (d) of 2.33×10^{-10}m, a molecular mass (m) of 6.65×10^{-27} kg, and a viscosity-temperature index (ω) of 0.66. Xenon has a molecular diameter of 5.74×10^{-10}m, a molecular mass of 218×10^{-27} kg, and a viscosity-temperature index of 0.85.

Although the diffuse wall temperatures at the top and bottom plates are kept at a constant temperature, there is a temperature jump of the order of Kn over these surfaces. First, due to the symmetric boundary conditions on the sides of this configuration, a heat transfer from the hot wall (the diffusive section of Plate B) towards the cold wall (Plate A) occurs for both helium and xenon gases. By setting $\alpha_{L,B} = 0$, where the lower plate has a specular-diffuse surface, a thermal discontinuity is created in the middle of Plate B. This results in a temperature jump, which in turn creates a thermal edge flow of the order $\text{Kn}^{0.5}$. This flow is shown by the dashed arrow along the lower plate in Fig. 6.51. The only cause of the counterclockwise flow is the thermal edge flow induced at the middle of the bottom wall (at the point between specular and diffuse reflection). Of course, to induce the thermal edge flow, the temperature difference between the top wall and the right half of the bottom wall is necessary.

Additionally, it is important to note that the direction of species diffusion streamlines differs from that of the thermal creep flow. The diffusive velocity, represented as $\vec{v}_D = \vec{U} - \vec{U}_{mix}$, refers to the average velocity of particles caused by random motion and collisions with other molecules, see Fig. 6.51b.

The mass flux of each component species can be described using the following equation:

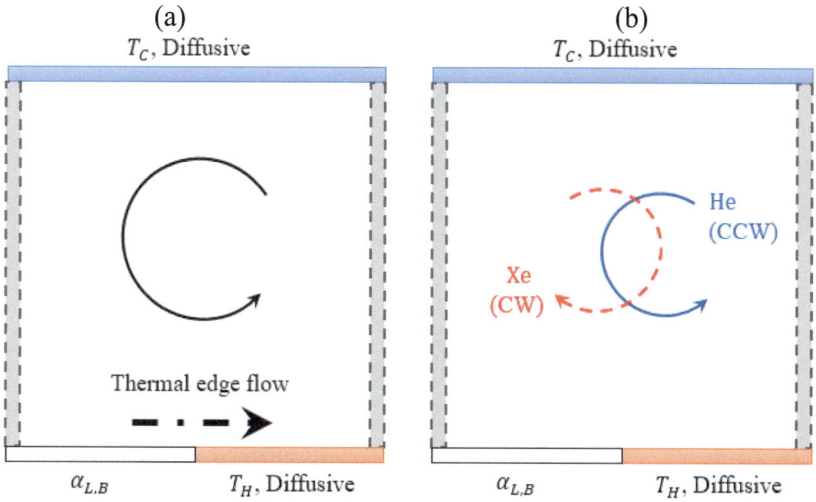

Fig. 6.51 Schematic of **a** flow mechanisms, **b** diffusive velocity streamline for helium and xenon in the investigated system (reproduced from Ref. [28])

$$J_{x_1,x_2} = C^{p}_{x_1,x_2}(\nabla p/p) + C^{\chi}_{x_1,x_2}(\nabla \chi/\chi) + C^{T}_{x_1,x_2}(\nabla T/T), \tag{6.126}$$

where x_1 and x_2 represent the two species that compose the mixture, and C are the coefficients for each driving force associated with gradients in pressure (∇p), mole fraction ($\nabla \chi$) and temperature (∇T). In this investigated system, the separation of components relies solely on the temperature gradient since no pressure or concentration gradients are applied. The mole fractions of helium and xenon are inversely related by the equation $\sum \chi_i = 1$, obtained directly from the definitions of χ_i. In this binary mixture, this has the implication that $\chi_{He} = 1 - \chi_{Xe}$ or $\nabla \chi_{He} = -\nabla \chi_{Xe}$.

Figure 6.52 compares the normalized average velocity magnitudes for helium, xenon, and the mixture at the investigated mixture Knudsen numbers. Each subplot illustrates the average velocities along the bottom wall, top wall, and inside the domain. The flow velocity remains minimal at lower Knudsen numbers (below 0.05), particularly the average normalized velocity inside the domain. As Knudsen numbers rise from 0.05 onward, rarefaction effects become more prominent. Specifically, thermal edge effects contribute to the formation of vortex structures due to temperature jump and thermal diffusion along the top and bottom surfaces of the domain.

In this Knudsen range (0.05 to 1), the normalized average velocity of helium is approximately 1.3 to 2 times higher than that of xenon. In contrast, the velocity of xenon is nearly equal to that of the mixture, see Fig. 6.52, where the average velocity of the mixture, $\vec{U}_{ave,mix}$, is calculated as:

$$\vec{U}_{avg,mix} = \frac{\chi_{He}.M_{He}.\vec{U}_{He} + \chi_{Xe}.M_{Xe}.\vec{U}_{Xe}}{\chi_{He}.M_{He} + \chi_{Xe}.M_{Xe}}, \tag{6.128}$$

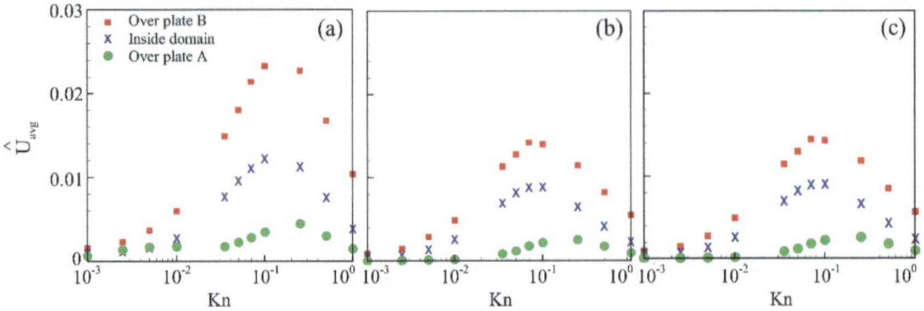

Fig. 6.52 Average normalized velocity magnitude at the investigated mixture Knudsen number for: **a** Helium, **b** Xenon, and **c** mixture, with $\hat{U} = |U|/U_{mp,mix}$. The simulation conditions are $\Delta T/T_C = 2.0$, $\alpha_{L,B} = 0$ and $L/D = 1.0$ (reproduced from Ref. [28])

where χ_{He} and χ_{Xe} are the mole fractions of helium and xenon, M_{He} and M_{Xe} are their molar masses (in g/mol), and $\vec{U}_{He}$ and $\vec{U}_{Xe}$ are their velocities.

Figure 6.53 reports normalized molar fraction change ($|\Delta\chi|/\chi_{inlet}$) of the binary gas mixture of helium-xenon at various Knudsen numbers, where χ_{inlet} represents the initial molar fraction of each gas at the inlet, set to 50% ($\chi_{inlet} = 0.5$). Any change in the molar fraction of helium ($\Delta\chi_{He}$) results in an equal and opposite change in xenon ($\Delta\chi_{Xe} = -\Delta\chi_{He}$). The value of $|\Delta\chi|$ is obtained from the difference between the maximum and minimum molar fractions, which for helium occur at the lower diffusive wall and the upper wall, respectively (this pattern is reversed for xenon.

As observed, the maximum separation of the two species occurs at Knudsen number 0.005. With an increase in the Knudsen number from 0.005 to 1, the separation percentage of both species decreases, reducing from 20.57 to 5.36%. This

Fig. 6.53 Normalized molar fraction percentage change of binary gas mixture for helium (equal to xenon) at the investigated mixture Knudsen number. The simulation conditions are $\Delta T/T_C = 2.0$, $\alpha_{L,B} = 0$ and $L/D = 1.0$ (reproduced from Ref. [28])

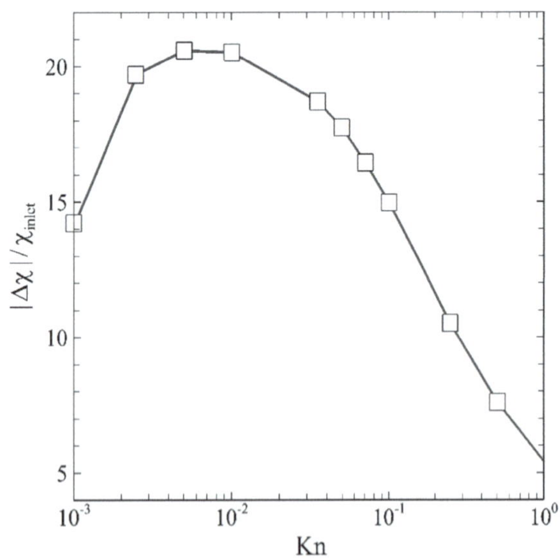

reduction may be due to increased molecular collisions with container walls instead of interacting with each other, leading to decreased molecular interactions and, consequently, decreased separation. In fact, with increasing Kn, the maximum x-velocity near the lower plate decreases. However, despite the thermal edge flow mechanism being of the order of $Kn^{0.5}$ in the middle of the lower wall, the significant increase in the temperature jump with higher Knudsen numbers leads to a decrease in temperature gradient. Consequently, beyond a Kn $= 0.1$, this does not necessarily result in improved separation. Moreover, as the Knudsen number decreases further into the continuum regime (below Kn approximately 10^{-2}), the separation percentage also decreases, consistent with the known behavior in the continuum limit, where both thermally-driven flows and thermally-induced separation diminish. This is due to the dominance of viscous effects and the lack of significant non-equilibrium phenomena.

Figure 6.54 shows the effect of the parameter $\alpha_{L,B}$ on the molar percentage change ($|\Delta\chi|/\chi_{inlet}$) for a constant Knudsen number of 0.05. As $\alpha_{L,B}$ increases, $|\Delta\chi|/\chi_{inlet}$ decreases, suggesting an inverse relationship between these variables under the given conditions. Additionally, as $\alpha_{L,B}$ approaches one, the thermal edge flow diminishes as the mechanism for flow induction weakens, i.e., there is no flow on the first frame in Fig. 6.54. At Kn $= 0.05$, the molar fraction variation remains stable within the range of 17.26–17.59% due to the temperature gradient, even in the absence of gas flow ($\alpha_{L,B} = 1$) and minimal vortex formation within ($0 < \alpha_{L,B} < 1$).

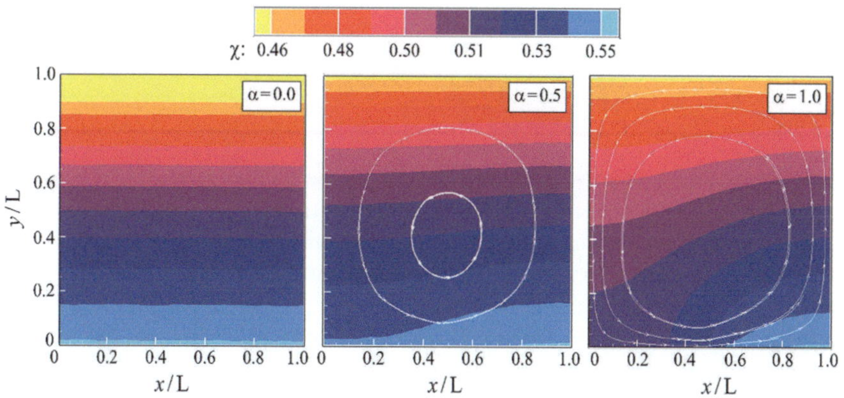

Fig. 6.54 Mole fraction of helium at various α of the bottom left wall at $Kn_{mix} = 0.05$, $\Delta T/T_C = 2.0$, and $L/D = 1.0$ (Reproduced from Ref. [28])

6.12 Thermally Driven Rarefied Flows Induced by a Partially Heated Diamond in a Channel

This section reports a radiometric pump design with a partially heated diamond in the middle of a cold channel [31]. The left side of the diamond is set to hot, and the right side is cold. Here, we aim to evaluate the effects of crucial parameters such as the accommodation coefficients of the channel's surfaces, the size of the diamond, the Knudsen number, and the imposed temperature affecting the gas flow characteristics in the channel. We show that the gas flows from the cold side to the hot side due to the radiometric flow induced by the temperature differences of the diamond. This flow is blocked by second-order thermal creep effects created due to temperature jump on the channel surfaces and centerline. We show that by making the channel surfaces specular, the latter is removed.

Figure 6.55 schematically shows the case of a diamond with the left (red) hot side T_h and right blue cold side T_c in the center of the channel. The length of the diamond and channel is set b and $4D$, and the height is a and $2D$, respectively, with $a/2D = 0.1$. The gas in the channel is nitrogen gas ($m = 46.5 \times 10^{-27}$ kg, $d = 4.17 \times 10^{-10}$ m, $T_{ref} = 273$ K), the rotational relaxation collision number is 5, and the viscosity-temperature index of ω is 0.74. We set the two sides of the channel as periodic to avoid the pressure boundary condition of the inlet and outlet to induce any flow.

6.12.1 *Mechanism of the Flow*

In the considered geometry, in addition to the radiometric flow induced at the middle tips of the diamond, the temperature jump and distribution on the top and bottom surface and at the centerline of the channels play a pivotal role, see Fig. 6.56. Thus, there are two types of flow: the radiometric flow induced from the cold to hot tips of the diamond and the second-order thermal creep flow generated from the temperature jump over the walls on top and bottom surfaces of the channel but removed by making the top and bottom walls specular.

To describe second-order thermal creep (that we have in Sect. 6.4 as well), consider asymptotic temperature expansion in the form of $T = T_0 + T_1\text{Kn} + T_2\text{Kn}^2 + \ldots$

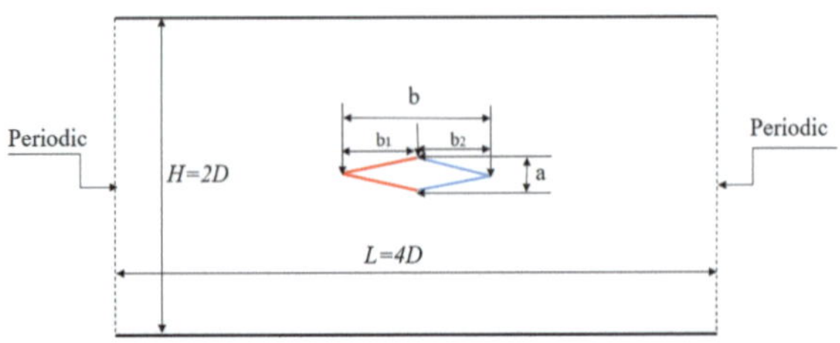

Fig. 6.55 Schematic of the geometry considered (Reproduced from Ref. [31])

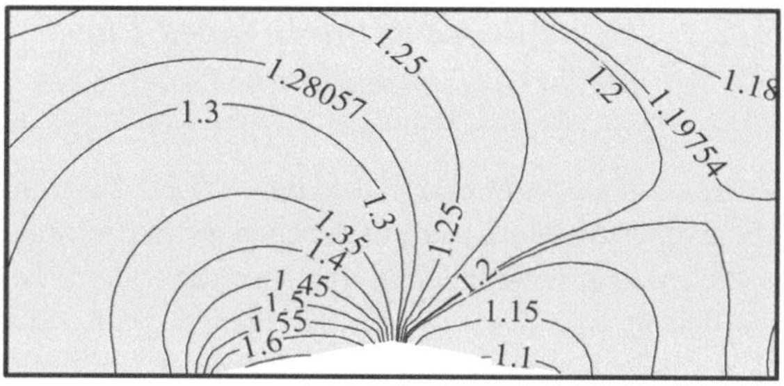

Fig. 6.56 Contour lines of temperature in the domain around the diamond, Kn = 1 (Reproduced from Ref. [31])

Consider the temperature of the gas (T_g) at the surface (T_s); thus $T_g = T_s + T_1\text{Kn} + T_2\text{Kn}^2 + \ldots$ The channel surfaces are held at a fixed temperature; accordingly, the first-order temperature distribution over them may not be uniform because of the temperature jump caused by the normal temperature gradient induced in the flow in the zeroth-order temperature, i.e., T_1 becomes non-uniform along the surface in such a way that temperature increases on the left sides of the top and bottom surfaces of the channel. When we plug this temperature change into the thermal creep velocity slip equation, that is:

$$(u_{s,c} - u_{wall}) \propto Kn\left(\frac{\partial T}{\partial x}\right), \tag{6.128}$$

where s is the velocity slip, and c stands for creep, we obtain:

$$(u_{s,c} - u_{wall}) \propto Kn^2\left(\frac{\partial T_1}{\partial x}\right) \tag{6.129}$$

In this way, the temperature gradient in the first order in Kn, originating from the temperature jump, results in the thermal creep in the second order in Kn. Although the local Knudsen number should replace Kn in Eq. (6.129), Eq. (6.129) gives an idea about the flows caused by the presence of the diamond with temperature distribution on the top and bottom surfaces.

The molecules hitting the hot side of the diamond reflect with a higher velocity than the reflection from the cold side because the reflection velocity of molecules is proportional to the surface temperature. That means the gas molecules have momentum from the cold side to the hot side. In other words, the gas molecules would receive a force from the cold side to the hot side. The force is radiometric forces that push the gas to flow from the cold side to the hot side in the channel. The strength of radiometric forces altered by the thermal jumps would induce different flow characteristics.

6.12.2 Dependence of the Induced Flow on the Accommodation Coefficient

If one set the case at $a/H = 0.1$, $b/L = 0.25$, $T_h = 600$ K, $T_c = 300$ K; the effect of the top and bottom walls of the channel with various accommodation coefficients is studied to observe the flow characteristic under different channel accommodation coefficients (σ). When $\sigma \neq 0$, the channel walls are set at T_c. The variation of σ affects reflection velocity and affects the gas's slip velocity along the surface. σ was changed between 0–1 to check the effect of the accommodation coefficient. When $\sigma =$ 1, the wall is fully diffuse. Figure 6.57 shows velocity contours and streamlines with different accommodation coefficients. In all cases, it is clearly indicated that the gas flows from the right side of the channel to the left side. At all σ, the maximum velocity is observed at the top and bottom tips of the diamond. The region of maximum is quite visible for higher σ cases. The tips are where the radiometric force is maximum and acts as a momentum source for the entire flow field.

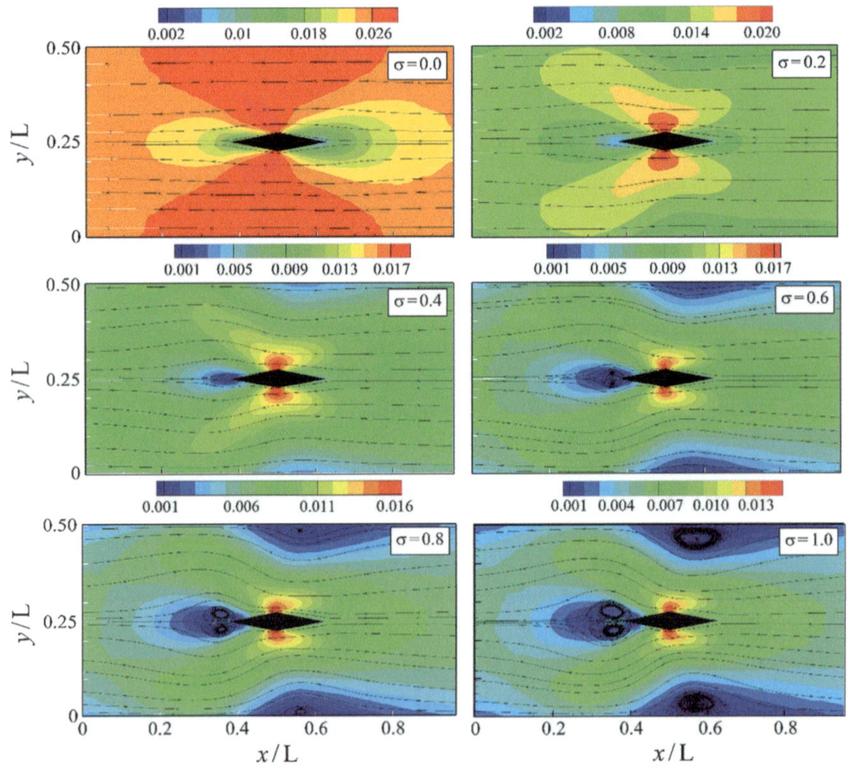

Fig. 6.57 Velocity contours and streamlines with different accommodation coefficients (Reproduced from Ref. [31])

As σ increases, the velocity magnitude decreases, and especially two vortices appear behind the diamond, i.e., when σ = 0.6, there appear two small eddies at the left tips of the diamond. When σ = 0.8, the vortices become bigger, and smaller vortices appear on the channel's upper and lower walls. When σ = 1, the vortices reach the maximum size at both positions. The reason of the formation of the eddies behind the diamond is as follows: the geometry and boundary conditions are symmetric, considering the temperature is higher at the centerline near the left tip of the diamond than the locations far from the left tip, thus, there is a temperature gradient over this symmetry line. The same mechanism as second-order thermal creep appears on the symmetry line and induces a rightward flow. This flow encounters the leftward radiometric flow appearing on the top of the diamond and moving left. When σ is small, the rightward flow is much weaker than the radiometric flow, but it becomes stronger with the increase of the σ, forming vortices at the left tip of the diamond.

The same second-order thermal creep mechanism appears on the top and bottom walls of the channel. The rightward flow on the top wall encounters the radiometric flow induced in the channel, and when the former becomes stronger, vortices appear on the top and bottom walls. Therefore, the case of diffuse channel walls experiences two pair of vortices on the top and bottom walls and behind the diamond; thus, the case of specular channel surface is considered for the rest of results provided in this paper.

6.12.3 Effect of Various Parameters on Induced Mass Flow Rate

Figure 6.58 shows the dependence of induced mass flow rate on various parameters. One can see that the mass flow rate decreases with the increase of the σ. The mass flow rate of σ = 0 is 5.2 times larger than σ = 1. This further justifies using a channel with specular surfaces for maximum throughput. The maximum mass flow rate appears for $a/H = 0.1$, a relatively thin diamond. Figure 6.58c shows the mass flow rate with various lengths. We can observe that the greatest mass flow rate is at where b/L around 0.1. The optimum case in terms of maximum throughput is the case with $b/L = 0.05$ and $a/H = 0.1$. Figure 6.58d shows the mass flow rate with Kn; where we can observe that the maximum mass flow rate appears at Kn = 0.8. Figure 6.58e shows the mass flow rate with temperature difference, which indicates it is almost proportional to the temperature but not in a linear way, in confirmation with the outcome of Ref. [25].

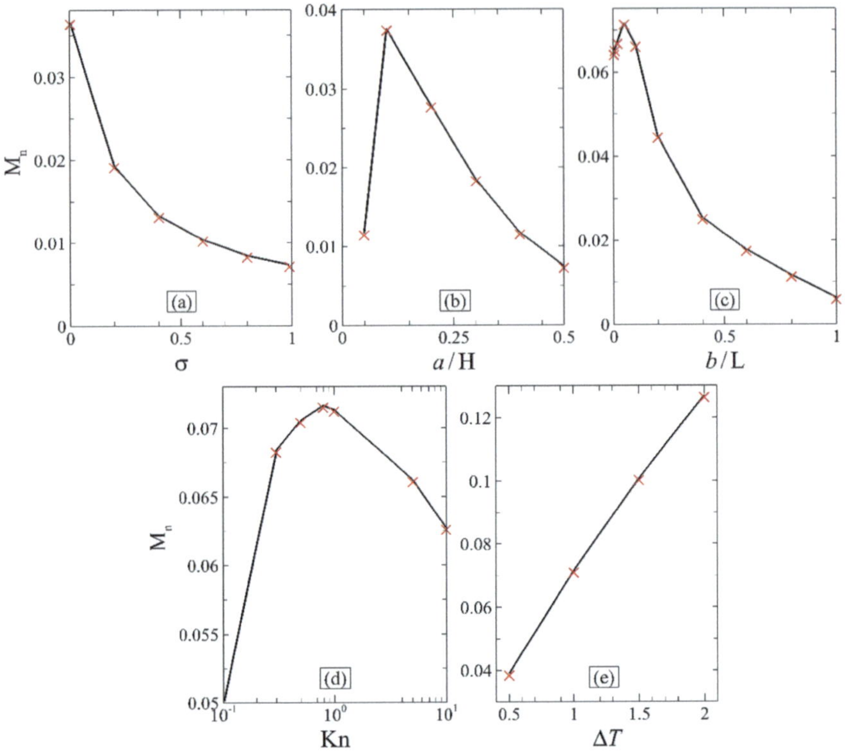

Fig. 6.58 Mass flow rate dependence on various parameters; **a** accommodation coefficient, **b** height, **c** length, **d** Knudsen, and **e** temperature difference (Reproduced from Ref. [31])

6.13 SNIT (Slow Non-Isothermal Thermal) Equations for Nonlinear Thermal Stress Flow

Adding the thermal stress term in the momentum equation of the continuum set of equations served as the foundation for constructing slow non-isothermal equations [32]. These equations describe the Boltzmann system as an asymptotic theory for small mean free time and small mean free path. The general behavior of the mildly rarefied gas is described by equations of the type used in fluid dynamics. Asymptotic analysis, based on Knudsen numbers, Kogan-Galkin-Friedlander (KGF) power series expansion, and evaluating the order of magnitude, is used to create the following second-order equations for continuum flows. Following KGF expansion, macro-variables like velocity and temperature are enlarged in terms of the Knudsen number. The time-independent Boltzmann equation's solution, however, can be split up into various length scales at minimal Knudsen numbers. Take into account the separation shown below [33]:

$$f = f_H + f_{kn}, \tag{6.130}$$

where f_H is the scale O(Kn^0) fluid-dynamic component of the velocity distribution function and f_{kn} is Knudsen-layer adjustment on the scale O(Kn). In a power series of Kn, the distribution function f_H and the macroscopic variables $h_H = \rho_H, u_{iH}$ and T_H can be expanded as follows:

$$f_H = f_{H0} + f_{H1}Kn + f_{H2}Kn^2 + \ldots, h_H = h_{H0} + h_{H1}Kn + h_{H2}Kn^2 + \ldots \quad (6.131)$$

We get the following system of equations of the fluid-dynamic type by substituting Eq. (6.131) into the Boltzmann equation and taking into account terms of the same order of Kn:

$$\frac{\partial p_{H0}}{\partial x_i} = 0 \quad (6.132)$$

$$\frac{\partial}{\partial x_i}\left(\frac{u_{iH1}}{T_{H0}}\right) = 0 \quad (6.133)$$

$$\frac{\partial p_{H1}}{\partial x_i} = 0 \quad (6.134)$$

$$\frac{\partial u_{iH1}}{\partial x_i} = \frac{\gamma_2}{2}\frac{\partial}{\partial x_i}\left(\sqrt{T_{H0}}\frac{\partial T_{H0}}{\partial x_i}\right) \quad (6.135)$$

$$\begin{aligned} &\frac{\partial}{\partial x_i}\left(\frac{u_{iH1}u_{jH1}}{T_{H0}}\right) - \frac{\gamma_1}{2}\frac{\partial}{\partial x_j}\left(\sqrt{T_{H0}}\left(\frac{\partial u_{iH1}}{\partial x_j} + \frac{\partial u_{jH1}}{\partial x_i} - \frac{2}{3}\frac{\partial u_{kH1}}{\partial x_k}\delta_{ij}\right)\right) \\ &- \frac{\gamma_7}{T_{H0}}\frac{\partial T_{H0}}{\partial x_i}\frac{\partial T_{H0}}{\partial x_j}\left(\frac{u_{jH1}}{\gamma_2\sqrt{T_{H0}}} - \frac{1}{4}\frac{\partial T_{H0}}{\partial x_j}\right) \\ &= -\frac{1}{2}\frac{\partial p^+_{H_2}}{\partial x_i} + \frac{p^2_{H_0F_{iH2}}}{T_{H0}} \end{aligned} \quad (6.136)$$

The energy and momentum equations are given in Eqs. (6.135) and (6.136), respectively, and P^+_{H2} is calculated as follows:

$$p^+_{H2} = p_{H0}p_{H2} + \frac{2\gamma_3}{3}\frac{\partial}{\partial x_k}\left(T_{H0}\frac{\partial T_{H0}}{\partial x_k}\right) - \frac{\gamma_7}{6}\left(\frac{\partial T_{H0}}{\partial x_k}\right)^2 \quad (6.137)$$

The correlations shown above are general to all molecular models. However, a specific choice of molecular model affects the transport coefficients. The gas-particle model can be used with the hard-sphere approach to calculate the following non-dimensional coefficients [32]:

$$\begin{aligned} \gamma_1 &= 1.270042427 \ \gamma_2 = 1.922284066 \\ \gamma_3 &= 1.947906335 \ \gamma_7 = 1.758705 \end{aligned} \quad (6.138)$$

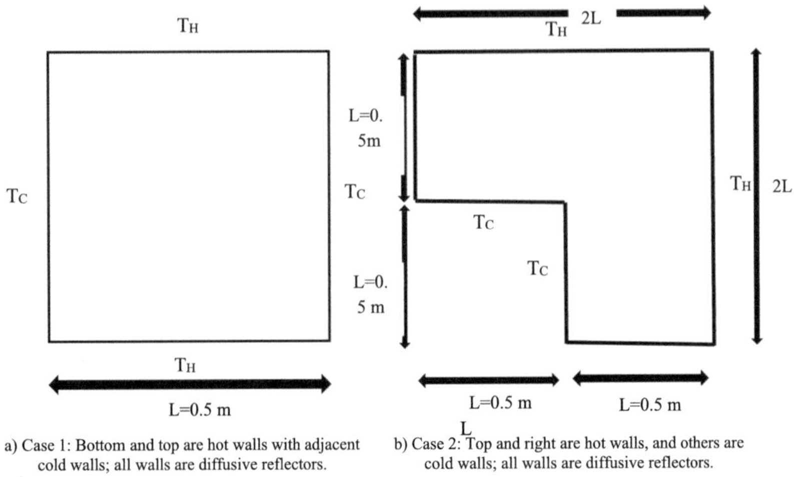

Fig. 6.59 Configurations of the square cavity and rectangular bend (Copied from open-source Ref. [32])

The nonlinear thermal term, which has coefficients of γ_3 and γ_7 on the left side of Eq. (6.136) and the right side of Eq. (6.137), are of Knudsen number order. The SNIT equations are developed at the limit of Kn approaching 0. As a result, these equations are only used at very low Kn to demonstrate the origin of the vortices in Fig. 6.59.

6.13.1 Geometry

Both geometries considered in [32] are shown in Fig. 6.59. In case 1, the cavity is a 2-D square shape with a side length of L. Opposite walls are at the same temperature, i.e., top and bottom sides are hot at 1500 K with adjacent cold walls at 300 K. Case 2 is a rectangular bend with top and right side hot with a temperature of 1500 K and other walls are cold with a temperature of 300 K. Both cases use a diffuse reflector for all borders as a boundary condition. the Knudsen number is λ/L.

6.13.2 Streamlines and Temperature Contours in the Square Cavity in Case 1

The streamlines in the cavity in case 1 are depicted in Fig. 6.60a by SNIT predictions at Kn = 0.01. The simulation results show that large vortices in the cavity are caused by nonlinear thermal stress flow. By solving the set of SNIT equations by setting the

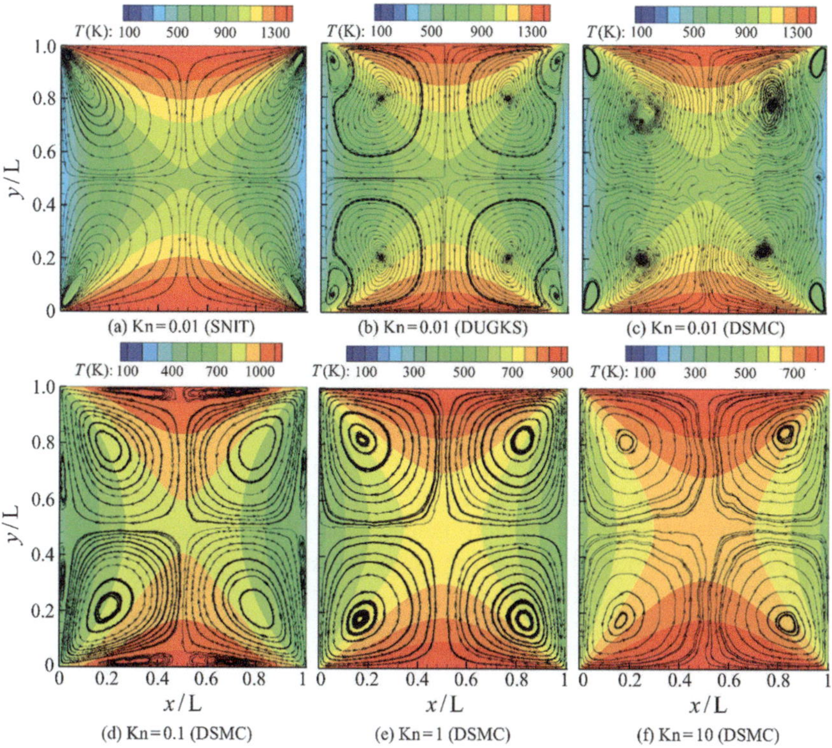

Fig. 6.60 Streamlines and temperature contours in the square cavity (Copied from open-source Ref. [32])

coefficients γ_3 and γ_7 on the left-hand side of Eqs. (6.136) and (6.137) to zero, no vortex or flow is observed in the domain; this shows that the nonlinear thermal stress factor is the parameter that causes the induced flow in the square under consideration. Four tiny vortices can be seen in DSMC forecasts around the four corners at Kn = 0.01 but not in the SNIT prediction. This may be a result of the SNIT equation's flaw: SNIT equation includes O(1) on temperature and O(Kn) effects in flow velocity; as a result, it is not anticipated that SNIT capture vortices close the corners which are O(Kn^2). The position and shape of the vortices with SNIT and the predictions of DSMC and DUGKS depicted in this figure differ as the vortices' centers are closer to corners in the solution of SNIT compared to DSMC and DUGKS. The DSMC and DUGKS streamline directions are from the cold to the hot zone.

The streamlines and temperature contours are shown in the same figure for Kn = 0.01, 0.1, 1, and 10 by DSMC and discrete unified gas kinetic scheme (DUGKS) solvers. At Kn = 1, 10, the realm is entirely dominated by four enormous vortices directed from the warm to the cold area, while at Kn = 0.01, both DSMC and DUGKS solvers' streamlines' direction are from the cold to warm zone. As the Knudsen number increases, the vortices' nature changes, causing the direction to reverse.

Since the isothermal curves are not straight and there is a significant temperature differential, primary vortices are likely the result of nonlinear thermal stress flow effects. This is compared to concentric elliptical cylinders discussed earlier in this chapter. The SNIT equations could capture these vast vortices, while the Navier–Stokes–Fourier (NSF) equations could not do so. It is evident that the nonlinear thermal stress caused these sizable vortices. Sone [1] asserts that the induced thermal field causes the nonlinear thermal stress flow in the domain. In contrast to the thermal stress slip and thermal creep flows, where the walls play an active part, the walls only play a secondary function in creating such a flow field. When temperature fluctuations are not small, thermal stress, a function of the temperature gradient, can be seen in Eq. (6.137). A force parallel to the temperature gradient pulls the gas in the opposite direction.

The primary distinction between the present study and previous work concentrating on thermally induced flows in elliptical cylinder geometries is the singular nature of the square geometry. Notwithstanding the fact the nonlinear thermal stress flow happens in a square, singularities may have the potential to influence and change it. The singularity at the sharp edges (points of the plate) is so strong that it dominates the overall flow field for both the thermal edge flow and the radiometric edge flow. Sharp corners in the square problem may have a less significant impact but still, compete with the nonlinear thermal stress flow to influence it. There are four tiny vortices near the square's corners at $\mathrm{Kn} = 0.01$ and eight vortices in the vicinity of all sides, though those disappeared at $\mathrm{Kn} = 1, 10$. In the current issue, there are four acute corners where the wall temperature is discontinuous; thus, the bottom/top and side walls close to the corner experience a significant temperature jump. A sharp temperature gradient develops in the gas along the bottom/top and side walls close to the corners. Due to the strong temperature gradient, flows are induced in the corners and along all sides of walls that move toward the corners at $\mathrm{Kn} = 0.1$ (the so-called thermal creep). At $\mathrm{Kn} = 0.01$ near the corner, these flows create four and eight vortices along all sides at $\mathrm{Kn} = 0.1$. When Kn is increased, vortices created by thermal creep vanish because as the number of Knudsen is increasing, vortices formed by nonlinear thermal stress become stronger and surround the entire cavity. Temperature increases as Kn rises, showing a more significant temperature jump along the bottom and top sides.

6.13.3 Streamlines and Temperature Contours in the Rectangular Bend in Case 2

The streamlines produced by the SNIT, DSMC, and DUGKS (discrete unified gas kinetic scheme) solvers are displayed in Fig. 6.61 for $\mathrm{Kn} = 0.01$, 0.1, 1, and 10. DUGKS was used to obtain the $\mathrm{Kn} = 0.01$, and the DSMC solver provided the others. At all the Knudsen numbers, the realm is entirely dominated by two enormous vortices. Due to the non-straight isothermal curves and the significant temperature

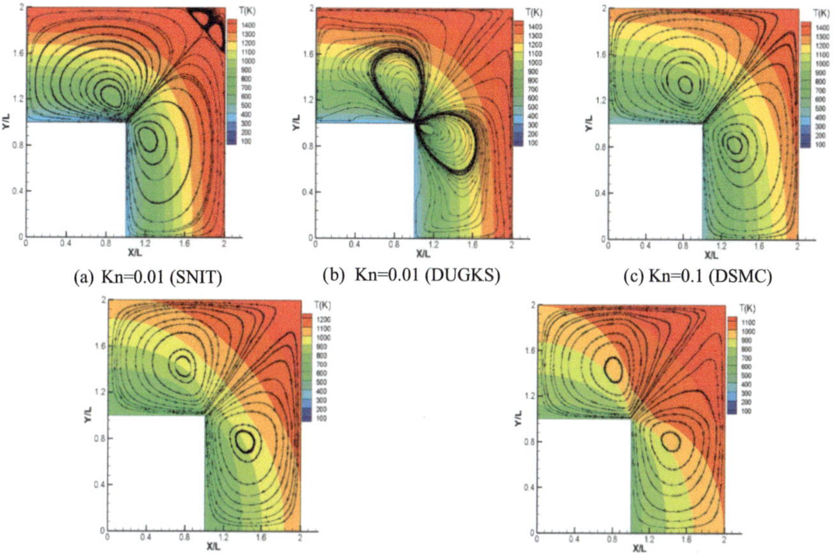

(a) Kn=0.01 (SNIT) (b) Kn=0.01 (DUGKS) (c) Kn=0.1 (DSMC)

(d) Kn=1 (DSMC) (e) Kn=10 (DSMC)

Fig. 6.61 a–d Streamlines and temperature contours in the rectangular bend in case 2, $0.01 \leq \text{Kn} \leq 10$ (Copied from open-source Ref. [32])

differential $|T_H - T_C|/T_C$, these vortices are most likely the result of nonlinear thermal stress effects.

In Case 2, the streamlines in the cavity from the SNIT predictions at Kn = 0.01 are depicted in Fig. 6.61a. Large vortices in the cavity are seen in the simulation results and are caused by nonlinear thermal stress flow. In the SNIT predictions, two small vortices appear close to the top right corner. These could be Moffatt vortices growing close to rectangular bend's corners. Due to the abrupt geometry change and boundary temperature near the corners, the SNIT equations are invalid, making the tiny vortices nonphysical.

6.14 Review of Knudsen Pumps

Based on the different mechanisms of the thermal-induced flows, different Knudsen pumps have been proposed and investigated in the literature. Here, different types of Knudsen pumps, i.e., thermal creep flow Knudsen pump, thermal edge flow Knudsen pump, and radiometric flow Knudsen pump, will be introduced and discussed. Also, different configurations with their performances will be reviewed [34].

6.14.1 Thermal Creep Flow (TCF) Knudsen Pump

In the early twentieth century, the first type of Knudsen pump was built based on the thermal creep flow by Knudsen [35, 36], see Fig. 6.62. A series of high-diameter and small-diameter pipes were connected. Some heaters are periodically located between the section and heat the gas. Rarefied gas through the small-diameter pipes flows from the cold section to the hot region based on the thermal creep effect. In high-diameter pipes, rarefaction effects are small and there is no induced thermal flow. However, a Poiseuille flow occurs in the high-diameter region from the heater toward the small-diameter pipe due to the pressure difference induced by the induced thermal creep flow. This Poiseuille flow weakens the thermal creep flow by weakening the temperature gradient distribution in the small-diameter region. The combined action of the two flows ultimately puts the Knudsen pump into dynamic equilibrium and the net mass flow rate is zero.

After introducing the basic concept of the Knudsen pump, different configurations for the thermal creep of Knudsen pumps have been investigated; see Fig. 6.63. A typical TCF Knudsen pump consists of a series of periodic elements.

The oldest type of TCF Knudsen pump consists of successive rectangular channels with decreasing and increasing wall temperature distributions, as shown in Fig. 6.63a [35–46]. The rectangular channel configuration exhibits one of the best performances among the TCF Knudsen pumps [47]. Figure 6.64 illustrates the effect of Knudsen number of flow patterns in a rectangular channel TCF pump.

Figure 6.63b illustrates a TCF pump that was first introduced by Aoki and co-workers [48, 49] and then investigated by different researchers [50–52]. This device is made of a simple smooth and curved channel with a periodic temperature distribution. It was shown that the straight-curved channels have the lowest performance among the TCF pump configurations [52, 53]. Figure 6.65 demonstrates the effect of the Knudsen number on the flow pattern of a straight-curved channel.

Double-curved TCF pump (Fig. 6.63c) was first introduced by Bond et al. [52]. It was shown that the mass flow rate for double-curved TCF pumps exceeds that of straight-channel TCF pumps by more than two and a half times; see Fig. 6.66. Compared to the other types of TCF pumps, double-curved pump leads to a steady

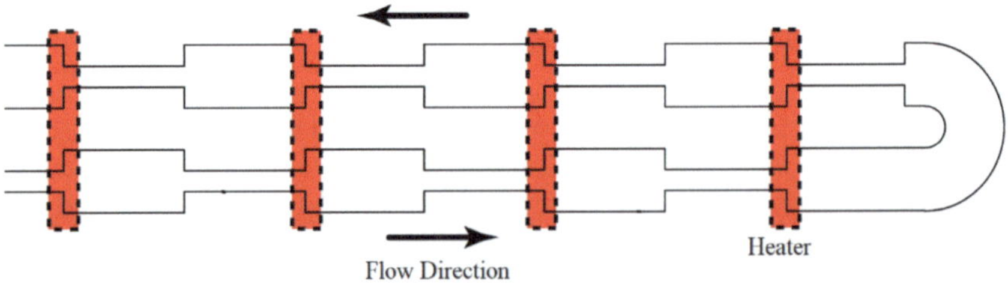

Fig. 6.62 The model of the early TCF Knudsen pump proposed by Knudsen (Reproduced from Ref. [35])

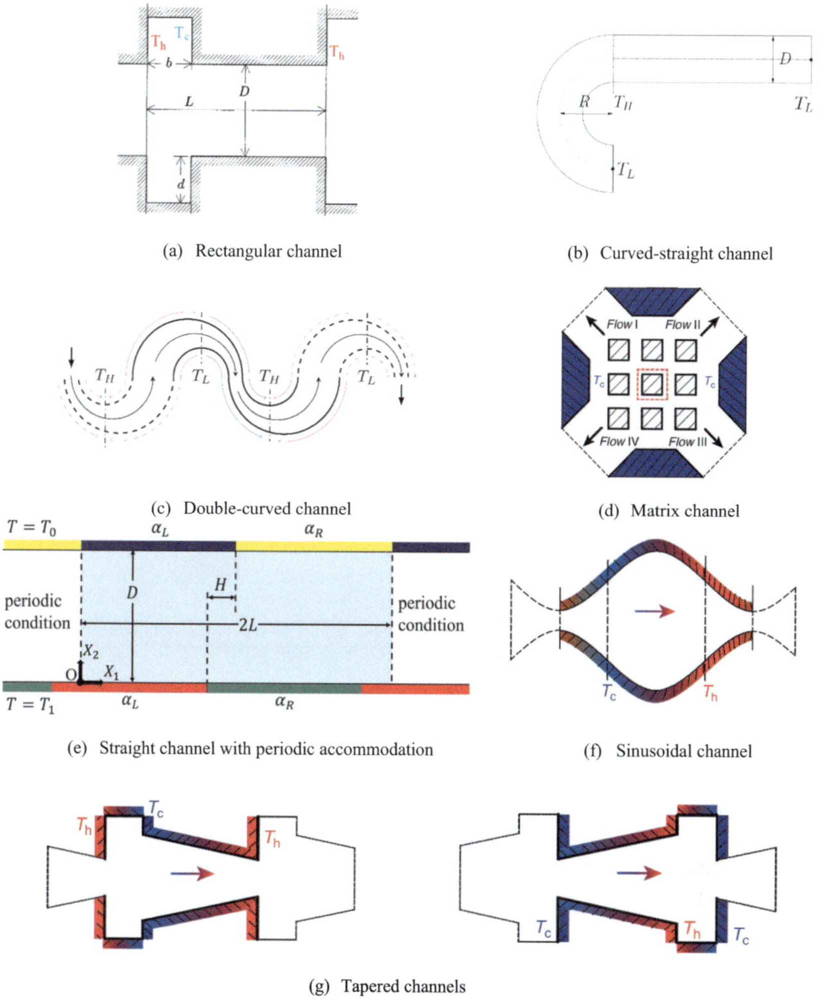

Fig. 6.63 Different proposed configurations of TCF Knudsen pumps (Copied with permission from Ref. [34])

flow. However, there is a higher complexity in manufacturing this pump because of the existing variable curvatures.

Bond et al. [53] introduced a matrix TCF Knudsen pump, as illustrated in Fig. 6.63d. The flow direction could be adjusted by changing the temperature profile over the individual elements (Fig. 6.67). It was shown that the performance of this type of TCF pump is the same as the rectangular channel type [53]. However, fabricating the matrix configuration with different temperature profiles for individual elements is very complicated.

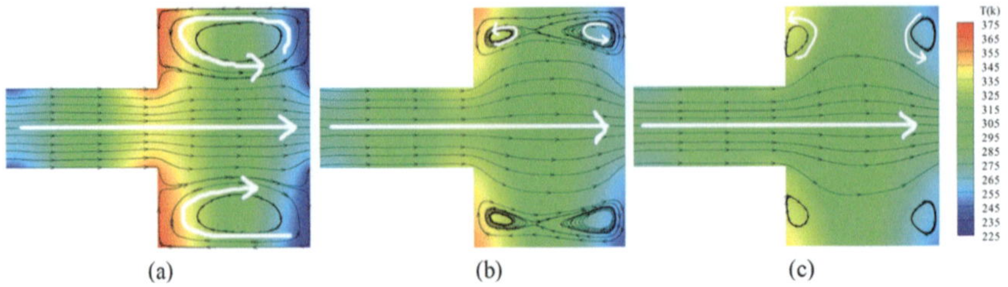

Fig. 6.64 Velocity streamlines and temperature contours for N_2 and O_2 mixture through a TCF Knudsen pump at different flow rarefactions of **a** Kn = 0.06, **b** 0.39, and **c** 3.87 (Copied from open access Ref. [45])

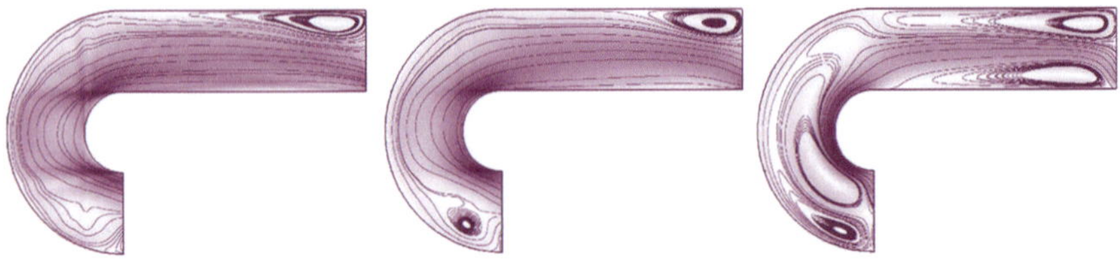

Fig. 6.65 Flow patterns through a curved-straight channel TCF pump at different Knudsen numbers of 1.0 (left), 0.5 (middle), and 0.1 (right) (Reproduced from Ref. [49])

Recently, Sugimoto and Sugimoto [54] introduced a new type of TCF pump based on a straight channel with a periodic distribution of accommodation coefficients over the walls, as indicated in Fig. 6.68. In this type of TCF pump, the flow changes periodically in response to the displacement of the discontinuity points of the accommodation coefficients, and in some cases, a one-way flow, the steady-state mass flux in the parallel to the flat plate generated between the plates, is generated. The one-way flow has a Knudsen number at which the flow rate becomes maximum and shows an approximately linear variation with the temperature difference between the two plates. On the other hand, the flow showed a quadratic variation in the accommodation coefficient difference (Fig. 6.68), indicating a nonlinear phenomenon that linear flow fields cannot express [54]. One period of the pump shown in Fig. 6.68a was used for species separation by Lotfian and Roohi [28], as discussed in Sect. 6.10.

The sinusoidal channel TCF pump (Fig. 6.63f) was first introduced by Bond et al. [53]. It was shown that the sinusoidal TCF pump improved the mass flow rate due to reductions in the tortuosity of the flow path and optimization of the re-circulation locality [53] (Fig 6.69).

The converging/diverging channels [6, 55, 56] have been employed as a type of TCF pump as indicated in Fig. 6.63g. It was shown that the mass flow rate is significantly higher when the tapered channel is perfused like a nozzle [55]. It can, therefore, be stated that under moderately rarefied conditions, micro-sized ducts having varying cross-sections act as a gas flow diode. The theoretically and experimentally analyzed diode effect increases with gaseous rarefaction [55]. A diverging channel has a larger

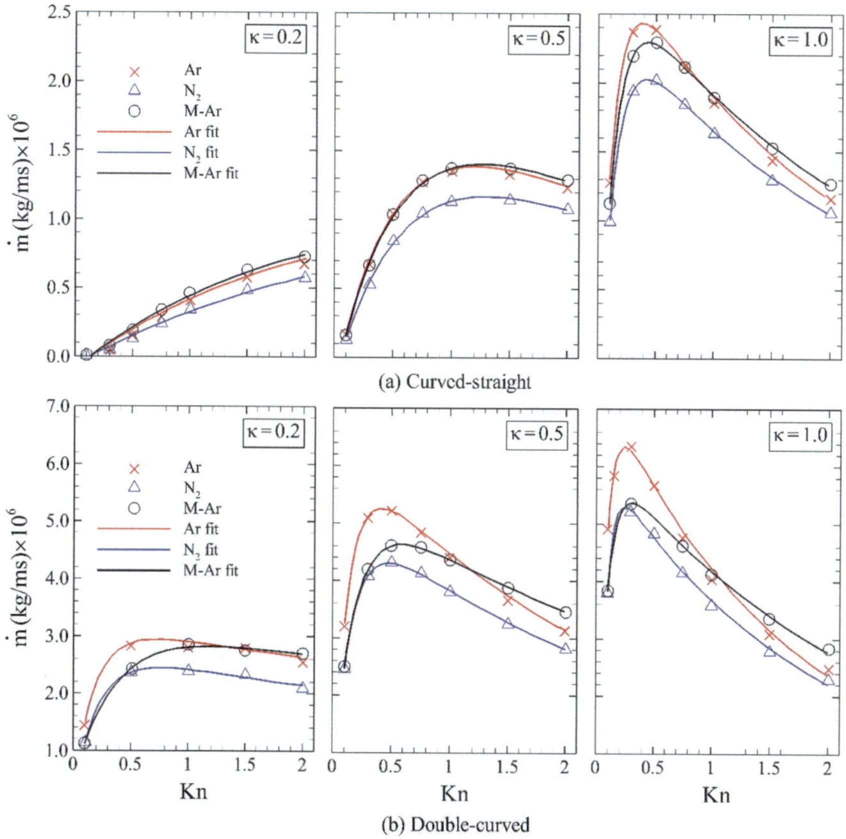

Fig. 6.66 Dimensional mass flow rates for the curved-straight (**a**) and double-curved (**b**) geometries (Reproduced from [52])

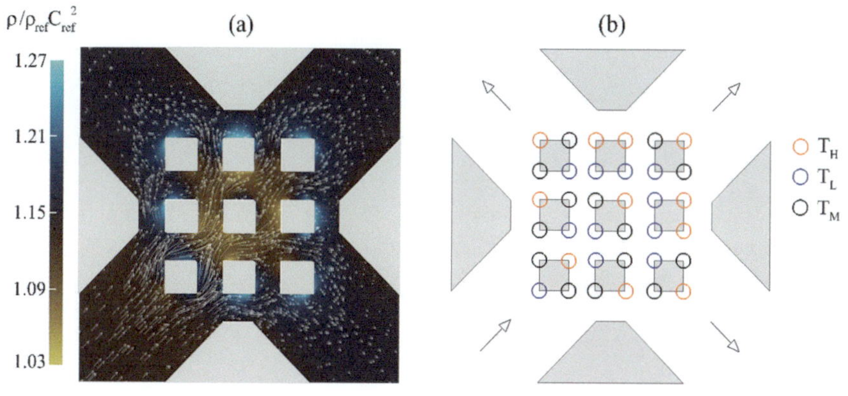

Fig. 6.67 Pressure gradient and flow vectors (**a**) and temperature profile for a 3 × 3 matrix TCF pump (Reproduced from [53])

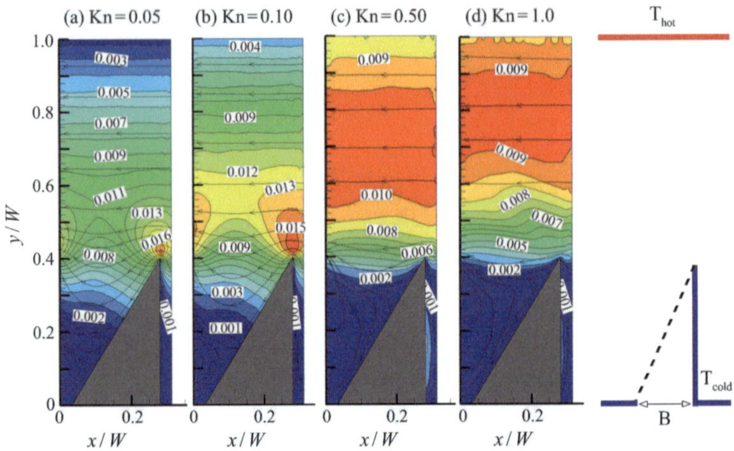

Fig. 6.68 Temperature fields and streamlines for Knudsen numbers of 0.5 (**a**) and 1.0 (**b**), and effect of accommodation difference on pump flow rate (**c**) (Reproduced from Ref. [54])

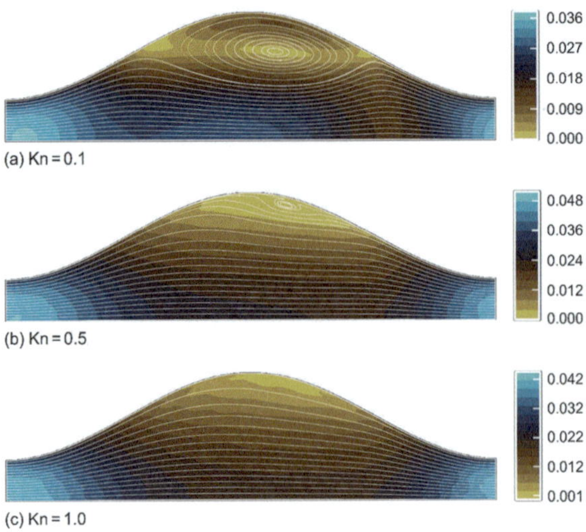

Fig. 6.69 Normalized velocity contours and streamlines for a sinusoidal TCF pump at different flow rarefaction (Reproduced from Ref. [53])

mass flow rate than a converging channel [6, 56]; see Fig. 6.70. The flow behavior in convergent/divergent channels was affected by area expansion effects, specified wall temperature gradient, and viscous forces [6].

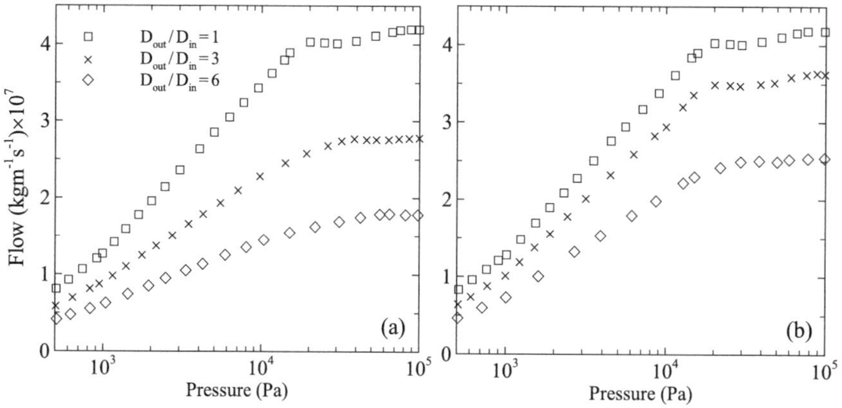

Fig. 6.70 Comparison of mass flow rates for converging (**a**) and diverging (**b**) channels with AR = 50 and zero pressure gradient under different outlet-to-inlet height ratios (Reproduced from Ref. [56])

6.14.2 Thermal Edge Flow (TEF) Knudsen Pump

Different proposed configurations for TEF Knudsen pumps have been listed in Fig. 6.71. All proposed arrangements involve ratchet-like patterns. This is because the best performance for induced thermal flow around a sharp edge occurs at zero thickness [14]. The mechanism of generated thermal flow in all configurations concerning thermal edge flow was previously described.

6.14.3 Radiometric Flow (RF) Knudsen Pump

The basis of the RF Knudsen pump is the radiometric flow which is previously described. Two types of RF Knudsen pumps have been proposed due to the edge existing between the adjacent surfaces. In the first type, adjacent surfaces have different temperatures [15, 25, 26, 31, 67–69]. In the second type, the adjacent surfaces have different accommodation coefficients [21, 22, 25, 57, 70]. The thermalization process is the flow mechanism in RF Knudsen pump. The gas molecules have weak or zero thermalization when they collide with cold or specular surfaces. In contrast, gas molecules are strongly thermalized with the surfaces when they collide with the hot or diffuse walls. As a result, gas flow occurs from the cold side to the hot side [47].

Different proposed configurations for RF Knudsen pumps have been listed in Fig. 6.72. The RF Knudsen pumps have been investigated based on the adjacent specular-diffusive walls, Fig. 6.72a, e, i, j, k, m, and adjacent walls with different temperatures, Fig. 6.72b, c, d, f, g, h, l. From the manufacturing viewpoint, case (i) is the simplest, and case (g) is the most complicated. It was shown that gas flow

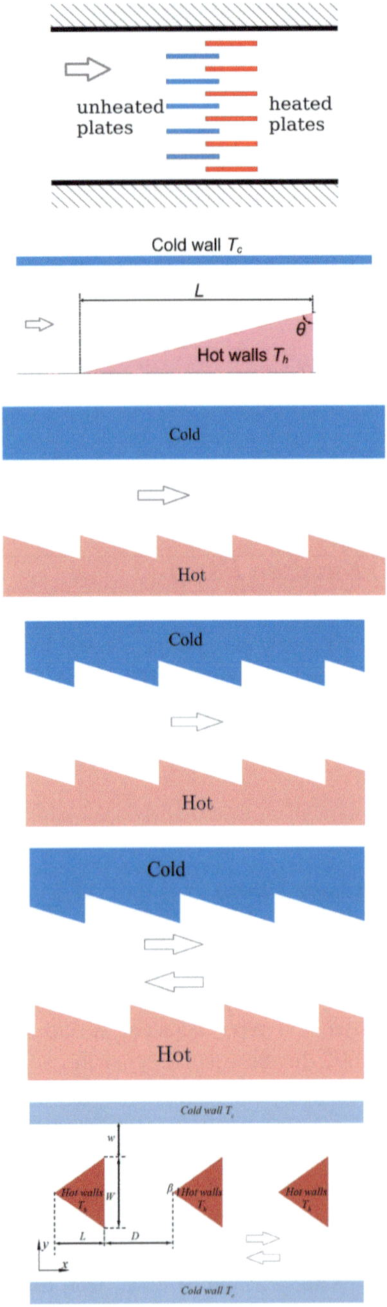

a) Heated and unheated plates [13]
- the flow direction from left-to-right
- the energy efficiency is improved 6 times [13] more than that of in rectangular TCF Knudsen pump (Fig. 6-63*a*)

b) Discrete ratchet-like patterns [22, 57]
- the flow direction from left-to-right
- the tip angle plays an important role in the pumping effect

c) Continues ratchet-like patterns [58-61]
- the flow direction from left to right
- more construction complexities

d) Double continues ratchet-like patterns [58-60]
- the flow direction from left to right
- enhanced thermal mass flow rate concerning case *c*

e) Double continues misaligned ratchet-like patterns [60, 62-64]
- flow direction from left to right at high Kn
- flow direction from right-to-left at low Kn
- an improved induced thermal flow rate could be achieved compared to the case *d*
- controlling the flow direction

f) Triangular obstacle [65]
- flow direction from left to right at high Kn
- flow direction from right-to-left at low Kn

Fig. 6.71 Different configurations of TEF Knudsen pumps (Copied with permission from Ref. [34])

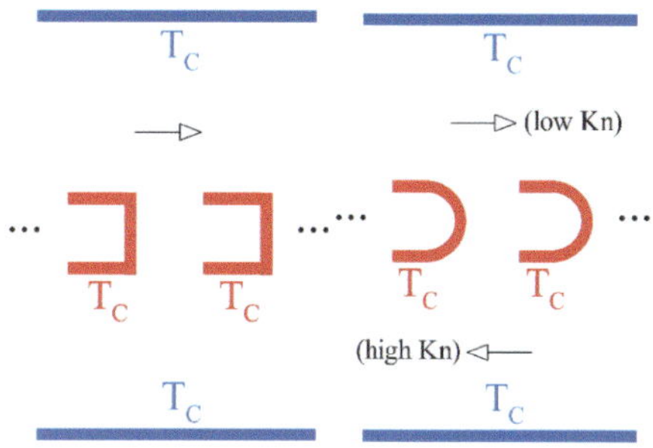

g) U-shaped obstacles [66]
- In squared type, the flow is left to right for all Knudsen numbers.
- In the curved type, the mass flow rate becomes reversed for high Knudsen numbers (Kn > 1) due to a kind of thermal edge flow (or a blunt edge flow) induced along the outer face of the curved part.
- The mass flow rate is generally more significant for curved types at the same Knudsen number.

Fig. 6.71 (continued)

rate is improved when the inclined surfaces are of low temperature or are specular, and the vertical surfaces are diffusive or high temperature [25, 47]. Also, channel configurations with both walls rachet types exhibit a higher mass flow rate than those with one surface ratchet wall [25].

Otic and Yonemura [67] considered 3 alternatives to ratchets, as shown in Fig. 6.72. Figure 6.73 shows the distribution of the net propulsive force in each of the microstructures at various temperature differences, at Kn = 1, where the temperature difference $T_h - T_c$ is changed while the middle value $(T_h + T_c)/2$ between the surface temperatures is kept at 400 K. As the temperature difference is increased, the net tangential Knudsen stress increases linearly, whose slope varies depending on the microstructure. The oblique plate has the largest slope, followed by the modified ratchet and the oblique ridge, while the conventional ratchet has the smallest slope.

Otic and Yonemura [68], in another attempt, tried to analyze the ratchet force using analytical molecular gas dynamics. They found momentum flux onto the upper surface brought by molecules leaving the lower surface without experiencing intermolecular collisions as:

$$J_{M,x}^{(wall)} = \int_{\phi_1}^{\phi_2}\int_0^{\pi}\int_0^{\infty} \exp\left(-\frac{L}{\lambda_c}\right) m n_{S_2} \sin^3\psi \sin\phi \cos\phi c^4 \left(\frac{m}{2\pi k_B T_{S_2}}\right)^{\frac{3}{2}} \cdot \exp\left(-\frac{m}{2k_B T_{S_2}} c^2\right) dc\, d\psi\, d\phi \tag{6.139}$$

The above equation was derived by considering the molecular interaction with the top wall of the ratchet configuration shown in Fig. 6.74. The notations used in Eq. (6.139) can be observed in this figure.

Momentum flux onto the upper surface brought by molecules coming from the bulk region after experiencing intermolecular collisions there was given by Eq. (7.141), see Fig. 6.75 for notation.

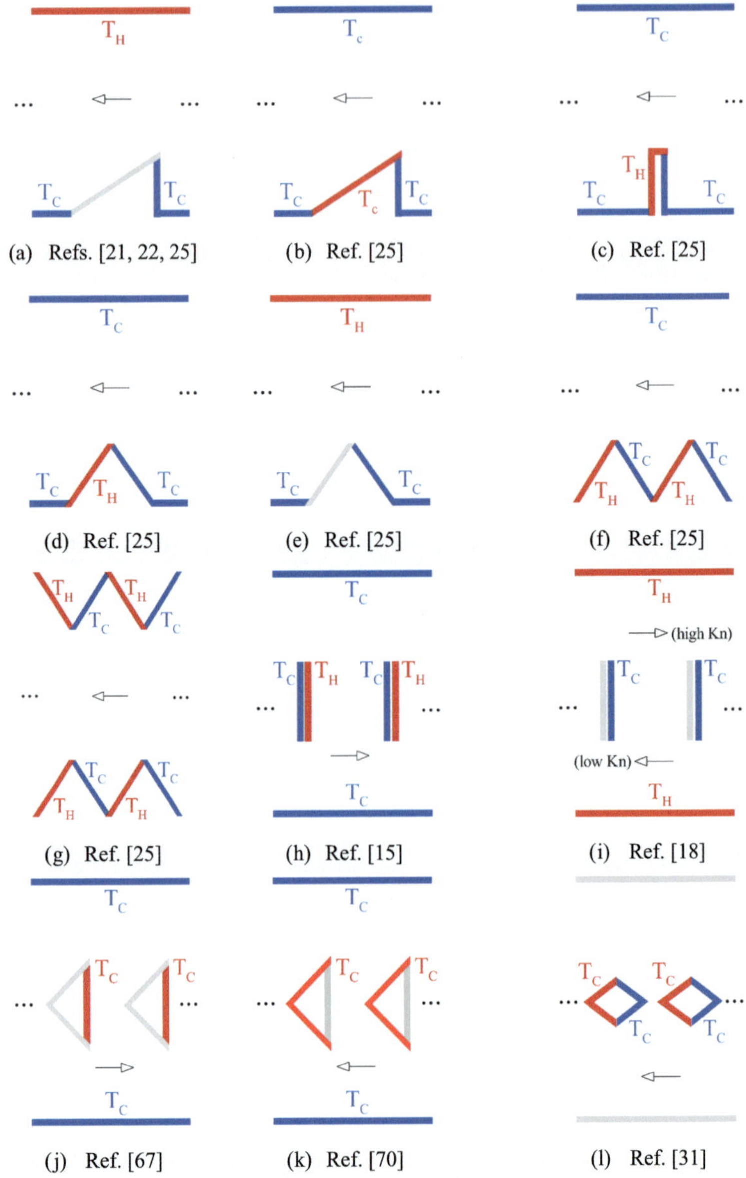

Fig. 6.72 Different configurations of RF Knudsen pumps (Reproduced from Ref. [34])

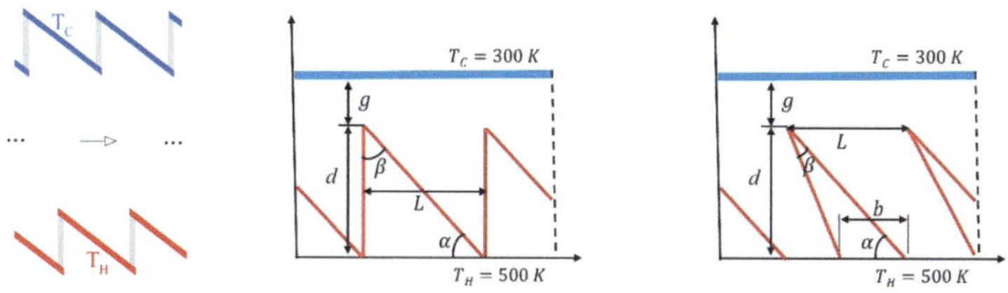

(m) Ref. [62] (n) Ref. [67] (o) Ref. [67]

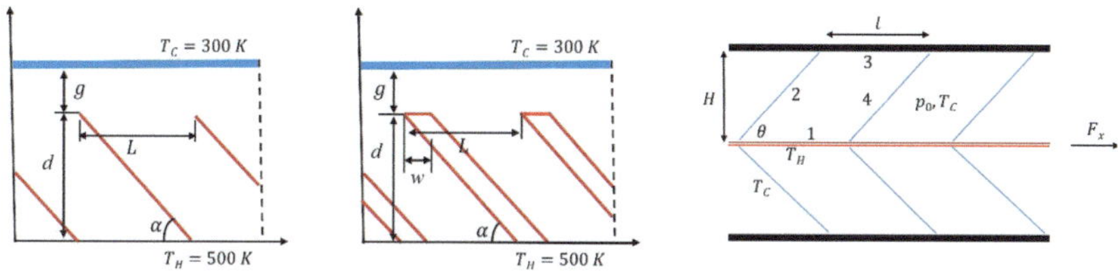

(p) Ref. [67] (q) Ref. [67] (r) Ref. [69]

Fig. 6.72 (continued)

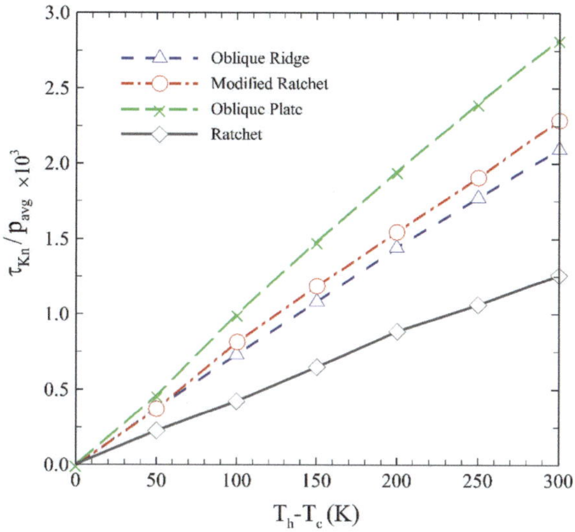

Fig. 6.73 Comparison of net tangential Knudsen stresses for different surface microstructures at different temperature differences in the case of Kn = 1 and $(T_h + T_c)/2 = 400$ K (Reproduced from Ref. [67])

$$J_{M,x}^{(gas)} = \int_0^{\pi}\int_0^{\pi}\int_0^{\infty}\left[\int_0^{L}\exp\left(-\frac{l}{\lambda_c}\right)dl\right] mn_g \sin^3\psi \sin\phi \cos\phi c^4 \left(\frac{m}{2\pi k_B T_g}\right)^{\frac{3}{2}}$$
$$.\exp\left(-\frac{m}{2k_B T_g}c^2\right) dc\, d\psi\, d\phi$$

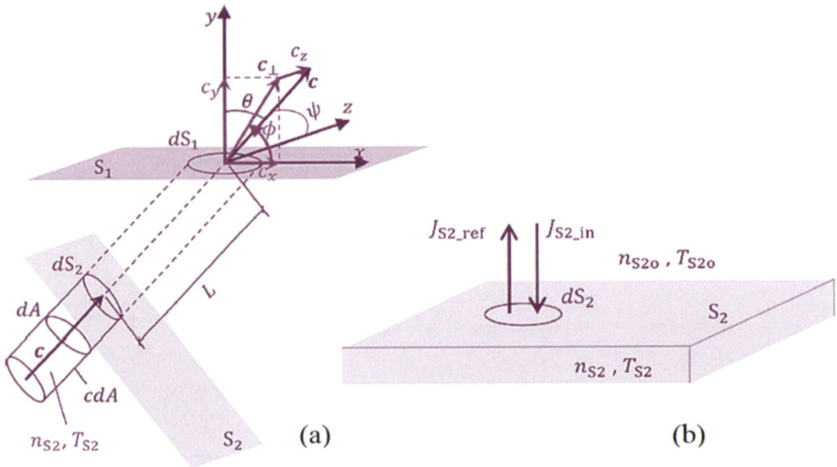

Fig. 6.74 Schematics of **a** the cylinder formed by all straight lines parallel to the velocity c that pass through the closed curve bounding dS1 on S1, and **b** the flux of gas molecules impinging on and leaving dS_2 on S_2 (Reproduced from Ref. [68])

$$= \int_0^{\pi}\int_0^{\pi}\int_0^{\infty}\left[1-\exp\left(-\frac{L}{\lambda_c}\right)\right] mn_g \sin^3\psi \sin\phi\cos\phi\, c^4\left(\frac{m}{2\pi k_B T_g}\right)^{\frac{3}{2}}$$
$$.\exp\left(-\frac{m}{2k_B T_g}c^2\right) dc\, d\psi\, d\phi \tag{6.140}$$

The integration for $J_{M,x}$ is done numerically.

According to Vincenti and Kruger [71] the viscous shear stress can be expressed by:

$$\tau_{flow} = \frac{1}{4} mn\overline{C}u\lambda_{ref} \tag{6.141}$$

Here, $\overline{C}$ is the mean molecular speed and u is the average x-velocity of molecules taken from the gas region located λ_{ref} below the upper surface. Reference [68] called this term as *flow effect*.

Thermal stress is given by:

$$\tau_{thermal} = p_{ref}\left(\frac{n_{ref}}{n}\frac{\lambda_{ref}^2}{T_{ref}}\right)\left(\gamma_3\frac{\partial^2 T}{\partial x\partial y} + \gamma_7\frac{1}{T}\frac{\partial T}{\partial x}\frac{\partial T}{\partial y}\right) \tag{6.142}$$

The variables with subscript "ref" are the reference values set in simulations, and the constants γ_3 and γ_7 are the thermal-stress transport coefficients, which for hard-sphere molecules have a value of 1.947906335 and 0.188106, respectively [72].

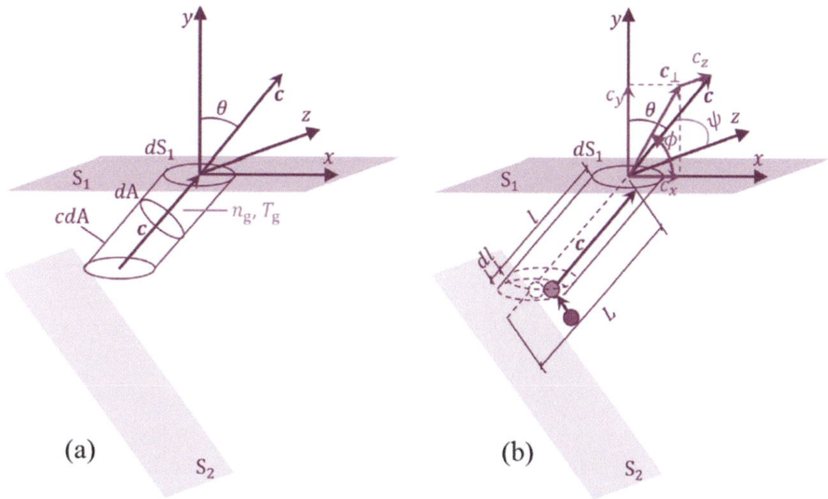

Fig. 6.75 Schematics of **a** the cylinder containing all of molecules of class c, which directly strike dS_1 within unit time unless they experience intermolecular collisions, and **b** the region where molecules of class c which directly strike dS1 experienced their latest intermolecular collision (Reproduced from Ref. [68])

Reference [67] chose the temperature at the gas region λ_{ref} below the upper surface for the calculation of the thermal stress. In Fig. 6.76, the distribution of the thermal stress is compared with the tangential stress obtained in the DSMC simulation minus the stress due to the flow effect. Here, the stress due to the flow effect is the one given by Eq. (6.141).

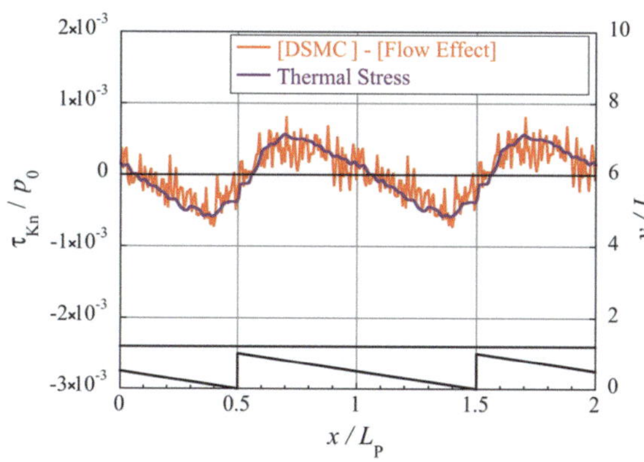

Fig. 6.76 Distribution of the thermal stress and that of the tangential Knudsen stress obtained by subtracting the flow effect from the DSMC results at Kn = 0.1. Here, p_{ref} is used as p_0 for the thermal stress, while the spatially averaged pressure pave is used as p_0 for the DSMC simulation and flow effect (Copied from open access Ref. [68])

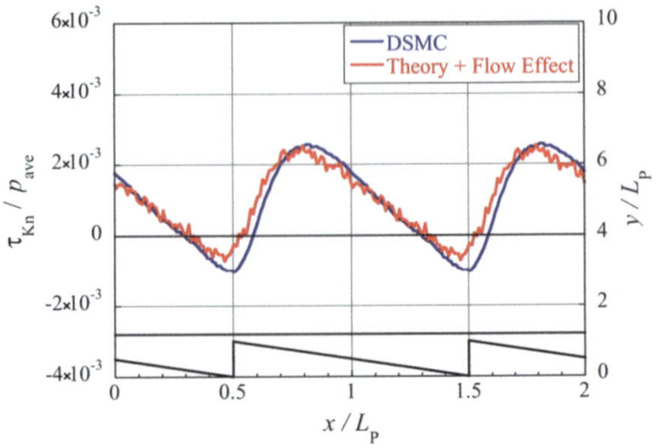

Fig. 6.77 Distribution tangential Knudsen stresses obtained by DSMC simulation and that obtained by the superposition of theoretical result and the contribution of the flow at Kn = 1 (Copied from open access Ref. [68])

At higher Knudsen numbers, the effects of thermal stress can be ignored. The reason is because the nondimensional temperature gradient with respect to the length scale of mean free path becomes smaller as the Knudsen number increases. Figure 6.77 compares the DSMC stress with theory (sum of Eqs. 6.139 and 6.140) and flow effects (6.142). There is a suitable agreement between the DSMC and analytical solution.

6.15 Summary

This chapter discusses various types of thermally driven flows in detail. Thermal creep flow, reverse thermal creep, thermal stress slip, nonlinear, thermal stress, thermal edge flow, and radiometric flow were introduced, with practical examples for each. Micropumps that operate based on radiometric flow were introduced, and several geometries of these pumps were examined. The possibility of gas mixture flow separation was also studied. Stokes flow was introduced, and Moffatt vortices and the conditions for their formation were discussed. Some details of various types of the thermally driven flow considered by authors were presented in detail. A comprehensive review of other researcher works on thermally-driven flow was provided.

Problems

1.A. This section aims to simplify the momentum equation for thermal creep flow in a channel. Simplify the following momentum equation for fully developed flow with no pressure gradient and low Reynolds number in a channel with thermal creep.

$$\rho u \frac{du}{dx} = -\frac{dP}{dx} + \mu \frac{\partial^2 u}{\partial y^2}$$

B. Show that the shear stress in steady thermal creep flow is zero.

C. Solve the simplified momentum equation with the following boundary condition and obtain the velocity profile. Assume a linear temperature distribution on the wall. Describe the resulting solution physically.

$$u_s = c_1 \frac{\partial u}{\partial y} + c_2 (\frac{\partial T}{\partial x})$$

D. The solution to the linearized Boltzmann equation for the velocity profile at very low Knudsen numbers (for example, 0.1, for which the answer to the previous section is valid) is as follows: (Kn $= k/\sqrt{\frac{\pi}{4}}$, the origin of the coordinate system is in the middle of the channel at height h).

$$u \sim k\left[1.726 - \left\{Y_1\left(\frac{1}{k}\left(0.5 + \frac{y}{h}\right)\right) + Y_1\left(\frac{1}{k}\left(0.5 - \frac{y}{h}\right)\right)\right\}\right],$$

where the function $Y_1(y)$ for any arbitrary value of the independent variable is shown in Fig. 6.74.

Show that the solution to this equation is similar to the one you obtained (Fig. 6.78).

2. In the solution of low Knudsen flow around a vane with a diffusive right side and a specular left side, the dimensionless isothermal lines (T/450, w = 4) are observed as Fig. 6.75:

A. Why the isothermal lines are perpendicular to the specular wall?
B. The isothermal lines are concentrated at the tip of the vane. For two horizontal points near the vane tip, calculate the temperature gradient. What

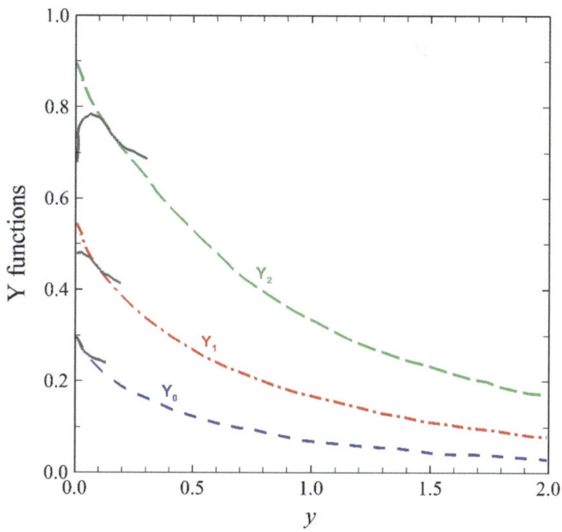

Fig. 6.78 Y functions (Reproduced from Ref. [5])

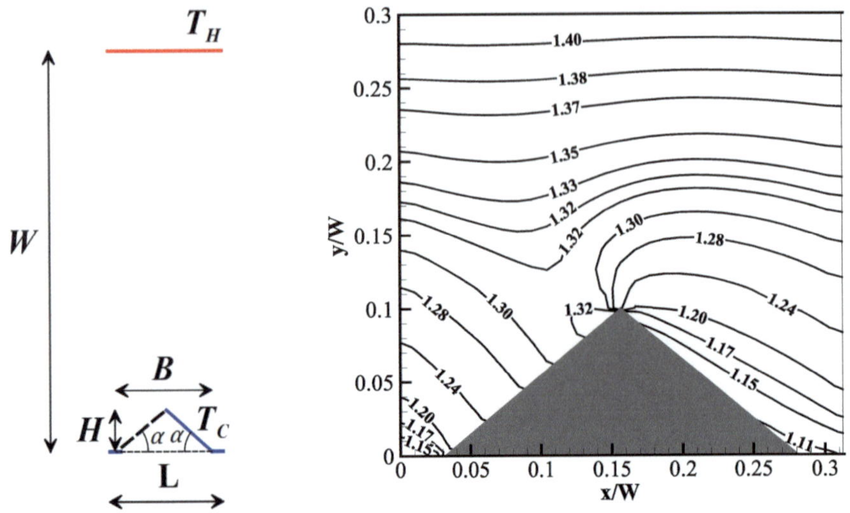

Fig. 6.79 Geometry of problem 2 (Copied from open access Ref. [21])

significance does the concentration of the lines near the tip have in generating the radiometric force? (Fig 6.79)

3. According to the figure below, the radiometric force at high Knudsen numbers (low pressures) is almost the same on all species. Why is there a difference in the radiometric force applied to different gas species at low Knudsen numbers (high pressures)? Justify your answer by mentioning the relations (Fig 6.80).

Fig. 6.80 Measured force values for different gas species in the Crookes radiometer (Reproduced from Ref. [16])

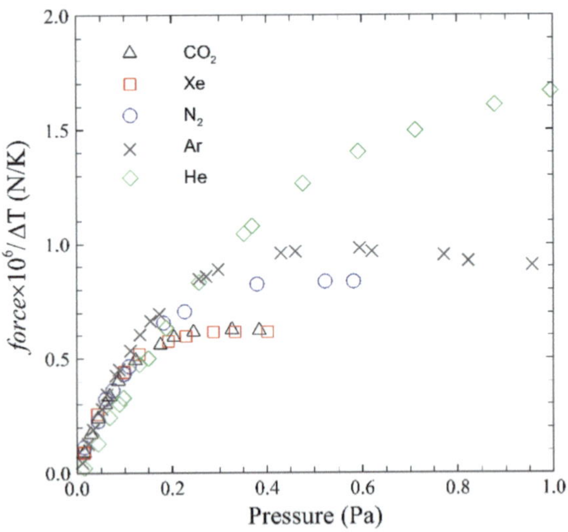

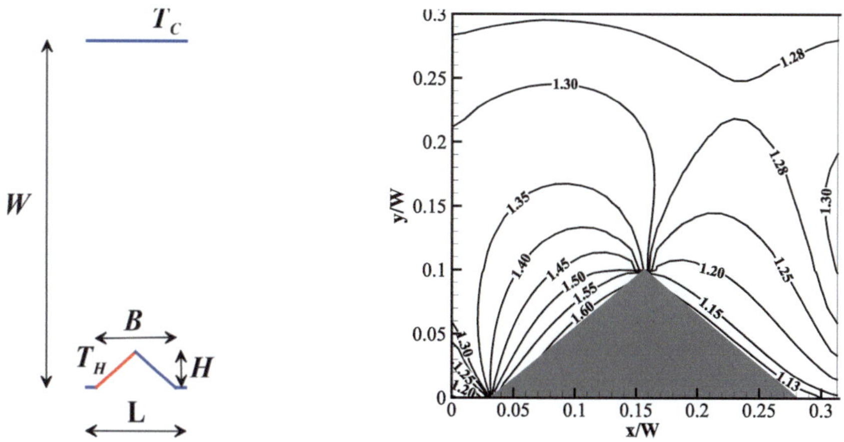

Fig. 6.81 Geometry of problem 4 (Copied from open access Ref. [21])

Hint: Consider the difference in free molecular mean free path lengths for different species.

4. In the solution of low Knudsen flow around a vane with a hot diffusive left side and a cold diffusive right side and lower horizontal sides, the dimensionless isotherms (T/450, w = 4) are observed as follows:

   A. Why do the isothermal lines not extend from the tip to the right corner and concentrate on the left? In your opinion, what is the reason for moving to the left?
   B. What kind of flow from the left horizontal wall towards the triangle? Explain. What will the final flow look like? (Fig 6.81)

5. Consider the velocity contours of the flow around the vane below. The inclined side of the blade is specular (Fig 6.82).

   A. Why is the maximum velocity limited to the tip of the vane at very low Knudsen numbers? Answer through the radiometric force mechanism.
   B. Why does a vortex form near the bottom wall? Why does the size of the formed vortex increase as the Knudsen number increases?
   C. Why does the velocity increase on the upper wall with an increase in the Knudsen number?

6.A. Define the phenomenon of thermophoresis.

B. How does the thermophoretic force depend on the Knudsen number?

C. The following relations give the thermophoretic force in two regimes: free molecular (first relation) and continuum (third relation). In these relations, R_p is the particle radius, and *c* is the most probable molecular velocity. Aside from the heat flux value, what is the difference in the force in these two regimes? What is the fundamental difference between the heat flux in these two regimes?

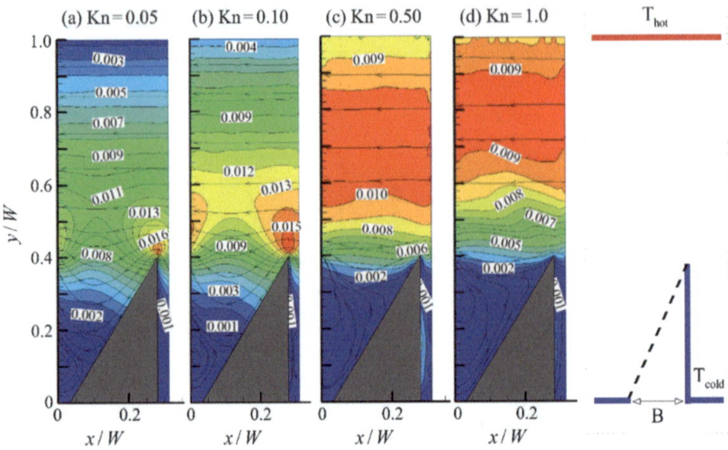

Fig. 6.82 Geometry of problem 5 (Copied from open access Ref. [21])

$$F_{FM} = \frac{3}{4}(\pi R_P^2)\frac{q}{c}$$

$$q_{FM} = -\frac{1}{2}P\bar{c}\frac{T_h - T_c}{T},\ T = \sqrt{T_h T_c}$$

$$F_{MX} = \left(\frac{32}{15\pi}\right)(\pi R_P^2)\frac{q}{c}$$

$$q_C = -k\frac{\partial T}{\partial z}$$

D. Propose a relation for the heat flux in all regimes.
E. Considering the right figure below, from a molecular point of view, what factor causes the movement of particles in the phenomenon of thermophoresis? (Fig 6.83)
F. The upper left figure shows the particle height based on the temperature of the upper wall. Why does the particle height decrease, and the particle fall with increasing temperature of the upper wall?
G. Assume the particle is modified so that its right half becomes a diffuse reflector, and its left half becomes a specular reflector. In which direction does the particle rotate? Why? (Hint: Consider the nature of the molecular collisions with the diffuse and specular walls.)
H. Sketch the reflection of particles from the specular and diffuse walls.

7. Consider a flat plate as follows:

   A. What conditions does a thermal edge flow form for T_{w+} and T_{w-}? Why?

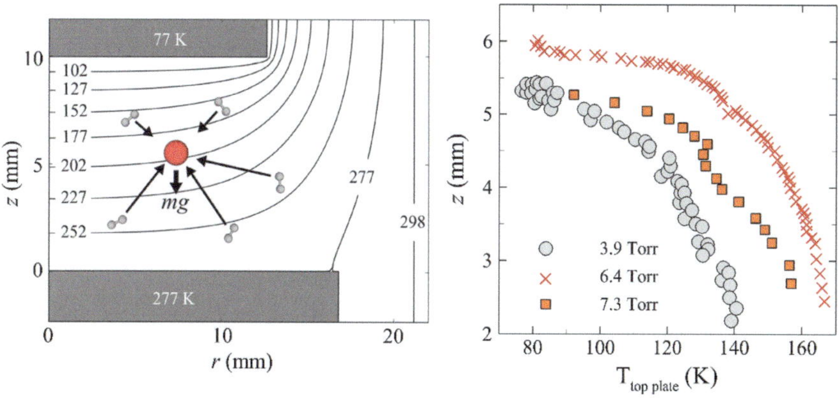

Fig. 6.83 A particle under the influence of thermophoretic force and plotting the forces as a function of the temperature difference between the upper and lower plates (Reproduced from Ref. [29])

Fig. 6.84 Flat plate with different thermal conditions

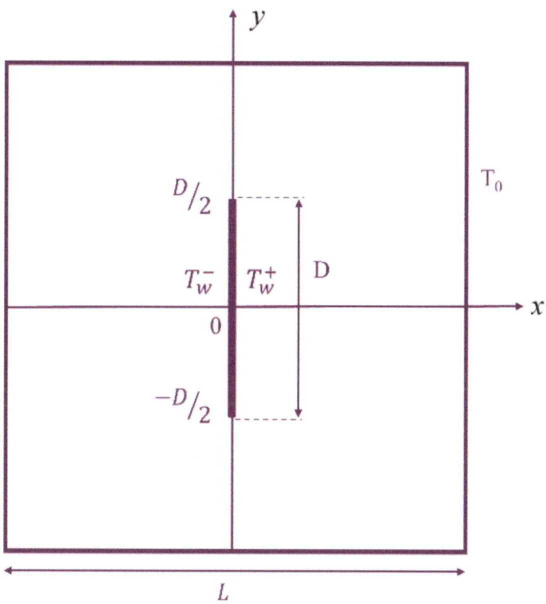

B. What conditions does a radiometric flow form for T_{w+} and T_{w-}? Why? (Fig 6.84)

8.A. Given the boundary condition:

$$U_S = \frac{1}{2}[(2-\sigma)U_\lambda + \sigma U_\omega] + U_C$$

where U_C is expressed as: $U_C = \frac{3\mu R}{4P}\frac{\partial T}{\partial S}$

Prove that:

$$\dot{M} = \frac{-h^3 P}{12\mu RT}\frac{dP}{dx}\left[1 + \sigma\frac{2-\sigma}{\sigma}\right] + \frac{3}{4}\frac{\mu h}{T}\frac{\partial T}{\partial x}$$

B. Based on part A of the question and the calculated $\dot{m}$, it is desired to prove the following relation:

$$P_1 - P_2 = \frac{9\mu^2 R}{h^2 \overline{P}}\frac{T_1 - T_2}{1 + 6\left(\frac{(2-\sigma)}{\sigma}(kn - kn^2)\right)}$$

9. (Stokes Flow) The problem of a circular cylinder in two-dimensional, inviscid incompressible weak shear flow can be described by the following PDE and boundary conditions (Fig 6.85):

$$\nabla^2\psi = \varepsilon\frac{U}{a}$$

$$\psi = 0 \text{ on } r = a$$

$$u = \left.\frac{\partial\psi}{\partial y}\right|_{r\to\infty} = U\left(1 + \varepsilon\frac{y}{a}\right)$$

or

$$\text{or } \psi(r)|_{r\to\infty} = U\left(r\sin\theta + \varepsilon\frac{r^2\sin^2\theta}{2a}\right)$$

$$U_\infty = U\left(1 + \varepsilon\frac{y}{a}\right)$$

Assume an asymptotic power series solution to this problem and, obtain the equation governing various orders (until ε^2). Solve the $O(1)$ equation and show:

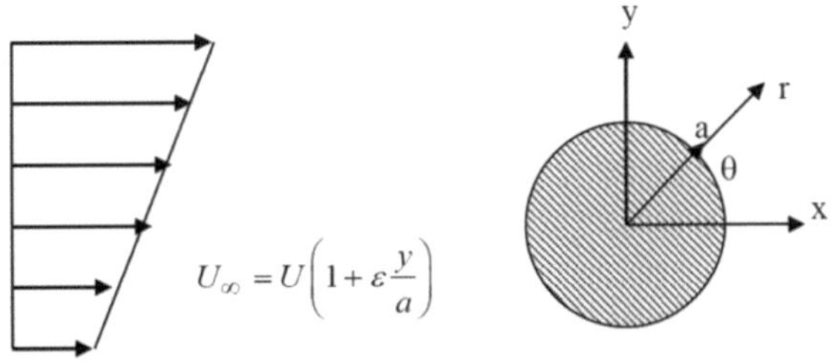

Fig. 6.85 Schematic of problem 9

$$\psi_0 = U\left(r - \frac{a^2}{r}\right)\sin\theta,$$

for $O(\varepsilon)$ and $O(\varepsilon^2)$, show that the solutions are as follows:

$$\psi_1 = \frac{Ua}{4}\left[\frac{r^2}{a^2}(1 - \cos 2\theta) + \frac{a^2}{r^2}\cos 2\theta - 1\right]$$

$$\psi_2 = 0$$

In fact all higher order terms of the stream function ψ are zero and the first three terms of the asymptotic expansion give the exact solution.

Hint:

$$\nabla^2\psi = \frac{1}{r}\frac{\partial}{\partial r}\left(r\frac{\partial\psi}{\partial r}\right) + \frac{1}{r^2}\frac{\partial^2\psi}{\partial\theta^2}$$

Hint: To obtain the solution in the general and particular form, use the boundary conditions and get an idea from the form of the answer.

References

1. Sone, Y.: Molecular Gas Dynamics: Theory, Techniques, and Applications. Birkhäuser, Boston (2007)
2. Karniadakis, G., Beskok, A., Aluru, N.: Microflows and Nanoflows: Fundamentals and Simulation, vol. 29. Springer Science & Business Media (2006)
3. Mohammadzadeh, A., Struchtrup, H.: Velocity dependent Maxwell boundary conditions in DSMC. Int. J. Heat Mass Transf. **87**, 151–160 (2015)
4. Ohwada, T., Sone, Y., Aoki, K.: Numerical analysis of the Poiseuille and thermal transpiration flows between two parallel plates on the basis of the Boltzmann equation for hard-sphere molecules. Phys. Fluids A **1**(12), 2042–2049 (1989)
5. Niimi, H.: Thermal creep flow of rarefied gas between two parallel plates. J. Phys. Soc. Jpn. **30**, 572–574 (1971)
6. Mozzafari, M.S., Roohi, E.: On the thermally-driven gas flow through divergent micro/ nanochannels. Int. J. Mod. Phys. C **28**(12), 1750143 (2017)
7. Takata, S., Funagane, H.: Poiseuille and thermal transpiration flows of a highly rarefied gas: overconcentration in the velocity distribution function. J. Fluid Mech. **669**, 242–259 (2011)
8. Su, W., Wang, P., Liu, H., Wu, L.: Accurate and efficient computation of the Boltzmann equation for Couette flow: influence of intermolecular potentials on Knudsen layer function and viscous slip coefficient. J. Comput. Phys. **378**, 573–590 (2019)
9. Lockerby, D.A., Reese, J.M., Emerson, D.R., Barber, R.W.: Velocity boundary condition at solid walls in rarefied gas calculations. Phys. Rev. E **70**(1), 017303 (2004)
10. Kogan, M.N., Galkin, V.S., Fridlender, O.G.: Stresses produced in gases by temperature and concentration inhomogeneities. New type of free convection. Soviet Physical Uspekhi **19**, 420–438 (1976)

11. Aoki, K., Sone, Y., Masukawa, N.: A rarefied gas flow induced by a temperature field. In: Harvey, J., Lord, G. (eds.) Rarefied Gas Dynamics, pp. 35–41. Oxford University Press, Oxford (1995)
12. Sone, Y., Yoshimoto, M.: Demonstration of a rarefied gas flow induced near the edge of a uniformly heated plate. Phys. Fluids **9**, 3530–3534 (1997)
13. Sugimoto, H., Sone, Y.: Vacuum pump without a moving part driven by thermal edge flow. In: Capitelli, M. (ed.) Rarefied Gas Dynamics, pp. 168–173. AIP, Melville, NY (2005)
14. Taguchi, S., Aoki, K.: Rarefied gas flow around a sharp edge induced by a temperature field. J. Fluid Mech. **694**, 191–224 (2012)
15. Taguchi, S., Aoki, K.: Motion of an array of plates in a rarefied gas caused by radiometric force. Phys. Rev. E **91**(6), 063007 (2015)
16. Ketsdever, A., Gimelshein, N., Gimelshein, S., Selden, N.: Radiometric phenomena: from the 19th to the 21st century. Vacuum **86**(11), 1644–1662 (2012)
17. Einstein, A.: Zur Theorie der Radiometerkrafte. Z. Angew. Phys. **27**, 1 (1924)
18. Baier, T., Hardt, S., Shahabi, V., Roohi, E.: Knudsen pump inspired by Crooks radiometer with a specular wall. Phys. Rev. Fluids **2**, 033401 (2017)
19. Schuster, A.: On the nature of the force producing the motion of a body exposed to rays of heat and light. Proc. R. Soc. Lond. **24**, 391–392 (1875)
20. Brüche, E., Littwin, W.: Experimental contributions to the radiometer question. Z. Angew. Phys. **52**, 318–335 (1928)
21. Shahabi, V., Baier, T., Roohi, E., Hardt, S.: Thermally induced gas flows in ratchet channels with diffuse and specular boundaries. Sci. Rep. **7**, 41412 (2017)
22. Donkov, A.A., Tiwari, S., Liang, T., Hardt, S., Klar, A., Ye, W.: Momentum and mass fluxes in a gas confined between periodically structured surfaces at different temperatures. Phys. Rev. E **84**, 016304 (2011)
23. Moffatt, H.K.: Viscous and resistive eddies near a sharp corner. J. Fluid Mech. **18**, 1–18 (1964)
24. Westphal, W.H.: Messungen am radiometer. Z. Angew. Phys. **23**, 92–100 (1920)
25. Lotfian, A., Roohi, E.: Radiometric flow in periodically patterned channels: fluid physics and improved configurations. J. Fluid Mech. **860**, 544–576 (2019)
26. Lotfian, A., Roohi, E.: Binary gas mixtures separation using microscale radiometric pumps. Int. Commun. Heat Mass Transf. **121**, 105061 (2021)
27. Panton, R.L.: Incompressible Flow. Wiley (2013)
28. Lotfian, A., Roohi, E.: Species separation induced by difference in accommodation coefficients. Int. Commun. Heat Mass Transf. **161**, 108524 (2025)
29. Fung, F., Usatyuk, M., DeSalvo, B.J., Chin, C.: Stable thermophoretic trapping of generic particles at low pressures. Appl. Phys. Lett. **110**(3), 034102 (2017)
30. Bird, G.A.: Molecular Gas Dynamics and the Direct Simulation of Gas Flows. Oxford University Press (1994)
31. Zhu, M., Roohi, E.: Thermally driven rarefied flows induced by a partially heated diamond in a channel. Int. Commun. Heat Mass Transf. **135**, 106095 (2022)
32. Mousivand, M., Roohi, E.: On the rarefied thermally-driven flows in cavities and bends. Fluids **7**(11), 354 (2022)
33. Rogozin, O.: Slow non-isothermal flows: numerical and asymptotic analysis of the Boltzmann equation. Comput. Math. Math. Phys. **57** (2017)
34. Akhlaghi, H., Roohi, E., Stefanov, S.: A comprehensive review on micro-and nano-scale gas flow effects: slip-jump phenomena, Knudsen paradox, thermally-driven flows, and Knudsen pumps. Phys. Rep. **997**, 1–60 (2023)
35. Knudsen, M.: Eine revision der gleichgewichtsbedingung der gase. Thermische molekularströmung. Annalen der Physik **336**(1), 205–229 (1909)
36. Knudsen, M.: Thermischer molekulardruck der gase in roehren und poroesen koerpern. Ann. Phys. **336**(3), 633–640 (1910)
37. Sone, Y., Waniguchi, Y., Aoki, K.: One-way flow of a rarefied gas induced in a channel with a periodic temperature distribution. Phys. Fluids **8**(8), 2227–2235 (1996)

38. Sone, Y., Sugimoto, H.: Vacuum pump without a moving part and its performance. In: AIP Conference Proceedings, vol. 663, no. 1, pp. 1041–1048. American Institute of Physics (2003)
39. Alexeenko, A.A., Gimelshein, S.F., Muntz, E.P., Ketsdever, A.D.: Kinetic modeling of temperature driven flows in short microchannels. In: International Conference on Nanochannels, Microchannels, and Minichannels, vol. 41855, pp. 483–491 (2005)
40. Aoki, K., Degond, P., Takata, S., Yoshida, H.: Diffusion models for Knudsen compressors. Phys. Fluids **19**(11), 117103 (2007)
41. Han, Y.L.: Thermal-creep-driven flows in Knudsen compressors and related nano/microscale gas transport channels. J. Microelectromech. Syst. **17**(4), 984–997 (2008)
42. Guo, X., Singh, D., Murthy, J., Alexeenko, A.A.: Numerical simulation of gas-phonon coupling in thermal transpiration flows. Phys. Rev. E **80**(4), 046310 (2009)
43. Han, Y.L.: Working gas temperature and pressure changes for microscale thermal creep-driven flow caused by discontinuous wall temperatures. Fluid Dyn. Res. **42**(4), 045505 (2010)
44. Dodulad, O.I., Kloss, Y.Y., Savichkin, D.O., Tcheremissine, F.G.: Knudsen pumps modeling with Lennard-Jones and ab initio intermolecular potentials. Vacuum **109**, 360–367 (2014)
45. Zhang, Z., Wang, X., Zhao, L., Zhang, S., Zhao, F.: Study of flow characteristics of gas mixtures in a rectangular Knudsen pump. Micromachines **10**(2), 79 (2019)
46. Ye, J., Shao, J., Xie, J., Zhao, Z., Yu, J., Zhang, Y., Salem, S.: The hydrogen flow characteristics of the multistage hydrogen Knudsen compressor based on the thermal transpiration effect. Int. J. Hydrog. Energy **44**(40), 22632–22642 (2019)
47. Wang, X., Su, T., Zhang, W., Zhang, Z., Zhang, S.: Knudsen pumps: a review. Microsyst. Nanoeng. **6**(1), 1–28 (2020)
48. Aoki, K., Degond, P., Mieussens, L., Takata, S., Yoshida, H.: A diffusion model for rarefied flows in curved channels. Multiscale Model. Simul. **6**(4), 1281–1316 (2008)
49. Aoki, K., Degond, P., Mieussens, L.: Numerical simulations of rarefied gases in curved channels: thermal creep, circulating flow, and pumping effect. Commun. Comput. Phys. **6**(5), 919 (2009)
50. Leontidis, V., Brandner, J.J., Baldas, L., Colin, S.: Numerical analysis of thermal creep flow in curved channels for designing a prototype of Knudsen micropump. In: Journal of Physics: Conference Series, vol. 362, no. 1, p. 012004. IOP Publishing (2012)
51. Leontidis, V., Chen, J., Baldas, L., Colin, S.: Numerical design of a Knudsen pump with curved channels operating in the slip flow regime. Heat Mass Transf. **50**(8), 1065–1080 (2014)
52. Bond, D.M., Wheatley, V., Goldsworthy, M.: Numerical investigation of curved channel Knudsen pump performance. Int. J. Heat Mass Transf. **76**, 1–15 (2014)
53. Bond, D.M., Wheatley, V., Goldsworthy, M.: Numerical investigation into the performance of alternative Knudsen pump designs. Int. J. Heat Mass Transf. **93**, 1038–1058 (2016)
54. Sugimoto, S., Sugimoto, H.: Thermal transpiration flows induced by differences in accommodation coefficients. Phys. Fluids **34**(4), 042005 (2022)
55. Graur, I., Veltzke, T., Méolans, J.G., Ho, M.T., Thöming, J.: The gas flow diode effect: theoretical and experimental analysis of moderately rarefied gas flows through a microchannel with varying cross section. Microfluid. Nanofluid. **18**(3), 391–402 (2015)
56. Tatsios, G., Lopez Quesada, G., Rojas-Cardenas, M., Baldas, L., Colin, S., Valougeorgis, D.: Computational investigation and parametrization of the pumping effect in temperature-driven flows through long tapered channels. Microfluid. Nanofluid. **21**(5), 1–17 (2017)
57. Han, F., Wang, X., Zhao, F., Zhang, S., Zhang, Z.: Numerical investigation of gas separation via thermally induced flows in ratchet-like patterned microchannels. Int. J. Therm. Sci. **172**, 107280 (2022)
58. Würger, A.: Leidenfrost gas ratchets driven by thermal creep. Phys. Rev. Lett. **107**(16), 164502 (2011)
59. Hardt, S., Tiwari, S., Baier, T.: Thermally driven flows between a Leidenfrost solid and a ratchet surface. Phys. Rev. E **87**(6), 063015 (2013)
60. Chen, J., Baldas, L., Colin, S.: Numerical study of thermal creep flow between two ratchet surfaces. Vacuum **109**, 294–301 (2014)

61. Zhu, L., Guo, Z.: Application of discrete unified gas kinetic scheme to thermally induced nonequilibrium flows. Comput. Fluids **193**, 103613 (2019)
62. Chen, J., Stefanov, S.K., Baldas, L., Colin, S.: Analysis of flow induced by temperature fields in ratchet-like microchannels by Direct Simulation Monte Carlo. Int. J. Heat Mass Transf. **99**, 672–680 (2016)
63. Rovenskaya, O.I.: Numerical investigation of thermally generated gas flow between saw-tooth like surfaces. Int. J. Heat Mass Transf. **147**, 118979 (2020)
64. Lan, J., Xie, J., Ye, J., Jiao, X., Peng, W.: Non-equilibrium evolution and characteristics of the serrated microchannel hydrogen Knudsen compressor. Int. J. Hydrog. Energy **47**(7), 4804–4813 (2022)
65. Wang, X., Zhang, Z., Zhang, W., Zhang, P., Zhang, S.: Numerical simulation of thermal edge flow in ratchet-like periodically patterned micro-channels. Int. J. Heat Mass Transf. **135**, 1023–1038 (2019)
66. Taguchi, S., Tsuji, T.: One-way flow over uniformly heated U-shaped bodies driven by thermal edge effects. Sci. Rep. **12**(1), 1–17 (2022)
67. Otic, C.J.C., Yonemura, S.: Effect of different surface microstructures in the thermally induced self-propulsion phenomenon. Micromachines **13**(6), 871 (2022)
68. Otic, C.J.C., Yonemura, S.: Mechanism of tangential Knudsen force at different Knudsen numbers. Phys. Fluids **34**(7) (2022)
69. Wadsworth, D.C., Muntz, E.P.: A computational study of radiometric phenomena for powering microactuators with unlimited displacements and large available forces. J. Microelectromech. Syst. **5**(1), 59–65 (1996)
70. Wang, R., Xu, X., Xu, K., Qian, T.: Onsager's cross coupling effects in gas flows confined to micro-channels. Phys. Rev. Fluids **1**(4), 044102 (2016)
71. Vincenti, W.G., Kruger, C.H.: Introduction to Physical Gas Dynamics. Krieger Publishing Company, Malabar, FL (1975)
72. Sone, Y.: Flows induced by temperature fields in a rarefied gas and their ghost effect on the behavior of a gas in the continuum limit. Annu. Rev. Fluid Mech. **32**, 779–811 (2000)

Chapter 7
Rarefied Couette Flow, Fourier Flow, and Cavity Flow

Abstract This chapter examines rarefied Couette and cavity flows and subsequently considers the Fourier flow. It presents analytical solutions across various rarefaction regimes, emphasizing these flow types' distinctive features and behaviors.

7.1 Couette Flow

The first part of this chapter examines the Couette flow, oscillatory Couette, and cylindrical Couette flow under rarefied conditions. These flows have been extensively studied in classical fluid mechanics, and this section aims to present analytical results for these flows in the slip, transitional, and free molecular regimes. The velocity profile of Couette flow in a continuous flow is linear. Exactly near the wall, the velocity is equal to the wall velocity, and as it gets closer to the center of the channel, the velocity approaches zero, see Fig. 7.1.

As the flow enters the rarefied regime, as shown in Fig. 7.2, the velocity profile deviates from linearity near the walls due to forming the Knudsen layer adjacent to the walls. The flow behaviour within this layer is entirely nonlinear. The velocity on the walls is not equal to the wall velocity because velocity slip occurs. On the lower wall, where the velocity gradient is negative, the flow velocity is less than the wall velocity. Conversely, on the upper wall, where the velocity gradient is positive, the flow velocity exceeds the wall velocity.

The governing equation for Couette flow based on the Navier–Stokes equation is obtained as follows:

$$\frac{dP_{xy}}{dy} = 0 \tag{7.1}$$

$$P_{xy} = -\tau_{xy} = -\mu \frac{\partial u}{\partial y} \tag{7.2}$$

Therefore, assuming a constant viscosity coefficient, the Couette flow equation in terms of velocity is obtained as a relation (7.3):

E. Roohi et al., *Advances in Direct Simulation Monte Carlo: From Micro-Scale to Rarefied Flow Phenomena*, https://doi.org/10.1007/978-981-96-8200-3_7

Fig. 7.1 Wall velocities in Couette flow of a channel

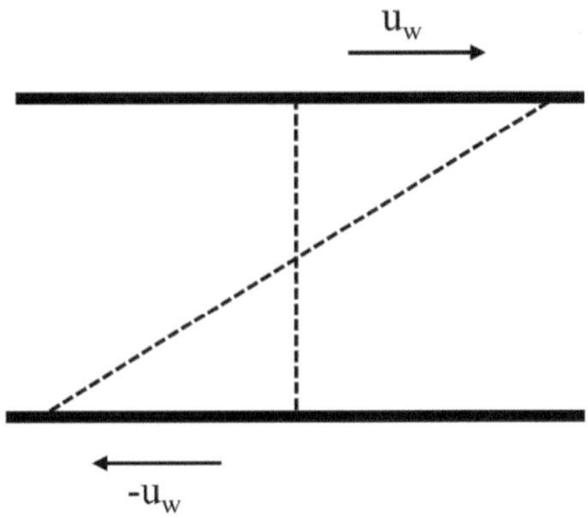

Fig. 7.2 Velocity profile in the Couette problem within the Knudsen layer and in the bulk flow

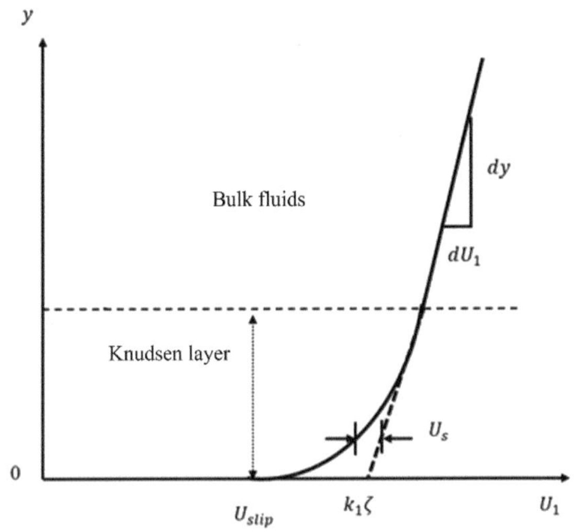

$$\frac{\partial^2 u(y)}{\partial y^2} = 0 \tag{7.3}$$

The first-order slip velocity boundary conditions are as follows:

$$u - u_w = \frac{2 - \sigma_v}{\sigma_v} \alpha \lambda \frac{\partial u}{\partial y} \tag{7.4}$$

If the boundary condition is Maxwell's, the value of α is 1. In another boundary condition, derived using the linearized Boltzmann equation, the equality $\alpha = 1.11$ holds. If the governing equation and boundary conditions are solved, the following velocity profile is obtained: ($Kn = \lambda/D$).

$$u_c = \frac{2u_w}{1 + 2\frac{2-\sigma_v}{\sigma_v}\alpha Kn}\frac{y}{D}, \quad -D/2 < y < D/2 \tag{7.5}$$

To extend the use of this relation for all Knudsen numbers, and not just the slip regime where this relation and the corresponding boundary condition hold, Karnidakis et al. [1] considered the coefficient α as a function of the Knudsen number as follows:

$$\alpha_m = \beta_0 + \beta_1 \tan^{-1}(\beta_2 Kn^{\beta_3})$$

They obtained the coefficients β by solving DSMC, as follows:

$$\beta_0 = 1.2977; \ \beta_1 = 0.71851; \ \beta_2 = -1.17488; \ \beta_3 = 0.58642$$

By using these coefficients, Eq. (7.5) provides good accuracy up to the range of Kn < 12. According to Fig. 7.3, the model proposed by Karniadakis et al. is accurate in other points except near the walls (Knudsen layer), where there is a slight error. As the Knudsen number increases, the slope of the velocity profile increases. In other words, the viscosity increases and the velocity changes decrease, see Fig. 7.4. Another crucial point is that these solutions are accurate for isothermal flows and M < 0.3; otherwise, the energy equation must be solved alongside the momentum equation.

The temperature behaviour is shown in Fig. 7.5, with the temperature distribution in Couette flow, plotted at different Mach numbers. As the Mach number increases, the temperature distribution deviates from linearity, and the thermal jump also increases.

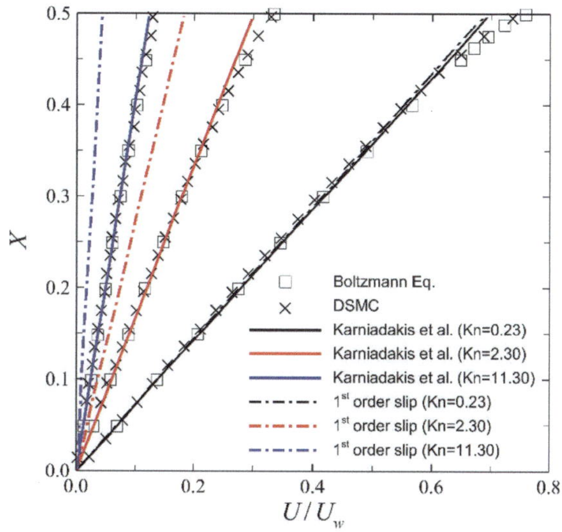

Fig. 7.3 Velocity diagram in the upper half of the Couette flow channel at different Knudsen numbers, comparison of the Karniadakis et al. model with the first-order slip equation, DSMC results, and the solution of the Boltzmann equation (Reproduced from Ref. [1])

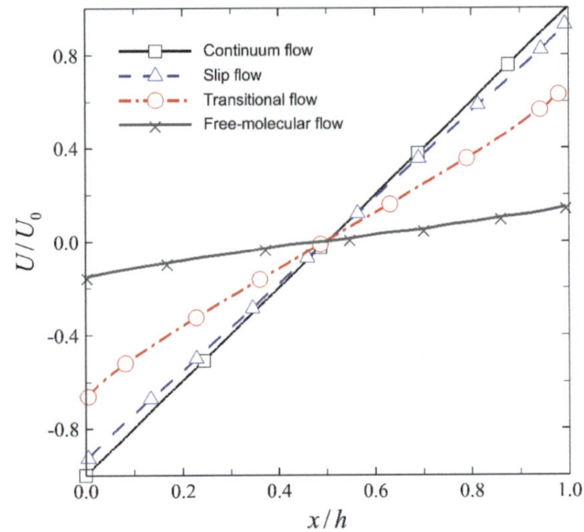

Fig. 7.4 Velocity diagram in Couette flow in the continuum, slip, and transient regimes (Reproduced from Ref. [2])

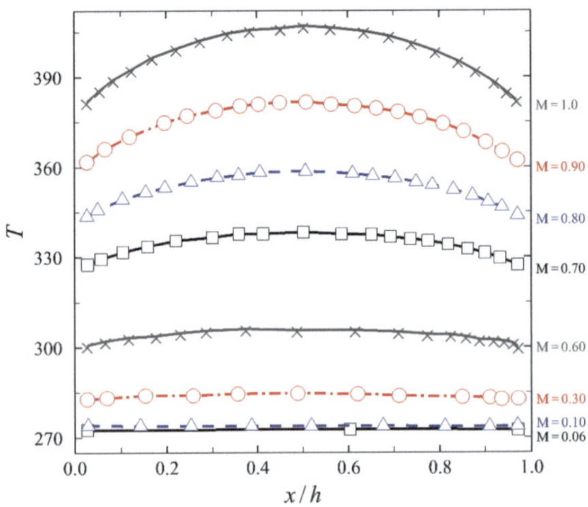

Fig. 7.5 Temperature distribution versus position at different Mach numbers (Reproduced from Ref. [1])

According to Fig. 7.6, by comparing the Karniadakis et al. [1] model and the DSMC solution, when the Knudsen number is constant and the Mach number is increasing, the proposed model deviates more from the DSMC solution. It is concluded that if M > 0.3, in addition to the temperature profile not being linear, the velocity profile of the Karniadakis et al. model loses its accuracy.

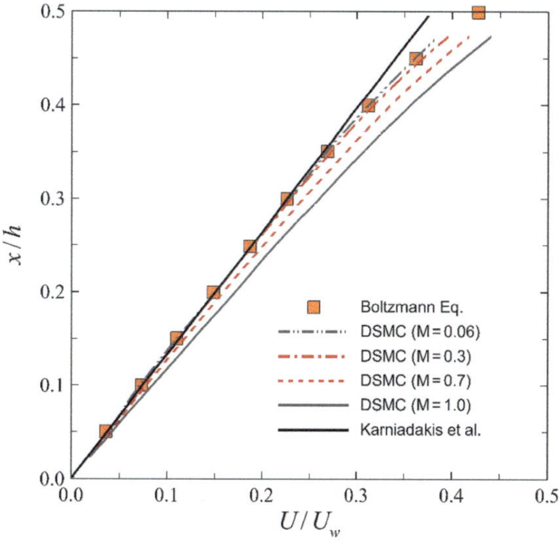

Fig. 7.6 Velocity diagram in the upper half of the channel at different wall Mach numbers, comparison of analytical results with the solution of the Boltzmann equation (Reproduced from Ref. [1])

7.2 Slip Length (l_s)

As shown in Fig. 7.7, if the linear velocity profile is extended linearly until the gas velocity equals the wall velocity, the added length is called the slip length. In essence, l_s is the length that is eliminated due to the non-continuum effects of the flow, and by adding this length to the channel thickness, the effect of the Knudsen layer can be neglected [1].

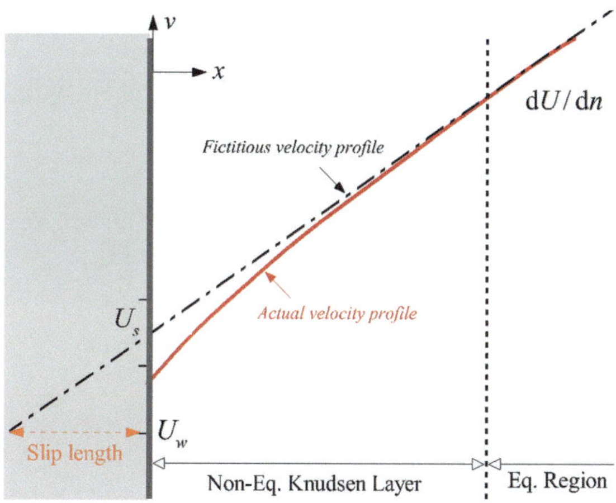

Fig. 7.7 Representation of slip length

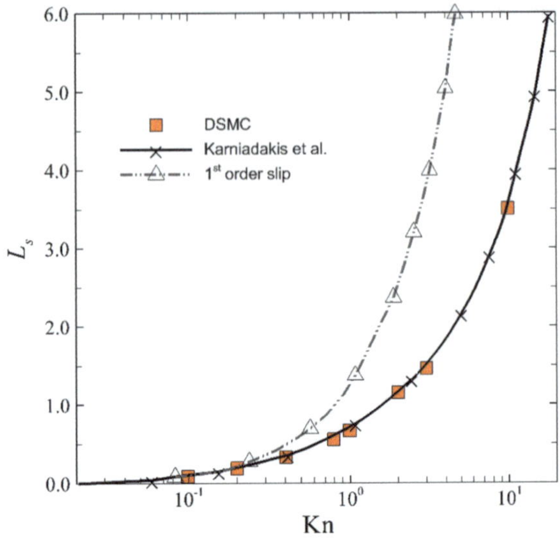

Fig. 7.8 Slip length as a function of Knudsen number (Reproduced from Ref. [1])

The slip length can be obtained from the following relation:

$$l_s = \frac{u(y = D/2) - U}{D\frac{\partial u(y)}{\partial y}}, \tag{7.6}$$

where D is the channel diameter, if this length is plotted against the Knudsen number, as shown in Fig. 7.8, it is observed that as the Knudsen number increases, the slip length also increases.

7.3 Incompressible Couette Flow Solution with the Energy Equation

This section aims to solve the momentum and energy equations governing Couette flow. The momentum equation is given by [3]:

$$\frac{dP_{xy}}{dy} = 0 \tag{7.7}$$

The energy equation is expressed as follows:

$$\frac{dq_y}{dy} + P_{xy}\frac{du}{dy} = 0 \tag{7.8}$$

Temperature and velocity boundary conditions:

$$u(y = \pm D) = \pm u_w \mp k_0 KnD \left.\frac{du}{dy}\right|_{y=\pm D} \tag{7.9}$$

$$T(y = \pm D) = T_0 \mp k_1 KnD \left.\frac{dT}{dy}\right|_{y=\pm D}, \tag{7.10}$$

where the channel height varies between $\pm D$. The temperature and velocity profiles are obtained by simultaneously solving the energy and momentum equations and applying the boundary conditions.

$$u = \frac{u_w}{1 + K_0 Kn} \frac{y}{D} \tag{7.11}$$

$$T = T_0 + \frac{2m}{15k}(1 + 2k_1 Kn)\left(\frac{u_w}{1 + K_0 Kn}\right)^2 - \frac{2m}{15k}\left(\frac{u_w}{1 + K_0 Kn}\right)^2 \left(\frac{y}{D}\right)^2 \tag{7.12}$$

The shear stress and heat flux are also obtained using relations (7.13) and (7.14).

$$P_{xy} = -\mu \frac{du}{dy} \tag{7.13}$$

$$q_y = -\lambda \frac{dT}{dy} \tag{7.14}$$

$$P_{xy} = -\mu \frac{u_w}{1 + K_0 Kn} \frac{1}{D} \tag{7.15}$$

$$q_y = \mu \left(\frac{u_w}{1 + K_0 Kn}\right)^2 \frac{y}{D^2} \tag{7.16}$$

The above results are valid if the fluid is Newtonian, and even if the Burnett equations are used, the above solutions remain unchanged; the only modification is that the horizontal flux term should be added to the heat flux expression:

$$q_x = \frac{\theta_5}{2} \frac{dv_x}{dy} \frac{dT}{dy} \tag{7.17}$$

$$\theta_5 = \frac{213}{8} \frac{\mu^2}{mn_0 T_0} \tag{7.18}$$

If the temperature and velocity obtained in the previous section are substituted, the Burnett term of the heat flux results in Eq. (7.19):

$$q_x = -\frac{213}{30} \frac{\mu^2}{n_0 k T_0} \left(\frac{V}{1 + K_0 Kn}\right)^3 \frac{y}{D^3} \tag{7.19}$$

7.4 Shear Stress in Couette Flow

According to Kogan's theory, the shear stress in the free molecular regime is proportional to the density and the velocity of the moving plates and can be expressed as follows [1]:

$$\tau_{xy,\infty} = -\rho_o U \sqrt{\frac{2RT_w}{\pi}} \tag{7.20}$$

Note that the shear stress in the free molecular regime is independent of the velocity gradient, and even if the velocity gradient tends towards zero, the above expression still holds and is non-zero.

The shear stress in the continuum regime is given by:

$$\tau_{xy,\infty} = -2\mu \frac{U}{D}, \tag{7.21}$$

where

$$\mu = \gamma_1 \sqrt{\frac{\pi RT}{8}} \rho_o \lambda = \sqrt{\frac{2RT_w}{\pi}} \rho_o \lambda \tag{7.22}$$

Cercignani proposed the following model for the shear stress in Couette flow [4]:

$$\frac{\tau_{xy}}{\tau_{xy,\infty}} = \pi_{xy} = \frac{aKn^2 + 2bKn}{aKn^2 + cKn + b}$$
$$a = 1.3056, \quad b = 2\pi, \quad c = 7.5939 \quad b = \pi \tag{7.23}$$

The coefficients of relation (7.23) are given for the hard sphere molecular model [4]. Sone et al. [5] derived the following relation for the range of very small Knudsen numbers:

$$\pi_{xy} = \frac{\gamma_1 \pi Kn}{2(1 + 2C_1 Kn)}$$
$$\gamma_1 = 1.270042, \quad C_1 = 1.1111 \tag{7.24}$$

Also, Bahukudumbi and Beskok [6] proposed the following relation for shear stress:

$$\frac{\tau_{xy}}{\tau_{xy,contiuum}} = \prod_{xy} = \frac{1}{2} \frac{aKn + 2b}{aKn^2 + cKn + b}$$
$$a = 0.5297, \quad b = 0.6030, \quad c = 1.6277 \tag{7.25}$$

The above coefficients are obtained based on least-squares fitting to the linearized Boltzmann solution data.

By taking the Knudsen number towards zero and infinity and expanding the Taylor series, we get the following relation [1]:

$$\pi_{xy} = 1 + \frac{2b - c}{a}\frac{1}{Kn} + \frac{-b - \left[\frac{-2b-c}{a}\right]c}{a}\frac{1}{Kn^2} + O\left(Kn^{-3}\right) \quad Kn \to \infty \tag{7.26}$$

$$\prod_{xy} = 1 + \frac{a - 2c}{2b}Kn + \frac{\left[\frac{2c-a}{b}\right]c - 2a}{2b}Kn^2 + O\left(Kn^3\right) \quad Kn \to 0 \tag{7.27}$$

$$\frac{a - 2c}{2b} = -2.2601 \approx -2\alpha$$
$$\begin{cases} if\ Kn \to \infty & \pi_{xy} \to 1 \\ if\ Kn \to 0 & \prod_{xy} \to 1 \end{cases} \tag{7.28}$$

In Fig. 7.9, the proposed shear stress from the model in relation (7.25) is compared and evaluated with the results of various models and different numerical methods. It is worth mentioning that these results were obtained under the condition of a wall Mach number equal to 0.05.

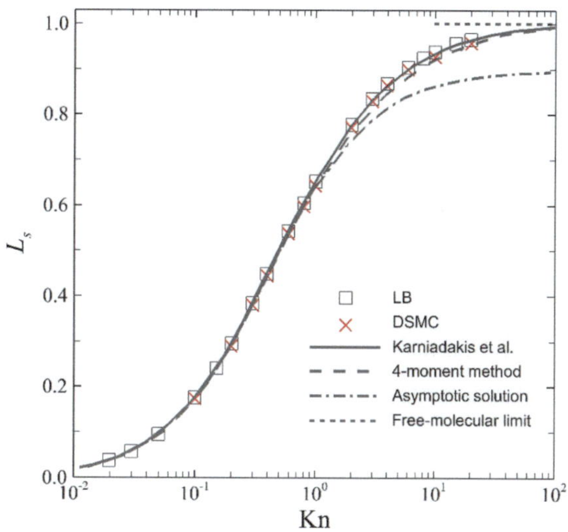

Fig. 7.9 Comparison of shear stress from the Bahukudumbi and Beskok (current) model with the Sone model, linearized Boltzmann results, the 4-moment method, free molecular regime results, and DSMC results (Reproduced from Ref. [1])

7.5 Behavior of Couette Flow Near the Wall

Different researchers have proposed relations to describe the velocity behavior in the Knudsen layer of Couette flow. Lockerby et al. [7], by fitting data, proposed the following relation for the velocity profile in the Knudsen layer:

$$U_s(x) \approx \frac{7}{20(1+x)^2} \tag{7.29}$$

Lilly and Sadr, by solving the linearized Boltzmann equation, proposed the following power-law relation [8]:

$$U_s(x) = U_s(0) - Cx^n, \tag{7.30}$$

where C is a constant in this relation, and the coefficient n was proposed to be 0.82. Based on the above relation, they stated that the slope of the velocity profile near the wall is indeterminate. In other words:

$$\frac{dU_s}{dx} \to \infty \quad x \to 0 \tag{7.31}$$

Takata and Funagane [9] analytically proved that the Couette flow velocity profile slope at smooth surfaces diverges logarithmically near the wall. Jiang and Luo [10], by solving the asymptotic form of the BGK equation, found that the Couette flow near the wall is described by the following power series:

$$U_s(x) = \sum_{n=0}^{N} \sum_{m=0}^{M} c_n, m^{x^n (x \ln x)^m}, x \to 0 \tag{7.32}$$

This again indicates that this relation's gradient near the wall is undefined.

7.6 Couette Flow Prediction by NCCR Equations

Myong derived the following nonlinear coupled constitutive relation (NCCR) equations from the Boltzmann equation [11, 12]:

$$\begin{aligned} &\rho \frac{D(\boldsymbol{\Pi}/\rho)}{Dt} + 2[\boldsymbol{\Pi} \cdot \nabla \boldsymbol{u}]^{(2)} + 2p[\nabla \boldsymbol{u}]^{(2)} = -\frac{p}{\eta} \boldsymbol{\Pi} q_{2nd}(\kappa), \\ &\rho \frac{D(\boldsymbol{Q}/\rho)}{Dt} + \frac{D\boldsymbol{u}}{Dt} \cdot \boldsymbol{\Pi} + \boldsymbol{Q} \cdot \nabla \boldsymbol{u} + \boldsymbol{\Pi} \cdot C_p \nabla T + pC_p \nabla T = -\frac{pC_p}{k} \boldsymbol{Q} q_{2nd}(\kappa), \end{aligned} \tag{7.33}$$

where the second-order dissipation term, q_{2nd}, and the first cumulant expansion term, κ, are given in terms of the hyperbolic sine form and a Rayleigh dissipation function [27], respectively;

$$q_{2nd}(\kappa) = \frac{\sin h\kappa}{\kappa}, \kappa = \frac{(mk_B)^{1/4}}{\sqrt{2}d}\frac{T^{1/4}}{p}\left(\frac{\boldsymbol{\Pi}:\boldsymbol{\Pi}}{2\eta} + \frac{\boldsymbol{Q}\cdot\boldsymbol{Q}}{kT}\right)^{1/2}.$$

In this expression, ρ, p, T, $\mathbf{u}$, E_t and D/Dt denote the density, the pressure, the temperature, the average velocity vector, the total energy, and the material time derivative, respectively. Higher-order non-conserved variables $\boldsymbol{\Pi}$ and $\boldsymbol{Q}$ are the shear stress tensor and the heat flux vector, respectively. The η, k represent the linear (Navier) viscosity and the linear (Fourier) thermal conductivity, respectively. The d, m, k_B denote the diameter of the molecule, the molecular mass, and the Boltzmann constant, respectively. The symbol $[]^{(2)}$ stands for a traceless symmetric part of the tensor and C_p represents the heat capacity per unit mass at constant pressure.

For the Couette flow, Myong [12] derived the following relation for the velocity profile:

$$u^*(s^*) = \Delta s^* \text{ from } \frac{du^*}{dy^*} = \frac{\Delta}{T^{*^n}} \text{ where } \Delta\left(\equiv \frac{du^*}{ds^*}\right) = -\Pi_0^* T_w^{*^n}. \tag{7.34}$$

The *velocity slope* Δ will be determined by combining the second-order NCCR constitutive and velocity slip models, repeated below:

$$u^*(s^* = 1/2) = 1 + \sigma_V \mathrm{Kn}\, \Pi^*]_{s^*=1/2} + \sigma_{VT}\frac{3}{4}\frac{(\gamma-1)}{\gamma/Pr\frac{k_w \Delta T/h}{pV} Q_x^*]_{s^*=1/2}}. \tag{7.35}$$

Note that a creeping term on the right-hand side appears since the tangential heat flux Q_x is generated by the driving force Q_{y_0}. Δ can be given by:

$$\frac{1}{2}\Delta = 1 + \Pi_{1/2}^*\left[\sigma_V \mathrm{Kn} + N_\delta^2 \sigma_{VT}\frac{3(\gamma-1)}{8\gamma}(1+\mathrm{Pr})\frac{\Delta^2}{q(\kappa_{1/2})T_w^{*^n}}\right], \tag{7.36}$$

where

$$\alpha_V \equiv \frac{1}{1 + 2\sigma_V \mathrm{Kn} T_w^{*^{-n}}/q(\kappa_{1/2})}, \tag{7.37}$$

$$A \equiv 1 + \sigma_{VT}\frac{9(\gamma-1)}{8\gamma}(1+\mathrm{Pr}),$$

$$\Delta = \left(r + \sqrt{g^3 + r^2}\right)^{1/3} + \left(r - \sqrt{g^3 + r^2}\right)^{1/3} + \frac{2}{3A},$$

where

$$g = \frac{1}{2\alpha_V A}\left(\frac{T_w^{*''}}{N_\delta/q(\kappa_{1/2})}\right)^2 - \left(\frac{2}{3A}\right)^2,$$

$$r = \frac{1}{2}\left(\frac{3}{A} - \frac{1}{\alpha_V A^2}\right)\left(\frac{T_w^{*''}}{N_\delta/q(\kappa_{1/2})}\right)^2 + \left(\frac{2}{3A}\right)^3$$

To validate the abnormal physical behaviors according to the new second-order theory in various regimes, DSMC solutions are obtained by using the Bird's Couette code with the diffusive wall condition. In the case of continuum theories, adjustable parameters of the velocity slip and temperature jump constants—chosen as best fit for the DSMC results—are assumed $\sigma_{V_C} = 0.8, \sigma_{T_C} = 1.2$ throughout for both of the Navier–Stokes–Fourier and second-order theories. As shown in Fig. 7.10, the velocity slope for a given Mach number decreases with respect to the Knudsen number in the transition regimes, but the velocity slope of the second-order theory (and the DSMC solutions of different Knudsen numbers 0.1, 0.5, 1.0, 2.5) remains greater than that of the Navier–Stokes–Fourier theory with varying coefficients and the difference increases with increasing Knudsen numbers, ultimately yielding the velocity gradient singularity. The ultimate reason behind this behavior is the shear-thinning property, that is, vanishing effective viscosity at high Knudsen number, embodied in the second-order constitutive relation.

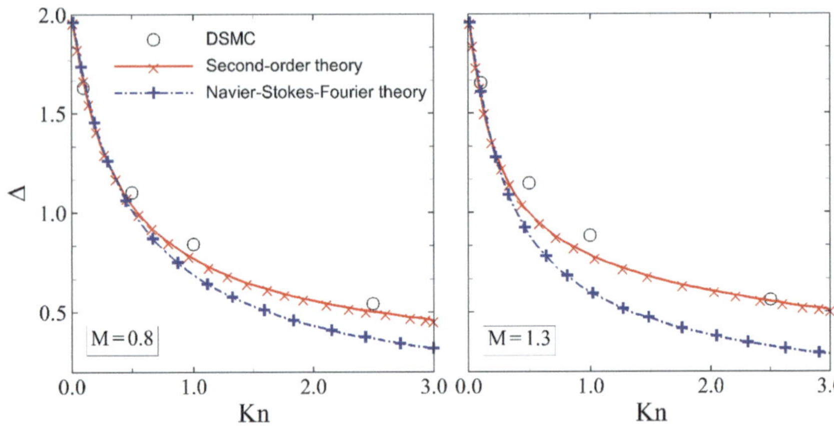

Fig. 7.10 The velocity slope Δ versus the Knudsen number in the plane Couette gas flow ($M = 0.8,\ 1.3,\ \sigma_{V_C} = 0.8,\ \sigma_{T_C} = 1.2$; Maxwellian). The present DSMC results obtained by the Bird's Couette code (Reproduced from Ref. [12])

7.7 Oscillatory Couette Flow

This section examines oscillatory Couette flow, which has multiple applications in micro-accelerometers, inertial sensors, and resonating filters. This problem can be expressed as a modified version of the second Stokes problem. In this flow, the upper wall has a velocity similar to Couette flow, which is oscillatory, see Fig. 7.11 [1].

Two dimensionless Knudsen and the Stokes numbers are influential in oscillatory Couette flow. The Stokes number represents the ratio of oscillatory forces to viscous forces and is defined by the relation (7.38):

$$\beta = \sqrt{\frac{h^2\omega}{\nu}} \tag{7.38}$$

The governing equation and slip boundary condition are:

$$\frac{\partial u}{\partial t} = \nu \frac{\partial^2 u}{\partial y^2} \tag{7.39}$$

$$u - u_w = C_1 \frac{2-\sigma_v}{\sigma_v} \lambda \frac{\partial u}{\partial y} \quad y = 0, h \tag{7.40}$$

If the upper wall velocity is $u = u_0 \sin \omega t$, the solution to the problem is the imaginary part of the function.

$$u(y,t) = Im\big[u(y)\exp(j\omega t)\big], \tag{7.41}$$

where we have $j^2 = -1$. Substituting this solution into the governing equation, we get:

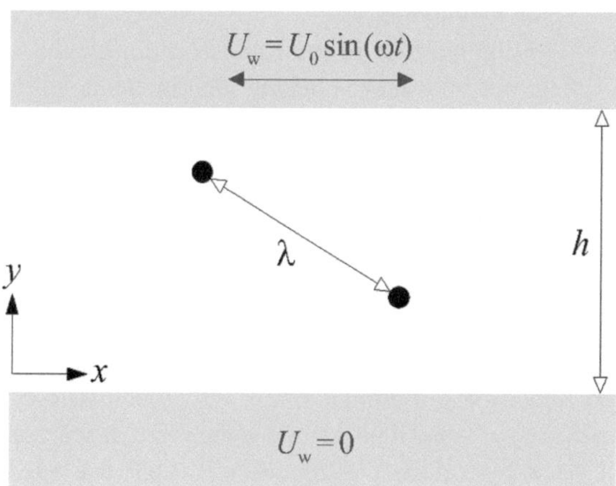

Fig. 7.11 Schematic of oscillatory Couette flow

$$j\omega u(y) = \nu \frac{\partial^2 u(y)}{\partial y^2} \tag{7.42}$$

Now, the resulting equation is only a function of y.

$$\frac{\partial^2 u(y)}{\partial y^2} - \psi^2 u(y) = 0 \tag{7.43}$$

$$\psi = \sqrt{\frac{j\omega}{\nu}} \tag{7.44}$$

By defining the solution $u(y)$ as follows:

$$u(y) = A_1 \sinh(\psi y) + A_2 \cosh(\psi y) \tag{7.45}$$

A_1 and A_2 are determined according to the boundary conditions. Therefore, the velocity profile for the imaginary part of the following function:

$$u(y,t) = \mathrm{Im}\left[\left(U_0 \frac{\sinh\left(\sqrt{j}\beta Y\right) + \sqrt{j}\beta C_1 Kn \cosh\left(\sqrt{j}\beta Y\right)}{\left(1 + j\beta^2 C_1^2 Kn^2\right)\sinh\left(\sqrt{j}\beta\right) + 2\sqrt{j}\beta C_1 Kn \cosh\left(\sqrt{j}\beta\right)}\right)\exp(j\omega t)\right] \tag{7.46}$$

where we have: $C_1 = 1.111$.

This is a general solution valid for all Stokes numbers in the continuum and slip regimes. Considering Fig. 7.12, the following points are noteworthy:

1. In the derived velocity function, as the Knudsen number increases, the velocity amplitude decreases because as the Knudsen number increases, the dissipation of the flow increases.
2. With the increase in the Knudsen number, the velocity profile inside the flow field becomes more uniform; that is, the amplitude of the velocity oscillations near the upper and lower walls will have little difference.
3. If $\beta < 0.25$, the analytical and DSMC solutions match perfectly across the entire range of Knudsen numbers.

In Fig. 7.13, all the velocity profiles pass through (0.5, 0.5). The so-called quasi-steady condition of oscillatory Couette flow is satisfied in these conditions. Under these conditions, the analytical solution provides a good accuracy up to a Knudsen number of less than 12, which can be confirmed in Fig. 7.12. In these conditions, the velocity distribution is linear with good accuracy, and the slip velocity for both stationary and oscillating walls is the same. If this condition is not met, the accuracy of the analytical Eq. (7.46) holds only up to a Knudsen number of 0.1.

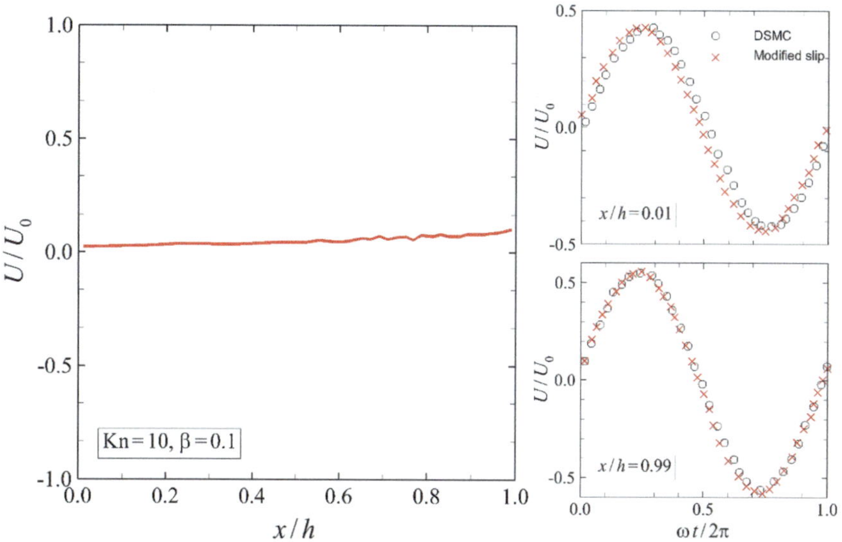

Fig. 7.12 Velocity diagram at a fixed location and time (Reproduced from Ref. [1])

Fig. 7.13 Velocity diagram versus location at different Knudsen numbers (Reproduced from Ref. [1])

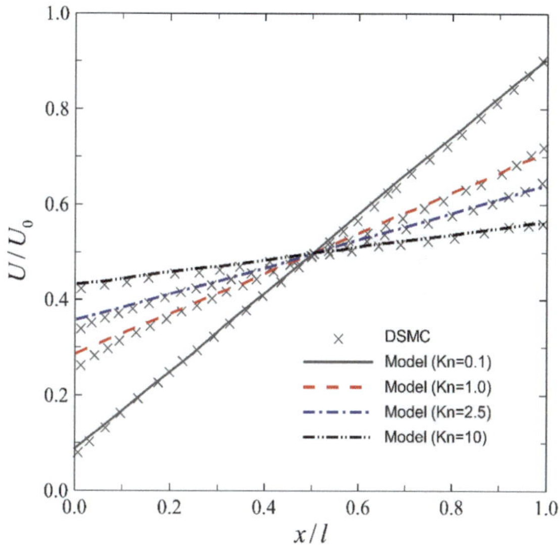

Figure 7.14 reports the dynamic response of the flow under the conditions Kn = 0.1 and b = 5. Figure 7.14a shows the velocity distribution at different times. Apart from the slip flow phenomenon, the dynamic behavior of the fluid resembles that of continuous flow. The velocity distribution predicted by the model of Eq. (7.46) aligns well with the DSMC results. The phase angle of the flow is defined as the time ratio in which the flow response lags or leads the applied wall velocity. A general

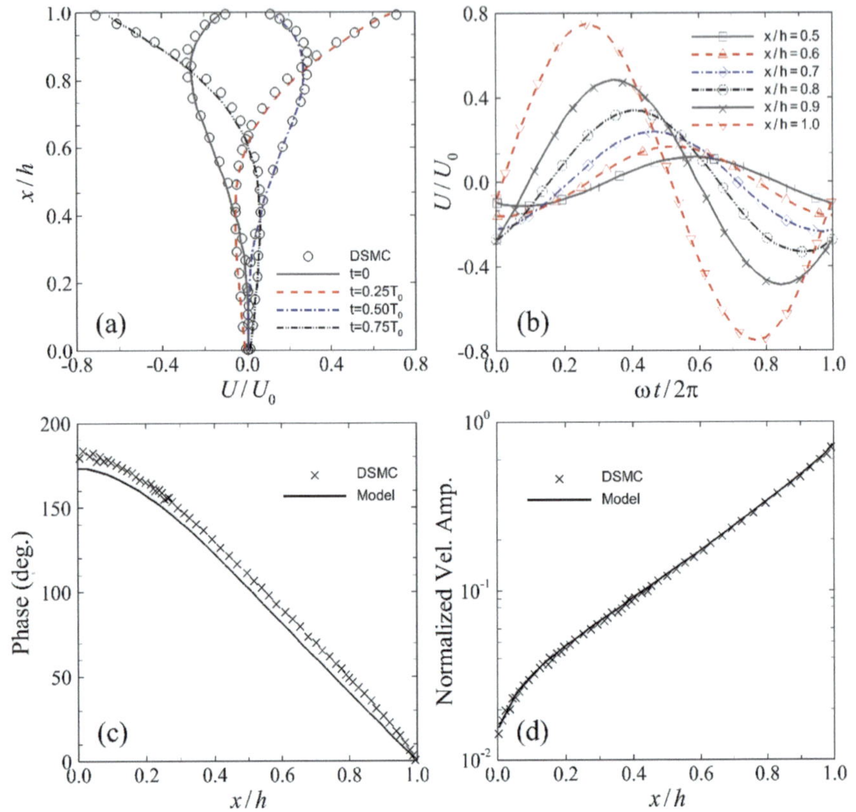

Fig. 7.14 Details of flow dynamics at Kn = 0.1 and β = 5; **a** dynamic response of medium, **b** velocity histories at various points, **c** phase variation throughout medium, and **d** log-plot of velocity amplitude (Reproduced from Ref. [1])

relation for determining the velocity response at any point Y is given by:

$$u(Y,t) = u_0\sin(\omega t + \Psi) \tag{7.47}$$

In this relation, Ψ is the phase angle. By expanding and rearranging Eq. (7.47), we get:

$$u(Y,t) = u_0[\sin(\omega t + \Psi) + \cos(\omega t)\sin\Psi = A(Y)\sin(\omega t) + B(Y)\cos(\omega t) \tag{7.48}$$

where

$$A(Y) = u_0\sin\Psi,\ B(Y) = u_0\cos\Psi$$

Therefore, the phase angle can be obtained as follows:

$$\Psi(Y) = \tan^{-1}\left[\frac{B(Y)}{A(Y)}\right] \tag{7.49}$$

Figure 7.14b shows the temporal history of the velocity at multiple points in the flow field. The velocity fields at different points exhibit a reduced amplitude compared to the wall velocity and multiple phases. Figure 7.14c compares the phase angle obtained from the present slip solution with the DSMC results. Both solutions show similar behavior. The changes in the phase angle at all points outside the Knudsen layer is linear. The wave propagation speed (phase velocity) can be extracted from the phase angle using the following relation:

$$c' = \omega / \left(\frac{\Delta\Psi}{\Delta Y}\right) \tag{7.50}$$

Therefore, the phase velocity is constant in regions where the phase angle changes linearly. The present slip model predicts a phase velocity of 1.770, while the DSMC method yields a value of 1.790. The classical Stokes solution predicts a phase velocity of 1.777. Near the wall, the phase velocity is not constant due to the presence of the boundary layer. Additionally, the phase velocity drops near stationary walls due to the interaction of the colliding and reflecting solutions. The dimensionless velocity amplitude on a logarithmic scale is shown in Fig. 7.14d. The results from the slip model and DSMC are in good agreement.

7.8 Cylindrical Couette Flow

In the cylindrical Couette problem, the flow is generated between two rotating cylinders, where the inner cylinder rotates with velocity ω_1 and the outer cylinder rotates with velocity ω_2 (see Fig. 7.15). The momentum equation is expressed as follows [13]:

$$\frac{d^2_{u_\theta}}{dr^2} + \frac{d}{dr}\left(\frac{u_\theta}{r}\right) = 0.0 \tag{7.51}$$

The slip boundary conditions are:

$$\begin{cases} u_\theta|_{r=R_1} = \omega_1 R_1 + \frac{2-\sigma_1}{\sigma_1}\lambda\left(\frac{du_\theta}{dr} - \frac{u_\theta}{r}\right)\Big)_{r=R1} \\ u_\theta|_{r=R_2} = \omega_2 R_2 - \frac{2-\sigma_2}{\sigma_2}\lambda\left(\frac{du_\theta}{dr} - \frac{u_\theta}{r}\right)\Big)_{r=R2} \end{cases}$$

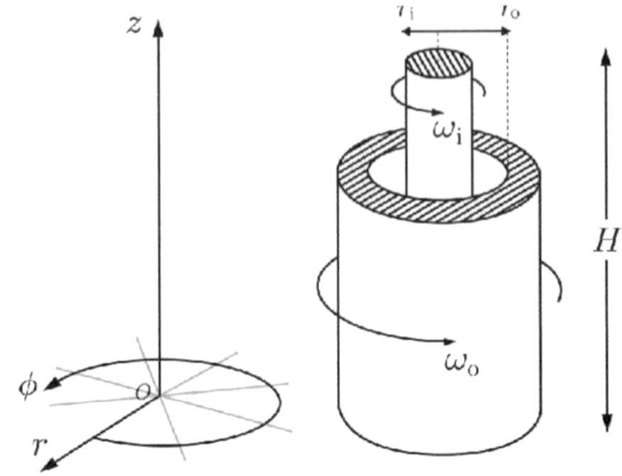

Fig. 7.15 Schematic diagram of the cylindrical Couette problem (usually considered two-dimensional $(r_o = R_2\, r_i = R_1)$)

The wall velocities are written as the term ωR, where ω is the angular velocity of the wall. The term inside the parentheses τ_{sn} corresponds to the shear stress in cylindrical coordinates, which is defined as $\tau_{sn} = \sigma_{r\theta} = \mu\left(\frac{1}{r}\frac{\partial u_r}{\partial \theta} + \frac{\partial u_\theta}{\partial r} - \frac{u_\theta}{r}\right)$. The term $\frac{1}{r}\frac{\partial u_r}{\partial \theta}$ is assumed to be zero as the flow is only angular. It is assumed that the momentum exchange coefficient of the inner and outer cylinders is not equal.

If we expand the momentum equation as follows, we obtain:

$$\frac{d^2 u_\theta}{dr^2} + \frac{1}{r}\frac{du_\theta}{dr} - \frac{u_\theta}{r^2} = 0$$

This equation is a Cauchy-Euler equation, which, by changing the variable $r = e^z$, its explanatory equation becomes:

$$D^n y = \frac{d^n u_\theta}{dz^n}$$

$$(D(D-1) + D - 1)u_\theta = 0 \Rightarrow D = \pm 1$$

$$u_\theta = ae^z + be^{-z} \overset{Z=Ln(r)}{\longrightarrow} u_\theta = ar + \frac{b}{r} \rightarrow$$

Therefore:

$$U(\theta) = ar + \frac{b}{r}$$

$$\begin{matrix} \frac{du}{dr} = a - \frac{b}{r^2} \\ -\frac{u(\theta)}{r} = -a - \frac{b}{r^2} \end{matrix} \Rightarrow \frac{du_\theta}{dr} - \frac{u_\theta}{r} = -\frac{2b}{r^2}$$

By applying boundary conditions:

$$\begin{cases} \omega_1 R_1 + \frac{(2-\sigma_1)}{\sigma_1}\lambda\left(-\frac{2b}{R_1^2}\right) = aR_1 + \frac{b}{R_1} \\ \omega_2 R_2 - \frac{(2-\sigma_2)}{\sigma_2}\lambda\left(-\frac{2b}{R_2^2}\right) = aR_2 + \frac{b}{R_2} \end{cases}$$

Thus:

$$\begin{cases} \omega_1 R_1 = aR_1 + \left(\frac{1}{R_1} + \left(\frac{2-\sigma_1}{\sigma_1}\right)\lambda\left(\frac{2}{R_1^2}\right)\right)b \\ \omega_2 R_2 = aR_2 + \left(\frac{1}{R_2} - \left(\frac{2-\sigma_2}{\sigma_2}\right)\lambda\left(\frac{2}{R_2^2}\right)\right)b \end{cases}$$

The solution is obtained using Cramer's rule:

$$a = \frac{\begin{vmatrix} \omega_1 R_1 & \frac{1}{R_1} + \left(\frac{2-\sigma_1}{\sigma_1} \cdot \frac{2\lambda}{R_1^2}\right) \\ \omega_2 R_2 & \frac{1}{R_2} - \left(\frac{2-\sigma_2}{\sigma_2} \cdot \frac{2\lambda}{R_2^2}\right) \end{vmatrix} / R_1 R_2}{\begin{vmatrix} R_1 & \frac{1}{R_1} + \left(\frac{2-\sigma_1}{\sigma_1} \times \frac{2\lambda}{R_1^2}\right) \\ R_2 & \frac{1}{R_2} - \left(\frac{2-\sigma_2}{\sigma_2} \cdot \frac{2\lambda}{R_2^2}\right) \end{vmatrix} / R_1 R_2}$$

The calculation of the denominator of the function a is:

$$\frac{\frac{R_1}{R_2} - R_1\left(\frac{2-\sigma_2}{\sigma_2} * \frac{2\lambda}{R_2^2}\right) - \frac{R_2}{R_1} + R_2\left(\frac{2-\sigma_1}{\sigma_1} * \frac{2\lambda}{R_1^2}\right)}{\mathrm{R_1 R_2}}$$

$$= \left[\frac{1}{R_2^2} - \left(\frac{1}{R_2} \times \left(\frac{2-\sigma_2}{\sigma_2} \times \frac{2\lambda}{R_2^2}\right)\right)\right] - \left[\frac{1}{R_1^2} + \left(\frac{1}{R_1}\left(\frac{2-\sigma_1}{\sigma_1} \times \frac{2\lambda}{R_1^2}\right)\right)\right]$$

$$\Rightarrow = \underbrace{\frac{1}{R_2^2}\left[1 - \left(\frac{2-\sigma_2}{\sigma_2} * \frac{2\lambda}{R_2}\right)\right]}_{(A)} - \underbrace{\frac{1}{R_1^2}\left[1 + \left(\frac{2-\sigma_1}{\sigma_1} * \frac{2\lambda}{R_1}\right)\right]}_{(B)}$$

The calculation of the numerator of the fraction a is:

$$\frac{R_1\omega_1 \times \left(\frac{1}{R_2} - \left(\frac{2-\sigma_2}{\sigma_2} \cdot \frac{2\lambda}{R_2^2}\right)\right) - R_2\omega_2 \times \left(\frac{1}{R_1} + \left(\frac{2-\sigma_1}{\sigma_1} \cdot \frac{2\lambda}{R_1^2}\right)\right)}{R_1 R_2}$$

$$= \omega_1 \times \underbrace{\frac{1}{R_2^2}\left(1 - \left(\frac{2-\sigma_2}{\sigma_2}\right) \cdot \left(\frac{2\lambda}{R_2}\right)\right)}_{A} - \omega_2 \times \underbrace{\frac{1}{R_1^2}\left(1 + \left(\frac{2-\sigma_1}{\sigma_1}\right) \cdot \left(\frac{2\lambda}{R_1}\right)\right)}_{B}$$

$$\omega_1 A - \omega_2 B \rightarrow a = \frac{\omega_1 A - \omega_2 B}{A - B}$$

The calculation of the b yields:

$$b = \frac{\begin{vmatrix} R_1 & \omega_1 R_1 \\ R_2 & w_2 R_2 \end{vmatrix}}{A\_B} \Big/ R_1 R_2 \rightarrow b = \frac{\omega_2 - \omega_1}{A - B} = \frac{\omega_1 - \omega_2}{B - A}$$

Note that the analytical solution obtained is valid for flow in the slip regime. With the increase in Knudsen number in the transition regime, alternative methods such as DSMC should be used. In Fig. 7.16, DSMC solution results are shown for a cylindrical Couette flow with an inner radius of 0.1875, a radius ratio of 5/3, a Knudsen number (based on the chamber radius) of 0.5, a Mach number (based on the inner cylinder speed) of 0.3, and a stationary outer cylinder. The results are plotted for two molecule-surface momentum exchange coefficients of 0.1 and 1.0 from three solvers. The difference in flow behavior for two cases is interesting.

At the momentum exchange coefficient of 0.1, the velocity on the inner wall is minimal and increases with a gentle slope towards the outer wall. This behavior, referred to as a special characteristic of rarefied flows in micro- and nano-dimensions, is called *velocity inversion.* However, at the momentum exchange coefficient of 1, the behavior is different, and as expected, the velocity decreases from the maximum value on the inner wall towards the outer wall. Figure 7.17 shows the velocity profile in the cylindrical Couette flow problem under conditions related to the previous figure, but both cylinders have a momentum exchange coefficient of 0.1. The radius of the inner cylinder is 3λ, and the radius of the outer cylinder is 5λ. The results for no-slip boundary conditions, using the common slip boundary condition (only including the term $\frac{du_\theta}{dr}$), using the complete slip boundary condition ($\frac{du_\theta}{dr} - \frac{u_\theta}{r}$), the analytical solution obtained in this section, and the DSMC results are presented [13].

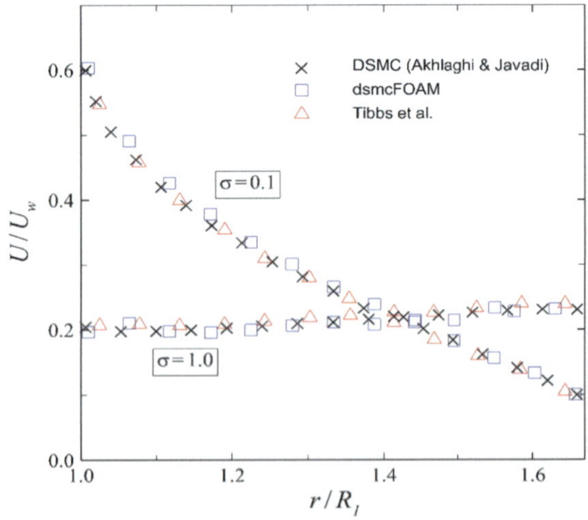

Fig. 7.16 Velocity distribution in the cylindrical Couette problem (Reproduced from Ref. [14])

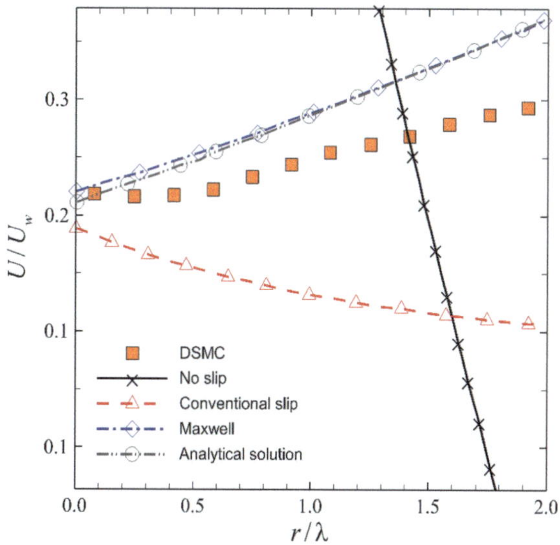

Fig. 7.17 Velocity distribution in the cylindrical Couette problem under different boundary conditions and the DSMC method (Reproduced from Ref. [13])

It is observed that the complete slip boundary condition can correctly predict the velocity inversion behavior correctly.

The maximum momentum exchange coefficient at which the velocity inversion phenomenon occurs is called the critical momentum exchange coefficient. Various researchers have proposed relations or results for calculating this coefficient. The analytical solution by Young et al. [13] from the Navier–Stokes equations with a slip boundary condition for the critical momentum exchange coefficient of the outer cylinder is given as follows:

$$\sigma_{2c} = 2\left(1 + \frac{R_2}{\lambda}\right)^{-1} \tag{7.52}$$

The above equation can be rewritten by defining the aspect ratio of the cylinders, $\chi = R_2/R_1$, as follows:

$$\sigma_{2c} = 2\left(1 + \frac{R_2}{\lambda}\right)^{-1} = 2\left(1 + \frac{\chi}{(\chi - 1)Kn}\right)^{-1} = \frac{2(\chi - 1)Kn}{(\chi - 1)Kn + \chi} \tag{7.53}$$

Aoki et al., using the linearized BGK model, proposed the following relation for the critical Knudsen number [15]:

$$\sigma_{2c} = \frac{\pi}{2}\gamma_1^* \frac{(\chi - 1)Kn}{\chi} \tag{7.54}$$

$\gamma_1^* = 1.270042$. The critical momentum exchange coefficient for the outer cylinder in the range $Kn < 0.5$ is plotted in Fig. 7.18, based on the linearized

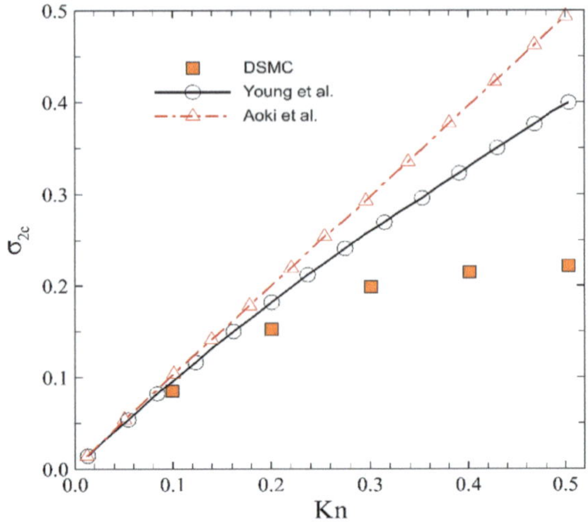

Fig. 7.18 Critical momentum exchange coefficient at Knudsen numbers less than 0.5 (Reproduced from Ref. [16])

Boltzmann solution, the analytical relation from Young et al., and the DSMC solution. According to the DSMC results, the critical exchange coefficient for the outer cylinder approaches 0.2 as the Knudsen number increases [16, 17].

7.9 Fourier Flow

The Fourier problem is one of the thermal flow problems caused by the temperature difference between the upper and lower plates of the channel [18]. It is one of the simplest conditions for studying the gas behavior under rarefaction conditions. In this flow, the gas is trapped between two infinite parallel plates separated by a distance of L. The plates are stationary, but their temperatures are unequal ($T_1 \neq T_2$). Starting from an arbitrary initial condition, the system reaches a steady state after a transient period in which the molecules have moved several times between the two walls. Once the steady state is achieved, there is a constant heat flux and the corresponding temperature gradient along the field. When the flow is in continuum conditions, according to Fourier's law, which is given by (7.55), the heat flux (q) is proportional to the temperature gradient in the central region of the solution domain.

$$q = -K(T)\frac{dT}{dx} \tag{7.55}$$

where $K(T)$ is the thermal conductivity coefficient as a function of temperature.

Figure 7.19 shows a schematic representation of the Fourier problem. Despite its apparent simplicity, the Fourier problem serves as a complex test for DSMC methods because the wall and field heat flux are related to the third-order moment of

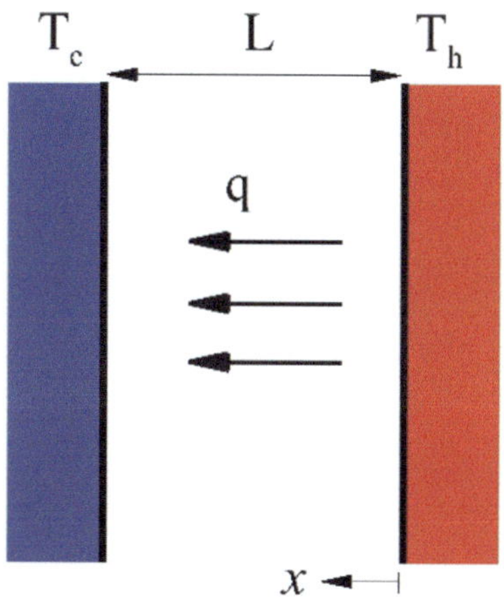

Fig. 7.19 Geometry and boundary conditions in Fourier flow

the fluctuating velocities of the simulated flow particles and converging to the correct value of heat flux in a low-speed flow is a challenging phenomenon.

In this section, the analytical solution of the Fourier flow is presented. Consider the following dimensionless equation [19, 20].

$$\frac{d}{dx^*}\left(k\frac{dT^*}{dx^*}\right) = 0 \tag{7.56}$$

The temperature T and the distance x are non-dimensionalized by T_h and L, respectively.

In the first step, assume that k is constant. We derive the analytical solution for the following boundary conditions. Where r is the temperature ratio of the cold wall to the hot wall $\left(r = \frac{T_c}{T_h}\right)$, and τ is the dimensionless temperature jump coefficient, which for a hard sphere gas model equals $\tau = 2.1904Kn$.

$$T^*(0) = 1 + \tau\frac{dT^*}{dx^*} \tag{7.57}$$

$$T^*(1) = r + \tau\frac{dT^*}{dx^*} \tag{7.58}$$

$$\frac{d}{dx^*}\left(k\frac{dT^*}{dx^*}\right) = 0 \to k = cte \to \frac{d^2T^*}{dx^{*2}} = 0 \to \frac{dT^*}{dx^*} = a \to T^* = ax^* + b$$

$$BC \to \begin{cases} T^*(0) = 1 + \tau\frac{dT^*}{dx^*} \to b = 1 + \tau a \\ T^*(1) = r + \tau\frac{dT^*}{dx^*} \to a + b = r + \tau a \end{cases} \to a = r - 1, b = \tau(r - 1) + 1$$

Relation (7.59) is obtained for the temperature distribution by substituting the values into the equation.

$$T^* = ax^* + b \rightarrow T^* = (r-1)x^* + \tau(r-1) + 1 = (r-1)\left(x^* + \tau\right) + 1 \quad (7.59)$$

If k is not constant and $k \propto T^{0.5}$ (hard sphere model):

$$k = \left(\frac{15}{32}\right) C_p \rho_0 \lambda c_m \sqrt{\pi} = \left(\frac{15}{32}\right)\left(\frac{\gamma k}{\gamma - 1}\right)\frac{\sqrt{RT}}{\sqrt{\pi} d^2}$$

The final solution, using boundary conditions (7.57) and (7.58), can be obtained as follows:

$$\frac{d}{dx^*}\left(\sqrt{T^*}\frac{dT^*}{dx^*}\right) = 0 \rightarrow \sqrt{T^*}\frac{dT^*}{dx^*} = C_1 \rightarrow \int \sqrt{T^*} dT^* = \int C_1 dx^*$$

$$\frac{{T^*}^{3/2}}{3/2} = C_1 x^* + C_2 \rightarrow {T^*}^{3/2} = \frac{3}{2C_1 x^*} + \frac{3}{2C_2} \rightarrow T^* = \left(ax^* + b\right)^{2/3} \rightarrow \frac{dT^*}{dx^*} = \frac{2}{3}a\left(ax^* + b\right)^{-\frac{1}{3}}$$

$$BC \rightarrow \begin{cases} T^*(0) = 1 + \tau\frac{dT^*}{dx^*} \\ T^*(1) = r + \tau\frac{dT^*}{dx^*} \end{cases} \rightarrow b = {T^*}^{\frac{3}{2}}(0),\, a = {T^*}^{\frac{3}{2}}(1) - {T^*}^{\frac{3}{2}}(0) \quad (7.60)$$

The obtained values are substituted into the above boundary conditions, and the following expressions are derived:

$$T^*(0) = 1 + \frac{2}{3}\tau\left({T^*}^{\frac{3}{2}}(1) - {T^*}^{\frac{3}{2}}(0)\right)\left({T^*}^{\frac{3}{2}}(0)\right)^{-\frac{1}{3}} \quad (7.61)$$

$$T^*(1) = r + \frac{2}{3}\tau\left({T^*}^{\frac{3}{2}}(1) - {T^*}^{\frac{3}{2}}(0)\right)\left({T^*}^{\frac{3}{2}}(1) - {T^*}^{\frac{3}{2}}(0) + {T^*}^{\frac{3}{2}}(0)\right)^{-\frac{1}{3}} \quad (7.62)$$

These are simplified as follows:

$$T^*(0) = 1 + \frac{2}{3}\tau\frac{T^*(1)^{3/2} - T^*(0)^{3/2}}{\sqrt{T^*(0)}} \quad (7.63)$$

$$T^*(1) = r + \sqrt{\frac{T^*(0)}{T^*(1)}}\left[1 - T^*(0)\right] \quad (7.64)$$

Therefore, under these conditions, the temperatures at the boundaries cannot be explicitly extracted, but they can be obtained using an iterative algorithm. The heat flux can be extracted from the following relation:

$$q = (5/16)\left(\frac{\gamma k}{\gamma - 1}\right)\frac{\sqrt{RT}}{\sqrt{\pi} d^2}\frac{a}{L}T_h\left(ax^* + b\right)^{-1/3} \quad (7.65)$$

Python code to solve analytical Fourier problem at low Kn is provided at the end of this chapter.

7.9.1 Comparison of the Analytical Fourier Prediction with the DSMC Solution

Here, we compare the DSMC solution with the analytical prediction of the Fourier equation [18]. We consider argon gas with the hard sphere molecular model under the conditions of $T_h = 373.15\text{K}$, $T_c = 173.15\text{K}$, $\Delta T = 200\text{K}$, $\text{Kn} = 0.001$. Figure 7.20 presents a comparison between the DSMC solution with the molecular collision method known as Simplified Bernoulli Trials (SBT) with one particle per cell in a mesh of 12,928 cells and the analytical solution (7.60). Since the Knudsen number is considered relatively small, the temperature jump value, and the size of the Knudsen layer region extend to a few mean free paths away from the surface. Considering these conditions, a very good agreement is observed between the molecular and the analytical solution.

Next, we increase the flow rarefaction of the flow. The one-dimensional Fourier problem is considered at $\text{Kn} = 0.024$ with $\Delta T = 100\text{K}$ ($T_c = 223.15\text{K}$, $T_h = 323.15\text{K}$). Figure 7.21 shows the temperature distribution from three different simulations is reported and compared with the analytical equation prediction (7.60). The temperature profile from the DSMC solution in the Knudsen layer region differs from that of the analytical solution. Furthermore, the discrepancy between the molecular and analytical solutions increases as we move toward the hot wall.

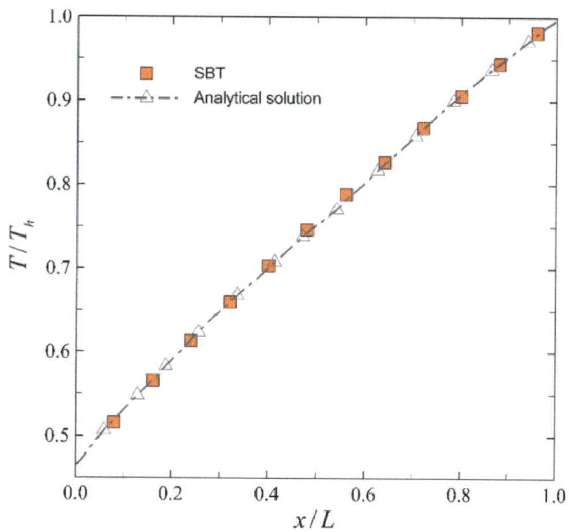

Fig. 7.20 Comparison between DSMC molecular method solution for wall temperature distribution at Kn = 0.001, ΔT = 200 K with analytical Fourier solution (Reproduced from Ref. [18])

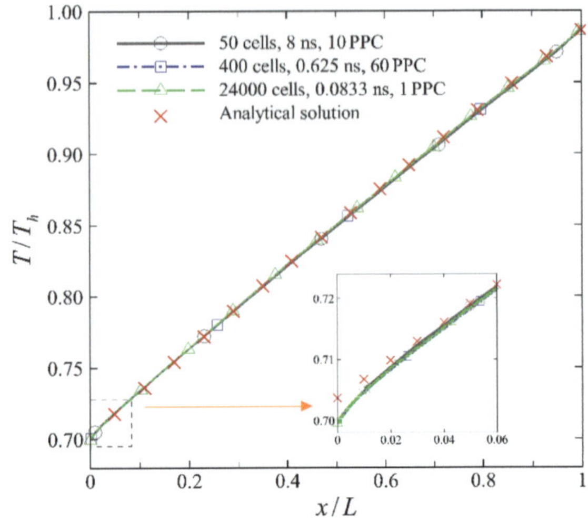

Fig. 7.21 Comparison between the DSMC molecular method solution for the wall temperature distribution at Kn = 0.024, ΔT = 200 K with the analytical Fourier solution (Reproduced from Ref. [18])

7.10 1D Steady Flow Between Two Parallel Plates in Free Molecular Regime

This section derives the density, temperature, and heat flux for Fourier flow in the free molecular regime. The schematic of the considered problem is shown in Fig. 7.22.

After detailed computations described in the appendix of this chapter, one derives the following non-dimensional $\left(T = \frac{T}{T_2},\ \ \rho = \frac{\rho}{\rho_{av}}\right)$ relations [21]:

$$\rho_1 = \left(\frac{T_2}{T_1}\right)^{1/2} \frac{(\alpha_1 + \alpha_2 - \alpha_1\alpha_2)}{\left[\alpha_2(2 - \alpha_1) + \alpha_1\left(\frac{T_2}{T_1}\right)^{1/2}(2 - \alpha_2)\right]} \tag{7.66}$$

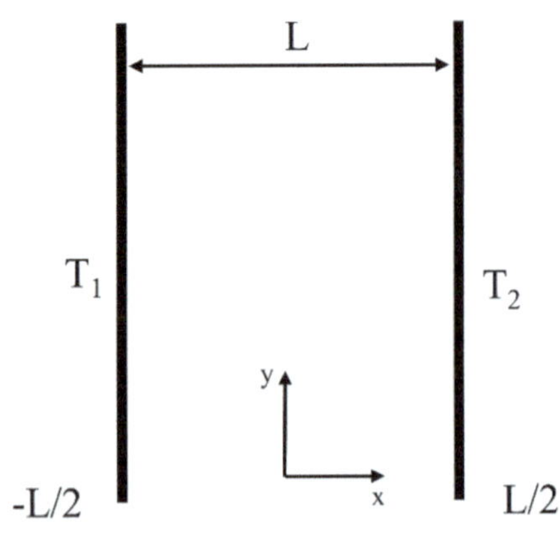

Fig. 7.22 Schematic of the Fourier flow problem considered at the free molecular regime

$$\rho_2 = \frac{(\alpha_1 + \alpha_2 - \alpha_1\alpha_2)}{\left[\alpha_2(2-\alpha_1) + \alpha_1\left(\frac{T_2}{T_1}\right)^{1/2}(2-\alpha_2)\right]} \tag{7.67}$$

$$T(x) = \frac{\left[\alpha_2(2-\alpha_1) + \alpha_1\left(\frac{T_1}{T_2}\right)^{1/2}(2-\alpha_2)\right]}{\left[\alpha_2(2-a_1) + a_1\left(\frac{T_2}{T_1}\right)^{1/2}(2-\alpha_2)\right]} \tag{7.68}$$

$$q(x) = \rho_{av}(2RT_2)^{3/2}\frac{1}{\sqrt{\pi}}\frac{a_1(2-\alpha_2)T_1 - \alpha_2(2-\alpha_1)T_2}{\alpha_2(2-a_1) + a_1(T_1T_2)^{1/2}(2-\alpha_2)}\frac{1}{T_2}, \tag{7.69}$$

where α_1 and α_2 are the accommodation coefficients of the right and left walls. Temperature and heat flux are constant in the domain of the free molecular regime. For $\alpha_1 = \alpha_2 = 1$, the above normalized relations simplify to:

$$\rho_1 = \left(\frac{T_2}{T_1}\right)^{1/2}\frac{1}{\left[1+\left(\frac{T_2}{T_1}\right)^{1/2}\right]} \tag{7.70}$$

$$\rho_2 = \frac{1}{\left[1+\left(\frac{T_2}{T_1}\right)^{1/2}\right]} \tag{7.71}$$

$$T(x) = \frac{\left[1+\left(\frac{T_1}{T_2}\right)^{\frac{1}{2}}\right]}{\left[1+\left(\frac{T_2}{T_1}\right)^{\frac{1}{2}}\right]} = \left(\frac{T_1}{T_2}\right)^{1/2} \tag{7.72}$$

$$q(x) = \rho_{av}(2RT_2)^{3/2}\frac{1}{\sqrt{\pi}}\frac{T_1 - T_2}{1+(T_1T_2)^{1/2}}\frac{1}{T_2} \tag{7.73}$$

Some other researchers considered heat transfer between parallel plates in free molecular regimes [21–25]. Sparrow and Kinney [22] suggested the following relation for the heat flux at free molecular regime:

$$q = \left[\frac{R}{2\pi}\right]^{1/2}\rho\frac{R(\gamma+1)}{(\gamma-1)}\frac{(1/\sigma_1 + 1/\sigma_2 - 1)}{(1/\alpha_1 + 1/\alpha_2 - 1)}(T_1 - T_2)$$
$$\left[\frac{2-\sigma_1}{\sigma_1}\left(T_2^*\right)^{-1/2} + \frac{2-\sigma_2}{\sigma_2}\left(T_1^*\right)^{-1/3}\right]^{-1} \tag{7.74}$$

$$T_1^* = \frac{[1-(1-\sigma_1)(1-\sigma_2)]\{\alpha_1[1-(1-\alpha_2)(1-\sigma_1)T_1 + \alpha_2(\sigma_1-\alpha_1)T_2\}}{\sigma_1\{[1-(1-\alpha_2)(1-\sigma_1)][1-(1-\alpha_1)(1-\sigma_2)] - (\sigma_1-\alpha_1)(\sigma_2-\alpha_2)\}}, \tag{7.75}$$

T_2^* has a similar relation with subscripts 1 and 2 interchanged. σ_1 and σ_2 are momentum accommodation coefficients and α_1 and α_2 are thermal accommodation coefficients. For fully diffuse/fully thermal accommodating conditions, the above relations simplify to:

$$q = \left[\frac{R}{2\pi}\right]^{1/2} \rho \frac{R(\gamma + 1)}{(\gamma - 1)}(T_1 - T_2)\left[\left(T_2^*\right)^{-1/2} + \left(T_1^*\right)^{-1/2}\right]^{-1} \tag{7.76}$$

$$T_1^* = T_1, T_2^* = T_2 \tag{7.77}$$

Fridlender [24] suggested the following relation:

$$\tau = (T_2/T_1)^{\frac{1}{2}},\ q_0 = \frac{1}{\sqrt{\pi}}\rho_0(2RT_1)^{\frac{3}{2}},\ q = q_0\tau|1 - \tau| \tag{7.78}$$

Or:

$$q = \frac{1}{\sqrt{\pi}}\rho_0(2RT_1)^{\frac{3}{2}}(T_2/T_1)^{\frac{1}{2}}[1 - (T_2/T_1)^{\frac{1}{2}}] \tag{7.79}$$

This is the same relation that Shen reported. Shen [25] reported a formula for the free-molecular heat flux between two parallel surfaces with complete thermal accommodation:

$$q = \frac{\rho(2R)^{\frac{3}{2}}\sqrt{T_1T_2}\left(\sqrt{T_1} - \sqrt{T_2}\right)}{\sqrt{\pi}}, \tag{7.80}$$

where $\rho = \rho_1 = \rho_2$. Figure 7.23 show the plots of Eq. (7.76) and (7.79) or (7.80). A comparison with the DSMC solution was also considered which shows excellent agreement with both graphs.

Figure 7.24 reports heat flux normalized to free molecular heat flux at various Kn magnitudes [26]. Interestingly, the maximum heat flux corresponds to free molecular regime.

7.11 Lid Driven Cavity

The driven cavity flow is a model problem with simple 2D geometry that can be related to many industrial applications [27, 28]. Lid driven cavity has been studied for different regimes by several authors [29–32]. Recently the attention of some researchers has been turned to non-classical thermal behaviors of lid-driven cavity

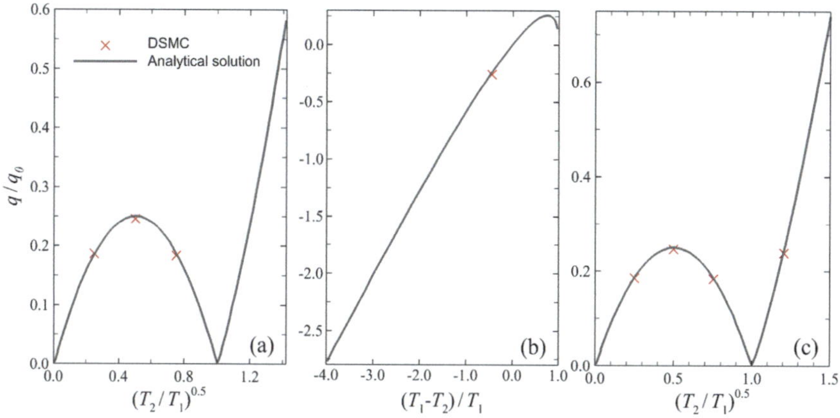

Fig. 7.23 Comparison of analytical free molecular equations with DSMC data; **a** plot of Eq. (7.73) and $T_1 = 700$ K, **b** plot of Eq. (7.76), and **c** plot of Eq. (7.79) or Eq. (7.80)

Fig. 7.24 Heat flow versus inverse Knudsen number in the linearized problem (Reproduced from Ref. [26])

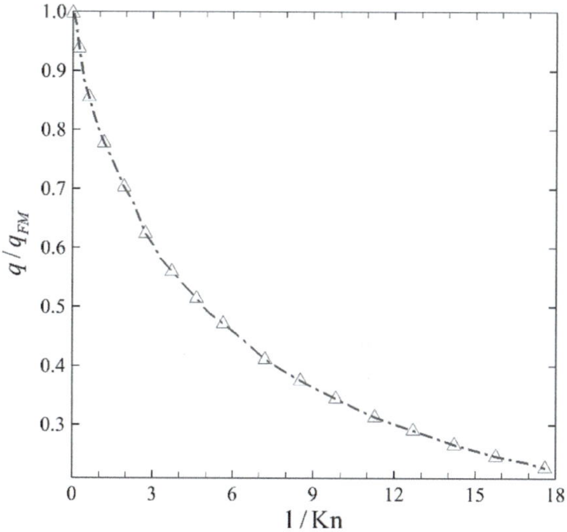

flow. Rana et al. [33] used regularized 13 moments equation to show that R13 equations are capable of capturing non-equilibrium thermal phenomena in the early transition regime inside a micro-cavity. John et al. [34] utilized the DSMC technique in the micro lid-driven cavity in the transition and free molecular regimes to show the cold-to-hot heat transfer. They attributed their observations to the non-equilibrium effects inside the flow field. John et al. [35] also studied the effect of incomplete surface accommodation on the heat transfer in the lid driven cavity. They reported that decreasing the thermal surface accommodation changes the isothermal wall cavity into the adiabatic wall cavity, and reduces the strength of sudden expansion in the driven cavity.

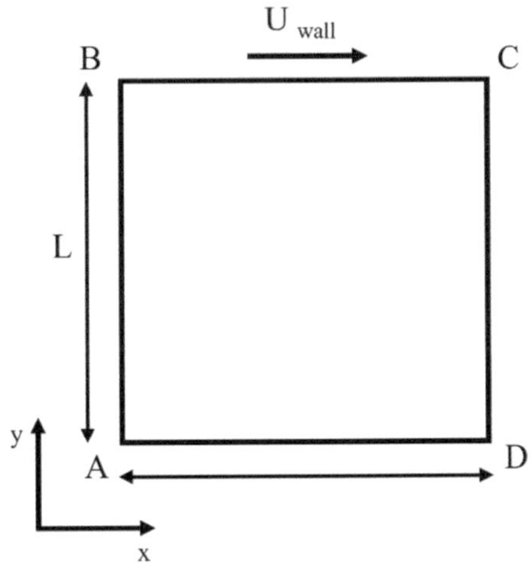

Fig. 7.25 Geometrical configuration of the micro- or nano-cavity

The micro driven cavity considered is shown in Fig. 7.25 [27]. Four corners of the cavity is denoted by A, B, C and D. The top driven lid moves in the positive x-direction at $U_{wall} = 100$ m/s. Cavity flow in early, middle and border of the slip regime, i.e., Kn = 0.005, 0.05 and 0.1 is considered in this study. The temperature of the walls are set equal to the reference temperature, $T_w = T_0 = 300$ K.

7.11.1 *Heat Flux Behavior*

Figure 7.26 shows the conductive heat flux lines overlaid on the temperature contours predicted by the Navier–Stokes (NS) and DSMC methods in three Knudsen numbers in the slip regime. Although these Knudsen numbers are in the "slip flow" range, according to the classification of flow regimes discussed in previous chapters, it does not mean that the NS with no-slip/jump boundary conditions give accurate results. The Knudsen numbers of 0.005, 0.05 and 0.1 are overall Knudsen numbers. However, only the local Knudsen number will define the validity of the NS or the continuum flow regime. In the considered cavities, chances are that the local Knudsen number will be higher than the overall Knudsen number. It is important to recall that the NS is valid when the flow is in thermodynamic equilibrium or it departs slightly from the thermodynamic equilibrium. As the Knudsen number increases, the departure from the thermodynamic equilibrium increases, and the first order shear stress and heat flux assumptions fail, where it is necessary to use the high order terms. As a result, the NS equations, even with traditional slip/jump boundary conditions, will not give good results at higher Kn cases for the investigated cavity.

For this particular cavity flow, the "driven force" of the flow is the top surface. Energy and momentum are transferred to the flow by the molecules that collide with the top surface. In an isothermal wall cavity, viscous dissipation serves as the primary mechanism for altering the gas temperature within the flow field. Acting consistently as a heat source, viscous dissipation is most pronounced in regions with strong shear stresses, particularly near the driven lid in the top-right corner of the cavity. Additionally, significant pressure variations near the top corners of the cavity influence the temperature distribution. Furthermore, non-equilibrium effects, which are dominant in the same regions, also impact the thermal field. The interplay of these factors collectively defines the thermal characteristics of the rarefied flow within the cavity.

In the early slip regime at Kn = 0.005, as shown in Fig. 7.26a, the predicted distribution of temperature obtained by the NS equations is in approximate agreement with the molecular approach. Both methods predict almost similar range of temperature variation in the flow field. In addition, the shapes of temperature contours are quite similar. The predicted direction of conductive heat flux by the DSMC is from hot to cold regions over the entire domain except for a small area near the top right corner. Temperature distribution in the middle slip regime obtained by the NS equations is shown in Fig. 7.26b. Maximum viscous dissipation occurs where the largest velocity gradients exist. As a result, the proximity of the top right corner, where the maximum shear stress occurs, is predicted as the hottest region by the NS equations. Figure 7.26b-right shows the temperature contour from the DSMC results at Kn = 0.05. It is seen that the maximum temperature for the DSMC solution is greater than the predicted value by the continuum approach. Moreover, the reduction in flow temperature near the left wall shows that the sudden expansion of rarefied flow in this region dominates the existing heat transfer mechanisms. Surprisingly, in the middle of slip regime, the conductive heat flux lines are from colder to the hotter regions in the upper half of the cavity. The assumed constitutive law of Fourier heat conduction, $q = -k\nabla T$, incorporated in the NS equations, cannot predict such direction for transfer of heat. This behavior is further enhanced in frame 7.26c at higher Kn, the left frame shows NS solution. Temperature distribution is very similar to the previous case; however, the maximum temperature is decreased. Smaller shear stress at Kn = 0.1 in comparison with Kn = 0.05 decreases viscous dissipation, the dominant heat generation mechanism in the NS equation, which in turn reduces the maximum temperature in the flow field.

Figure 7.26c-right shows the conductive heat flux lines in the cavity from the DSMC solution at Kn = 0.1. The conductive heat flux lines and temperature contours are quite similar to the previous test case; however, the minimum temperature is decreased by one degree that demonstrates stronger non-equilibrium effects in the top left corner of the cavity.

Clearly, the simple constitutive law of Fourier heat conduction in the NS equations fails to predict thermal characteristics of the flow even in the middle slip regime. As a result, the capability of the extended continuum-based methods in capturing non-equilibrium heat transfer phenomena was considered. The solution of the full regularized 13 moments (R13) equations for the driven cavity flow at Kn = 0.05. Description of the R13 governing equations is given in Ref. [33].

Figure 7.27a shows the conductive heat flux lines overlaid on the temperature distribution predicted by the R13 method. It is seen that the full R13 equations

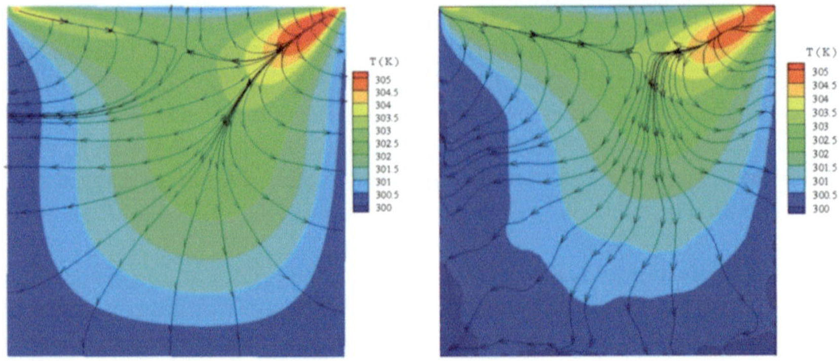

(a) Kn=0.005

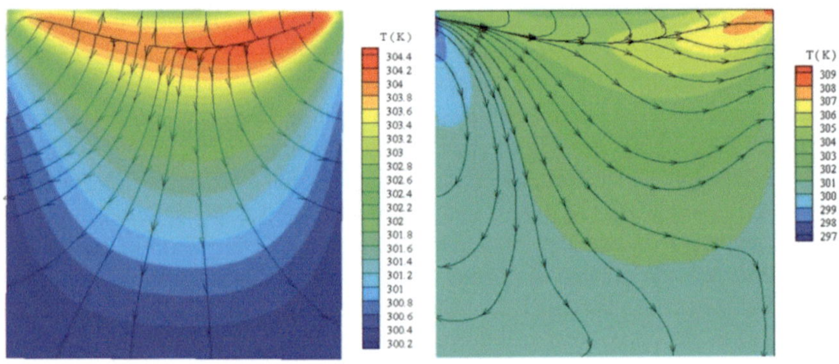

(b) Kn=0.05

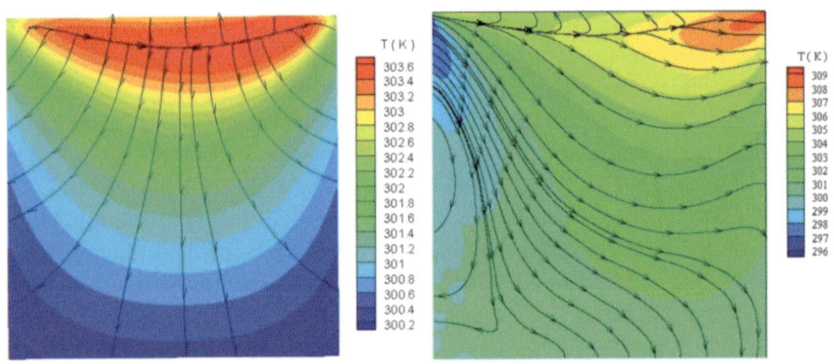

(c) Kn=0.1

Fig. 7.26 Conductive Heat flux lines overlaid on the temperature contour, left row: NS, right row: DSMC (Copied with permission from Ref. [27])

can capture cold-to-hot heat flux phenomenon in the middle of slip regime. This figure also illustrates that the predicted temperature distribution by the regularized 13 moment method are similar to the DSMC solution in Fig. 7.6. Figure 7.27b shows the comparison of non-dimensional vertical heat flux and temperature profile along Y/L = 0.8 and X/L = 0.8, respectively. In this figure, heat flux is non-dimensionalized by its maximum positive value along the horizontal line. It is well observed that the predicted temperature by the R13 methods is in good agreement with the DSMC solution. Although the heat flux profile exhibits a small error in the magnitude, the heat flux direction is correctly predicted. It is expected that employing higher moment methods, i.e., R26, provide closer agreement with DSMC solution [36, 37].

The sharp corners in the cavity geometry induce a bend in the velocity profile near the top corners, leading to an increase in the second derivative of velocity in these regions. The interaction between the temperature gradient and the second derivative of velocity dictates the direction of conductive heat flux within the cavity. When the flow temperature variation is minimal, owing to the relatively low lid velocity, the heat flux associated with the second derivative of velocity—cold-to-hot—dominates over the conventional Fourier heat flux. However, as the Knudsen number decreases, intermolecular collisions become more frequent, causing the diffusive heat flux, governed by the temperature gradient, to surpass the contribution from the velocity's second derivative. These observations align with the weakly nonlinear form of the Boltzmann equation, which attributes higher-order heat flux terms to the second derivative of velocity.

We also consider the relations derived for the heat flux in the asymptotic theory of the Boltzmann equation [5], presented by Sone. To this end, the velocity and temperature fields obtained using the DSMC method at Kn = 0.05 are employed to examine the capability of this method in capturing the cold-to-hot transfer phenomenon.

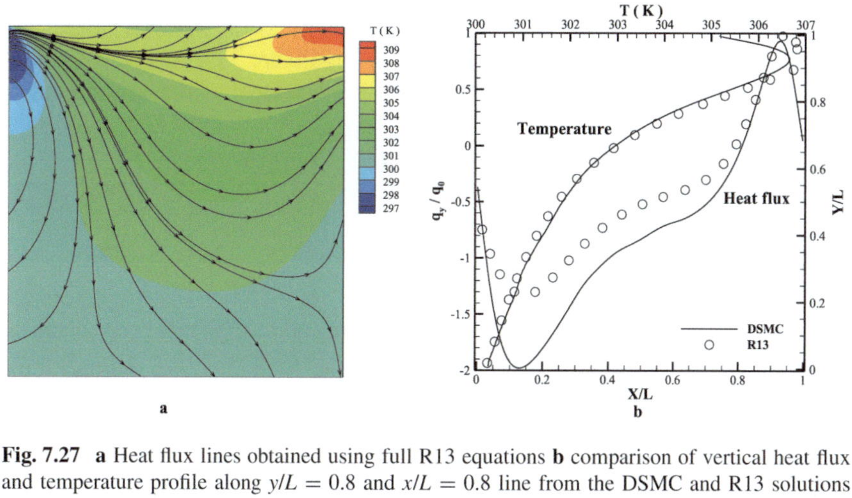

Fig. 7.27 **a** Heat flux lines obtained using full R13 equations **b** comparison of vertical heat flux and temperature profile along $y/L = 0.8$ and $x/L = 0.8$ line from the DSMC and R13 solutions (Copied with permission from Ref. [27])

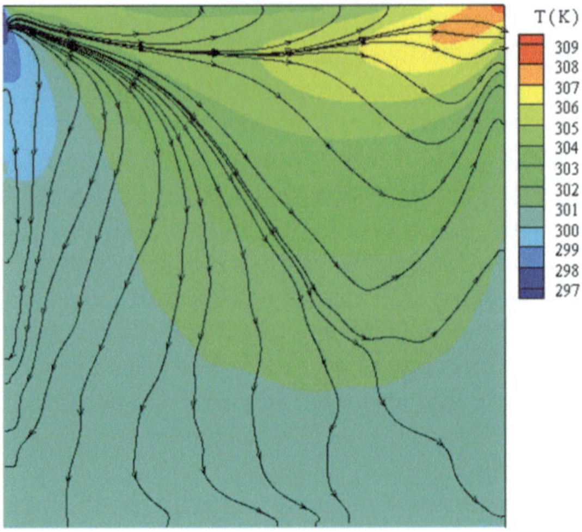

Fig. 7.28 Heat transfer lines obtained at Kn = 0.05 based on the heat flux relation in the asymptotic theory of the Boltzmann equation of Sone [5]

The equations are given by:

$$q_{is} = Kn^{**} q_{is1} + Kn^{**2} q_{is2} + Kn^{**3} q_{is3} \tag{7.81}$$

$$q_{is1} = 0 \tag{7.82}$$

$$q_{is2} = -5/4\delta_2 \frac{\partial T_{s1}^{**}}{\partial x_i} \tag{7.83}$$

$$q_{is3} = -5/4\delta_2 \frac{\partial T_{s2}^{**}}{\partial x_i} - 5/4\delta_5 T_{s1}^{**} \frac{\partial T_{s1}^{**}}{\partial x_i} + 0.5\gamma_3 \frac{\partial^2 u_{is1}}{\partial x_j^2}, \tag{7.84}$$

where $Kn^{**} = Kn\sqrt{\pi}/2$, $T^{**} = (T - T_0)/T_0$, and $h_s = h_{s1}Kn^{*} + h_{s2}Kn^{*2} + \ldots$ is a thermodynamics property expanded in terms of Kn. Figure 7.28 illustrates the heat transfer lines obtained using this method. As observed, these equations can predict the cold-to-hot heat transfer and accurately forecast heat transfer lines similar to those predicted by the DSMC method.

7.12 Summary

This chapter introduced the Couette flow in micro- and nano-scales, i.e., rarefied Couette flow. This flow's slip length, analytical solution in different rarefaction regimes, and shear stress were presented. Oscillatory Couette flow and cylindrical Couette flow were also reviewed. The rarefied Fourier flow was also reported, and a comparison between DSMC method predictions and analytical solutions at continuum limit and free molecular limit was provided. The end of this chapter considered cavity flow with an emphasis on cold to hot transfer and emphasis that

R13 and asymptotic theory of the Boltzmann equation of Sone can predict this phenomenon.

Problems

1. For the Kramer problem, the differential equations are:

   a. $\tau = -\mu \frac{dv}{dx} - \frac{2}{3}\frac{\mu^3}{P\rho}\frac{d^3v}{dx^3}$

   b. $$\tau = -\mu\frac{dv}{dx} - \frac{2}{5}\frac{\mu}{P}\frac{dq}{dx}$$
   $$q = B\frac{\mu^2}{P\rho}\frac{d^2q}{dx^2}$$

   Solve with the following boundary conditions:

$$v_{x=0} = -\lambda\frac{\tau}{\mu}\sqrt{\frac{2}{\pi}}$$
$$\frac{dv}{dx}_{x\to 0} = -1.7\frac{\tau}{\mu}$$
$$\frac{dv}{dx}_{x\to\infty} = -\frac{\tau}{\mu}$$

Note that the stress value is constant. The velocity boundary condition at infinity is obtained from the DSMC solution (Fig. 7.29). Show that the solution is equal to:

$$v = -\frac{\tau}{\mu}\left[x + \lambda\sqrt{\frac{2}{\pi}} + \frac{7}{10C}\lambda\left(1 - e^{-Cx/\lambda}\right)\right]$$

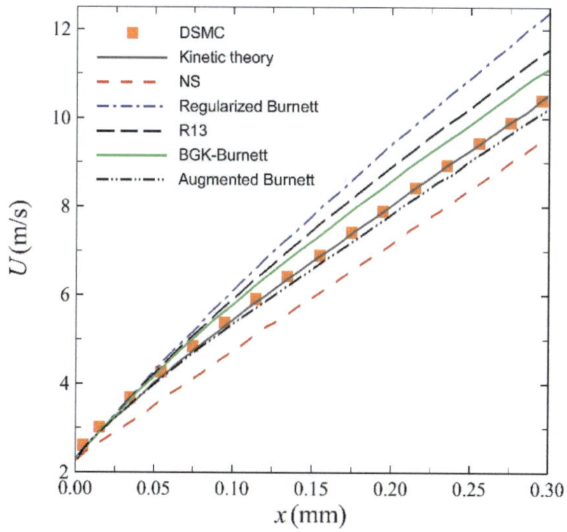

Fig. 7.29 Velocity distribution in the Kramer problem using different methods (Reproduced from Ref. [38])

2.A. The goal of the first section is to linearize the conservation equations. Consider the following dimensionless parameters. The dimensionless quantities are denoted by superscript (^).

$$T = T_0\left(1+\hat{T}\right) \quad v_i = \sqrt{RT_0}\hat{v}_i \quad \sigma_{ij} = \rho_0 RT_0 \hat{\sigma}_{ij} \quad q_i = \rho_0 \sqrt{RT_0}^{3} \hat{q}_i$$
$$= \rho_0(1+\hat{\rho})$$

$$x_i = L\hat{x}_i \quad t = \frac{L}{\sqrt{RT_0}}\hat{t} \quad \text{Kn} = \frac{\mu_0}{\rho_0\sqrt{RT_0}L}$$

The conservation equations are expressed as follows:

$$\frac{\partial\rho}{\partial t} + \nabla\cdot(\rho\boldsymbol{V}) = 0$$
$$\frac{\partial(\rho\boldsymbol{V})}{\partial t} + \nabla\cdot(\boldsymbol{V}(\rho\boldsymbol{V})) + \nabla p + \nabla\cdot\boldsymbol{\sigma} = 0$$
$$\frac{\partial(\rho E)}{\partial t} + \nabla\cdot(\boldsymbol{V}(\rho E)) + \nabla\cdot(\boldsymbol{V}p) + \nabla\cdot(\boldsymbol{\tau}\cdot\boldsymbol{V}) + \nabla\cdot\boldsymbol{q} = 0$$

where $E = e + \frac{|V^2|}{2}$.

Show that the above equations are linearized for a monatomic gas ($e = c_v T = 3/2\ T$) as follows. Use the guidance file at the end of the exercise and substitute the pressure in the momentum equation with temperature and density from the equation of state $p = \rho RT$. The superscript (^) has been removed from the dimensionless quantities for simplicity.

$$\frac{\partial\rho}{\partial t} + \nabla\cdot(\boldsymbol{V}) = 0.$$
$$\frac{\partial(\boldsymbol{V})}{\partial t} + \nabla\rho + \nabla T + \nabla\cdot\boldsymbol{\sigma} = 0$$
$$\frac{3}{2}\frac{\partial(T)}{\partial t} + \nabla\cdot(\boldsymbol{V}) + \nabla\cdot\boldsymbol{q} = 0$$

B. This section aims to use the linearized conservation equations and their analytical solutions in simple flows. Assume that the following relations give the linearized stress tensor and heat flux vector:

$$\sigma_{ij} = \sigma_{ij}^{(0)} + \text{Kn}\sigma_{ij}^{(1)} + \text{Kn}^2\sigma_{ij}^{(2)} + \text{Kn}^3\sigma_{ij}^{(3)} + \ldots$$
$$q_i = q_i^{(0)} + \text{Kn}q_i^{(1)} + \text{Kn}^2 q_i^{(2)} + \text{Kn}^3 q_i^{(3)} + \ldots$$
$$\sigma_{ij}^{(0)} = q_i^{(0)} = 0$$
$$\sigma_{ij}^{(1)} = -2\frac{\partial v_{<i}}{\partial x_{j>}} \quad q_i^{(1)} = -\frac{15}{4}\frac{\partial T}{\partial x_i}$$

$$\sigma_{ij}^{(2)} = -2\frac{\partial^2 \rho}{\partial x_{<i}\partial x_{j>}} + \frac{\partial^2 T}{\partial x_{<i}\partial x_{j>}} \quad q_i^{(2)} = -\frac{13}{4}\frac{\partial^2 v_k}{\partial x_k x_i} + \frac{3}{2}\frac{\partial^2 v_i}{\partial x_k \partial x_k}$$

$$\sigma_{ij}^{(3)} = \frac{5}{3}\frac{\partial^2}{\partial x_{<i}\partial x_{j>}}\frac{\partial v_k}{\partial x_k} - \frac{4}{3}\frac{\partial^2}{\partial x_k \partial x_i}\frac{\partial v_{<i}}{\partial x_{j>}}$$

$$q_i^{(3)} = -\frac{157}{16}\frac{\partial^3 T}{\partial x_i \partial x_k \partial x_k} - \frac{5}{8}\frac{\partial^3 \rho}{\partial x_i \partial x_k \partial x_k}$$

The angled brackets denote the symmetric, trace-free part of a second-order tensor. Considering the zeroth-order terms of stress and heat flux leads to the Euler equations, the zeroth and first terms of the Navier–Stokes equations, the zeroth, first, and second terms of the Brenner equations, all terms from zeroth to third of the Super-Burnett equations are obtained. Note that an index in the bracket indicates the deviatoric of that tensor. If the third terms are replaced with the following terms, the augmented Burnett equations are obtained:

$$\sigma_{ij}^{(a)} = \frac{1}{3}\frac{\partial^2}{\partial x_k x_k x_{j>}}\frac{\partial v_{<i}}{\partial x_{j>}} \quad q_i^{(a)} = \frac{11}{6}\frac{\partial^3 T}{\partial x_i x_k x_k} - \frac{5}{8}\frac{\partial^3 \rho}{\partial x_i x_k x_k}$$

C. For the one-dimensional Kramer problem with velocity $v(x)$, where the flow is isothermal and at constant pressure, show that the shear stress relation for the Super-Burnett and the augmented Burnett equations simplifies as follows:

$$\sigma_{xy} = -\mu\frac{dv}{dx} - \frac{2}{3}\frac{\mu^3}{\rho p}\frac{d^3 v}{dx^3}$$

$$\sigma_{xy} = -\mu\frac{dv}{dx} + \frac{1}{6}\frac{\mu^3}{\rho p}\frac{d^3 v}{dx^3}$$

D. Solve the above stress equation using the three boundary conditions given in the 2005 paper by Lockerbie, Rees, and Gliss.

$$v_{x=0} = -\lambda\frac{\tau}{\mu}\sqrt{\frac{2}{\pi}} \quad \frac{dv}{dx} = -\frac{\tau}{\mu} \;\; (x \to \infty) \quad \frac{dv}{dx} = -1.7\frac{\tau}{\mu} \;\; (x \to 0)$$

E. For the Couette flow, where only the component $v(x)$ is non-zero, show that the linearized conservation equations simplify as follows:

$$\frac{\partial \rho}{\partial x} + \frac{\partial T}{\partial x} + \frac{\partial \sigma_{11}}{\partial x} = 0 \quad \frac{\partial \sigma_{12}}{\partial x} = 0 \quad \frac{\partial q_1}{\partial x} = 0$$

Therefore:

$$\sigma_{12} = cte \quad q_1 = cte$$

The stresses for the Navier–Stokes equations are obtained as follows:

$$\sigma_{11} = 0 \quad \sigma_{12} = -\mathrm{Kn}\frac{\partial v}{\partial x} \quad q_1 = -\frac{15}{4}\mathrm{Kn}\frac{\partial T}{\partial x} \quad q_2 = 0$$

As a result, the solution to the conservation equations for the Navier–Stokes conditions is obtained as follows:

$$v = v_0 - \sigma_{12}\frac{x}{\mathrm{Kn}} \quad T = T_0 - \frac{4}{15}q_1\frac{x}{\mathrm{Kn}} \quad \rho = \rho_0 - \frac{4}{15}q_1\frac{x}{\mathrm{Kn}}$$

F. For the Burnett equations, the stresses are simplified as follows:

$$\sigma_{11} = -\frac{4}{3}Kn^2\frac{\partial^2 \rho}{\partial x^2} + \frac{2}{3}Kn^2\frac{\partial^2 T}{\partial x^2} \quad \sigma_{12} = -Kn\frac{\partial v}{\partial x}$$

$$q_1 = -\frac{15}{4}\mathrm{Kn}\frac{\partial T}{\partial x} \quad q_2 = \frac{3}{2}\mathrm{Kn}^2\frac{\partial^2 v}{\partial x^2}$$

Show that the velocity and temperature profiles are exactly as before are given by the following expression, but the density and stress profiles are different:

$$v(x) = v_0 - \sigma_{12}\frac{x}{\mathrm{Kn}} \quad T(x) = T_0 - \frac{4}{15}q_1\frac{x}{\mathrm{Kn}} \quad q_2 = 0$$

$$\rho = \rho_0 + \frac{4}{15}\frac{q_1}{\mathrm{Kn}}x + A\sinh\left[\frac{\sqrt{3}}{2}\frac{x-\frac{1}{2}}{\mathrm{Kn}}\right] + B\cosh\left[\frac{\sqrt{3}}{2}\frac{x-\frac{1}{2}}{\mathrm{Kn}}\right]$$

$$\sigma_{11} = -A\sinh\left[\frac{\sqrt{3}}{2}\frac{x-\frac{1}{2}}{\mathrm{Kn}}\right] - B\cosh\left[\frac{\sqrt{3}}{2}\frac{x-\frac{1}{2}}{\mathrm{Kn}}\right]$$

G. For the Super-Burnett equations, we have:

$$\sigma_{11} = -\frac{4}{3}\mathrm{Kn}^2\frac{\partial^2 \rho}{\partial x^2} + \frac{2}{3}\mathrm{Kn}^2\frac{\partial^2 T}{\partial x^2} \quad \sigma_{12} = -\mathrm{Kn}\frac{\partial v}{\partial x} - \frac{2}{3}\mathrm{Kn}^3\frac{\partial^3 v}{\partial x^3}$$

$$q_1 = -\frac{15}{4}\mathrm{Kn}\frac{\partial T}{\partial x} - \frac{157}{16}Kn^3\frac{\partial^3 T}{\partial x^3} - \frac{5}{8}Kn^3\frac{\partial^3 \rho}{\partial x^3}$$

$$q_2 = \frac{3}{2}\mathrm{Kn}^2\frac{\partial^2 v}{\partial x^2}$$

Also, for the augmented Burnett, we have:

$$\sigma_{12,aB} = -\mathrm{Kn}\frac{\partial v}{\partial x} + \frac{1}{6}Kn^3\frac{\partial^3 v}{\partial x^3}$$
$$q_{1,aB} = -\frac{15}{4}\mathrm{Kn}\frac{\partial T}{\partial x} + \frac{11}{6}\mathrm{Kn}^3\frac{\partial^3 T}{\partial x^3} - \frac{5}{8}\mathrm{Kn}^3\frac{\partial^3 \rho}{\partial x^3}$$

Show that, as a result, the velocity profile for the super-Burnett and augmented Burnett is as follows:

$$v_{SB} = v_0 - \sigma_{12}\frac{x}{\mathrm{Kn}} + A\cos\left[\frac{\sqrt{3}}{2}\frac{x-\frac{1}{2}}{\mathrm{Kn}}\right] + B\sin\left[\frac{\sqrt{3}}{2}\frac{x-\frac{1}{2}}{\mathrm{Kn}}\right]$$
$$v_{aB} = v_0 - \sigma_{12}\frac{x}{\mathrm{Kn}} + A\cosh\left[\frac{\sqrt{3}}{2}\frac{x-\frac{1}{2}}{\mathrm{Kn}}\right] + B\sinh\left[\frac{\sqrt{3}}{2}\frac{x-\frac{1}{2}}{\mathrm{Kn}}\right]$$

Show that the temperature profile for the super-Burnett and augmented Burnett is as follows:

$$T_{SB} = T_0 - \frac{4}{15}q_1\frac{x}{\mathrm{Kn}} + C\cos\left[\lambda_1\frac{x-\frac{1}{2}}{\mathrm{Kn}}\right] + D\sin\left[\lambda_1\frac{x-\frac{1}{2}}{\mathrm{Kn}}\right]$$
$$+F\cosh\left[\lambda_2\frac{x-\frac{1}{2}}{\mathrm{Kn}}\right] + G\sinh\left[\lambda_2\frac{x-\frac{1}{2}}{\mathrm{Kn}}\right]$$
$$\lambda_1 = \sqrt{\frac{3\sqrt{53129}-201}{1216}} = 0.635 \quad \lambda_2 = \sqrt{\frac{3\sqrt{53129}+201}{1216}} = 0.857$$
$$T = T_0 - \frac{4}{15}q_1\frac{x}{\mathrm{Kn}} + C\cosh\left[\gamma_1\frac{x-\frac{1}{2}}{\mathrm{Kn}}\right] + D\sinh\left[\gamma_1\frac{x-\frac{1}{2}}{\mathrm{Kn}}\right]$$
$$+F\cosh\left[\gamma_2\frac{x-\frac{1}{2}}{\mathrm{Kn}}\right] + G\sinh\left[\gamma_2\frac{x-\frac{1}{2}}{\mathrm{Kn}}\right]$$
$$\gamma_1 = \sqrt{\frac{3\sqrt{5081}+303}{128}} = 2.009 \quad \gamma_2 = \sqrt{\frac{303-3\sqrt{5081}}{1216}} = 0.835$$

Appendix 1: Linearization of the Euler Equations

If it is assumed that the flow is adiabatic and inviscid, and only small perturbations are propagating in the medium, the nonlinear governing equations for the unsteady one-dimensional flow simplify. This section extracts the governing equation for the propagation of weak waves, which will be derived by linearizing the mass, momentum, and energy conservation equations.

The conservation equations for the unsteady one-dimensional flow in differential form are as follows:

$$\begin{cases} \rho_t + (\rho u)_x = 0 \\ \rho u_t + \rho u u_x = -P_x T ds = de + P dv \\ \rho h_{0,t} + \rho u h_{0,x} = P_t \end{cases} \tag{7.85}$$

In this section, these equations will be linearized. For the linear solution, the small perturbation method is used. Here, the small changes in density are defined as $\tilde{S}$, which is equivalent to ε in the mathematical small perturbation method.

$$\tilde{S} = \frac{\Delta\rho}{\rho_1} \tag{7.86}$$

$$\begin{aligned} \rho &= \rho_1\left(1 + \tilde{S}\right) \\ \tilde{S} &= \frac{\rho - \rho_1}{\rho_1}, \tilde{S} \ll 1 \end{aligned} \tag{7.87}$$

It should be noted that in isentropic compressible flows, a relation between pressure and density can be obtained by using the definition of the speed of sound; therefore, there is no need to use the energy equation.

$$a = \sqrt{\frac{\gamma P}{\rho}} = \sqrt{\gamma RT} \tag{7.88}$$

$$P = P(\rho, s) \tag{7.89}$$

where s is the entropy, however, since the flow is isentropic, Eq. (8.56) simplifies to:

$$P = P(\rho) \tag{7.90}$$

$$\Rightarrow \frac{1}{\rho} P_x = \frac{1}{\rho} \frac{\partial P}{\partial \rho} \frac{\partial \rho}{\partial x} = \frac{a^2 \tilde{S}_x}{1 + \tilde{S}} \tag{7.91}$$

If we summarize the assumptions made:
28) Isentropic flow

$$2)\ \rho = \rho_1 + \Delta\rho \quad \Delta\rho \ll \rho_1$$

$$3)\ P = P_1 + \Delta P \quad \Delta P \ll P_1$$

$$4)\ O(u) = O\left(a_1 \tilde{S}\right) \quad \tilde{S} = \varepsilon$$

The continuity and momentum equations can be rewritten in terms of the parameter $\tilde{S}$ as follows:

$$\begin{cases} \tilde{S}_t + (1+\tilde{S})u_x + u\tilde{S}_x = 0 \\ u_t + uu_x + \frac{a^2}{1+\tilde{S}}\tilde{S}_x = 0 \end{cases} \quad (7.92)$$

The two continuity and momentum equations in the above relation are the transformed versions of the original equations, simplified under the assumption of isentropic flow and small perturbations for the motion of a weak moving wave.

Notes

From the expansion of the isentropic speed of sound relation, we have:

$$28)\ a^2 = \left.\frac{dP}{d\rho}\right)_s \approx \left.\frac{dP}{d\rho}\right|_1 + [\rho - \rho_1]\left.\frac{d^2P}{d^2\rho}\right| + \cdots \approx a_1^2 + O(\tilde{S}) + \cdots$$

Since $\tilde{S}$ is very small:

$$\Rightarrow a^2 \approx a_1^2$$

$$2)\ \frac{1}{1+\tilde{S}} \approx (1 - \tilde{S} + \cdots)$$

The third term of the momentum equation: $\frac{a^2}{1+\tilde{S}}\tilde{S}_x = a_1^2(1-\tilde{S})\tilde{S}_x \approx a_1^2\tilde{S}_x$

$$3)\text{Assume} \rightarrow u_1 = 0 \quad \Delta u = u \quad \frac{u\tilde{S}_x}{u_x} = \frac{u\frac{d\tilde{S}}{dx}}{\frac{du}{dx}} \approx \frac{u}{u}\tilde{S} \approx O(\tilde{S})$$

$$u_1 = 0 \rightarrow u\tilde{S}_x \ll u_x$$

$$4)\frac{uu_x}{a_1^2\tilde{S}_x} = \frac{u\Delta u}{a_1^2\Delta\tilde{S}} = \frac{u}{a_1}\frac{\Delta\left(\frac{u}{a}\right)}{\Delta\tilde{S}} = O(\tilde{S})$$

$$\Rightarrow \begin{cases} \text{Continuity}: \ \tilde{S}_t + u_x = 0 \\ \text{Momentum}: \ u_t + a_1^2\tilde{S}_x = 0 \end{cases}$$

The above equations are the linearized forms of the continuity and momentum equations.

Appendix 2: Details of Free Molecular Flow Derivation for Fourier Flow

The velocity distribution function (VDF) is given by:

$$f(x,\xi)=\frac{\rho}{(2\pi RT)^{3/2}}\exp(-\frac{\xi_x^2+\xi_y^2+\xi_z^2}{2RT}) \tag{7.93}$$

On the surfaces, we have the following VDF:
For positive wall $\xi_x > 0$:

$$f(-L/2,\xi)=\alpha_1\frac{\rho_1}{(2\pi RT_1)^{3/2}}\exp\left(-\frac{\xi_x^2+\xi_y^2+\xi_z^2}{2RT_1}\right) \\ +(1-\alpha_1)f\left(-L/2,-\xi_x+\xi_y+\xi_z\right) \tag{7.94}$$

For negative wall $\xi_x < 0$, we have

$$f(L/2,\xi)=\alpha_2\frac{\rho_2}{(2\pi RT_2)^{3/2}}\exp\left(-\frac{\xi_x^2+\xi_y^2+\xi_z^2}{2RT_2}\right) \\ +(1-\alpha_2)f\left(L/2,-\xi_x+\xi_y+\xi_z\right) \tag{7.95}$$

$$\rho_1=\left(\frac{2\pi}{RT_1}\right)^{1/2}\int\limits_{\xi_x>0}\xi f(-L/2,\xi)d\xi \tag{7.96}$$

$$\rho_2=\left(\frac{2\pi}{RT_2}\right)^{1/2}\int\limits_{\xi_x<0}\xi f(L/2,\xi)d\xi \tag{7.97}$$

The parameter α is the corresponding accommodation coefficient in the Maxwell diffuse-specular boundary condition. It means α_1 at the left boundary and α_2 at the right boundary. The first term α_1 is for diffuse reflection, second $(1-\alpha_1)$ is for specular reflection. Let non-dimensional quantities as follows:

$$x=\frac{x}{L},\ T=\frac{T}{T_2},\ \rho=\frac{\rho}{2\rho_{av}},\ \boldsymbol{\zeta}=\frac{\boldsymbol{\xi}}{(2RT_2)^{1/2}},\ f=\frac{(2RT_2)^{3/2}}{\rho_{av}}f$$

Then, we have:

$$f(-1/2,\zeta)=\alpha_1\frac{\rho_1}{\pi^{3/2}}\left(\frac{T_2}{T_1}\right)^{3/2}\exp\left(-\frac{T_2}{T_1}\left(\zeta_x^2+\zeta_y^2+\zeta_z^2\right)\right) \\ +(1-\alpha_1)f\left(-1/2,-\zeta_x+\zeta_y+\zeta_z\right) \tag{7.98}$$

$$f(1/2,\zeta) = \alpha_2 \frac{\rho_2}{\pi^{3/2}} \exp\Big(-\big(\zeta_x^2 + \zeta_y^2 + \zeta_z^2\big)\Big) + (1-\alpha_2) f\big(1/2, -\zeta_x + \zeta_y + \zeta_z\big), \tag{7.99}$$

and:

$$\rho_1 = \pi^{1/2} \left(\frac{T_2}{T_1}\right)^{1/2} \int\limits_{\xi_x > 0} \zeta f(-1/2,\zeta) d\xi \tag{7.100}$$

$$\rho_2 = \pi^{1/2} \int\limits_{\xi_x < 0} \zeta f(1/2,\zeta) d\xi \tag{7.101}$$

It is more straightforward to transform to the polar coordinates:

$$\xi_r = \sqrt{\xi_x^2 + \xi_y^2},\ \xi_x = \xi_r \cos\theta_\xi,\ \xi_y = \xi_r \sin\theta_\xi,\ \theta = \tan^{-1}\left(\frac{\xi_y}{\xi_x}\right)$$

Then, we introduce new variables:

$$F_1 = \frac{\rho_1}{\pi^{3/2}} \left(\frac{T_2}{T_1}\right)^{3/2} \exp\left(-\frac{T_2}{T_1}\big(\zeta_x^2 + \zeta_y^2 + \zeta_z^2\big)\right) \tag{7.102}$$

$$F_2 = \frac{\rho_2}{\pi^{3/2}} \exp\Big(-\big(\zeta_x^2 + \zeta_y^2 + \zeta_z^2\big)\Big) \tag{7.103}$$

$$F_1^- = f\big(-1/2, -\zeta_x, \zeta_y, \zeta_z\big) \tag{7.104}$$

$$F_2^- = f\big(1/2, -\zeta_x, \zeta_y, \zeta_z\big) \tag{7.105}$$

We have the equation system:

$$F_1^- = \alpha_2 F_2 + (1-\alpha_2) F_2^- \tag{7.106}$$

$$F_2^- = \alpha_1 F_1 + (1-\alpha_1) F_1^- \tag{7.107}$$

We solve:

$$\begin{aligned}
F_1^- &= \alpha_2 F_2 + (1-\alpha_2)\big(\alpha_1 F_1 + (1-\alpha_1) F_1^-\big) = \\
&= \alpha_2 F_2 + (1-\alpha_2)\alpha_1 F_1 + (1-\alpha_2)(1-\alpha_1) F_1^- \rightarrow \\
F_1^-&(1-(1-\alpha_2)(1-\alpha_1)) = \alpha_2 F_2 + (1-\alpha_2)\alpha_1 F_1 \\
F_2^- &= \alpha_1 F_1 + (1-\alpha_1)\alpha_2 F_2 + (1-\alpha_1)(1-\alpha_2) F_2^-
\end{aligned}$$

$$F_1^- = \frac{\alpha_2 F_2 + (1-\alpha_2)\alpha_1 F_1}{(\alpha_1 + \alpha_2 - \alpha_1\alpha_2)} \tag{7.108}$$

$$F_2^- = \frac{\alpha_1 F_1 + (1-\alpha_1)\alpha_2 F_2}{(\alpha_1 + \alpha_2 - \alpha_1\alpha_2)} \tag{7.109}$$

In the polar coordinate, the two plates are vertical with respect to the coordinate system, therefore, the integrals will be from $-\pi/2$ to $\pi/2$.

$$\begin{aligned}\rho_1 &= \pi^{1/2}\left(\frac{T_2}{T_1}\right)^{1/2} \int\limits_{|\theta_\zeta|<\pi/2} \zeta_r^2 \cos\theta_\zeta f(-1/2,\zeta)d\zeta_r d\theta_\zeta d\zeta_z \\ &= \rho_1 = \pi^{1/2}\left(\frac{T_2}{T_1}\right)^{1/2} \int\limits_{|\theta_\zeta|<\pi/2} \zeta_r^2 \cos\theta_\zeta F_1^- d\zeta_r d\theta_\zeta d\zeta_z\end{aligned} \tag{7.110}$$

Next, we calculate the integrals $\int \zeta_x F_1 d\vec{\zeta}$ and $\int \zeta_x F_2 d\vec{\zeta}$:

$$\begin{aligned}&\int\limits_{\pi/2\le\theta_\zeta<-\pi/2} \zeta_r^2 \cos\theta_\zeta F_2 d\zeta_r d\theta_\zeta d\zeta_z \\ &= \int\limits_{\pi/2\le\theta_\zeta<-\pi/2} \cos\theta_\zeta d\theta_\zeta . \frac{\rho_2}{\pi^{3/2}} \int_{-\infty}^{0} \zeta_r^2 \exp(-\zeta_r^2)d\zeta_r \int_{-\infty}^{\infty} \exp(-\zeta_z^2)d\zeta_z \\ &= \frac{\rho_2}{2\pi^{1/2}}\end{aligned}$$

$$\begin{aligned}&\int\limits_{|\theta_\zeta|<\pi/2} \zeta_r^2 \cos\theta_\zeta F_1 d\zeta_r d\theta_\zeta d\zeta_z \\ &= \frac{\rho_1}{\pi^{3/2}}\left(\frac{T_2}{T_1}\right)^{3/2} \int\limits_{|\theta_\zeta|<\pi/2} \zeta_r^2 \cos\theta_\zeta \exp\left(-\frac{T_2}{T_1}(\zeta_r^2+\zeta_z^2)\right)d\zeta_r d\theta_\zeta d\zeta_z = \\ &= \frac{2\rho_1}{\pi^{\frac{3}{2}}}\left(\frac{T_2}{T_1}\right)^{\frac{3}{2}} \int_0^{\infty} \zeta_r^2 \exp\left(-\frac{T_2}{T_1}\zeta_r^2\right)d\zeta_r \int_{-\infty}^{\infty} \exp\left(-\frac{T_2}{T_1}\zeta_z^2\right)d\zeta_z \int_{|\theta_\zeta|<\pi/2} \cos\theta_\zeta d\theta_\zeta = \\ &= \frac{2\rho_1}{\pi^{3/2}}\left(\frac{T_2}{T_1}\right)^{3/2} \frac{\pi^{1/2}}{4\left(\frac{T_2}{T_1}\right)^{3/2}} 2\frac{\pi^{1/2}}{2\left(\frac{T_2}{T_1}\right)^{1/2}} = \frac{\rho_1}{2\pi^{1/2}}\left(\frac{T_1}{T_2}\right)^{1/2}\end{aligned}$$

In Cartesian coordinate, we will have:

$$\int \zeta_x F_2 d\zeta_x d\zeta_y d\zeta_z$$

$$= \frac{\rho_2}{\pi^{3/2}} \overbrace{\int_{-\infty}^{0} \zeta_x \exp\left(-\zeta_x^2\right) d\zeta_x}^{(-)} \overbrace{\int_{\infty}^{-\infty} \exp\left(-\zeta_y^2\right) d\zeta_y}^{(-)} \int_{-\infty}^{\infty} \exp\left(-\zeta_z^2\right) d\zeta_z = \frac{\rho_2}{2\pi^{1/2}}$$

$$\int \zeta_x F_1 d\zeta_x d\zeta_y d\zeta_z$$

$$= \frac{\rho_1}{\pi^{3/2}} \left(\frac{T_2}{T_1}\right)^{3/2} \int_{0}^{\infty} \zeta_x \exp\left(-\frac{T_2}{T_1}\zeta_x^2\right) d\zeta_x \int_{-\infty}^{\infty} \exp\left(-\frac{T_2}{T_1}\zeta_y^2\right) d\zeta_y \int_{-\infty}^{\infty} \exp\left(-\frac{T_2}{T_1}\zeta_z^2\right) d\zeta_z$$

$$= \frac{\rho_1}{2\pi^{1/2}} \left(\frac{T_1}{T_2}\right)^{1/2}$$

Replacing to (7.110), we will find:

$$\rho_1 = \pi^{1/2} \left(\frac{T_2}{T_1}\right)^{1/2} \left(\frac{\alpha_2}{(\alpha_1 + \alpha_2 - \alpha_1\alpha_2)} \frac{\rho_2}{2\pi^{1/2}} + \frac{(1-\alpha_2)\alpha_1}{(\alpha_1 + \alpha_2 - \alpha_1\alpha_2)} \frac{\rho_1}{2\pi^{1/2}} \left(\frac{T_1}{T_2}\right)^{1/2} \right)$$

$$\rho_1 = \frac{(1-\alpha_2)\alpha_1\rho_1}{2(\alpha_1 + \alpha_2 - \alpha_1\alpha_2)} + \frac{\alpha_2\rho_2}{2(\alpha_1 + \alpha_2 - \alpha_1\alpha_2)} \left(\frac{T_2}{T_1}\right)^{1/2}$$

$$\rho_1(\alpha_1 + \alpha_2 - \alpha_1\alpha_2) - (1-\alpha_2)\alpha_1\rho_1 = \alpha_2\rho_2 \left(\frac{T_2}{T_1}\right)^{1/2} \tag{7.111}$$

$$\rho_1 = \rho_2 \left(\frac{T_2}{T_1}\right)^{1/2}$$

For $\zeta < 0$, we have:

$$f(x,\zeta) = \alpha_2 \frac{\rho_2}{\pi^{3/2}} \exp\left(-\left(\overbrace{\zeta_x^2 + \zeta_y^2}^{\zeta_r^2} + \zeta_z^2\right)\right) + (1-\alpha_2) f\left(1/2, -\zeta_x, \zeta_y, \zeta_z\right) =$$

$$= \alpha_2 F_2 + (1-\alpha_2) \frac{\alpha_1 F_1 + (1-\alpha_1)\alpha_2 F_2}{(\alpha_1 + \alpha_2 - \alpha_1\alpha_2)} =$$

$$= \frac{(\alpha_1 + \alpha_2 - \alpha_1\alpha_2)\alpha_2 F_2 + (1-\alpha_2)(\alpha_1 F_1 + (1-\alpha_1)\alpha_2 F_2)}{(\alpha_1 + \alpha_2 - \alpha_1\alpha_2)}$$

$$= \frac{\alpha_2 F_2 + \alpha_1(1-\alpha_2) F_1}{(\alpha_1 + \alpha_2 - \alpha_1\alpha_2)}$$

$$f^-(x,\zeta) = \frac{\rho_2}{\pi^{3/2}} \left[\frac{\alpha_2 \exp\left(-\left(\zeta_x^2 + \zeta_y^2 + \zeta_z^2\right)\right)}{\alpha_1 + \alpha_2 - \alpha_1\alpha_2} \right.$$

$$+\frac{\alpha_1(1-\alpha_2)\left(\frac{T_2}{T_1}\right)^2\exp\left(-\frac{T_2}{T_1}\left(\zeta_x^2+\zeta_y^2+\zeta_z^2\right)\right)}{\alpha_1+\alpha_2-\alpha_1\alpha_2}\Bigg] \tag{7.112}$$

For $\zeta > 0$

$$\begin{aligned}
f(x,\zeta) &= \alpha_1\frac{\rho_1}{\pi^{3/2}}\left(\frac{T_2}{T_1}\right)^{3/2}\exp\left(-\frac{T_2}{T_1}\left(\zeta_x^2+\zeta_y^2+\xi_z^2\right)\right)\\
&\quad + (1-\alpha_1)f(-1/2,-\zeta_x,\zeta_y,\zeta_z)\\
&= \alpha_1F_1 + (1-\alpha_1)F_1^- = \alpha_1F_1 + (1-\alpha_1)\frac{\alpha_2F_2+(1-\alpha_2)\alpha_1F_1}{(\alpha_1+\alpha_2-\alpha_1\alpha_2)}\\
&= \frac{\alpha_1F_1+\alpha_2(1-\alpha_1)F_2}{(\alpha_1+\alpha_2-\alpha_1\alpha_2)}\\
f^+(x,\zeta) &= \frac{\rho_2}{\pi^{3/2}}\left[\frac{\alpha_1\left(\frac{T_2}{T_1}\right)^2\exp\left(-\frac{T_2}{T_1}\left(\zeta_x^2+\zeta_y^2+\zeta_z^2\right)\right)}{\alpha_1+\alpha_2-\alpha_1\alpha_2}\right.\\
&\quad \left.+\frac{\alpha_2(1-\alpha_1)\exp\left(-\left(\zeta_x^2+\zeta_y^2+\zeta_z^2\right)\right)}{\alpha_1+\alpha_2-\alpha_1\alpha_2}\right]
\end{aligned} \tag{7.113}$$

In order to fully describe the DVF, it is left to calculate ρ_1 and ρ_2 (densities of the molecules scattered diffusely from the walls 1 and 2, respectively).

The first relationship is already found as: $\rho_1 = \rho_2\left(\frac{T_2}{T_1}\right)^{1/2}$.

It is also known that the average density should be equal to 1 (in non-dimensional form):

$$\rho_{av} = \frac{1}{L}\int_{-L/2}^{L/2}\rho dx = 1 \Rightarrow \rho = 1 \tag{7.114}$$

Density:

$$\begin{aligned}
\rho(x) &= \int f\left(x,\vec{\zeta}\right)d\vec{\zeta} =\\
&= \frac{\rho_2}{\pi^{3/2}(\alpha_1+\alpha_2-\alpha_1\alpha_2)}\left[\alpha_1\left(\frac{T_2}{T_1}\right)^2 J_1+\alpha_2(1-\alpha_1)J_2\right]+\\
&\quad + \frac{\rho_2}{\pi^{3/2}(\alpha_1+\alpha_2-\alpha_1\alpha_2)}\left[\alpha_2J_2+\alpha_1(1-\alpha_2)\left(\frac{T_2}{T_1}\right)^2 J_1\right]
\end{aligned}$$

$$J_1 = \iint\int \exp\left(-\frac{T_2}{T_1}\left(\zeta_x^2 + \zeta_y^2 + \zeta_z^2\right)\right) d\zeta_x d\zeta_y d\zeta_z =$$

$$= \int_0^\infty \exp\left(-\frac{T_2}{T_1}\zeta_x^2\right) d\zeta_x \int_{-\infty}^\infty \exp\left(-\frac{T_2}{T_1}\zeta_y^2\right) d\zeta_y \int_{-\infty}^\infty \exp\left(-\frac{T_2}{T_1}\zeta_z^2\right) d\zeta_z =$$

$$= \int_0^\infty \exp\left(-\frac{T_2}{T_1}\zeta_x^2\right) d\zeta_x \int_{-\infty}^\infty \exp\left(-\frac{T_2}{T_1}\zeta_y^2\right) d\zeta_y \int_{-\infty}^\infty \exp\left(-\frac{T_2}{T_1}\zeta_z^2\right) d\zeta_z =$$

$$= \frac{\sqrt{\pi}}{2\left(\frac{T_2}{T_1}\right)^{1/2}} 2 \frac{\sqrt{\pi}}{2\left(\frac{T_2}{T_1}\right)^{1/2}} 2 \frac{\sqrt{\pi}}{2\left(\frac{T_2}{T_1}\right)^{1/2}} = \frac{\pi^{3/2}}{2}\left(\frac{T_1}{T_2}\right)^{3/2}$$

$$J_2 = \iint\int \exp\left(-\left(\zeta_x^2 + \zeta_y^2 + \zeta_z^2\right)\right) d\zeta_x d\zeta_y d\zeta_z =$$

$$= \int_0^\infty \exp(-\zeta_x^2) d\zeta_x \int_{-\infty}^\infty \exp\left(-\zeta_y^2\right) d\zeta_y \int_{-\infty}^\infty \exp(-\zeta_z^2) d\zeta_z =$$

$$= \frac{\sqrt{\pi}}{2} 2 \frac{\sqrt{\pi}}{2} 2 \frac{\sqrt{\pi}}{2} = \frac{\pi^{3/2}}{2}$$

$$\rho(x) = \int f\left(x, \vec{\zeta}\right) d\vec{\zeta} =$$

$$= \frac{\rho_2}{\pi^{3/2}(\alpha_1 + \alpha_2 - \alpha_1\alpha_2)}\left[\alpha_1\left(\frac{T_2}{T_1}\right)^2 \frac{\pi^{3/2}}{2}\left(\frac{T_1}{T_2}\right)^{3/2} + \alpha_2(1-\alpha_1)\frac{\pi^{3/2}}{2}\right] +$$

$$+ \frac{\rho_2}{\pi^{3/2}(\alpha_1 + \alpha_2 - \alpha_1\alpha_2)}\left[\alpha_2\frac{\pi^{3/2}}{2} + \alpha_1(1-\alpha_2)\left(\frac{T_2}{T_1}\right)^2 \frac{\pi^{3/2}}{2}\left(\frac{T_1}{T_2}\right)^{3/2}\right]$$

$$\rho(x) = \frac{\rho_2}{(\alpha_1 + \alpha_2 - \alpha_1\alpha_2)}\left[\alpha_2(2 - a_1) + a_1\left(\frac{T_2}{T_1}\right)^{1/2}(2 - \alpha_2)\right] \tag{7.115}$$

To calculate ρ_2:

$$\rho(x) = \frac{\rho_2}{(\alpha_1 + \alpha_2 - \alpha_1\alpha_2)}\left[\alpha_2(2 - a_1) + a_1\left(\frac{T_2}{T_1}\right)^{1/2}(2 - \alpha_2)\right] = 1$$

$$\rho_2 = \frac{(\alpha_1 + \alpha_2 - \alpha_1\alpha_2)}{\left[\alpha_2(2 - a_1) + a_1\left(\frac{T_2}{T_1}\right)^{1/2}(2 - \alpha_2)\right]} \tag{7.116}$$

To calculate ρ1:

$$\rho_1 = \rho_2 \left(\frac{T_2}{T_1}\right)^{1/2}$$

$$\rho_1 = \left(\frac{T_2}{T_1}\right)^{1/2} \frac{(\alpha_1 + \alpha_2 - \alpha_1\alpha_2)}{\left[\alpha_2(2 - \alpha_1) + \alpha_1\left(\frac{T_2}{T_1}\right)^{1/2}(2 - \alpha_2)\right]} \tag{7.117}$$

Heat flux:

$$q(x) = \int\int\int \zeta_x \vec{\zeta}^2 f\left(x, \vec{\zeta}\right) d\vec{\zeta} =$$

$$q(x) = \int\int\int \zeta_x\left(\zeta_x^2 + \zeta_y^2 + \zeta_z^2\right) f\left(x, \zeta_x, \zeta_y, \zeta_z\right) d\zeta_x d\zeta_y d\zeta_z =$$

$$= \frac{\rho_2}{\pi^{3/2}(\alpha_1 + \alpha_2 - \alpha_1\alpha_2)}\left[\alpha_1\left(\frac{T_2}{T_1}\right)^2 J_1 + \alpha_2(1 - \alpha_1)J_2\right]$$

$$+ \frac{\rho_2}{\pi^{3/2}(\alpha_1 + \alpha_2 - \alpha_1\alpha_2)}\left[\alpha_2 J_2 + \alpha_1(1 - \alpha_2)\left(\frac{T_2}{T_1}\right)^2 J_1\right]$$

$$J_1 = \iint\int \zeta_x\left(\zeta_x^2 + \zeta_y^2 + \zeta_z^2\right)\exp\left(-\frac{T_2}{T_1}\left(\zeta_x^2 + \zeta_y^2 + \zeta_z^2\right)\right) d\zeta_x d\zeta_y d\zeta_z =$$

$$= \int_0^\infty \zeta_x^3 \exp\left(-\frac{T_2}{T_1}\zeta_x^2\right) d\zeta_x \int_{-\infty}^\infty \exp\left(-\frac{T_2}{T_1}\zeta_y^2\right) d\zeta_y \int_{-\infty}^\infty \exp\left(-\frac{T_2}{T_1}\zeta_z^2\right) d\zeta_z +$$

$$+ \int_0^\infty \zeta_x \exp\left(-\frac{T_2}{T_1}\zeta_x^2\right) d\zeta_x \int_{-\infty}^\infty \zeta_y^2 \exp\left(-\frac{T_2}{T_1}\zeta_y^2\right) d\zeta_y \int_{-\infty}^\infty \exp\left(-\frac{T_2}{T_1}\zeta_z^2\right) d\zeta_z +$$

$$+ \int_0^\infty \zeta_x \exp\left(-\frac{T_2}{T_1}\zeta_x^2\right) d\zeta_x \int_{-\infty}^\infty \exp\left(-\frac{T_2}{T_1}\zeta_y^2\right) d\zeta_y \int_{-\infty}^\infty \zeta_z^2 \exp\left(-\frac{T_2}{T_1}\zeta_z^2\right) d\zeta_z =$$

$$= \frac{1}{2\left(\frac{T_2}{T_1}\right)^2} \frac{2\sqrt{\pi}}{2\left(\frac{T_2}{T_1}\right)^{1/2}} \frac{2\sqrt{\pi}}{2\left(\frac{T_2}{T_1}\right)^{1/2}} + \frac{1}{2\left(\frac{T_2}{T_1}\right)} \frac{2\sqrt{\pi}}{4\left(\frac{T_2}{T_1}\right)^{3/2}} \frac{2\sqrt{\pi}}{2\left(\frac{T_2}{T_1}\right)^{1/2}}$$

$$+ \frac{1}{2\left(\frac{T_2}{T_1}\right)} \frac{2\sqrt{\pi}}{2\left(\frac{T_2}{T_1}\right)^{1/2}} \frac{2\sqrt{\pi}}{4\left(\frac{T_2}{T_1}\right)^{3/2}} =$$

$$= \frac{\pi}{2}\left(\frac{T_1}{T_2}\right)^3 + \frac{\pi}{4}\left(\frac{T_1}{T_2}\right)^3 + \frac{\pi}{4}\left(\frac{T_1}{T_2}\right)^3 = \pi\left(\frac{T_1}{T_2}\right)^3$$

$$J_2 = -\iint\int \zeta_x\left(\zeta_x^2+\zeta_y^2+\zeta_z^2\right)\exp\left(-\left(\zeta_x^2+\zeta_y^2+\zeta_z^2\right)\right)d\zeta_x d\zeta_y d\zeta_z =$$

$$= -\int_0^{\infty}\zeta_x^3\exp(-\zeta_x^2)d\zeta_x\int_{-\infty}^{\infty}\exp\left(-\zeta_y^2\right)d\zeta_y\int_{-\infty}^{\infty}\exp(-\zeta_z^2)d\zeta_z -$$

$$-\int_0^{\infty}\zeta_x\exp(-\zeta_x^2)d\zeta_x\int_{-\infty}^{\infty}\zeta_y^2\exp\left(-\zeta_y^2\right)d\zeta_y\int_{-\infty}^{\infty}\exp(-\zeta_z^2)d\zeta_z -$$

$$-\int_0^{\infty}\zeta_x\exp(-\zeta_x^2)d\zeta_x\int_{-\infty}^{\infty}\exp\left(-\zeta_y^2\right)d\zeta_y\int_{-\infty}^{\infty}\zeta_z^2\exp(-\zeta_z^2)d\zeta_z =$$

$$= -\left(\frac{1}{2}2\frac{\sqrt{\pi}}{2}2\frac{\sqrt{\pi}}{2}+\frac{1}{2}2\frac{\sqrt{\pi}}{4}2\frac{\sqrt{\pi}}{2}+\frac{1}{2}2\frac{\sqrt{\pi}}{2}2\frac{\sqrt{\pi}}{4}\right) = -\left(\frac{\pi}{2}+\frac{\pi}{4}+\frac{\pi}{4}\right) = -\pi$$

$$q(x) = \frac{\rho_2}{\pi^{3/2}(\alpha_1+\alpha_2-\alpha_1\alpha_2)}\left[\alpha_1\left(\frac{T_2}{T_1}\right)^2\pi\left(\frac{T_1}{T_2}\right)^3 - \alpha_2(1-\alpha_1)\pi\right] +$$

$$+\frac{\rho_2}{\pi^{3/2}(\alpha_1+\alpha_2-\alpha_1\alpha_2)}\left[-2\pi+\alpha_1(1-\alpha_2)\left(\frac{T_2}{T_1}\right)^2\pi\left(\frac{T_1}{T_2}\right)^3\right] =$$

$$\begin{aligned} q(x) &= \frac{\rho_2}{\sqrt{\pi}(\alpha_1+\alpha_2-\alpha_1\alpha_2)}\left[-\alpha_2(2-\alpha_1)+\alpha_1\frac{T_1}{T_2}(2-\alpha_2)\right] \\ q(x) &= \frac{\rho_2}{\sqrt{\pi}(\alpha_1+\alpha_2-\alpha_1\alpha_2)T_2}[\alpha_1(2-\alpha_2)T_1-\alpha_2(2-\alpha_1)T_2] \end{aligned} \tag{7.118}$$

Replace ρ_2 with (7.84):

$$\begin{aligned} q(x) &= \frac{\frac{(\alpha_1+\alpha_2-\alpha_1\alpha_2)}{\left[\alpha_2(2-\alpha_1)+\alpha_1\left(\frac{T_2}{T_1}\right)^{1/2}(2-\alpha_2)\right]}}{\sqrt{\pi}(\alpha_1+\alpha_2-\alpha_1\alpha_2)}[\alpha_1(2-\alpha_2)T_1-\alpha_2(2-\alpha_1)T_2] \\ q(x) &= \frac{[\alpha_1(2-\alpha_2)T_1-\alpha_2(2-\alpha_1)T_2]}{\sqrt{\pi}T_2\left[\alpha_2(2-\alpha_1)+\alpha_1\left(\frac{T_2}{T_1}\right)^{1/2}(2-\alpha_2)\right]} \end{aligned} \tag{7.119}$$

In dimensional form for $\alpha_1 = 1,\ \alpha_2 = 1$: $q(x) = \rho_{av}\left(\frac{2kT_2}{m}\right)^{3/2}\frac{1}{\sqrt{\pi}}\frac{T_1-T_2}{1+(T_2/T_1)^{1/2}}\frac{1}{T_2}$.
For temperature

$$
\begin{aligned}
T(x) &= \frac{2}{3\rho}\int\left(\zeta_x^2+\zeta_y^2+\zeta_z^2\right)f\left(x,\vec{\zeta}\right)d\vec{\zeta} \\
&= \frac{2}{3\rho}\left\{\frac{\rho_2}{\pi^{3/2}(\alpha_1+\alpha_2-\alpha_1\alpha_2)}\left[\alpha_1\left(\frac{T_2}{T_1}\right)^2 J_1+\alpha_2(1-\alpha_1)J_2\right]\right. \\
&\quad \left.+\frac{\rho_2}{\pi^{3/2}(\alpha_1+\alpha_2-\alpha_1\alpha_2)}\left[\alpha_2 J_2+\alpha_1(1-\alpha_2)\left(\frac{T_2}{T_1}\right)^2 J_1\right]\right\}
\end{aligned}
$$

where:

$$
\begin{aligned}
J_1 &= \iint\int\left(\zeta_x^2+\zeta_y^2+\zeta_z^2\right)\exp\left(-\frac{T_2}{T_1}\left(\zeta_x^2+\zeta_y^2+\zeta_z^2\right)\right)d\zeta_x d\zeta_y d\zeta_z = \\
&= \int_0^\infty \zeta_x^2\exp\left(-\frac{T_2}{T_1}\zeta_x^2\right)d\zeta_x\int_{-\infty}^\infty \exp\left(-\frac{T_2}{T_1}\zeta_y^2\right)d\zeta_y\int_{-\infty}^\infty \exp\left(-\frac{T_2}{T_1}\zeta_z^2\right)d\zeta_z+ \\
&+\int_0^\infty \exp\left(-\frac{T_2}{T_1}\zeta_x^2\right)d\zeta_x\int_{-\infty}^\infty \zeta_y^2\exp\left(-\frac{T_2}{T_1}\zeta_y^2\right)d\zeta_y\int_{-\infty}^\infty \exp\left(-\frac{T_2}{T_1}\zeta_z^2\right)d\zeta_z+ \\
&+\int_0^\infty \exp\left(-\frac{T_2}{T_1}\zeta_x^2\right)d\zeta_x\int_{-\infty}^\infty \exp\left(-\frac{T_2}{T_1}\zeta_y^2\right)d\zeta_y\int_{-\infty}^\infty \zeta_z^2\exp\left(-\frac{T_2}{T_1}\zeta_z^2\right)d\zeta_z = \\
&= \frac{\sqrt{\pi}}{4\left(\frac{T_2}{T_1}\right)^{3/2}}2\frac{\sqrt{\pi}}{2\left(\frac{T_2}{T_1}\right)^{1/2}}2\frac{\sqrt{\pi}}{2\left(\frac{T_2}{T_1}\right)^{1/2}}+\frac{\sqrt{\pi}}{2\left(\frac{T_2}{T_1}\right)^{1/2}}2\frac{\sqrt{\pi}}{4\left(\frac{T_2}{T_1}\right)^{3/2}}2\frac{\sqrt{\pi}}{2\left(\frac{T_2}{T_1}\right)^{1/2}}+ \\
&+\frac{\sqrt{\pi}}{2\left(\frac{T_2}{T_1}\right)^{1/2}}2\frac{\sqrt{\pi}}{2\left(\frac{T_2}{T_1}\right)^{1/2}}2\frac{\sqrt{\pi}}{4\left(\frac{T_2}{T_1}\right)^{3/2}} = \frac{\pi^{3/2}}{4}\left(\frac{T_1}{T_2}\right)^{5/2}+\frac{\pi^{3/2}}{4}\left(\frac{T_1}{T_2}\right)^{5/2} \\
&+\frac{\pi^{3/2}}{4}\left(\frac{T_1}{T_2}\right)^{5/2}
\end{aligned}
$$

$$J_1 = \frac{3\pi^{3/2}}{4}\left(\frac{T_1}{T_2}\right)^{5/2}$$

$$J_2 = \iint\int\left(\zeta_x^2+\zeta_y^2+\zeta_z^2\right)\exp\left(-\left(\zeta_x^2+\zeta_y^2+\zeta_z^2\right)\right)d\zeta_x d\zeta_y d\zeta_z =$$

$$= \int_0^\infty \zeta_x^2\exp(-\zeta_x^2)d\zeta_x \int_{-\infty}^\infty \exp\left(-\zeta_y^2\right)d\zeta_y \int_{-\infty}^\infty \exp(-\zeta_z^2)d\zeta_z +$$

$$+ \int_0^\infty \exp(-\zeta_x^2)d\zeta_x \int_{-\infty}^\infty \zeta_y^2\exp\left(-\zeta_y^2\right)d\zeta_y \int_{-\infty}^\infty \exp(-\zeta_z^2)d\zeta_z +$$

$$+ \int_0^\infty \exp(-\zeta_x^2)d\zeta_x \int_{-\infty}^\infty \exp\left(-\zeta_y^2\right)d\zeta_y \int_{-\infty}^\infty \zeta_z^2\exp(-\zeta_z^2)d\zeta_z =$$

$$= \frac{\sqrt{\pi}}{4}2\frac{\sqrt{\pi}}{2}2\frac{\sqrt{\pi}}{2} + \frac{\sqrt{\pi}}{2}2\frac{\sqrt{\pi}}{4}2\frac{\sqrt{\pi}}{2} + \frac{\sqrt{\pi}}{2}2\frac{\sqrt{\pi}}{2}2\frac{\sqrt{\pi}}{4} = \frac{3\pi^{3/2}}{4}$$

$$T(x) = \frac{2}{3\rho}\frac{\rho_2}{\pi^{3/2}(\alpha_1+\alpha_2-\alpha_1\alpha_2)}\left[\alpha_1\left(\frac{T_2}{T_1}\right)^2 J_1 + \alpha_2(1-\alpha_1)J_2\right] +$$

$$+ \frac{2}{3\rho}\frac{\rho_2}{\pi^{3/2}(\alpha_1+\alpha_2-\alpha_1\alpha_2)}\left[\alpha_2 J_2 + \alpha_1(1-\alpha_2)\left(\frac{T_2}{T_1}\right)^2 J_1\right] =$$

$$= \frac{2}{3\rho}\frac{\rho_2}{\pi^{3/2}(\alpha_1+\alpha_2-\alpha_1\alpha_2)}\left[\alpha_1\left(\frac{T_2}{T_1}\right)^2\frac{3\pi^{3/2}}{4}\left(\frac{T_1}{T_2}\right)^{5/2} + \alpha_2(1-\alpha_1)\frac{3\pi^{3/2}}{4}\right] +$$

$$+ \frac{2}{3\rho}\frac{\rho_2}{\pi^{3/2}(\alpha_1+\alpha_2-\alpha_1\alpha_2)}\left[\alpha_2\frac{3\pi^{3/2}}{4} + \alpha_1(1-\alpha_2)\left(\frac{T_2}{T_1}\right)^2\frac{3\pi^{3/2}}{4}\left(\frac{T_1}{T_2}\right)^{5/2}\right] =$$

$$= \frac{2}{3\rho}\frac{3\rho_2}{4(\alpha_1+\alpha_2-\alpha_1\alpha_2)}\left[\alpha_1\left(\frac{T_1}{T_2}\right)^{1/2} + \alpha_2 - \alpha_1\alpha_2\right]$$

$$+ \frac{2}{3\rho}\frac{3\rho_2}{4(\alpha_1+\alpha_2-\alpha_1\alpha_2)}\left[\alpha_2 + \alpha_1\left(\frac{T_1}{T_2}\right)^{1/2} - \alpha_1\alpha_2\left(\frac{T_1}{T_2}\right)^{3/2}\right] =$$

$$= \frac{1}{2\rho}\frac{\rho_2}{(\alpha_1+\alpha_2-\alpha_1\alpha_2)}\left[2\alpha_1\left(\frac{T_1}{T_2}\right)^{1/2} + 2\alpha_2 - \alpha_1\alpha_2 - \alpha_1\alpha_2\left(\frac{T_1}{T_2}\right)^{1/2}\right]$$

$$T(x) = \frac{1}{2\rho}\frac{\rho_2}{(\alpha_1+\alpha_2-\alpha_1\alpha_2)}\left[\alpha_2(2-\alpha_1) + \alpha_1\left(\frac{T_1}{T_2}\right)^{1/2}(2-\alpha_2)\right]$$

Replace ρ and ρ_2 with (7.116):

$$T(x) = \frac{1}{2\frac{\rho_2}{2(\alpha_1+\alpha_2-\alpha_1\alpha_2)}\left[\alpha_2(2-a_1)+a_1\left(\frac{T_2}{T_1}\right)^{1/2}(2-\alpha_2)\right]} \frac{\rho_2}{(\alpha_1+\alpha_2-\alpha_1\alpha_2)}$$

$$\left[\alpha_2(2-\alpha_1)+\alpha_1\left(\frac{T_1}{T_2}\right)^{1/2}(2-\alpha_2)\right]$$

$$T(x) = \frac{\left[\alpha_2(2-\alpha_1)+\alpha_1\left(\frac{T_1}{T_2}\right)^{1/2}(2-\alpha_2)\right]}{\left[\alpha_2(2-a_1)+a_1\left(\frac{T_2}{T_1}\right)^{1/2}(2-\alpha_2)\right]} \tag{7.120}$$

Extend for Gas Mixture

Formulation of the problem:

Total density:

$$\rho = \sum_{a=1}^{N} n_a m_a \tag{7.121}$$

where: $a = 1, 2...N$ number of species.

The total number density:

$$n = n_1 + n_2 \tag{7.122}$$

The molecular concentration is defined as:

$$C = \frac{n_1}{n_1 + n_2} = \frac{n_1}{n} \tag{7.123}$$

Average molecular mass of the mixture:

$$m(C) = Cm_1 + (1 - C)m_2 \tag{7.124}$$

The mass average velocity [Bird 1.43]:

In the free molecular regime each species conducts the heat independently on each other:

$$q'_x = q'_{1x} + q'_{2x} \tag{7.125}$$

In equilibrium state: $P_0 = nkT_0$.

Characteristic molecular speed of the mixture: $u_0 = \left(\frac{2kT_0}{m}\right)^{1/2}$.

The dimensionless heat flux q related to the dimension one: $q'_x = qP_0u_0\frac{\Delta T}{T_0}$.

The heat flux through one species (Bird p155): $q'_x = \frac{n_a m_a}{2\sqrt{\pi}}\left(\frac{2kT_0}{m_a}\right)^{3/2}\frac{\Delta T}{T_0}$ (for diffuse gas-surface interaction: accommodation coefficient $\alpha = 1$).

We calculate the dimensionless heat flux using all the above equations:

$$qP_0u_0\boxed{\frac{\Delta T}{T_0}} = \frac{n_1m_1}{2\sqrt{\pi}}\left(\frac{2kT_0}{m_1}\right)^{3/2}\boxed{\frac{\Delta T}{T_0}} + \frac{n_2m_2}{2\sqrt{\pi}}\left(\frac{2kT_0}{m_2}\right)^{3/2}\boxed{\frac{\Delta T}{T_0}}$$

$$q\underbrace{(nkT_0)}_{P_0}\overbrace{\left(\frac{2kT_0}{m}\right)^{1/2}}^{u_0} = \frac{(2kT_0)^{3/2}}{2\sqrt{\pi}}\left(\frac{n_1m_1}{m_1^{3/2}} + \frac{n_2m_2}{m_2^{3/2}}\right)$$

$$qn\sqrt{2}\frac{(kT_0)^{3/2}}{\sqrt{m}} = \frac{2^{3/2}(kT_0)^{3/2}}{2\sqrt{\pi}}\left(\frac{n_1m_1}{m_1^{3/2}} + \frac{n_2m_2}{m_2^{3/2}}\right)$$

$$q = \frac{\sqrt{m}}{n\sqrt{\pi}}\left(\frac{n_1}{\sqrt{m_1}} + \frac{n_2}{\sqrt{m_2}}\right) = \frac{1}{\sqrt{\pi}}\left(\frac{n_1}{n}\frac{\sqrt{m}}{\sqrt{m_1}} + \frac{n_2}{n}\frac{\sqrt{m}}{\sqrt{m_2}}\right)$$

$$q = \frac{1}{\sqrt{\pi}}\left(C\sqrt{\frac{m}{m_1}} + (1-C)\sqrt{\frac{m}{m_2}}\right)$$

$\left(\text{For single gas C} = 1,\ \text{m} = \text{m}_1 = \text{m}_2, q = \frac{1}{\sqrt{\pi}}\right)$

Appendix 3: Python Code for Analytical Solution of the Fourier Problem at Low Kn

The Python code to solve Fourier equations given in Sect. 7.9 is given below:

```
import numpy as np
import matplotlib.pyplot as plt
import pandas as pd

# Constants
# Boltzmann constant (J/K)
kb = 1.38064852e-23
# Ratio of specific heats (monoatomic gas)
k = 5. / 3.
# Prandtl number
pr = 2. / 3.
# Temperature ratio T_min / T_max
r = 223.15 / 323.15
# Knudsen number
kn = 0.0237
# A parameter related to frequency, seems to be large
fr = 1e9
# A derived heat transfer parameter
q = 1.996 * (k / (k + 1)) * (kn / pr)

# Initial guess for the iterative solution for dimensionless temperatures at the boundaries
# t1_0 is at the hot boundary (x=1), t0_0 is at the cold boundary (x=0)
t1_0 = 1 + (2 / 3) * q * ((r**(3/2) - 1.))
t0_0 = r * (1 - (2 / 3) * q * (r**(3/2) - 1) / (r**(3/2)))

# Iteratively solve for the boundary temperatures until convergence
eps = 1.
while eps > 1e-12:
    t1 = 1 + (2. / 3.) * q / np.sqrt(t1_0) * (t0_0**(3./2.) - t1_0**(3./2.))
    t0 = r + np.sqrt(t1_0 / t0_0) * (1. - t1_0)
    # Check for convergence
    eps = max(abs(t0 - t0_0), abs(t1 - t1_0))
    # Update values for the next iteration
    t0_0 = t0
    t1_0 = t1

# Parameters for calculating the profiles based on the converged boundary temperatures
a = t1**(3./2.) - t0**(3./2.)
b = t0**(3./2.)
c = -2. / fr

# Generate profiles across the domain
N = 101 # Number of points
z_vals = np.linspace(0, 1, N) # Dimensionless position from 0 to 1
```

```
# Calculate Temperature profile
tt = a * z_vals + b
T_vals = tt ** (2. / 3.)

# Calculate Heat Flux profile
# Note: This is a dimensionless flux
flux = (2. / 3.) * a * (tt ** (-1. / 3.)) * np.sqrt(T_vals)

# Calculate Density profile
ro = np.exp(-(2./3.) * np.log(tt) + 3 * tt ** (1./3.) * c / a)
# Normalize the density profile
ro1 = ro * N / np.sum(ro)

# Constants for converting dimensionless heat flux to physical units (qqq2)
Th = 373.15  # Reference temperature (K)
gc = 208.1   # Gas constant for Argon (J/kgK)
dm = 3.63e-10 # Atomic diameter (m)
vth = np.sqrt(2 * gc * Th) # Thermal velocity (m/s)
rmass = 6.63e-26 # Mass of Argon atom (kg)
rl = 1e-3 # Characteristic length (m)
rl0 = kn * rl # Mean free path (m)
rn0 = 1 / (np.sqrt(2.) * np.pi * dm**2 * rl0) # Number density (1/m^3)
rho = rmass * rn0 # Mass density (kg/m^3)
cp = 523.1 # Specific heat at constant pressure (J/kgK)
# Thermal conductivity
lambdah = (15. / 32.) * cp * vth * rl0 * rho * np.sqrt(np.pi)
# Physical heat flux (W/m^2)
qqq2 = lambdah * flux * Th / rl

# Create a pandas DataFrame for the results
df = pd.DataFrame({
    "x": z_vals,
    "Density": ro1,
    "Temperature": T_vals,
    "HeatFlux": qqq2
})

# Print the DataFrame to the console
print("--- Heat Conduction Profiles ---")
print(df)

# Plotting the results
plt.figure(figsize=(14, 10))
```

```
# Plot Density
plt.subplot(3, 1, 1)
plt.plot(z_vals, ro1, label='Density', color='blue')
plt.ylabel("Density")
plt.title("Calculated Profiles for Heat Conduction")
plt.grid(True)
plt.legend()

# Plot Temperature
plt.subplot(3, 1, 2)
plt.plot(z_vals, T_vals, label='Temperature $T/T_{max}$', color='orange')
plt.ylabel("Temperature")
plt.grid(True)
plt.legend()

# Plot Heat Flux
plt.subplot(3, 1, 3)
plt.plot(z_vals, qqq2, label='Conductive Heat Flux', linestyle='-', color='green')
plt.xlabel("x/L")
plt.ylabel("Heat Flux (W/m^2)")
plt.grid(True)
plt.legend()

# Display the plots
plt.tight_layout()
plt.show()
```

References

1. Karniadakis, G., Beskok, A., Aluru, N.: Microflows and Nanoflows: Fundamentals and Simulation, vol. 29. Springer Science & Business Media (2006)
2. Agrawal, A., Kushwaha, H.M., Jadhav, R.S.: Microscale Flow and Heat Transfer. Springer International Publishing (2020)
3. Marques, Jr.W., Kremer, G.M., Sharipov, F.M.: Couette flow with slip and jump boundary conditions. Continuum Mech. Thermody **12**(6), 379–386 (2000)
4. Cercignani, C.: Plane Poiseuille flow and Knudsen minimum effect. In: Laurmann, J.A. (ed.) Proceedings of the Third International Symposium on Rarefied Gas Dynamics, vol. 1, pp. 92–101. Academic Press, New York (1963)
5. Sone, Y.: Molecular Gas Dynamics: Theory, Techniques, and Applications. Birkhäuser, Boston (2007)
6. Bahukudumbi, P., Beskok, A.: A phenomenological lubrication model for the entire Knudsen regime. J. Micromech. Microeng. **13**(6), 873–884 (2003)
7. Lockerby, D.A., Reese, J.M.: On the modelling of isothermal gas flows at the microscale. J. Fluid Mech. **604**, 235–261 (2008)
8. Lilley, C.R., Sader, J.E.: Velocity gradient singularity and structure of the velocity profile in the Knudsen layer according to the Boltzmann equation. Phys. Rev. E **76**, 026315 (2007)
9. Takata, S., Funagane, H.: Singular behaviour of a rarefied gas on a planar boundary. J. Fluid Mech. **717**, 30–47 (2013)

10. Jiang, S., Luo, L.S.: Analysis and accurate numerical solutions of the integral equation derived from the linearized BGKW equation for the steady Couette flow. J. Comput. Phys. **316**, 416–434 (2016)
11. Myong, R.S.: On the high Mach number shock structure singularity caused by overreach of Maxwellian molecules. Phys. Fluids 26(5).c
12. Myong, R.S.: Theoretical description of the gaseous Knudsen layer in Couette flow based on the second-order constitutive and slip-jump models. Phys. Fluids **28**(1) (2016)
13. Yuhong, S., Barber, R.W., Emerson, D.R.: Inverted velocity profiles in rarefied cylindrical Couette gas flow and the impact of the accommodation coefficient. Phys. Fluids **17**(4), 047102 (2005)
14. Akhlaghi, H., Javadi, K.: Torque data on non-isothermal rarefied cylindrical Couette flow. Vacuum **121**, 56–63 (2015)
15. Aoki, K., Yoshida, H., Nakanishi, T., Garcia, A.L.: Inverted velocity profile in the cylindrical Couette flow of a rarefied gas. Phys. Rev. E **68**(1), 016302 (2003)
16. Stefanov, S.K., Barber, R.W., Emerson, D.R., Reese, J.M.: The critical accommodation coefficient for velocity inversion in rarefied cylindrical Couette flow in the slip and near free-molecular regimes. In: Proceedings of the 25th International Symposium on Rarefied Gas Dynamics, St. Petersburg, Russia, pp. 1146–1151 (2006)
17. Emerson, D.R., Gu, X.J., Stefanov, S.K., Yuhong, S., Barber, R.W.: Nonplanar oscillatory shear flow: From the continuum to the free-molecular regime. Phys. Fluids **19**(10), 107105 (2007)
18. Taheri, E., Roohi, E., Stefanov, S.: On the convergence of the simplified Bernoulli Trial collision scheme in rarefied Fourier flow. Phys. Fluids **29**(6), 062003 (2017)
19. Stefanov, S., Roussinov, V., Cercignani, C.: Rayleigh-Bénard flow of a rarefied gas and its attractors. I. Convection regime. Phys. Fluids **14**, 2255 (2002)
20. Manela, A., Frankel, I.: On the Rayleigh-Bénard problem in the continuum limit. Phys. Fluids **17**(3), 036101 (2005)
21. Meskos, S.: Mathematical models and stochastic simulation of gas mixtures and mixing processes in microscale. Ph.D. Thesis, Bulgarian Academy of Science (2020)
22. Sparrow, E.M., Kinney, R.B.: Free molecule heat transfer between parallel plates. Phys. Fluids **7**(3), 473–475 (1964)
23. Springer, G.S.: Heat transfer in rarefied gases. In: Advances in Heat Transfer, vol. 7, pp. 163–218. Elsevier (1971)
24. Fridlender, O.G.: Heat transfer in a very rarefied gas. Fluid Dyn. **15**(1), 165–168 (1980)
25. Shen, C.: Rarefied Gas Dynamics. Springer, Berlin (2005)
26. Ohwada, T.: Heat flow and temperature and density distributions in a rarefied gas between parallel plates with different temperatures. Finite-difference analysis of the nonlinear Boltzmann equation for hard-sphere molecules. Phys. Fluids **8**(8), 2153–2160 (1996)
27. Mohammadzadeh, A., Roohi, E., Niazmand, H., Stefanov, S., Myong, R.S.: Thermal and second-law analysis of a micro-or nanocavity using direct-simulation Monte Carlo. Phys. Rev. E Stat. Nonlinear Soft Matter Phys. **85**(5), 056310 (2012)
28. Oztop, H.F., Dagtekin, I.: Mixed convection in two-sided lid-driven differentially heated square cavity. Int. J. Heat Mass Transf. **47**(8–9), 1761–1769 (2004)
29. Naris, S., Valougeorgis, D.: The driven cavity flow over the whole range of the Knudsen number. Phys. Fluids **17**(9) (2005)
30. Auld, D., Lan, Y.: Simulation of lid-driven cavity flow by parallel DSMC method. In: 24th AIAA Applied Aerodynamics Conference. p. 3328 (2006)
31. Mizzi, S., Emerson, D.R., Stefanov, S.K., Barber, R.W., Reese, J.M.: Effects of rarefaction on cavity flow in the slip regime. J. Comput. Theor. Nanosci. **4**(4), 817–822 (2007)
32. Perumal, D.A., Krishna, V., Sarvesh, G., Dass, A.K.: Numerical simulation of gaseous microflows by lattice Boltzmann method. Int. J. Recent Trends Eng. **1**(5), 15 (2009)
33. Rana, A., Torrilhon, M., Struchtrup, H.: A robust numerical method for the R13 equations of rarefied gas dynamics: application to lid driven cavity. J. Comput. Phys. **236**, 169–186 (2013)
34. John, B., Gu, X.J., Emerson, D.R.: Investigation of heat and mass transfer in a lid-driven cavity under nonequilibrium flow conditions. Numer. Heat Transf. Part B Fundam. **58**(5), 287–303 (2010)

35. John, B., Gu, X.J., Emerson, D.R.: Effects of incomplete surface accommodation on non-equilibrium heat transfer in cavity flow: a parallel DSMC study. Comput. Fluids **45**(1), 197–201 (2011)
36. Gu, X.J., Emerson, D.R.: A high-order moment approach for capturing non-equilibrium phenomena in the transition regime. J. Fluid Mech. **636**, 177–216 (2009)
37. Gu, X.J., John, B., Tang, G.H., Emerson, D.R.: Heat and mass transfer of a rarefied gas in a driven micro-cavity. In: International Conference on Micro/Nanoscale Heat Transfer, vol. 43901, pp. 559–568 (2009)
38. Lockerby, D.A., Reese, J.M., Gallis, M.A.: The usefulness of higher-order constitutive relations for describing the Knudsen layer. Phys. Fluids **17**(10) (2005)

Chapter 8
Airfoil Aerodynamics at Rarefied Flow Regimes

Abstract This chapter explores airfoil aerodynamics in rarefied environments, presenting a semi-analytical method for assessing lift and drag in free-molecular flow. Parametric studies investigate the influence of rarefaction on aerodynamic performance, considering variables such as Mach number, Reynolds number, angle of attack, camber, and airfoil thickness.

8.1 Introduction

The airfoil's aerodynamic performance studies at free-stream conditions have largely relied on intensive numerical calculations, such as Direct Simulation Monte Carlo (DSMC), Boltzmann equation modeling, or CFD solvers based on the continuum limit equations. This section examines the aerodynamic characteristics under extremely rarefied flow conditions at large Knudsen numbers in a wind tunnel and presents a semi-analytical solution in a free-molecular flow regime. This section investigates the rarefied effects of airfoil's aerodynamic characteristics. Also, a parametric study is conducted to examine the influence of Mach number, Reynolds number, thickness, angle of attack, and camber on rarefaction effects for airfoils.

8.2 Statement of the Problem

Figure 8.1 provides a schematic of the problem setup. The inlet and outlet sections of the channel are assumed to be linked to the perfect monatomic gas reservoirs at thermodynamic equilibrium with constant temperatures and densities.

Gas particles entering the channel are distributed by the equilibrium Maxwellian. The Maxwell boundary condition models the interactions between the plate and particles and the channel's boundaries.

E. Roohi et al., *Advances in Direct Simulation Monte Carlo: From Micro-Scale to Rarefied Flow Phenomena*, https://doi.org/10.1007/978-981-96-8200-3_8

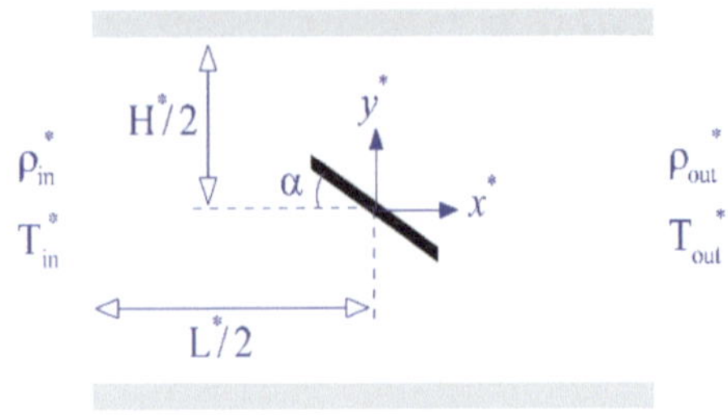

Fig. 8.1 Schematic of the problem: a thin flat plate in a channel with length L^* and width H^* [1]

$$f\left(\mathrm{r}_b^*, \xi^* \cdot \hat{\mathrm{n}} > 0\right) = \beta \frac{\rho_b^*\left(\mathrm{r}_b^*\right)}{\pi^{3/2} U_{\mathrm{mp}}^{*3}} \exp\left[-\frac{\xi^{*2}}{U_{\mathrm{mp}}^{*2}}\right] + (1-\beta) f\left(\mathrm{r}_b^*, \xi^* - 2\left(\xi^* \cdot \hat{\mathrm{n}}\right)\hat{\mathrm{n}}\right) \tag{8.1}$$

At each boundary point $r_b^* = \left(x_b^*, y_b^*\right)$, a portion of the gas molecules, β, are diffusely emitted, and the rest, $(1-\beta)$, are emitted specularly. $\rho_b^*\left(r_b^*\right)$ is an unknown function associated with the particle's mass flux released from the boundary and the molecular most probable speed is calculated as $\sqrt{2\mathcal{R}^* T_b^*}$. The solid boundaries are supposed to be isothermal, maintaining the same temperature as the inlet reservoir, $T_b^* = T_{\mathrm{in}}^*$.

8.3 Free-Molecular Limit

The state of gas molecules is determined by the probability density function $f = f(\mathrm{r}, \xi)$ in the two-dimensional steady configuration, representing the probability of finding a gas molecule at position $\mathrm{r} = (x, y)$ with velocity $\xi = \left(\xi_x, \xi_y, \xi_z\right)$. Under free-molecular conditions, this function follows the collisionless Boltzmann equation and extends with the boundary condition of scaled Maxwell to account for the reflected particles at the surface.

$$f(\mathrm{r}_b, \xi \cdot \hat{\mathrm{n}} > 0) = \beta \frac{\rho_b(\mathrm{r}_b)}{\pi^{3/2}} \exp\left[-\xi^2\right] + (1-\beta) f(\mathrm{r}_b, \xi - 2(\xi \cdot \hat{\mathrm{n}})\hat{\mathrm{n}}) \tag{8.2}$$

8.3.1 *Diffuse Reflecting Walls*

When $\beta = 1$ is applied in Eq. (8.2), each gas particle state depends on its last interaction with the solid wall or free boundaries. The overall solution is subsequently provided by Shapiro and Manela [1]:

$$f(\mathrm{r}, \xi) = \frac{\tilde{\rho}(\mathrm{r}, \xi)}{\pi^{3/2}\tilde{T}^{3/2}} \exp\left[-\frac{\xi^2}{\tilde{T}}\right] \tag{8.3}$$

Here, $\tilde{T}$ is the defined temperature of the emitting boundary. Based on the particle's location and velocity components (ξ_x, ξ_y), the surface from which the particle was last emitted can be uniquely identified. The calculation of $\tilde{\rho}(x, y, \mathrm{c})$ is performed by applying the impermeability condition.

$$\int\limits_{\xi\cdot\hat{\mathrm{n}}>0} (\xi \cdot \hat{\mathrm{n}}) f(\mathrm{r}_b, \xi) \mathrm{d}\xi + \int\limits_{\xi\cdot\hat{\mathrm{n}}<0} (\xi \cdot \hat{\mathrm{n}}) f(\mathrm{r}_b, \xi) \mathrm{d}\xi = 0 \tag{8.4}$$

The first integral accounts for the effect of outgoing particles, while the second integral accounts for the impact of incoming particles on the total velocity normal to the surface. The particles arriving may have various origins along the channel boundaries; each source's influence must be analyzed separately.

Referring to Fig. 8.2a, at the specified point $(x_b, H/2)$, the walls' impermeability condition is imposed, and using Eq. (8.2), the contribution of the particles in Eq. (8.4) is determined.

$$\int\limits_{\xi\cdot\hat{\mathrm{n}}>0} (\xi \cdot \hat{\mathrm{n}}) f_{\mathrm{uw}} \,\mathrm{d}\xi = \frac{\rho_{\mathrm{uw}}(x_b)}{2\sqrt{\pi}} \tag{8.5}$$

The subscript "uw" signifies the function's value at a point located on the upper wall. A more complex computation is needed to calculate the effect of incoming particles in Eq. (8.4). The dashed blue lines in Fig. 8.2a represent the kinematic boundaries and define the regions within the channel for particles arriving at $(x_b, H/2).(x_b, H/2)$.

We derive the following result by considering the effect of reflected particles from the lower wall.

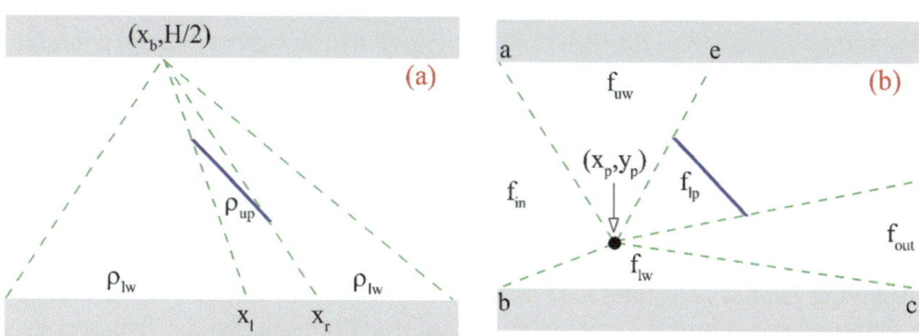

Fig. 8.2 Division of a diffuse-wall channel for the **a** imposition of the impermeability condition at a point $(x_b, H/2)$ along the upper tunnel wall, and **b** calculation of the hydrodynamic fields within the channel at a point (x_p, y_p) (Reproduced from Ref. [1])

$$\int_{\xi\cdot\hat{\mathbf{n}}<0} (\xi \cdot \hat{\mathrm{n}}) f_{\mathrm{lw}}\, \mathrm{d}\xi = -\frac{H^2}{4\sqrt{\pi}}\left(\int_{-L/2}^{x_l} \frac{\rho_{\mathrm{lw}}(x)\mathrm{d}x}{\left[H^2+(x-x_b)^2\right]^{3/2}} + \int_{x_r}^{L/2} \frac{\rho_{\mathrm{lw}}(x)\mathrm{d}x}{\left[H^2+(x-x_b)^2\right]^{3/2}}\right) \tag{8.6}$$

On the other hand, the contribution of reflected particles from the upper surface of the plate is as follows [1]:

$$\int_{\xi\cdot\hat{\mathrm{n}}<0} (\xi \cdot \hat{\mathrm{n}}) f_{\mathrm{up}}\, \mathrm{d}\xi == \frac{-1}{4\sqrt{\pi}} \int_{-\frac{1}{2}}^{\frac{1}{2}} \rho_{up}(z) \frac{(z \sin\alpha + H/2)(x_b \sin\alpha + (H \cos\alpha)/2)}{\left[\left(z \sin\alpha + \frac{H}{2}\right)^2 + (x_b - z\cos\alpha)^2\right]^{\frac{3}{2}}} dz \tag{8.7}$$

The parameter z in Eq. (8.7) is the integration variable at the upper surface. At large angles of attack, particularly near the left edge, some parts of the upper surface might be blocked, causing particles to reach from the lower plate surface. In these cases, the terms f_{up} and ρ_{up} in Eq. (8.7) are substituted with f_{lp} and ρ_{lp}, respectively.

The impact of particles reaching from the inlet or outlet regions is expressed as follows:

$$\int_{\xi\cdot\hat{\mathrm{n}}<0} (\xi \cdot \hat{\mathrm{n}}) f_{\mathrm{in}}\, \mathrm{d}\xi = \frac{1}{4\sqrt{\pi}}\left(\frac{\frac{L}{2}+x_b}{\sqrt{H^2+\left(\frac{L}{2}+x_b\right)^2}} - 1\right) \tag{8.8}$$

$$\int_{\xi\cdot\hat{\mathrm{n}}<0} (\xi \cdot \hat{\mathrm{n}}) f_{\mathrm{out}}\, \mathrm{d}\xi = \frac{\rho_{\mathrm{out}}\sqrt{T_{\mathrm{out}}}}{4\sqrt{\pi}}\left(\frac{L/2-x_b}{\sqrt{H^2+(L/2-x_b)^2}} - 1\right) \tag{8.9}$$

By summing Eqs. (8.5) through (8.9) into Eq. (8.4), the impermeability condition is derived at the point $(x_b, H/2)$ [1].

$$= -\left(\frac{\frac{L}{2}+x_b}{\sqrt{H^2+\left(\frac{L}{2}+x_b\right)^2}} - 1\right) - \rho_{\mathrm{out}}\sqrt{T_{out}}\left(\frac{\frac{L}{2}-x_b}{\sqrt{H^2+\left(\frac{L}{2}-x_b\right)^2}} - 1\right) \tag{8.10}$$

Applying similar reasoning will calculate the impermeability conditions for the remaining solid surfaces. By combining this information with Eq. (8.10), we create a system of interconnected equations that calculate how a substance's density changes at the system's boundaries.

By discretizing the fluxes and using the trapezoidal rule along the boundaries to approximate the integral terms, the impermeability conditions were solved numerically for any valid combination of $0 \leq \rho_{\mathrm{out}} \leq 1$ and $0 < T_{\mathrm{out}} \leq 1$. This produced non-homogeneous, linear equations that were solved using MATLAB code, requiring only a minor computational cost compared to the DSMC solution.

The probability density function in Eq. (8.3) can be determined after the fluxes have been calculated, and the hydrodynamic fields can then be computed using the proper quadratures over the velocity space. In particular, the density, as well as velocity components, can be calculated as:

$$\rho(x,y)=\int_{-\infty}^{\infty} f\,\mathrm{d}\xi,\quad u_x(x,y)=\frac{1}{\rho(x,y)}\int_{-\infty}^{\infty}\xi_x f\,\mathrm{d}\xi,\quad \text{and}\quad u_y(x,y)=\frac{1}{\rho(x,y)}\int_{-\infty}^{\infty}\xi_y f\,\mathrm{d}\xi \tag{8.11}$$

respectively, and the normal (P_{xx}, P_{yy}, and P_{zz}) and shear $\left(P_{xy}\right)$ stresses are computed by:

$$\begin{aligned} P_{xx}(x,y)&=\int_{-\infty}^{\infty}(\xi_x-u_x)^2 f\,\mathrm{d}\xi,\quad P_{yy}(x,y)=\int_{-\infty}^{\infty}\left(\xi_y-u_y\right)^2 f\,\mathrm{d}\xi\\ P_{zz}(x,y)&=\int_{-\infty}^{\infty}\xi_z^2 f\,\mathrm{d}\xi,\quad \text{and}\quad P_{xy}(x,y)=\int_{-\infty}^{\infty}(\xi_x-u_x)\left(\xi_y-u_y\right)f\,\mathrm{d}\xi \end{aligned} \tag{8.12}$$

The pressure is derived by:

$$p(x,y)=\frac{2}{3}\left(P_{xx}+P_{yy}+P_{zz}\right) \tag{8.13}$$

Also, the temperature is calculated by the scaled state equation of ideal gas $T=p/\rho$.

8.3.2 Specular Reflecting Walls

Unlike the diffuse reflecting walls, the impermeability conditions are automatically satisfied at a specular wall ($\beta=0$). Figure 8.3b illustrates the division of the channel into regions where particles originate from and arrive at $(x,y)=(0.7,0.2)$. This process must be repeated for all relevant (x,y) locations to obtain the solution for (x,y) in the five-dimensional space. Then, the hydrodynamic fields can be determined using Eqs. (8.11) and (8.12).

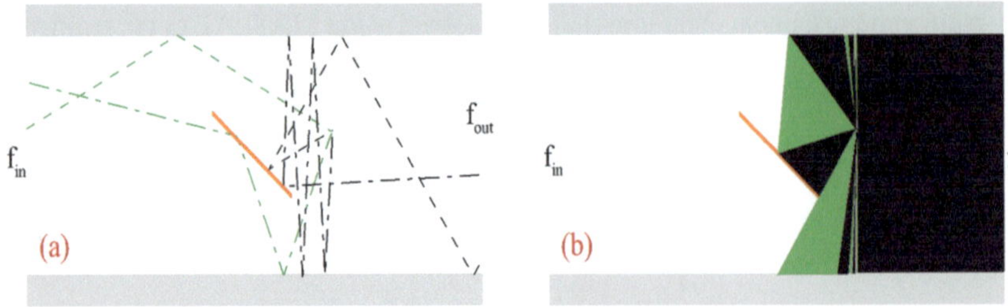

Fig. 8.3 **a** Specular-wall particle trajectories originating at the inlet and outlet sections and arriving at $(x, y) = (0.7, 0.2)$. **b** Division of a specularly reflecting channel into sections of particles originating at the tunnel inlet (red) and outlet (blue) sections (Reproduced from Ref. [1])

8.4 Results for Density-Driven Flow

Figure 8.4 shows the contours of free molecular for an airfoil with the angle of attack $\alpha = \pi/4$ within a diffuse-wall channel where the scaled gas density at the outlet reservoir is $\rho_{\text{out}} = 0.2$ [1].

Under free-molecular flow conditions, despite the large angle of attack, the flow remains attached and adheres to the impermeability condition at the plate's surface. The gas speed peaks near the tips of the airfoil and at the outlet of the channel, corresponding to a subsonic Mach number of approximately $\approx 0.27\sqrt{2/\gamma} \approx 0.3$ for $\gamma = 5/3$. The flow remains mostly stagnant on the lower surface, with a noticeable slip at the airfoil's upper surface.

Considering Fig. 8.4c, it is observed that the flow is almost isothermal, resulting in pressure and density distributions that are nearly the same. A pressure difference of approximately $\Delta p = p_{\text{lower}} - p_{\text{upper}} \approx 0.35$ (in $\rho^*_{\text{in}} U^{*2}_{\text{mp in}}$ units) is observed across

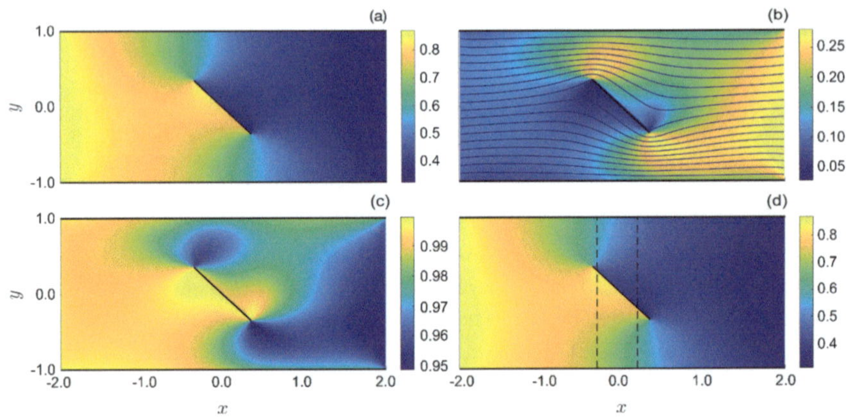

Fig. 8.4 Contours of the free-molecular **a** density, **b** velocity magnitude, **c** temperature, and **d** pressure fields for an airfoil in a diffuse-wall tunnel (Copied from open access Ref. [1])

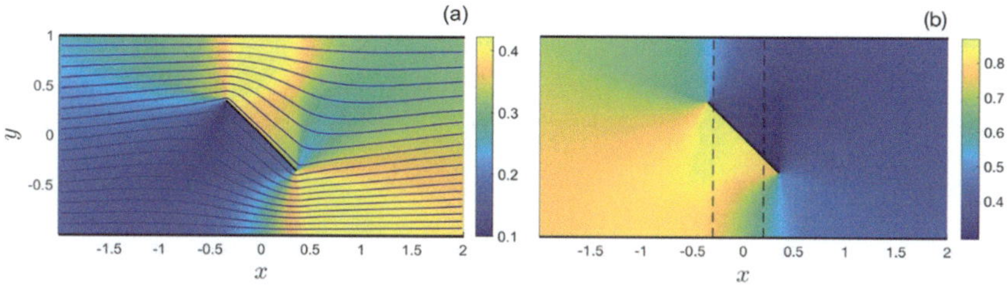

Fig. 8.5 Colormaps of the free-molecular **a** velocity magnitude and **b** pressure fields for an airfoil in a specular-wall tunnel (Copied from open access Ref. [1])

the plate, indicating higher density upstream and lower density downstream in the channel (Fig. 8.4d).

Figure 8.5 analyzes the impact of surface and tunnel conditions on the system's behaviour, using specular-wall conditions instead of diffuse-wall conditions while maintaining the geometry and reservoir state from Fig. 8.4.

In contrast to Fig. 8.4b, Fig. 8.5 shows significantly higher speeds ($\approx$0.42, Mach number $\approx$0.46) and greater velocity slip along the tunnel and airfoil walls, a result of specular-wall interactions, where the solid walls do not absorb kinetic energy, unlike in the diffuse-wall setup. The specular-wall configuration results in higher pressure jumps, approximately $\Delta p \approx 0.5$, and these differences are also evident in the changes in the aerodynamic loading on the plate. The specular-wall configuration shows sharper flow gradients near the airfoil tips, caused by the confined region influenced by the emission of particles from the surface of the airfoil, unlike the more uniform dispersion seen in the diffuse-wall setup.

Figure 8.6 evaluates the validity of the free-molecular model by comparing it to the DSMC solution at finite Knudsen numbers, using sections at x $= -0.3$ and x $= 0.2$ (the dashed lines in Figs. 8.4d and 8.5b). The figure shows distributions of pressure and x-velocity for both specular and diffuse-wall cases, with dashed lines linking data from sides of the airfoil, highlighting the discontinuous "jumps".

The ballistic and DSMC solution align well at Kn $=$ 10 for the diffuse-wall results and slightly decrease as Kn decreases. However, the free-molecular results align closely with DSMC data at Kn $=$ 1, indicating that molecular collisions have a negligible impact on Kn $\gtrsim$ 1 in the narrow (H $=$ 2) channel. For the specular-wall case, the free-molecular model shows noticeable discrepancies at Kn $\sim$ O(1), though DSMC results approach the collisionless limit as Kn increases. Overall, the good match between ballistic results and DSMC estimations at Kn $=$ 10 for both wall conditions is a positive outcome.

The tunnel and plate walls influence Fig. 8.6's results and their agreement with DSMC data. An empty channel setup with varying Knudsen numbers was analyzed to evaluate the free-molecular solution's accuracy (Fig. 8.7). The results confirm that this model is valid for high Knudsen numbers and presents a good result for Kn $\gtrsim$ 1, with a deviation of around 4.5%.

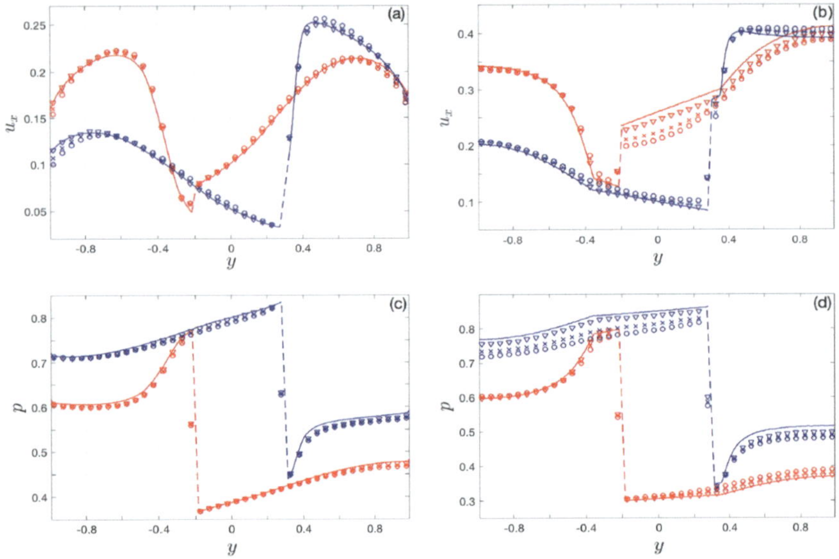

Fig. 8.6 Comparison between free-molecular (solid lines) and DSMC predictions (symbols) for the x-velocity **a** and **b** and pressure **c** and **d** fields along $x = -0.3$ (in blue) and $x = 0.2$ (in red). **a** and **c** show results for a diffuse-wall channel, and **b** and **d** present data for a specular tunnel. The triangles, crosses, and circles show DSMC data at Kn = 10, 2, and 1, respectively (Copied from open access Ref. [1])

Fig. 8.7 Comparison between free-molecular (pluses) and DSMC predictions (other symbols) for the x-velocity field along $x = -0.3$ (in blue) and $x = 0.2$ (in red) in a diffuse-wall channel in the absence of a plate. The triangles, crosses, and circles show DSMC data at Kn = 10, 2, and 1, respectively (Copied from open access Ref. [1])

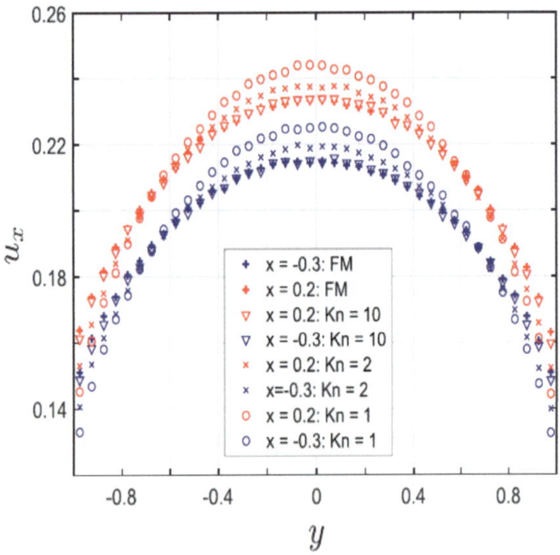

In this setup, the lift force in the y-direction per unit span on the plate is scaled by the term $\rho_{\text{in}}^* U_{\text{mp}_{\text{in}}}^{*2} c^*$, is given by:

$$L = \int_{-0.5}^{0.5} \left[\left(P_{s_\perp s_\perp}\left(s_\parallel, s_\perp = 0^-\right)\cos\alpha - P_{s_\parallel, s_\perp}\left(s_\parallel, s_\perp = 0^-\right)\sin\alpha \right) - \left(P_{s_\perp s_\perp}\left(s_\parallel, s_\perp = 0^+\right)\cos\alpha - P_{s_\parallel, s_\perp}\left(s_\parallel, s_\perp = 0^+\right)\sin\alpha \right) \right] ds_\parallel \quad (8.14)$$

In this context, $s_\parallel$ and $s_\perp$ represent the parallel and normal body coordinates originating at the plate's midchord. The stress components in Eq. (8.14) include the normal stress $\left(P_{s_\perp s_\perp}\right)$ and the shear stress $\left(P_{s_\parallel s_\perp}\right)$, calculated using Eq. (8.12), with the x and y indices replaced by $s_\parallel$ and $s_\perp$, respectively.

The drag force is calculated by:

$$D = \int_{-0.5}^{0.5} \left[\left(P_{s_\perp s_\perp}\left(s_\parallel, s_\perp = 0^-\right)\sin\alpha + P_{s_\parallel s_\perp}\left(s_\parallel, s_\perp = 0^-\right)\cos\alpha \right) - \left(P_{s_\perp s_\perp}\left(s_\parallel, s_\perp = 0^+\right)\sin\alpha + P_{s_\parallel s_\perp}\left(s_\parallel, s_\perp = +\right)\cos\alpha \right) \right] ds_\parallel \quad (8.15)$$

In Fig. 8.8a, b, the free-molecular and DSMC results are closely aligned at Kn = 10. However, similar to the findings in Fig. 8.6, the gap between the collisionless and DSMC data increases as Kn decreases, particularly in the specular-wall setup, where differences at Kn = 2 are evident.

Figure 8.8a shows that the lift force changes non-monotonically with the plate's inclination angle (α), peaking at an intermediate angle. At = 0 and $\alpha = \pi/2$, the lift force is zero because of symmetry. The lift becomes positive and negatively antisymmetric between these angles for $\pi/2 \leq \alpha \leq \pi$.

The lift curves are almost symmetrical around $\alpha = \pi/4$ in the diffuse-wall case, where the peak lift is observed. For the specular-wall setup, the peak shifts to a lower α. As the density difference between reservoirs increases, the lift force grows, reaching a limit at conditions of expansion-to-vacuum ($\rho_{\text{out}} = 0$), shown by the thin magenta curve, the same as the $\rho_{\text{out}} = 0.01$ case. As predicted, the lift force is higher in the specular wall compared to the diffuse-wall case.

In Fig. 8.8b, the drag force steadily increases with α from 0 to $\pi/2$, peaking at $\alpha = \pi/2$, then symmetrically decreases for $\pi/2 \leq \alpha \leq \pi$. The drag is zero at $\alpha = 0$ due to the absence of shear stress on the smooth surface $P_{xy}\left(s_\perp = 0^\pm; \alpha = 0\right) \equiv 0$ in the specular-wall case. However, in the diffuse-wall setup, a non-zero drag at $\alpha = 0$ is attributed to the energy transfer between the gas and the plate at $\beta = 1$.

The drag force is typically smaller in the diffuse-wall than in the specular-wall case, except small values of α ($\alpha/\pi \leq 0.1$), where the drag in the specular-wall setup becomes lower than in the diffuse-wall setup due to the absence of shear stress at very small α. $P_{xy}\left(s_\perp = 0^\pm; \alpha \ll 1\right)$.

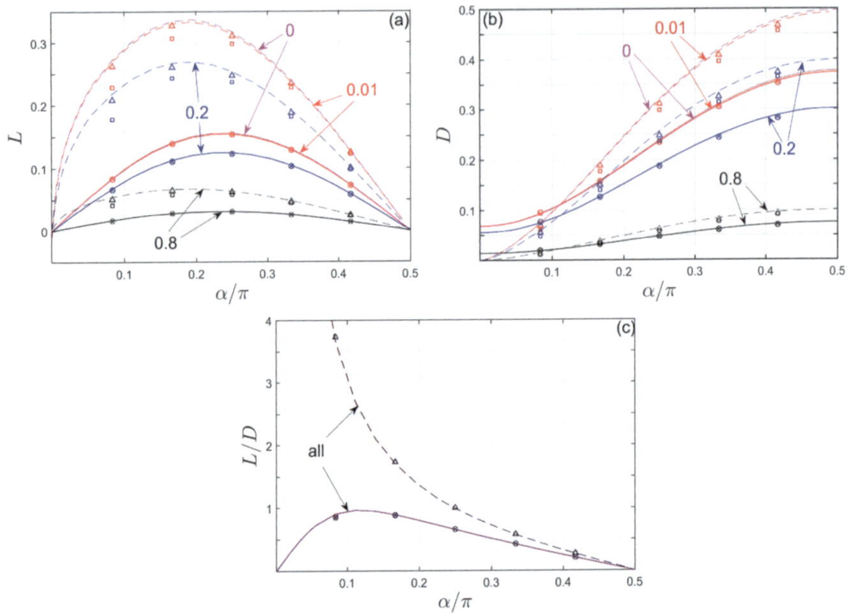

Fig. 8.8 Variations of the free-molecular **a** lift, **b** drag, and **c** lift-to-frag force ratio for diffuse-(solid lines) and specular-wall (dashed curves) tunnels. The crosses and circles mark DSMC results in a diffuse-wall tunnel with Kn = 10 and 2, respectively, whereas the triangles and squares denote the counterpart data in a specular-wall configuration (Copied from open access Ref. [1])

The lift-to-drag ratio (L/D) in the specular-wall configuration decreases steadily from an infinite value to zero in $0 \leq \alpha \leq \pi/2$, however the L/D ratio value is zero at $\alpha = 0$ and $\alpha = \pi/2$ for diffuse-wall, peaking at $L/D \approx 1$ around $\alpha \approx 0.11\pi$. For both setups, all L/D curves align on a single line due to the dependence of Maxwellian distributions on density, which cancels out when calculating the ratio. The primary distinction between the two configurations is in how L/D behaves for small $\alpha \ll 1$: it diverges in the specular case and vanishes in the diffuse case due to the absence of P_{xy} at $\alpha = 0$ $\left(P_{xy}\left(s_{\perp} = 0^{\pm}; \alpha = 0\right)\right)$ in the specular-wall case, leading to a singular L/D value as $\alpha \to 0$.

8.5 Results for Temperature-Driven Flow

For a simple channel with specular walls and no airfoil, the mass flow rate is given by the equation (excluding an airfoil) [1]:

$$\dot{m}^{(\text{no-airfoil,spec})} = \frac{H}{2\sqrt{\pi}}\left(1 - \rho_{\text{out}}\sqrt{T_{\text{out}}}\right) \tag{8.16}$$

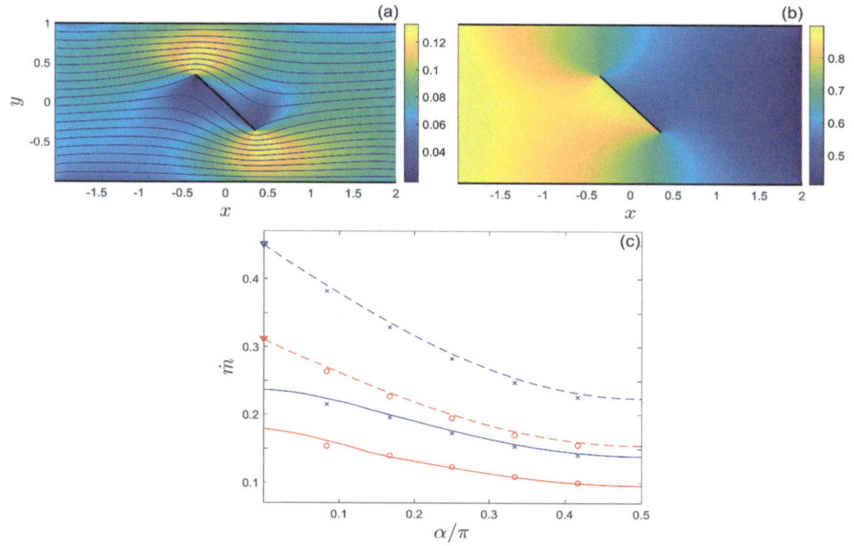

Fig. 8.9 Colormaps of the free-molecular **a** velocity magnitude and **b** pressure field. **c** Variations of the free-molecular tunnel mass flow rate with the airfoil angle of attack in diffuse- (solid lines) and specular-wall (dashed lines) channels. The blue and red curves show results for density-driven ($\rho_{\text{out}} = 0.2, T_{\text{out}} = 1$) and temperature-driven ($\rho_{\text{out}} = 1, T_{\text{out}} = 0.2$) setups, respectively, and the triangles at $\alpha = 0$ mark the free-molecular mass flow rates in a specular channel with no airfoil. The crosses and circles present respective DSMC results at Kn $= 10$ (Copied from open access Ref. [1])

According to Fig. 8.9, at the same pressure ratio as in previous experiments ($p^*_{\text{out}}/p^*_{\text{in}} = 0.2$), the tunnel experiences significantly lower speeds and pressure differences in a temperature-driven flow setup ($T_{\text{out}} = 0.2$, $\rho_{\text{out}} = 1$) than density-driven counterparts.

At high Knudsen numbers, temperature-driven channel flows result in less mass flow rates than density-driven flows. As shown in Fig. 8.9, at $\alpha = 0$, the mass flow rate matches the value from Eq. (8.16) in a specular channel, and the results are strongly validated by the DSMC data presented at Kn $= 10$. As α increases from 0 to $\pi/2$, the obstruction from the airfoil reduces the mass flow rate, with lower flow rates observed in temperature-driven setups than density-driven ones. Also, in diffuse-wall setups, the mass flow rate is lower compared to specular ones, as the flow's interaction with the walls leads to a transfer of energy, reducing its momentum, with $\approx$50% reduction in specular-wall configurations and $\approx$35% reduction in diffuse-wall setups.

In Fig. 8.10a, the lift force is lower in the temperature-driven case than the density-driven setup. Figure 8.10b reveals that, unlike the density-driven case where all curves merge into one, the lift-to-drag ratio changes with varying T_{out}, gradually aligning with the density-driven curve as the outlet temperature decreases. This behavior illustrates the nonlinear relation between Maxwellian distributions and temperature, and despite these changes, the optimal angle of attack for aerodynamic efficiency is

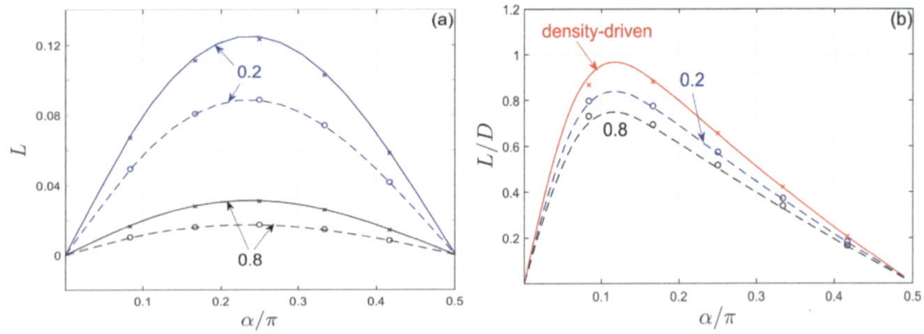

Fig. 8.10 Variations of the free-molecular **a** lift and **b** lift-over-drag ratio with the plate angle of attack in a diffuse-wall tunnel for density-driven (solid lines) and temperature-driven (dashed curves) flows. The numbers indicate the values of ρ_{out} (= 0.8, 0.2 with $T_{\text{out}} = 1$) or T_{out} (= 0.8, 0.2 with $\rho_{\text{out}} = 1$). The symbols denote DSMC predictions at Kn = 10 in the respective density-driven (crosses) and temperature-driven (circles) setups (Copied from open access Ref. [1])

slightly greater than $\alpha \approx 0.1\pi$. Furthermore, the density-driven setup produces higher aerodynamic forces at a fixed pressure ratio due to a larger mass flow rate than the temperature-driven configuration.

8.6 The Free-Stream Limit

To examine the correlation of the calculations with the free-stream results at the high Knudsen number flow regimes, first, the collisionless free-stream aerodynamic loading expressions for a plate with an angle α in a uniform flow are considered. For a diffuse-reflecting plate, the normal (F_N) and parallel (F_S) forces in the free-stream regime are described by [1]:

$$F_N^{(\text{FS diff})} = \frac{1}{2}\left[\left(2U_N^2 + 1\right)\text{erf}(U_N) + \frac{2}{\sqrt{\pi}}U_N \exp\left[-U_N^2\right] + \sqrt{\pi}U_N\right] \tag{8.17}$$

$$F_S^{(\text{FS,diff})} = U_S\left[U_N \text{erf}(U_N) + \frac{1}{\sqrt{\pi}}\exp\left[-U_N^2\right]\right] \tag{8.18}$$

Here, $U_N = U_0 \sin\alpha$, $U_S = U_0 \cos\alpha$, with $\text{erf}(s) = 2\pi^{-1/2}\int_0^s \exp\left[-q^2\right]dq$ representing the Gauss error function. U_0 is the dimensionless speed of flow. The lift and drag forces are then expressed as:

$$L = F_N \cos\alpha - F_S \sin\alpha \text{ and } D = F_N \sin\alpha + F_S \cos\alpha \tag{8.19}$$

The setup in this study differs significantly from the free-stream description. It involves a "blow-down" tunnel where the flow consists of Maxwellians of vanishing macroscopic speeds resulting from particles originating from the entry and exit points

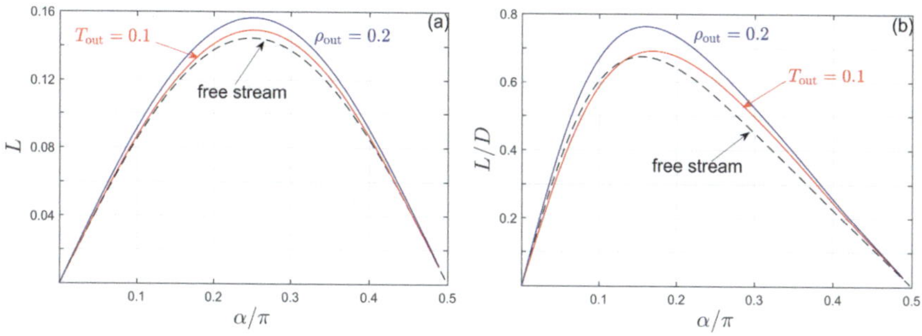

Fig. 8.11 Variations with plate inclination of the **a** lift force and **b** lift-to-drag ratio for collisionless free-stream flow (dashed curves) and tunnel flow (blue and red lines). The free-stream results are for a diffuse plate set at uniform x-directed flow of speed $U_0 = 0.2$. The blue and red curves mark the predictions for density-driven (with $\rho_{out} = 0.2$) and temperature-driven (with $T_{out} = 0.1$) flows in a diffuse tunnel of width $H = 10$, respectively (Copied from open access Ref. [1])

of the tunnel. Unlike blow-down tunnels under continuum conditions, the current setup lacks molecular collisions, and the hydrodynamic description may not converge to the free-stream limit, especially when the tunnel walls are far apart.

According to Fig. 8.11, despite the obvious distinctions between the blow-down tunnel and the free stream, the results show that the differences in aerodynamic forces are relatively small. Both setups, where density or temperature differences drive flow, show similar trends in how lift and lift-to-drag ratio change. However, the temperature-driven setup shows better agreement with the free-stream results, with less than a 3% discrepancy at the peak. This better correlation is due to the Maxwellian distribution dependency on T_{out} and velocity U_0, compared to the linear dependence on outlet density ρ_{out} in the density-driven case.

To examine the convergence of our approach to the free-stream limit, we replace identical Maxwellians that have a velocity U_0 with the pressure gradient between reservoirs in the inlet and outlet.

$$f_{in} = f_{out} = \frac{1}{\pi^{3/2}} \exp\left[-|\xi - U_0 \hat{\mathbf{x}}|^2\right] \tag{8.20}$$

As the channel width increases (H $\gg$ 1), the system is expected to approach the non-confined model, offering insight into how the tunnel walls influence the deviation from free-stream conditions in the high Knudsen number regime.

According to Fig. 8.12, the lift force at a channel is lower than in the free stream due to flow momentum loss from wall collisions, but as the channel width increases, the solution converges to the free-stream results. At H = 10, the walls no longer significantly impact the flow, and the results closely match the free-stream limit.

When H becomes very large, the effects of the tunnel's surfaces become insignificant, and the particles approach the plate surface. Due to the asymmetry of the problem, an x-direction flow is created, and the free-molecular forces on the plate can be calculated in a closed integral form.

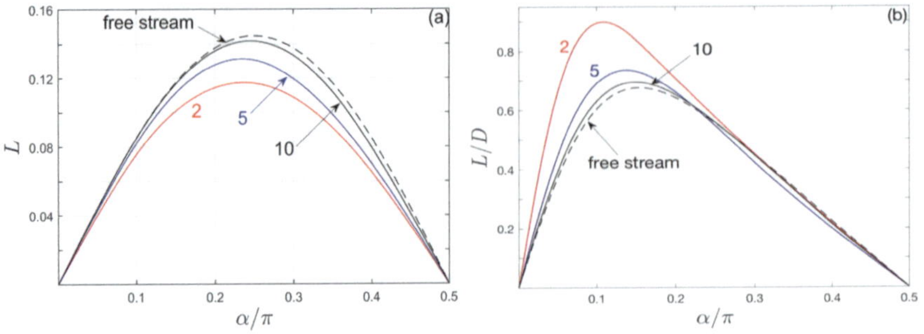

Fig. 8.12 Convergence of the free-molecular **a** lift force and **b** lift-to-drag ratio to the free-stream limit in a diffuse-wall tunnel with $U_0 = 0.2$. The red, blue, and black curves show variations with the plate angle of attack at the indicated $H = 2,\ 5$, and 10 tunnel width values, respectively. The dashed lines show the counterpart free-stream results for a diffuse reflecting plate (Copied from open access Ref. [1])

Figure 8.13a demonstrates the convergence of drag force across a specular plate as H increases, with $\rho_{out} = 0.2$, $T_{out} = 1$, and specified angles of attack α. Convergence occurs when $H \gtrsim 20$ and for $H = 2$, as discussed in previous sections, the results depend on H to highlight the impact of channel reflections and boundaries, with a large H reducing this impact by increasing the geometrical decay of the signal released from the walls.

Also, in Fig. 8.13a, the results are further supported by analyzing the large-H limit of the velocity reservoir setup. As with the thermodynamic reservoir setup, convergence is observed for $H \gtrsim 20$, reaching the free-stream force values listed in Eqs. (8.17)–(8.19). From the findings in Fig. 8.13, it is clear that the results for the narrower $H = 2$ tunnel are influenced by sidewall reflections, aligning with the objective of this work to study the coupling effect of the walls in a rarefied gas wind

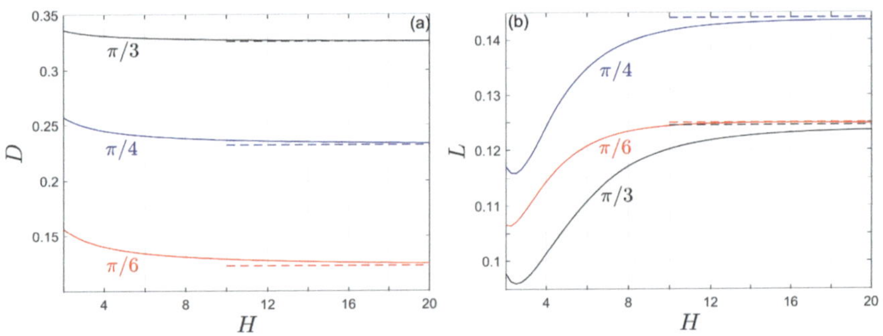

Fig. 8.13 Variation with H of the free-molecular **a** drag force in a specular-wall tunnel with $\rho_{out} = 0.2$, $T_{out} = 1$; **b** lift force in a diffuse-wall tunnel with $U_0 = 0.2$. The numbers indicate the values of α, the plate angle of attack, and the dashed lines mark the large-H asymptotes calculated in Appendix B for **a** and Eqs. (24)–(26) for **b** (Copied from open access Ref. [1])

tunnel. Tunnel confinement impacts are minimal at high Knudsen numbers when the channel walls are positioned over ≈10 chord-lengths from the plate surfaces.

8.7 Aerodynamic Characteristics

Shoja-Sani et al. [2] investigated the simulation of rarefied supersonic and subsonic gas flows around a NACA 0012 airfoil using both continuum and particle methods. The schematic of NACA 0012 has been illustrated in Fig. 8.14 and its coordinated is formulated as:

$$y_t(x) = 0.6\left[0.2969\sqrt{x} - 0.1260x - 0.3516x^2 + 0.2843x^3 - 0.1015x^4\right] \quad (8.21)$$

where, y_t is the half thickness at normalized location of x. The Navier–Stokes equations with first-order slip/jump boundary conditions are solved using OpenFOAM, while the DSMC solver (dsmcFOAM) is enhanced with a newly introduced "Simplified Bernoulli Trial (SBT)" scheme for inter-molecular collision modeling. Figure 8.14 shows the C_p distribution calculated by DSMC and NS slip flow at $\alpha = 0°$ at $Kn = 0.26$. Significant differences are observed between the two methods in predicting the pressure coefficient.

Figure 8.15 illustrates the impact of the rising Knudsen number on the slip velocity and pressure coefficient (C_p). In Fig. 8.15a, C_p remains constant on the upper surface across all cases, while on the lower surface, it increases as Kn rises. This occurs because, with higher Kn, the initial energy of the molecules is transformed into increased viscous forces and pressure. For an airfoil at an angle of attack, the stagnation point is on the lower surface, where viscous forces become more influential on

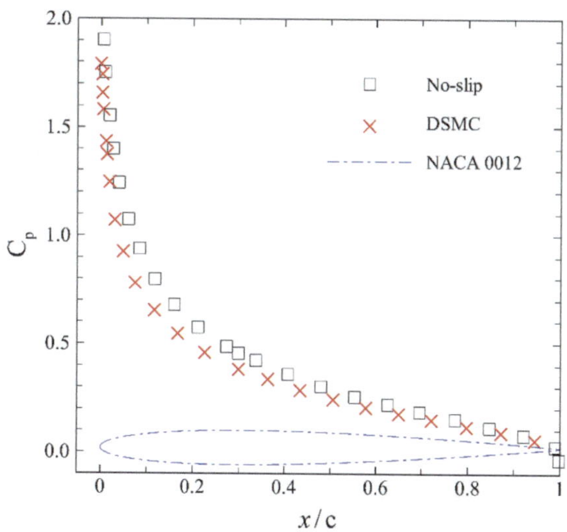

Fig. 8.14 Comparison of surface pressure distributions on the upper surface of the airfoil at Mach = 2, $\alpha = 0°$ and $Kn = 0.26$ (Reproduced from Ref. [2])

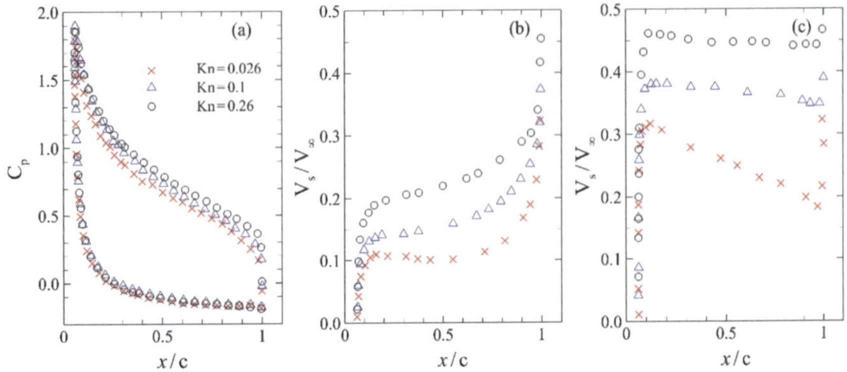

Fig. 8.15 Effect of Knudsen number on the **a** pressure coefficient, **b** slip velocity distributions at the upper surface, and **c** slip velocity distributions at the lower surface, Mach = 2, α = 20° (Reproduced from Ref. [2])

velocity, and an increase in the Knudsen number leads to higher viscosity. Conversely, viscous forces have less effect on the upper surface, where velocity is higher, making C_p relatively independent of viscosity and nearly constant.

Figure 8.15b shows that slip velocity increases significantly on the airfoil's upper surface. This is due to the rarer flow on the upper surface, where fewer particles are present, preventing them from reaching equilibrium with the surface at higher Knudsen numbers.

Figure 8.16 compares the pressure coefficient along the stagnation line in front of the airfoil at $\alpha = 0°$. The figure highlights the shock formation location, which starts further away from the airfoil as Kn increases. Higher viscous forces result in a wider and more diffusive shock wave. Furthermore, stagnation pressure changes slightly with increasing Kn.

Figure 8.17 demonstrates how the shock's shape and curvature depend on the Knudsen number. In highly rarefied regimes, the shock wave transitions into gradual diffusive waves propagating upstream and appears more curved. In contrast, at lower Kn numbers, the free stream's higher flow density and greater total kinetic energy result in sharper shock waves closer to the leading edge.

Figure 8.18 shows that the shear stress is highest at the leading edge on the airfoil's upper surface, followed by a sharp decrease in all Kn numbers. As the Kn number increases, lower shear stress values are observed, attributed to fewer particle–surface interactions at higher rarefaction levels.

Figure 8.19 illustrates the aerodynamic performance of the NACA 0012 airfoil in rarefied flow conditions. Figure 8.19a, b compare the lift (C_L) and drag (C_D) coefficients with predictions from linearized theory. As shown in Fig. 8.19a, C_L remains unchanged with increasing Kn. In contrast, C_D increases with Kn, as illustrated in Fig. 8.19b. For small angles of attack (up to $\alpha = 10°$), the lift coefficient matches well with the linearized theory, but the drag coefficient deviates. This deviation highlights that the linearized theory fails to accurately predict C_D due to neglecting viscous

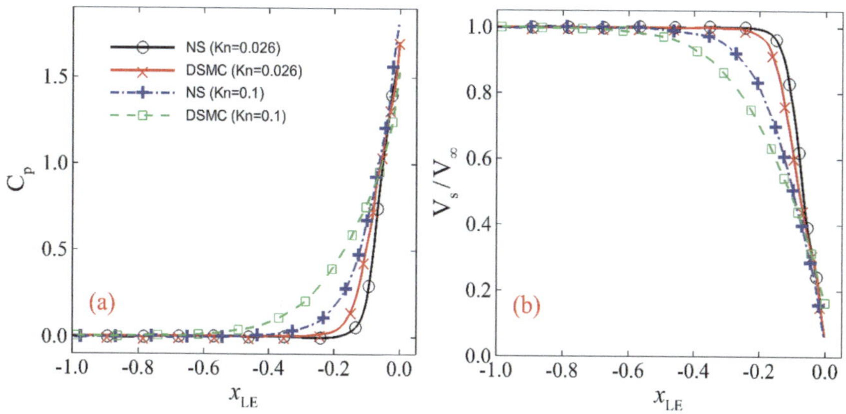

Fig. 8.16 **a** Pressure coefficient distribution and **b** Normalized velocity on the stagnation line of the airfoil, Mach = 2, α = 0° (Reproduced from Ref. [2])

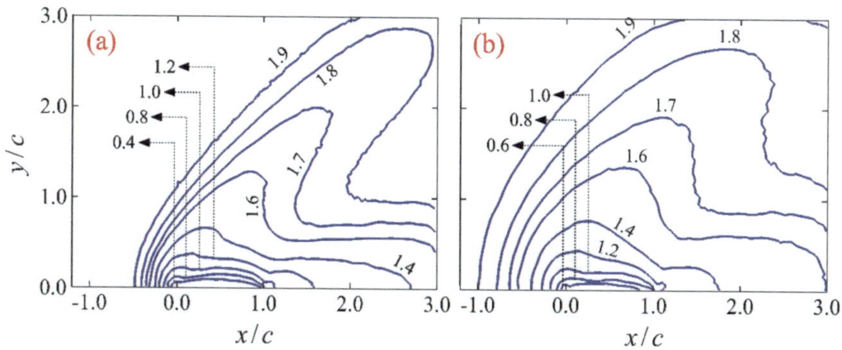

Fig. 8.17 Mach number contour from DSMC solution at Mach = 2, α = 0° **a** Kn = 0.1 **b** Kn = 0.26 (Reproduced from Ref. [2])

forces. Figure 8.19c shows that the maximum value of the airfoil's drag polar (C_L/C_D) occurs at α = 25° for all cases. As the Kn number increases, the value of C_L/C_D decreases due to the drag coefficient increase while the lift coefficient remains nearly constant, as seen in Fig. 8.19a, b. Lower Kn numbers are advantageous since they result in a higher C_L/C_D at the same angle of attack. Figure 8.19d shows that the lift coefficient's rate of change to the angle of attack is less than the supersonic linearized theory predicts at low Kn numbers but increases with higher Kn. The drag coefficient, however, is higher than the linearized value and continues to increase as Kn rises.

Tai [3] used Bird's Direct Simulation Monte Carlo (DSMC) method to investigate low-density aerodynamic characteristics around airfoils in the transition regime. Numerical simulations are conducted for a NACA-0009 airfoil at Mach 4 and 5, with an angle of attack of 1.25°, at altitudes of 56 and 62 miles above sea level. The

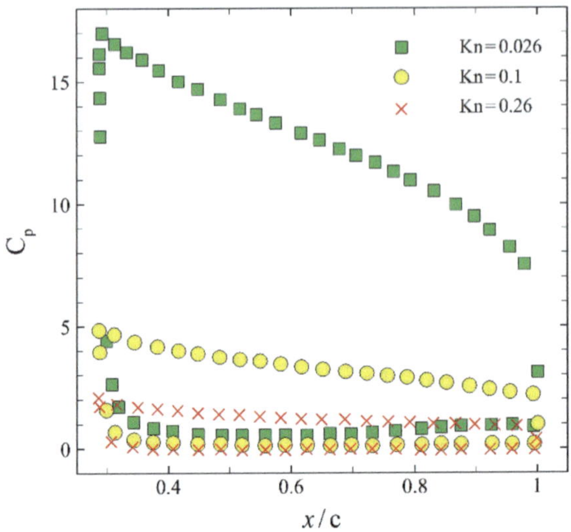

Fig. 8.18 Wall shear stress on the airfoil surface at different Kn numbers, Mach = 2, α = 45°, DSMC calculations (Reproduced from Ref. [2])

Knudsen number ranges from 0.47 to 1.15, and the Reynolds number varies between 15 and 114, significantly departing from continuum theory.

Figure 8.20 presents the lift coefficient for the NACA-0009 airfoil as a function of Reynolds number. The lift coefficient slightly increases as the Reynolds number decreases or the Knudsen number increases.

Salehin [4] analyzed supersonic flow over symmetric NACA airfoils (NACA 0012, NACA 0010, and NACA 0007) in transitional rarefied conditions using the Direct Simulation Monte Carlo (DSMC) method. The simulations are conducted with the dsmcFOAM solver in the OpenFOAM platform. The Mach number ranges from 2 to 4, and the Knudsen number varies from 0.5 to 5.0, representing a transition between rarefied and continuum flow. The angle of attack for all the simulations is kept at 0°, and the focus is on comparing the aerodynamic performance and heat load across three different airfoils.

The relationship between fluid density and drag coefficient is explored in this study. It is found that while drag increases with fluid density, the drag coefficient decreases as fluid density increases. This is because the drag coefficient is inversely proportional to fluid density when referenced in terms of drag force, which accounts for the density. Conversely, the drag coefficient increases exponentially at higher altitudes, where air viscosity and density are reduced.

The study also investigated the effect of the Knudsen number on aerodynamic drag. For fixed Mach numbers, drag increases with an increasing Knudsen number. The relationship between drag coefficient and velocity is also analyzed, showing that the drag coefficient decreases as velocity increases. However, for constant Knudsen numbers, a linear decrease in drag is observed as the Mach number increases. These

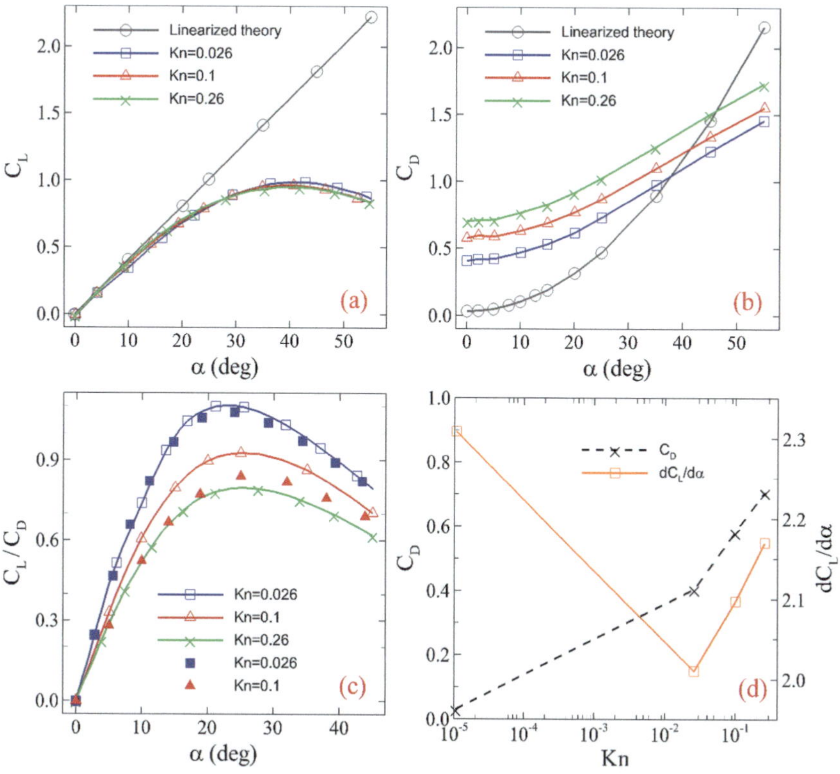

Fig. 8.19 Aerodynamics performance of the airfoil at the supersonic regime **a** C_L comparison **b** C_D comparison **c** Drag polar **d** Lift slop and drag coefficient at $\alpha = 0°$ (Reproduced from Ref. [2])

Fig. 8.20 Lift coefficient versus Reynolds number for NACA-0009 airfoil at h = 56 and 62 miles (Reproduced from Ref. [3])

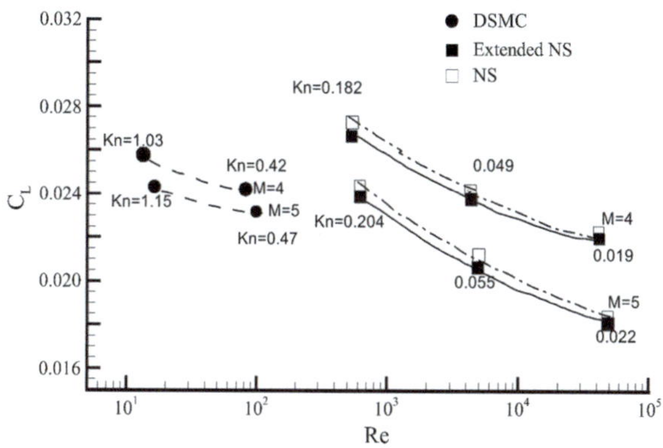

findings indicate the significant impact of rarefaction effects on the drag force in transitional flow conditions (Fig. 8.21; Tables 8.1, 8.2 and 8.3).

Furthermore, aerothermodynamic performance is compared for the three airfoils (NACA 0012, NACA 0010, and NACA 0007), noting that despite similar performance in terms of pressure coefficient, density, and shear stress, airfoil thickness

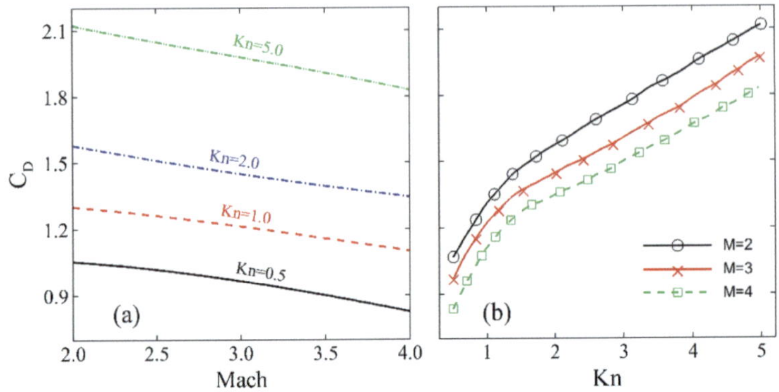

Fig. 8.21 Representation of drag coefficient as the function of **a** Mach number and **b** Knudsen number for NACA 0012 at AOA = 0° (Reproduced from Ref. [4])

Table 8.1 Drag coefficient for NACA 0012 airfoil [4]

Kn	M = 2	M = 3	M = 4
0.5	1.06	0.97	0.83
1.0	1.31	1.22	1.11
2.0	1.58	1.45	1.35
5.0	2.12	1.98	1.84

Table 8.2 Drag coefficient for NACA 0010 airfoil [4]

Kn	M = 2	M = 3	M = 4
0.5	1.02	0.92	0.79
1.0	1.26	1.18	1.07
2.0	1.54	1.42	1.32
5.0	2.08	1.94	1.80

Table 8.3 Drag coefficient for NACA 0007 airfoil [4]

Kn	M = 2	M = 3	M = 4
0.5	0.95	0.83	0.71
1.0	1.16	1.08	0.98
2.0	1.45	1.30	1.24
5.0	1.97	1.78	1.68

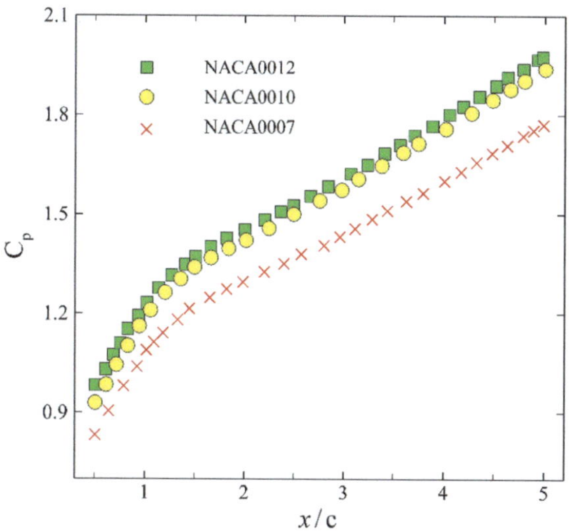

Fig. 8.22 Airfoil thickness effect on drag coefficient at $M_\infty = 3$ (Reproduced from Ref. [4])

plays a crucial role in determining drag and heat flux. Among these, the NACA 0007 airfoil is identified as the most optimized for drag reduction and heat flux performance under supersonic flow in transitional rarefied conditions. On the other hand, the NACA 0012 airfoil, due to its larger thickness, experiences higher drag and heat flux, acting more like a blunt body at supersonic speeds (Fig. 8.22).

8.8 Generalized Hydrodynamic Model

The Newtonian framework, which relies on linear uncoupled constitutive relations such as Fourier's law for thermal conduction and Newton's law for viscous stress, leads to the Navier–Stokes-Fourier (NSF) equations. Xiao and Wang [5] investigated the numerical analyses of rarefied gas flows around an airfoil using the generalized hydrodynamic (GH) model, an implicit non-NSF framework that extends beyond the limitations of traditional NSF equations.

Figures 8.23 and 8.24 show that both cases have a sharp pressure drop near the leading edge. However, the pressure increases at the trailing edge for $M = 0.8$ and $Kn = 0.2$ cases, a behavior not observed in other cases where expansion and a corresponding pressure drop occur. This unique recompression at the trailing edge in high Knudsen conditions is absent in NSF solutions. In near-equilibrium states, NSF solutions predict higher normal stress in compression regions than what occurs, leading to exaggerated recompression. These findings suggest that conventional airfoils are not optimal for rarefied gas flow conditions.

Figure 8.25a presents the lift (C_L) and drag (C_D) coefficients for flows over a microscale NACA 0012 airfoil with a fixed angle of attack ($\alpha = 20°$). Both C_L and C_D increase as Kn increases. However, the lift coefficient shows only a slight increase

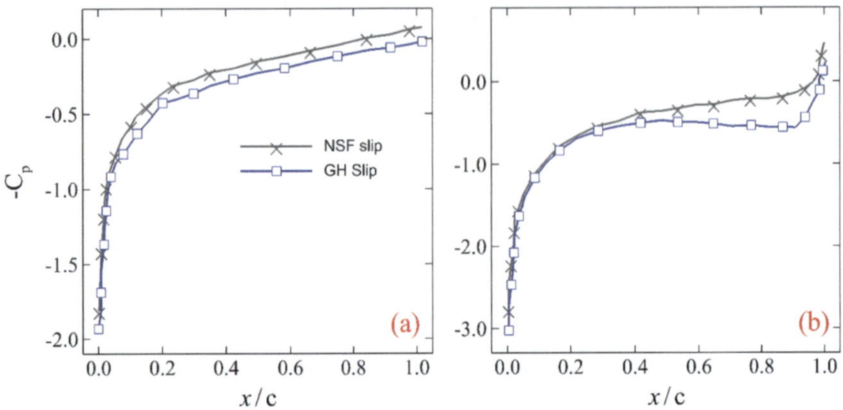

Fig. 8.23 Surface pressure distributions at M = 0.8 for **a** Kn = 0.02 and **b** Kn = 0.2 (Reproduced from Ref. [5])

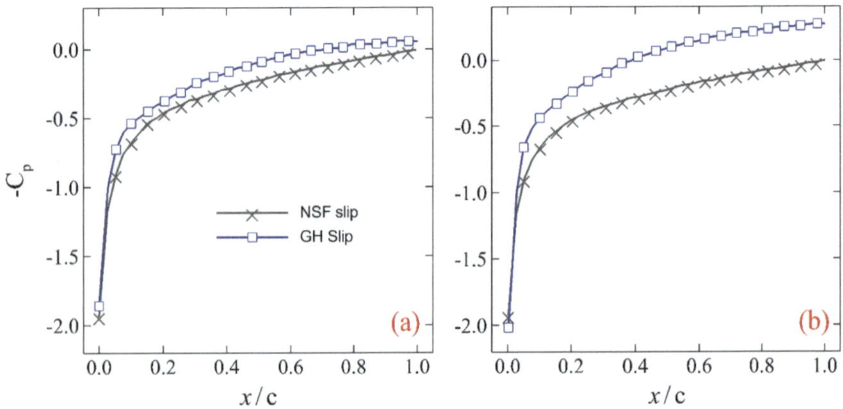

Fig. 8.24 Surface pressure distribution at M = 2.0 for **a** Kn = 0.02 and **b** Kn = 0.2 (Reproduced from Ref. [5])

compared to the more rapid rise in the drag coefficient. Consequently, the lift-to-drag ratio reduces and falls below 1 at high Knudsen states, as illustrated in Fig. 8.25b.

The nonequilibrium state of the flow depends on the Knudsen number (Kn) and Mach number (Ma). Figures 8.26 and 8.27 show the lift and drag coefficients, along with the lift-to-drag ratio, as functions of the angle of attack. Increasing the angle of attack results in a higher lift-drag ratio. However, the trends differ between subsonic and supersonic conditions. In subsonic conditions, the NSF framework overpredicts the lift-drag ratio, while in supersonic conditions, it underpredicts the ratio.

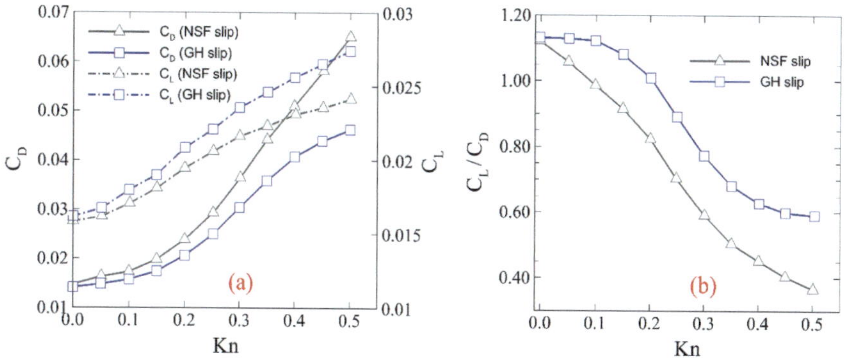

Fig. 8.25 **a** Lift and drag coefficients and **b** Lift-drag ratio at M = 2.0 and α = 20 (Reproduced from Ref. [5])

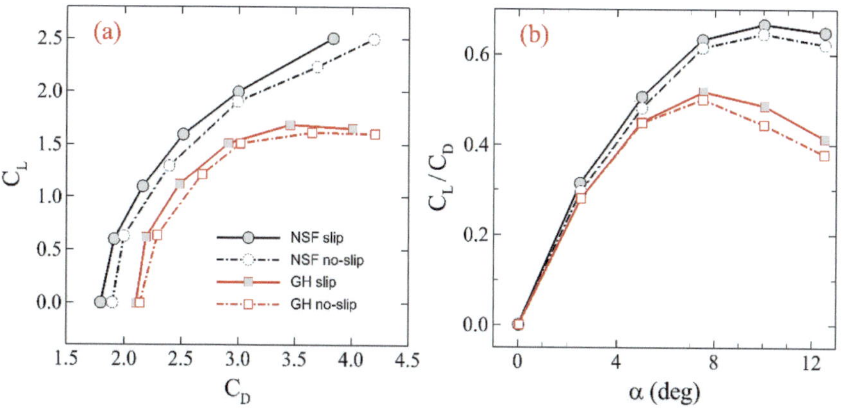

Fig. 8.26 C_L and C_L/C_d at M = 0.8 and Kn = 0.5 (Reproduced from Ref. [5])

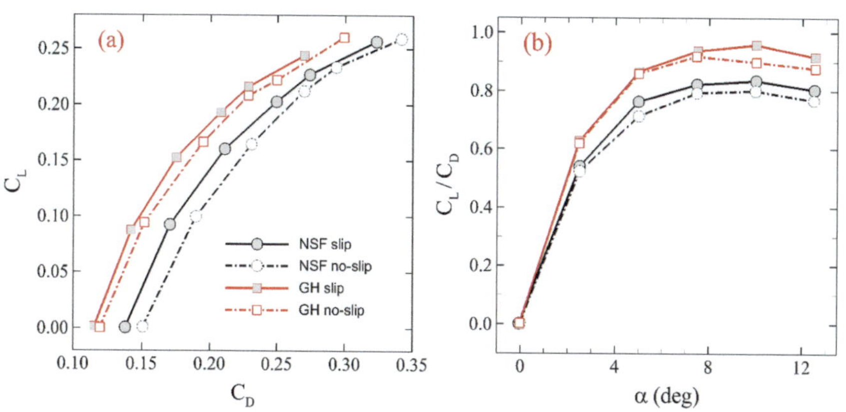

Fig. 8.27 C_L and C_L/C_d at M = 2.0 and Kn = 0.5 (Reproduced from Ref. [5])

In supersonic cases, the viscous stress predicted by the GH model is lower than that of the NSF framework in compression regions, leading to weaker compression at the leading edge and reduced drag. Conversely, in subsonic cases, the near-equilibrium state causes the GH model to predict higher viscous stress than the NSF framework in compression regions, resulting in stronger compression at the leading edge and higher drag. These findings highlight that rarefied effects on an airfoil's aerodynamic characteristics differ significantly between far-from-equilibrium and near-equilibrium conditions.

8.9 Information Preservation Method

Sun and Boyd [6] investigated microscale flows over zero-thickness flat plates using Information Preservation (IP) and continuum with slip boundary conditions methods. The (IP) method, based on the Direct Simulation Monte Carlo (DSMC) method, is implemented to predict behavior flows at low Reynolds numbers ($0.1 \leq Re \leq 100$). The IP method effectively reduces statistical scatter in low-speed, constant-density flow systems by preserving macroscopic and microscopic information. The results suggest that continuum methods may not fully capture rarefied impacts.

Figure 8.28 illustrates that the Information Preservation (IP) method aligns with free-molecular theory ($C_D = 1.35/M$) at very low Reynolds numbers ($Re < 0.2$) and matches numerical solutions of the Navier–Stokes equations for incompressible flows at higher Reynolds numbers ($Re > 10$). DSMC results are also included but show greater statistical scatter at lower Reynolds numbers. Tamada and Miura's theoretical prediction for the drag coefficient in slip flow, based on the Oseen-Stokes equations, aligns with the IP method for Reynolds numbers around 1–2. Schaaf and Sherman's experimental results, displayed in Fig. 8.28, were mostly collected with a Mach number of approximately 0.2. However, they used a slightly lower Mach number (0.167) for the experiment with a Reynolds number of 3.15. Experimental data match well with IP results, except for a slight discrepancy at $Re = 3.15$, attributed to Mach number effects. Incompressible experimental data from Janour show minimal differences from results when $Re > 10$. The Blasius boundary layer theory is invalid for low Reynolds number flows, but the continuum approach with slip boundary conditions provides accurate predictions at higher Reynolds numbers ($Re > 10$). Overall, the IP method demonstrates good agreement with experimental and theoretical results, making it effective for studying drag in rarefied and low Reynolds number flows.

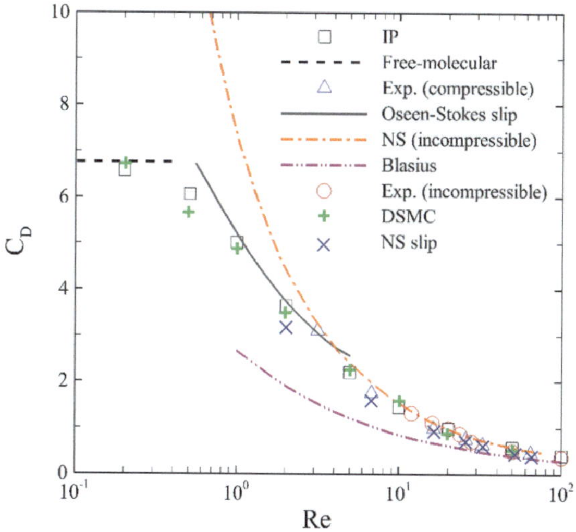

Fig. 8.28 Drag coefficient of a finite plate at low Reynolds numbers (Reproduced from Ref. [6])

8.10 Gurney Flap

Amini et al. [7] numerically investigated rarefied gas flow around a NACA 0012 airfoil with an attached Gurney flap at Mach 2 and three Knudsen numbers (0.026, 0.1, and 0.26). The study examines the effects of angle of attack and Gurney flap height on the aerodynamic characteristics of the airfoil, using the Navier–Stokes equations with slip boundary conditions and the DSMC method. The DSMC simulations are performed using dsmcFoam, an OpenFOAM solver, implementing the Variable Hard Sphere (VHS) collision model and Larsen-Borgnakke internal energy model.

Figure 8.29a–c shows that the pressure coefficient increases with higher Knudsen numbers at the leading edge. At the trailing edge, the pressure rises with the height of the Gurney flap (GF), contributing to an increased lift coefficient, and the pressure coefficient for a 5% flapped airfoil increases as Kn rises. This figure's top lines correspond to the lower surface, while the bottom lines represent the airfoil's upper surface. The pressure coefficient reduces across the surface, reaching a minimum at the trailing edge on the lower surface of a clean airfoil. However, adding a Gurney flap causes the pressure coefficient to initially decrease and then increase sharply near the flap. In contrast, the upper surface pressure distribution remains largely unaffected for both clean and Gurney flap configurations.

Figure 8.30 demonstrates that, for all Gurney flap heights, the drag coefficient increases with higher Knudsen numbers, regardless of the angle of attack. Additionally, at each Knudsen number, the drag coefficient increases as the angle of attack rises.

Figure 8.31 reveals that increasing the Gurney flap height significantly raises the lift coefficient. However, in all Knudsen numbers, C_L decreases beyond an angle

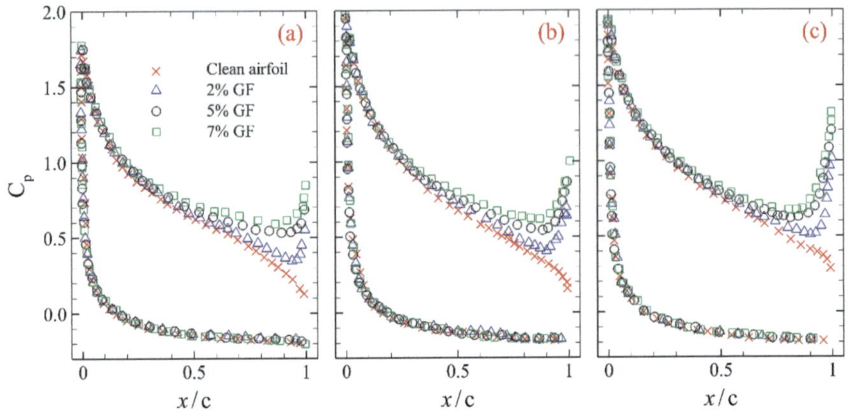

Fig. 8.29 Pressure coefficients distribution on the airfoil surface for clean and flapped airfoils for $M = 2$ and AOA = 20° at **a** Kn = 0.026, **b** Kn = 0.1, and **c** Kn = 0.26 (Reproduced from Ref. [7])

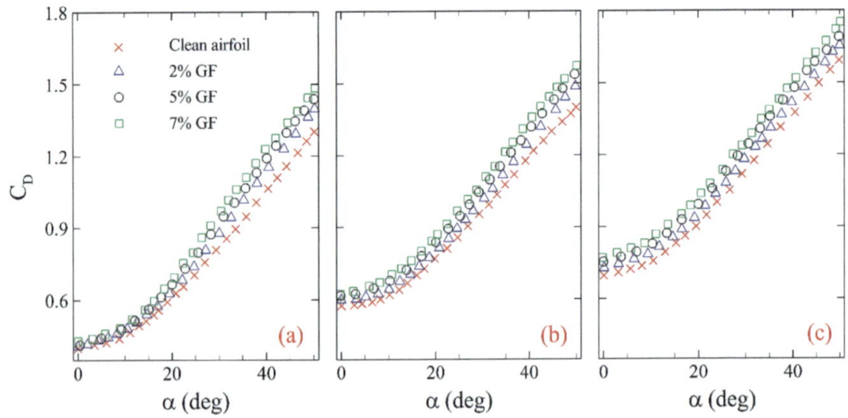

Fig. 8.30 Drag coefficient at **a** Kn = 0.026, **b** Kn = 0.1, and **c** Kn = 0.26 (Reproduced from Ref. [7])

of attack of about 40°. Additionally, C_L slightly decreases as the Knudsen number increases.

Figure 8.32 illustrates that, for Kn = 0.026 and Kn = 0.1, Gurney flaps with heights of 5% and 7% significantly enhance the C_L/C_D at angles of attack below 30°, while the 2% Gurney flap has minimal effect. At angles of attack above 30°, for Kn = 0.026, C_L/C_D decreases; for Kn = 0.1, C_L/C_D values for all Gurney flap heights are nearly identical and greater than those of a clean airfoil. At Kn = 0.26, all Gurney flap heights increase C_L/C_D across all angles, with the 5% and 7% flaps having a much stronger effect than the 2% flap.

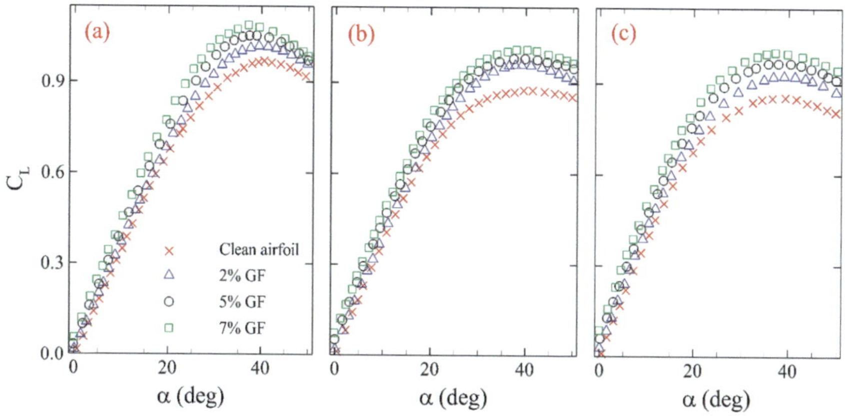

Fig. 8.31 Lift coefficient at **a** Kn = 0.026, **b** Kn = 0.1, and **c** Kn = 0.26 (Reproduced from Ref. [7])

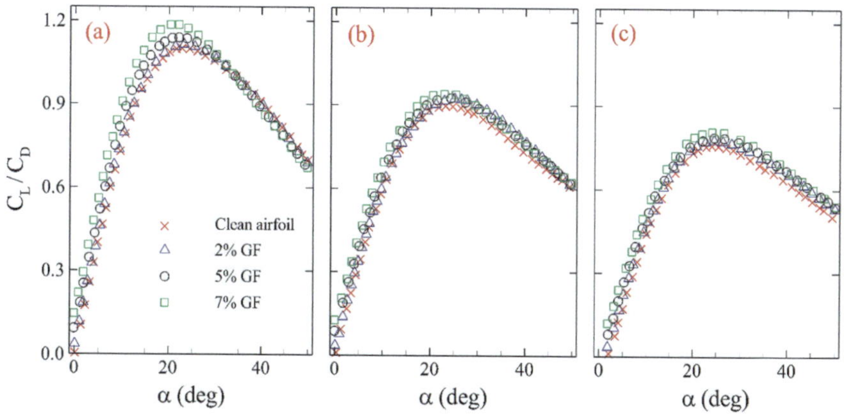

Fig. 8.32 Lift to-drag coefficient ratio at **a** Kn = 0.026, **b** Kn = 0.1, and **c** Kn = 0.26 (Reproduced from Ref. [7])

8.11 Rarefaction Effects for Transonic Flows at Low Reynolds Numbers

Pekarda and Alexeenko [8] analyzed the rarefaction effects on airfoils in the low Reynolds transonic flow regime ($M_\infty = 0.8$) using the continuum Navier–Stokes (NS) equations, the ellipsoidal statistical Bhatnagar–Gross–Krook (ES-BGK) kinetic equation, and the Direct Simulation Monte Carlo (DSMC) method with the variable soft sphere (VSS) model. A parametric study is conducted to examine the influence of Mach number, Reynolds number, thickness, angle of attack, and camber on rarefaction effects for symmetric (NACA 0012 and 0007) and cambered (NACA 2407) airfoils. With and without slip corrections, the ANSYS Fluent solver is used for NS simulations, and DSMC simulations use Bird's molecular collision parameters.

8.11.1 Rarefaction Effects with Reynolds Number

Rarefaction effects in the transonic regime are quantified for the NACA 0012 airfoil at $M_\infty = 0.8$, with Reynolds numbers ranging from 10^1 to 10^4, with $\alpha = 0°$ for different. Horizontal velocity and local Knudsen number contours (Fig. 8.33a–d) show that rarefaction effects become significant as the Reynolds number decreases from 10^4 to 10^3, with the flow becoming fully rarefied at $Re_\infty = 10$. These effects are influenced not only by wall slip but also by regions of high gradients. Table 8.4 summarizes drag differences between ES-BGK, DSMC, and NS solutions, showing that local Knudsen numbers significantly influence rarefaction effects, even when the free-stream Knudsen number suggests continuum behavior. The results highlight that continuum solvers are less accurate in capturing high-gradient flow regions near the airfoil, even at higher Reynolds numbers.

8.11.2 Rarefaction Effects with Mach Number

The impact of Mach number on continuum breakdown is analyzed for a NACA 0012 airfoil at $\alpha = 10°$, comparing $M_\infty = 0.147$ and $M_\infty = 0.8$ ($Kn_\infty = 1.19 \times 10^{-3}$). Velocity magnitudes and local Knudsen numbers (Fig. 8.34) reveal that increasing M_∞ from 0.147 to 0.8 enlarges the flow separation region, creating a high-gradient region where local Knudsen numbers exceed 0.01. This stronger gradient region causes a 21% difference in drag at $M_\infty = 0.8$, compared to only 1% at $M_\infty = 0.15$.

At $M_\infty = 0.8$, Mach number contours (Fig. 8.34c) show that the separation region and wake vortices generate slip regions due to high-velocity gradients. These effects are confirmed by local Knudsen number contours (Fig. 8.34d), highlighting the slip regime near the airfoil and within the vortex core. Conversely, the lower Mach number $M_\infty = 0.15$ exhibits minimal rarefaction effects and negligible drag differences. The findings are summarized in Table 8.5. And show that the lift is less sensitive to rarefaction, with a difference of less than 3% between ES-BGK and NS results.

8.11.3 Rarefaction Effects with Thickness

The effect of airfoil thickness on fluid rarefaction is studied by comparing drag results for NACA 0012 and NACA 0007 airfoils at $M_\infty = 0.\ 8$, $\alpha = 0$, and $Kn_\infty = 1.19 \times 10^{-3}$. Velocity contours (Fig. 8.35a, b) reveal that the NACA 0012 airfoil exhibits higher maximum velocities due to its larger thickness and gradients. These

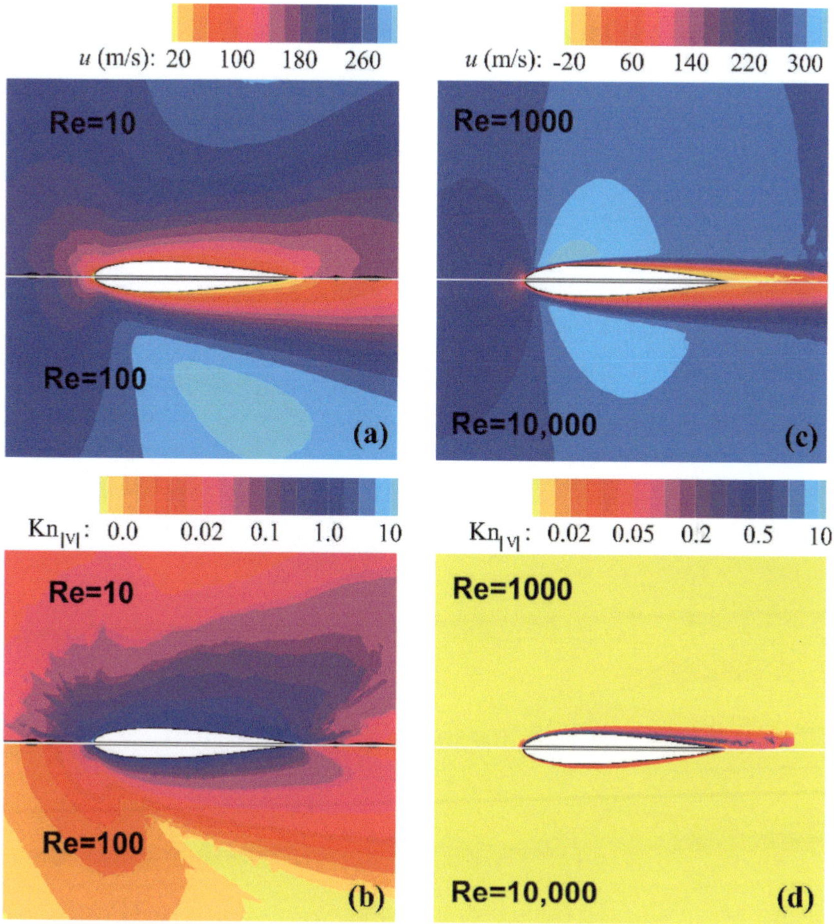

Fig. 8.33 Contours of x-velocity (**a**, **c**) and local Knudsen number (**b**, **d**) with ES-BGK method for $M_\infty = 0.8$, $\alpha = 0°$, and at different Reynolds numbers (Reproduced from Ref. [8])

Table 8.4 Drag values for NACA 0012 at $M_\infty = 0.8$; $\alpha = 0°$ with NS and ES-BGK solvers [8]

Re_∞	Kn_∞	$D_{ES\text{-}BGK}$	D_{DSMC} (mN)	D_{NS}	$(\lvert D_{NS}\text{-}D_{ES\text{-}BGK}\rvert / D_{ES\text{-}BGK}) \times 100$
10^1	1.19×10^{-1}	27.33 mN	27.72	72.01 mN	163.51
10^2	1.19×10^{-2}	107.21 mN	105.2	155.09 mN	45.52
10^3	1.19×10^{-3}	344.7 mN	–	418.41 mN	18.03
10^4	1.19×10^{-4}	1.28 N	–	1.35 N	5.01

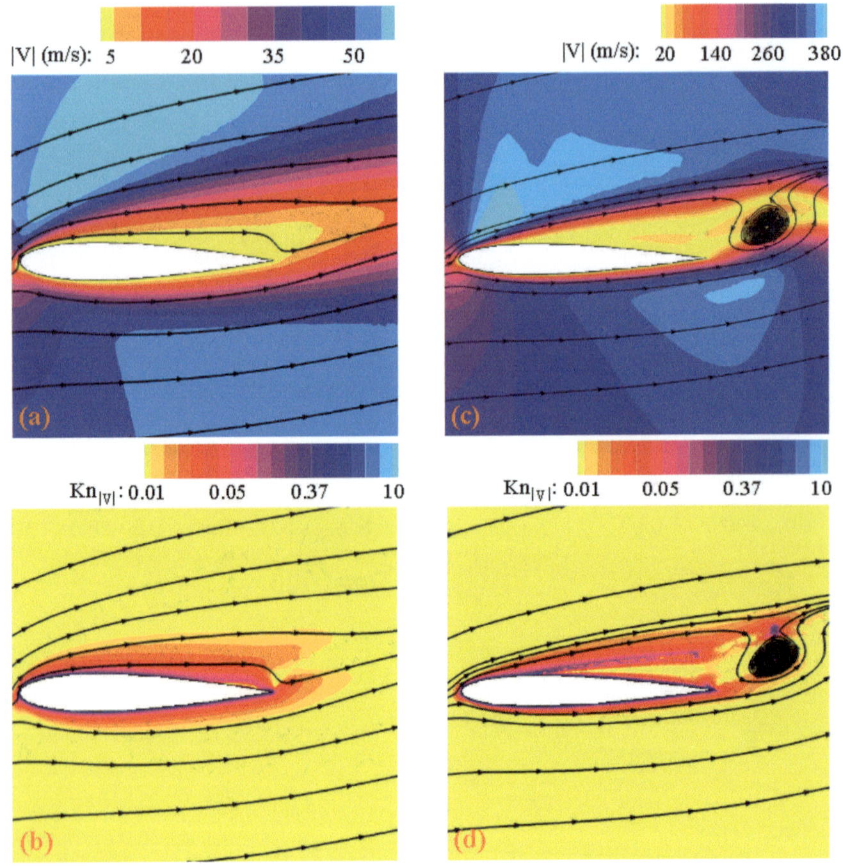

Fig. 8.34 Velocity magnitude and local Knudsen number with ES-BGK method for $M_\infty = 0.15$ (**a**, **b**) and 0.8 (**c**, **d**) at $Kn_\infty = 1.19 \times 10^{-3}$ and $\alpha = 10°$ (Reproduced from Ref. [8])

Table 8.5 Effect of speed at $\alpha = 10°$ and $Kn_\infty = 1.19 \times 10^{-3}$with the NS and ES-BGK solvers for NACA0012 [8]

M_∞	$D_{ES\text{-}BGK}$ (N)	D_{NS} (N)	$L_{ES\text{-}BGK}$ (N)	L_{NS} (N)	$(\lvert D_{NS}-D_{ES\text{-}BGK}\rvert/D_{ES\text{-}BGK}) \times 100$
0.15	0.01	0.011	0.034	0.035	1
0.80	0.43	0.52	1.51	1.53	21

high gradients, particularly from the stagnation point to the suction point, make the NACA 0012 results less accurate when using NS simulations, as confirmed by local Knudsen number contours (Fig. 8.35c, d). The larger separation region on the NACA 0012 airfoil creates a larger slip region from the suction point to the trailing edge, amplifying rarefaction effects. Additionally, using a thinner airfoil, such as the NACA 0007 instead of the NACA 0012, reduces rarefaction effects.

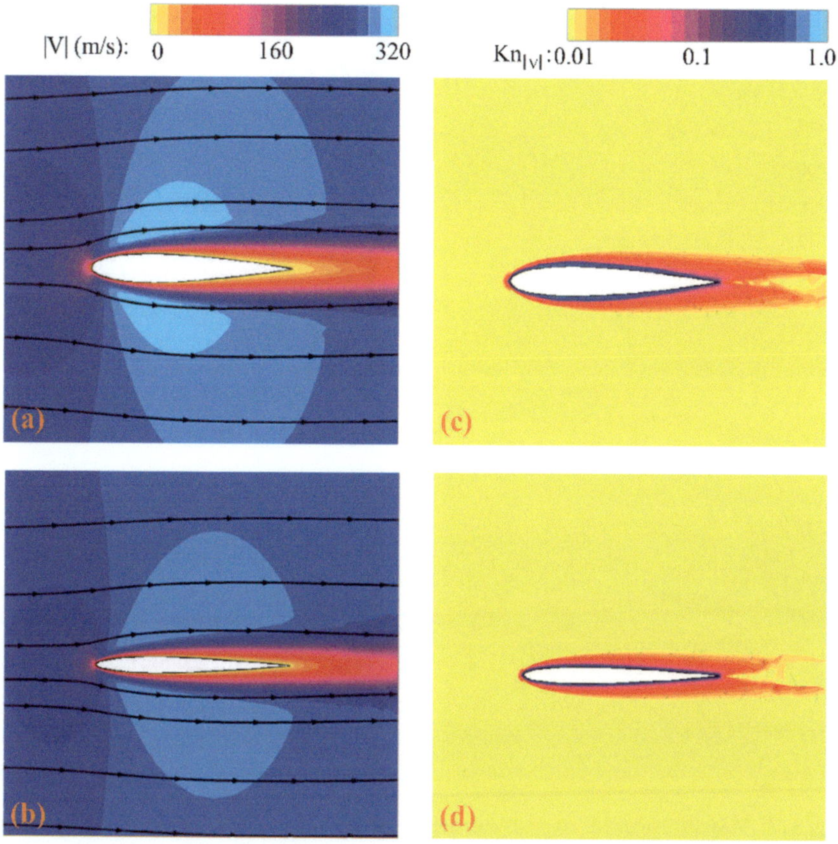

Fig. 8.35 Velocity magnitude contours and streamlines (**a**, **b**), and local Knudsen number contours (**c**, **d**) obtained by ES-BGK method for NACA 0012 (**a**, **c**) and NACA 0007 (**b**, **d**) at $M_\infty = 0.8$; $Kn_\infty = 1.19 \times 10^{-3}$, $\alpha = 0°$ (Reproduced from Ref. [8])

8.11.4 Rarefaction Effects with Camber

The effect of camber on drag and continuum breakdown is analyzed by comparing the NACA 0007 and NACA 2407 airfoils at $M_\infty = 0.8$, $\alpha = 0°$, and $Kn_\infty = 1.19 \times 10^{-3}$. Velocity contours in Fig. 8.36a, b show that the NACA 2407 has a larger separation region, making its drag and lift more susceptible to breakdown. This results in a 5% greater rarefaction difference in drag between NS and ES-BGK solutions. Therefore, camber-induced separation amplifies rarefaction effects for a given angle of attack. Thus, slip effects dominate at low Reynolds numbers, particularly for thicker airfoils like NACA 2407, where increased thickness and flow separation enhance rarefaction effects.

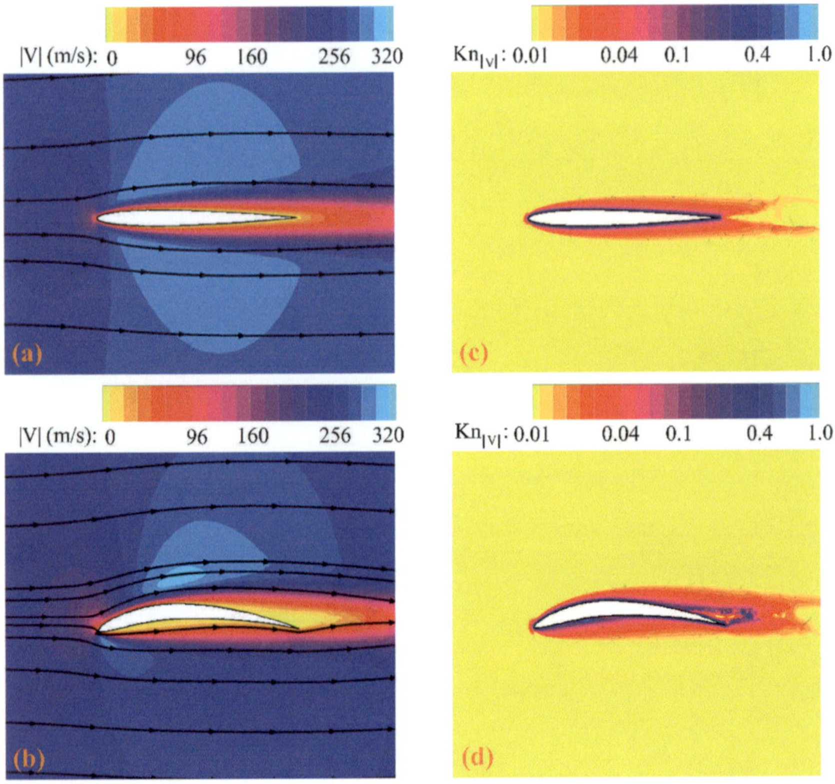

Fig. 8.36 Velocity magnitude contours and streamlines (**a**, **b**), and local Knudsen number contours (**c**, **d**) obtained by ES-BGK method for NACA 0007 (**a**, **c**) and NACA 2407 (**b**, **d**) at $M_\infty = 0.8$, $Kn_\infty = 1.19 \times 10^{-3}$, $\alpha = 0°$ (Reproduced from Ref. [8])

8.11.5 Rarefaction Effects with Angle of Attack

The influence of angle of attack on drag and continuum breakdown is analyzed for the NACA 0012 airfoil at Mach 0.8 and angles of attack 0° and 10°. Figure 8.37a, b show velocity contours, indicating that the separation region generates a greater slip zone extending from the suction point to the trailing edge, amplifying rarefaction effects. Increasing the angle of attack by 10° enhances rarefaction effects by approximately 3%.

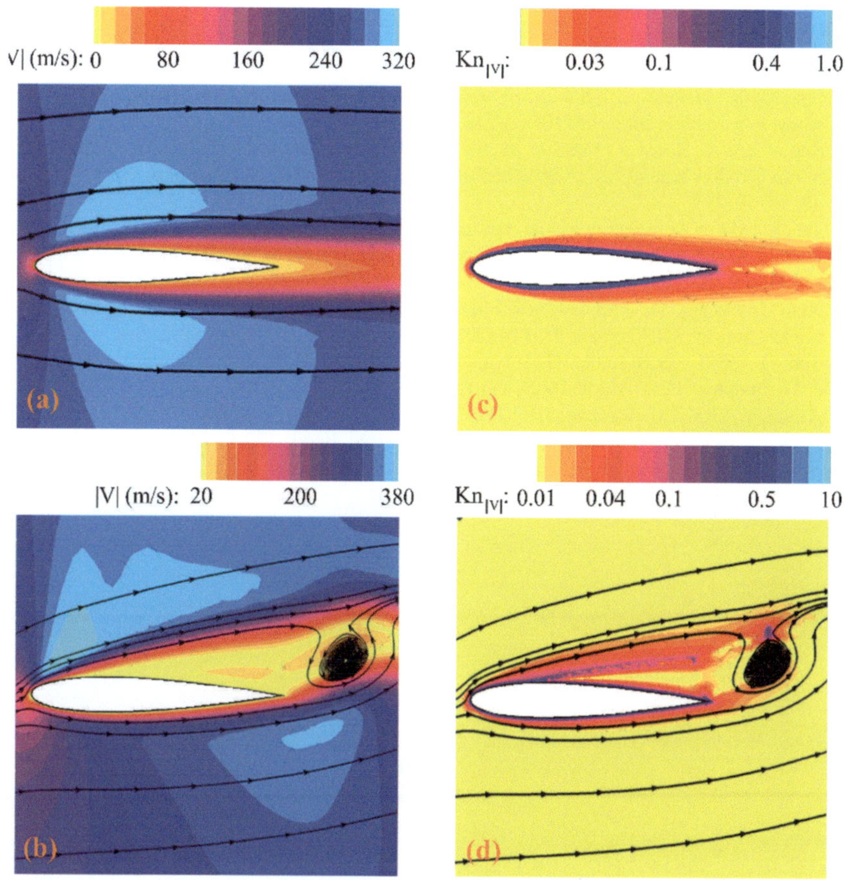

Fig. 8.37 Velocity magnitude contours and streamlines (**a**, **b**), and local Knudsen number contours (**c**, **d**) obtained by ES-BGK method for NACA0012 at $M_\infty = 0.8$; $Kn_\infty = 1.19 \times 10^{-3}$ and at different angles of attacks of 0° (**a**, **c**) and 10° (**b**, **d**) (Reproduced from Ref. [8])

8.12 Summary

The study investigates the aerodynamic performance of airfoils in rarefied flow conditions, showing that lift coefficients slightly increase with increasing Knudsen number (Kn), compared to the more rapid rise in the drag coefficient, and the lift-to-drag ratio decreases with increasing Kn, highlighting the effects of rarefaction on drag. Overall, rarefied flow significantly alters aerodynamic characteristics, and as the Reynolds number decreases, rarefaction becomes more significant. Increasing the Mach number expands flow separation, enhancing rarefaction effects. Airfoil thickness and camber significantly amplify rarefaction effects and impact drag coefficient, while thicker airfoils have larger separation regions. Finally, increasing the angle of attack also intensifies rarefaction effects.

References

1. Shapiro, R., Manela, A.: Thin-airfoil aerodynamics in a rarefied gas wind tunnel: a theoretical study. Phys. Fluids **36**(5), 057107 (2024)
2. Shoja-Sani, A., Roohi, E., Kahrom, M., Stefanov, S.: Investigation of aerodynamic characteristics of rarefied flow around NACA 0012 airfoil using DSMC and NS solvers. Eur. J. Mech.-B/Fluids **48**, 59 (2014)
3. Tai, T.C.: Direct simulation of low-density flow over airfoils. J. Aircr. **29**(5), 806 (1992)
4. Salehin, M.: Aerothermal analysis of supersonic flow around an airfoil in transitional rarefied conditions. (2020)
5. Xiao, H., Wang, D.: Rarefied airfoil aerodynamics based on the generalized hydrodynamic model. Aerosp. Sci. Technol. **92**, 148 (2019)
6. Sun, Q., Boyd, I.D., Candler, G.V.: Numerical simulation of gas flow over microscale airfoils. J. Thermophys. Heat Transf. **16**(2), 171 (2002)
7. Amini, Y., Liravi, M., Izadpanah, E.: The effects of Gurney flap on the aerodynamic performance of NACA 0012 airfoil in the rarefied gas flow. Comput. Fluids **170**, 93 (2018)
8. Pekardan, C., Alexeenko, A.: Rarefaction effects for transonic airfoil flows at low Reynolds numbers. AIAA J. **56**(2), 765 (2018)

Chapter 9
Rayleigh–Bénard Flow of a Rarefied Gas

This chapter delves into non-equilibrium flows marked by instabilities and transitions between distinct attractors. It centers on the Rayleigh–Bénard convection problem—a classic framework for investigating self-organization, chaos, and turbulence. The chapter introduces alternative dimensionless parameters in rarefied regimes, where traditional Rayleigh and Prandtl numbers become inadequate. Emphasis is placed on kinetic-based analysis of the conduction-to-convection transition in a two-dimensional system, focusing on hard-sphere gases and molecular-level modeling.

9.1 Introduction

The previous Chaps. 5–7 consider examples of rarefied gas flows at conditions ensuring the evolution of the flow system to a stable, steady-state final regime. In such kinds of flows, the development of any type of flow instability cannot be observed. In this chapter, the analysis of rarefied gas flows will be extended by a study of non-equilibrium gas flows developing flow instability leading to various final states (attractors). The development and formation of convective flow in a simple thermal system heated from below is known in hydrodynamics as the Rayleigh-Bénard (RB) problem (see Ref. [1]). It studies thermal convection in a fluid between two horizontal parallel plates under the gravity force and with the bottom plate heated. In RB convection, buoyancy from density differences due to temperature variations drives convection and is counteracted by diffusion as an equilibrating fluid property. It is commonly used to explore hydrodynamic instability, self-organization, chaotic behaviour, and fluid turbulence transition. The system is typically analyzed using the Oberbeck-Boussinesq approximation for small temperature variations within continuum fluid

E. Roohi et al., *Advances in Direct Simulation Monte Carlo: From Micro-Scale to Rarefied Flow Phenomena*, https://doi.org/10.1007/978-981-96-8200-3_9

dynamics and is characterized by Rayleigh and Prandtl numbers. In rarefied slip-flow regimes, convection requires high-temperature differences, requiring alternative dimensionless parameters beyond Ra and Pr. When studying a Rayleigh-Bénard system with larger temperature and density differences, the compressibility of the fluid must be considered. For a compressible heat-conducting viscous fluid, it has been found that coexisting attractors can arise. The long-term behavior of the Rayleigh-Bénard model can lead to complex attractors, including heat conduction, stable vortical convection, or oscillating (periodic or chaotic) flow. From the molecular dynamics viewpoint, the problem has an essential meaning because the molecular description allows one to investigate the onset of the convection from a kinetic viewpoint with due account for the gas rarefaction. The first results of the RB convection study of a rarefied gas were reported in Refs. [2–4]. This study examines rarefaction effects on the transition from conduction to convection in a 2D RB system of a "hard sphere" gas, defining the convection zone.

9.2 Convection Regime

Stefanov et al. [5] examined the Rayleigh-Bénard problem for a rarefied gas, where the molecular description allows investigation of convection onset from a kinetic viewpoint with due account for the gas rarefaction. They investigated the Rayleigh-Bénard flow for nondimensional parameters in the ranges Kn $\in [1.0 \times 10^{-3}, 4 \times 10^{-2}]$, Fr $\in [1.0 \times 10^{-1}, 1.5 \times 10^{3}]$ and the temperature ratio r $= T_c/T_h = 0.1$, corresponding to a large temperature difference for which the RB system is believed to reach most of the possible attractors. They explored RB instability in a rarefied gas using two numerical approaches for small Knudsen number flows: direct simulation Monte Carlo (DSMC) (the molecular model) and Navier–Stokes finite difference (FD) method, i.e., continuum approach based on the model of compressible viscous heat conducting gas with state-dependent transport coefficients from the Chapman–Enskog expansion.

9.2.1 Problem Configuration

The schematic view of the two-dimensional Rayleigh-Bénard (RB) system is given in Fig. 9.1. The monatomic hard-sphere model gas is confined by two diffusively reflecting horizontal hot and cold isothermal walls and periodic boundary conditions at two vertical planes.

The Bénard instability zone is studied for temperature ratio $r = 0.1$ and Kn ≥ 0.005, using FD for the smallest Knudsen number (Kn $= 0.001$) due to the high computational cost of DSMC, and the DSMC runs were done for Fr $= 1.0$ and Fr $= 2.0$.

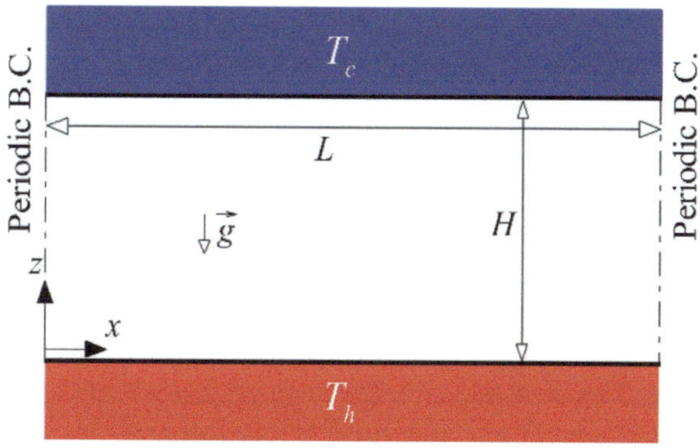

Fig. 9.1 Schematic of the 2D Rayleigh-Bénard flow configuration, with the aspect ratio of the domain $\Lambda = \frac{L}{H} = 2$

9.2.2 *Exact Solution for a Pure-Conduction State*

The gravitational instability of the RB system is analyzed using the exact solution of the pure-conduction state, leading to two analytical conditions for large and small Froude numbers as exterior bounds of the convection zone. The exact solution for a hard sphere rarefied gas to find a necessary condition for the stability of a pure-conduction state is as follows [5]:

$$
\begin{aligned}
T(z) &= (az+b)^{2/3} \\
\rho(z) &= \frac{1}{\rho_0}\exp\left[-\frac{2}{3}\ln|az+b| + \frac{3c}{a}(az+b)^{1/3}\right]
\end{aligned}
\tag{9.1}
$$

where $a = T(1)^{3/2} - T(0)^{3/2}$, $b = T(0)^{3/2}$, and $c = -\frac{2}{Fr}$.

Using the heat transfer equation and the boundary conditions, unknown temperatures $T(0)$ and $T(1)$ are obtained following an iterative scheme, i.e. (0 and 1 are non-dimensional and correspond to hot and cold walls):

$$T(0) = 1 + \frac{2}{3}\tau\frac{T(1)^{3/2} - T(0)^{3/2}}{\sqrt{T(0)}} \tag{9.2}$$

$$T(1) = r + \sqrt{\frac{T(0)}{T(1)}}[1 - T(0)] \tag{9.3}$$

ρ_0 is obtained as follows:

$$\rho_0 = \int_0^1 \rho(z)dz \tag{9.4}$$

By using this analytical solution, Manela and Frankel [6] formulated and studied the linearized perturbation problem about the pure-conduction reference state described by (9.1)–(9.4).

9.2.3 The Zone of the Bénard Instability

Data analysis was performed within the parameter range $0.001 \le \mathrm{Kn} \le 0.03$, $0.5 \le \mathrm{Fr} \le 1500$, and fixed $r = 0.1$, separating convection states from pure conduction. The model's evolution is sensitive to initial conditions, so the Bénard instability zone was identified by running both DSMC and FD codes starting from equilibrium conditions. FD solutions, unlike DSMC, lacked fluctuations, and for $\mathrm{Kn} \ge 0.01$, round-off errors were insufficient to trigger convection, requiring artificial temperature disturbances in the initial condition for a very large Fr number near the right boundary of the instability zone.

The computed instability zone, shown in Fig. 9.2 as a gray shaded area, indicates that the onset of Bénard instability occurs at $\mathrm{Kn} = 0.029$ for Monte Carlo simulations and $\mathrm{Kn} = 0.028$ for finite difference computations, demonstrating good agreement between the two approaches. This onset occurs at small Knudsen numbers where the continuum approach is sufficiently accurate. Overall, the comparison between the continuum (FD) and molecular (DSMC or MC) approaches shows good agreement within the stable convection zone ($\mathrm{Kn} \ge 0.005$), except in the largest Kn zone, where the Bénard instability area from DSMC is slightly larger than that from FD. The obtained numerically boundaries of the convection domain are convincingly confirmed by the linear analysis of the problem conducted in Ref. [6].

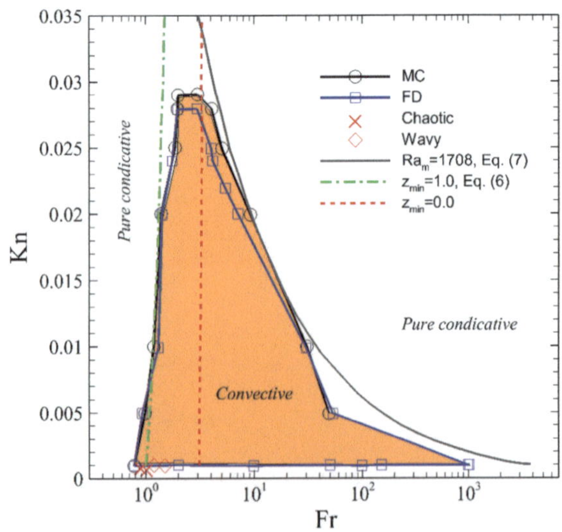

Fig. 9.2 The convection zone computed by DSMC and FD methods. The bold solid and dashed lines are analytical estimates bounding the zone (Reproduced from Ref. [5])

According to Table 9.1, for Kn ≥ 0.01, a two-roll convection flow is observed for all internal points of the Bénard zone. Varying Fr from large to small values shows that the vortex centers move from above to below the midplane of the computational domain. The vortical motion reaches maximum intensity near the left boundary of the instability zone. For Kn $= 0.005$ and Kn $= 0.001$, the final configurations differ: four- and six-roll patterns are seen on the left boundary (Figs. 9.3 and 9.4), while a two-roll pattern appears on the right boundary, similar to those for larger Knudsen numbers. Data analysis shows that the two-roll pattern persists until Fr approaches the left boundary of the instability zone.

In the two-roll configuration, vortices occupy the entire domain with vortex size equal to the distance L between walls. For smaller Knudsen and Froude numbers, stronger stratification near the hot wall propels vortices toward the cold wall, transforming the two-roll pattern into four or six rolls and reducing the vortex size.

The DSMC and FD averaged density and temperature profiles, along with the pure-conduction exact solution, are shown in Figs. 9.5 and 9.6. The Monte Carlo temperature is slightly lower than the FD solution, and for large Froude numbers, convection flattens density and temperature within the internal flow layer.

Table 9.1 The vortex number at points of the curves bounding the zone of convection flow shown in Fig. 9.3, computed by both DSMC and FD methods [5]

Left branch				Right branch			
Kn	*Fr*	Rolls (MC)	Rolls (FD)	Kn	*Fr*	Rolls (MC)	Rolls (FD)
0.03		0	0	0.03		0	0
0.029	2.0	2	0	0.029	3.0	2	0
0.028	2.0	2	2	0.028	3.0		2
				0.028	4.0	2	
0.025	1.9	2	2	0.025	4.0		2
				0.025	5.0	2	
0.024	1.8		2	0.024	4.2		2
				0.022	5.0		2
0.02	1.4	2	2	0.02	7.0		2
				0.02	9.0	2	
0.01	1.2	2	2	0.01	30.0	2	2
0.005	0.95	4	4	0.005	50.0	2	
				0.005	55.0		2
0.001	0.84		4	0.001	1000.0		2
0.001	0.8		6				

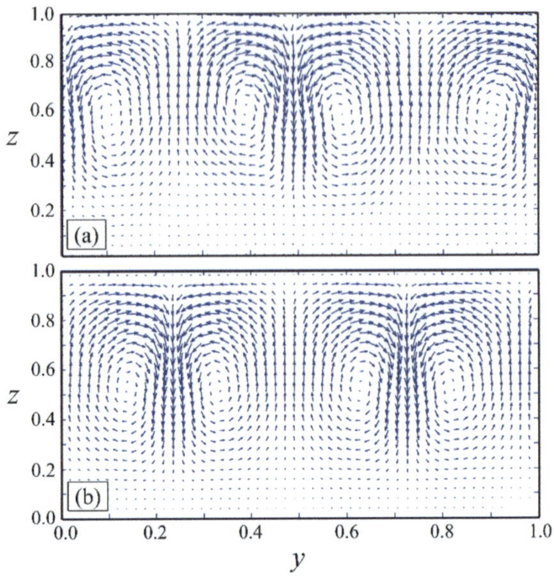

Fig. 9.3 The final four-roll state computed by DSMC (**a**) and FD (**b**) methods for Kn = 0.005, Fr = 0.95, and $r = 0.1$ (Reproduced from Ref. [5])

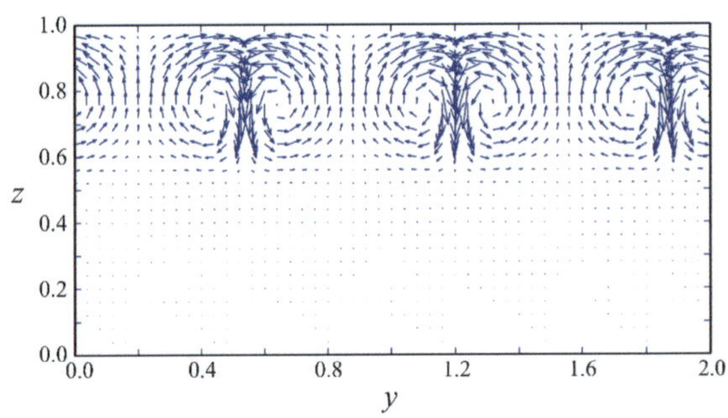

Fig. 9.4 The final six-roll state computed by the FD method for Kn = 0.001, Fr = 0.8, and $r = 0.1$ (Reproduced from Ref. [5])

9.2.4 Stability Analysis of the Pure-Conduction State

The convection flow zone is estimated using stability analysis of the pure-conduction regime, with approximate boundary estimates based on the instability mechanism and pure-conduction solution. Density profiles for the pure-conduction regime are analyzed at two Knudsen numbers for a fixed temperature ratio r = 0.1.

The continuum model of a compressible viscous heat-conducting hard sphere gas reveals a simple exact solution for the pure-conduction state, showing explicit density and temperature profiles between the hot and cold walls. According to Fig. 9.7, the DSMC results slightly differ from the pure-conduction exact solution, especially for temperature at small Fr. The analytical solution for temperature, under boundary conditions, is independent of Fr. In contrast, the one-dimensional FD solution for

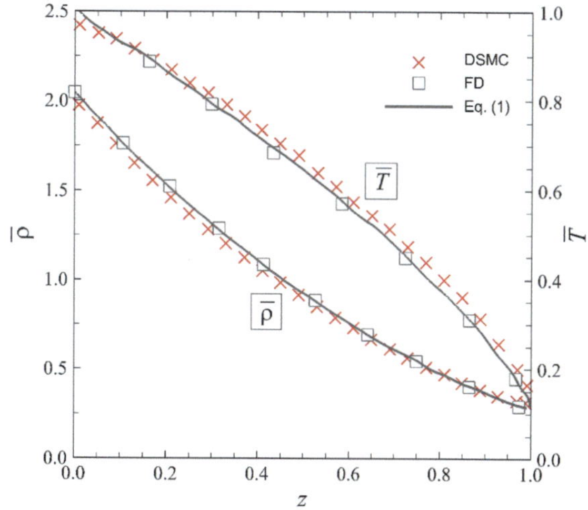

Fig. 9.5 The averaged density and temperature profiles computed by DSMC and FD methods compared with the pure conduction solution, for Kn = 0.005, Fr = 1.0, and $r = 0.1$ (Reproduced from Ref. [5])

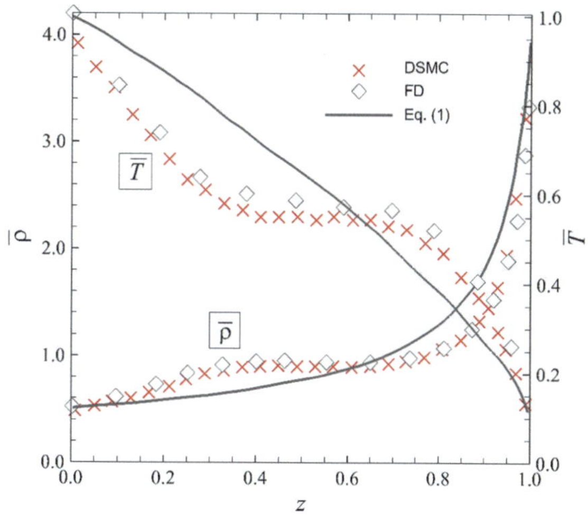

Fig. 9.6 The averaged density and temperature profiles computed by DSMC and FD methods compared with the pure conduction solution, for Kn = 0.005, Fr = 50.0, and $r = 0.1$ (Reproduced from Ref. [5])

pure conduction matches the analytical one. The agreement between DSMC and the analytical solution is excellent for lower Kn, as shown in Fig. 9.8.

Figures 9.7 and 9.8 show three types of density profiles between hot and cold walls. The existence of one extremum (minimum) of density that falls within the interval $z_{min} \in (0, 1)$ for a certain interval of Froude numbers proves this statement in a strict way.

$$\frac{d\rho(z)}{dz} = 0 \rightarrow z_{min} = -\left(\frac{b}{a}\right) + \left(\frac{2}{3c}\right)^3 a^2 \tag{9.5}$$

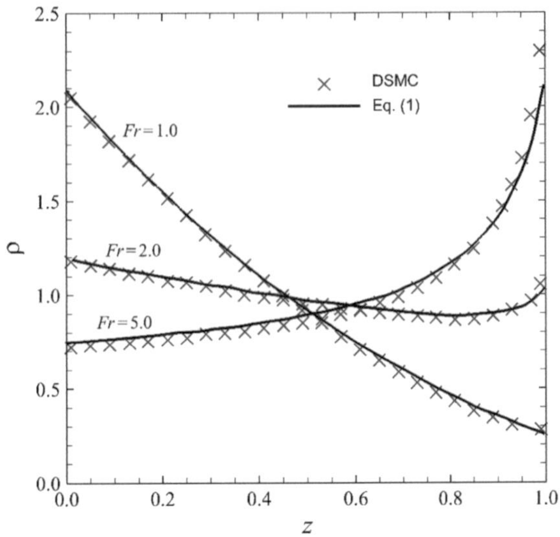

Fig. 9.7 Comparison of density profiles obtained by the DSMC method and the analytical solution for three values of the Froude number Fr = 1.0, 2.0, 5.0 and Kn = 0.02 (Reproduced from Ref. [5])

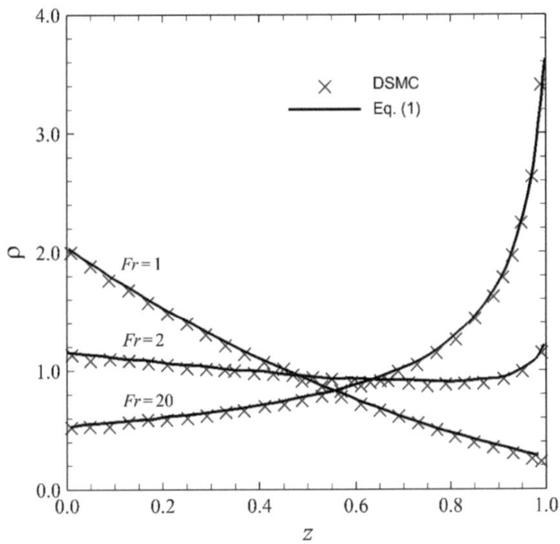

Fig. 9.8 Comparison of density profiles obtained by the DSMC method and the analytical solution for three values of the Froude number Fr = 1.0, 2.0, 20.0 and Kn = 0.005 (Reproduced from Ref. [5])

First, the density profile is monotonic for small Froude numbers, with heavier gas layers below lighter ones, preventing Bénard instability. Second, as the Froude number increases, the density profile becomes nonmonotonic, with a minimum forming near the cold wall, allowing instability to occur for small Knudsen numbers. This condition is a necessary criterion for convection transition and is shown in Fig. 9.2 by the solid line.

$$z_{\min} = -\left(\frac{b}{a}\right) + \left(\frac{2}{3c}\right)^3 a^2 = 1 \tag{9.6}$$

Third, for large Froude numbers, the minimum z_{min} disappears through the left boundary (hot wall), and the density profile becomes monotonic again but with a positive slope, unlike the case for small Froude numbers (the dotted line in Fig. 9.2).

In this condition, gravitational instability manifests under critical conditions, which can be estimated using a modified Rayleigh number, as suggested by Golshtein and Elperin [7]. Ra depends on (r, Kn, Fr) for a hard-sphere gas, and the critical value $Ra_m = 1708$ is estimated to be closely matched to our numerical results. This condition, $Ra_m = 1708$ defines the dashed curve along the right boundary of the computed instability zone, corresponding to large Fr values (Fig. 9.2).

$$Ra_m = \frac{2048}{75\pi}\frac{(1-r)}{(1+r)^2}FrKn^2 \tag{9.7}$$

For Kn > 0.02, the computed Bénard zone is smaller than predicted by analytical conditions. Good agreement is observed for Kn ∈ (0.02, 0.01), while deviations for Kn = (0.005, 0.001) from the analytical estimate at large Froude numbers may be due to nonlinear effects or hysteresis loops, explaining why the computed left boundary exceeds the analytical limit. However, the explanation might also be connected with the hysteresis loops of the final regimes with the back and forth variation of the conditions in the gas system.

9.2.5 Hysteresis Phenomena in Rayleigh–Bénard Convection

The analysis of the instability zone revealed two issues related to the sensitive dependence of the final state on initial conditions. First, near the critical values of the governing parameters (e.g., Kn = 0.005, Fr = 1.0, r = 0.1), a convection flow was excited when starting from the equilibrium initial condition, but not from the perturbed pure-conduction regime. This resulted in two different final states (convection or pure conduction) for the same set of parameters (Kn, Fr, r). Second, coexisting attractors for a combination of parameters (Kn, Fr, r) inside the instability zone were observed. It is shown that cyclic transitions between two states (attractors) occur, i.e., for the case of a compressible viscous heat conducting "hard sphere" gas, we have obtained a hysteresis loop.

Figures 9.9 and 9.10 show the hysteresis loops along two paths, one on the left branch and the other on the right branch of the convection zone (Fig. 9.2). The peak vertical velocity $\max|w_i(y, z)|$ is used to characterize a given state Si(Fr). Near the considered points, $\max|w_i(y, z)| > 0$ corresponds to a stable two-roll final regime, while $\max|w_i(y, z)| \approx 0$ corresponds to a pure-conduction regime.

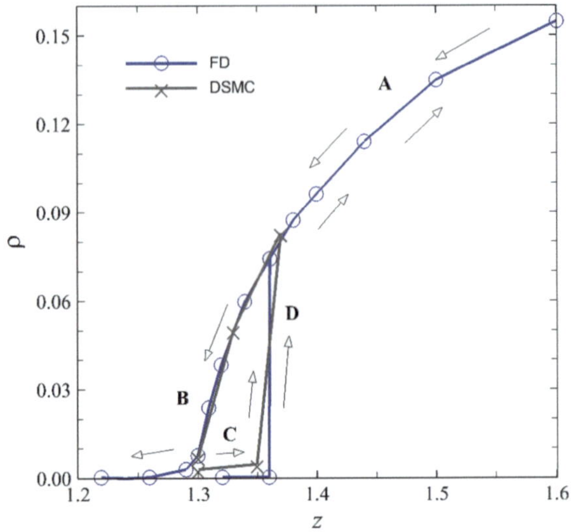

Fig. 9.9 The hysteresis loop computed by DSMC and FD methods for small Fr numbers and a fixed Kn = 0.02 (Reproduced from Ref. [5])

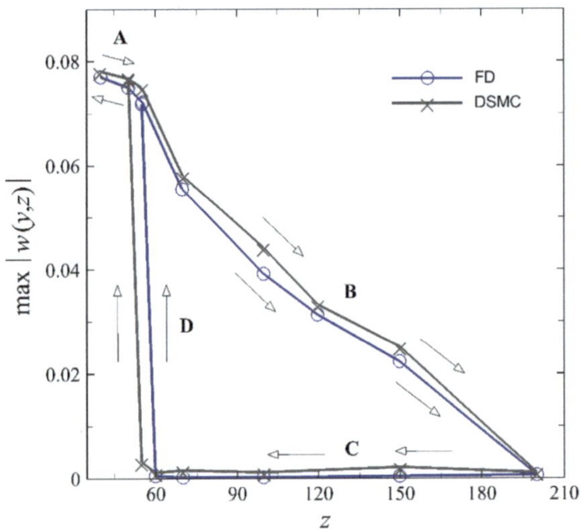

Fig. 9.10 The hysteresis loop computed by DSMC and FD methods for large Fr numbers and a fixed Kn = 0.005 (Reproduced from Ref. [5])

The path starting from a two-roll state towards interval B leads to states with decreasing intensity until a critical point (Fr ≈ 1.22 for the left branch) where convection and pure conduction become indistinguishable (between Fr = 150.0 and Fr = 200.0 for the right branch). Reversing Fr does not repeat the previous sequence; instead, it starts from the pure-conduction state (interval C) until a second critical point (D), where the state jumps back to two-roll convection (interval A), with increasing intensity in subsequent states. The analysis reveals a hysteresis loop with a jump from pure conduction to convection, where two coexisting attractors (convection in B and pure conduction in C) exist for a given combination of parameters. The resulting state depends on the path, indicating memory properties in the RB gas.

The differences between the analytically and numerically computed instability zone boundaries arise from the hysteresis properties of the gas system. For Kn < 0.01, convection can be excited for Froude numbers outside the analytical prediction when starting from an equilibrium initial state. This occurs because waves traveling between the walls periodically create a reverse density slope near the cold wall, allowing enough time for vortex self-organization. The transition period and wave effects are avoided when starting from the exact pure-conduction solution, preventing convection outside the analytical prediction. The hysteresis graphics in Figs. 9.9 and 9.10 show that once excited by a special initial condition, the roll pattern can be maintained outside the instability zone predicted by the analytical conditions. Therefore, the instability zone cannot be precisely determined without defining the initial conditions.

9.3 Chaotic and Periodic Convective Regimes

In the second part of their study, Stefanov et al. [8] analyze convection regimes for near-continuum conditions (Kn = 0.001) in the Rayleigh–Bénard flow of a rarefied gas. This approach allows observation of initial perturbations and the stages of self-organization in RB convection, along with nondimensional heat fluxes at the hot and cold walls, averaged along the relevant wall as integral characterizations of the instantaneous flow state.

*9.3.1 Near-Continuum Regimes (*Kn = *0.001)*

To simplify the analysis of the RB continuum solution, which is embedded in an infinite-dimensional phase space in terms of nonlinear dynamics, averaged heat fluxes at the walls are used to reduce the phase space to two dimensions while preserving key attractor properties. This approach helps investigate unsteady-state solutions, and the nondimensional fluxes transform into the classical Nusselt number under linear temperature variation and no-slip boundary conditions.

The time evolution of the nondimensional heat fluxes at the walls for the continuum model, computed by the FD method, is shown in Fig. 9.11. DSMC results for Fr = 1.0 are not included due to numerical statistical fluctuations comparable to those in the chaotic heat transfer process. However, averaged values of DSMC and FD heat fluxes and macro characteristics of the flow show good agreement. Figure 9.11 shows that the transition period strongly depends on the Froude number, becoming longer for Froude numbers near the convection zone's boundary. The shortest transition occurs for $5.0 \leq \text{Fr} \leq 10.05$, where the nondimensional heat flux reaches its maximum.

Interestingly, the heat flux for a pure conduction state does not depend on the Froude number, and for fixed Kn, the heat flux remains the same for all Froude numbers outside the convection zone.

According to Fig. 9.11a, b, the transition from the equilibrium state to the flow attractor occurs in three stages: (1) Convergence of heat fluxes after separation due to initial cooling and gravity, with large oscillations and one-dimensional flow. The convergence leads to a pure-conduction state, accompanied by large regular oscillations with decreasing amplitude. This represents the contribution of the initially induced wave traveling between the cold and hot walls. (2) Instability excitation and formation of convection flow patterns, leading to a two-dimensional one with a vivid spatiotemporal structure. (3) Establishment of the final state, where the flow achieves a vortical structure or unsteady spatiotemporal behavior and approaches the final attractor.

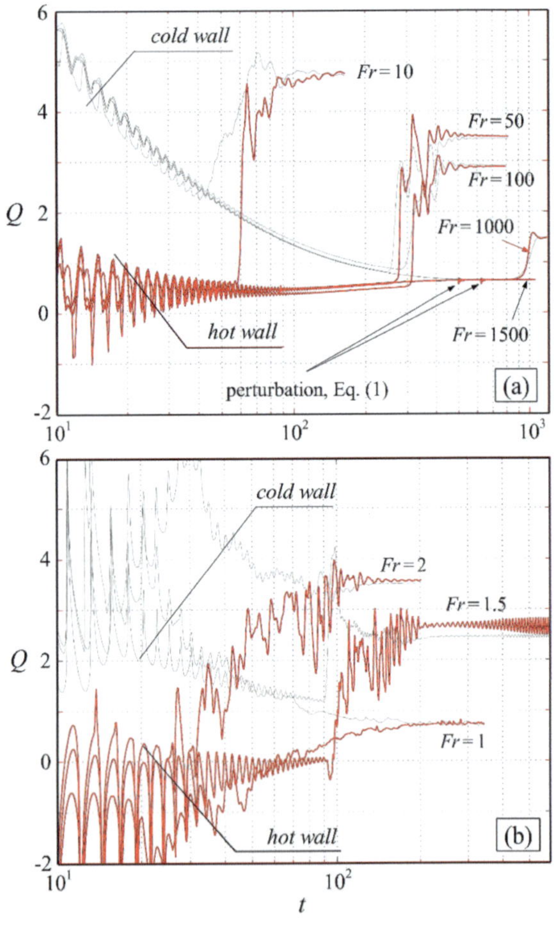

Fig. 9.11 The heat flux behavior for a set of **a** large and **b** small Froude numbers. The Knudsen number and the temperature ratio are fixed (Kn = 0.001 and r = 0.1) (Reproduced from Ref. [8])

The flow field was perturbed, with instability excited for Fr = 1000 but not Fr = 1500. For Fr < 1000, instability was caused by round-off errors during the first stage. For $1.2 \leq \text{Fr} \leq 2.0$ which the final state changes from a stable two-roll configuration (Fr = 2.0) through a periodic (Fr = 1.5) to a quasiperiodic (Fr = 1.2, not shown in Fig. 9.11b) behavior, the first stage of the transition transforms into a mixed (first and second) and a third stage observed. For Fr = 1.0, the transition is smooth into a chaotic state after a very short first stage, without passing through well-separated stages, as observed for larger Froude numbers.

9.3.2 Transition to Stable Two-Roll State

For Froude numbers $2.0 \leq \text{Fr} \leq 10^3$, the transition period consists of three separated stages, leading to a stable two-roll final configuration. Figure 9.12 presents the FD calculation for Fr = 50.0, while DSMC computations are not performed for this case. Figure 9.12a–i show the evolution of the velocity vector field, where the horizontal arrow represents a reference velocity vector. Before the time marked in Fig. 9.12a, the flow maintains a one-dimensional character, typical of the first stage, until $t \approx 120.0$. A two-dimensional perturbation, caused by round-off errors, grows slowly in vortex dipoles near the upper wall. This aligns with the theoretical prediction that instability first arises near the cold wall due to the largest positive density slope. Until $t \leq 230.0 \div 240.0$, the vortex dipoles grow slowly and do not significantly affect heat transfer through the walls. The second stage of the transition begins with a rapid acceleration in the growth of vortex dipoles, which move towards the hot wall. During this stage, the cold wall experiences the influence of the excited convection flow, causing the heat flux to increase rapidly, but the rolls have not yet reached the hot wall (Fig. 9.12c). The vortex configuration is symmetric about the midline, and the heat flux at the hot wall begins to rise as the vortices reach the bottom.

As the vortex interaction intensifies, the heat transfer decreases, and a local minimum appears in the heat flux profiles (Fig. 9.12d). The vortex configuration shifts to a more stable four-roll structure, and heat transfer increases again (Fig. 9.12e). This vortex interaction repeats (Fig. 9.12f, h), transforming the convection flow into a final two-roll configuration (Fig. 9.12i). The aspect ratio A = 2.0 allows the eight-roll configuration to transform into a four-roll one while maintaining some symmetry, but transitioning to a two-roll configuration requires a complete loss of symmetry (Fig. 9.12f, g illustrate it clearly). The third stage culminates in the final symmetric two-roll state (Fig. 9.12i), which is significantly shifted compared to the previous configuration (Fig. 9.12h).

The final state's averaged density and temperature profiles are shown in Fig. 9.13, compared with the pure conduction solution. The FD computed profiles are flattened in a wide area between the walls. This effect is also observed for larger Froude numbers and higher Knudsen numbers. The heat flux at the cold wall deviates from the typical first stage behavior earlier than at the hot wall (Fig. 9.11b) due to the development of additional temporal structures near the cold wall. In Fig. 9.14, the

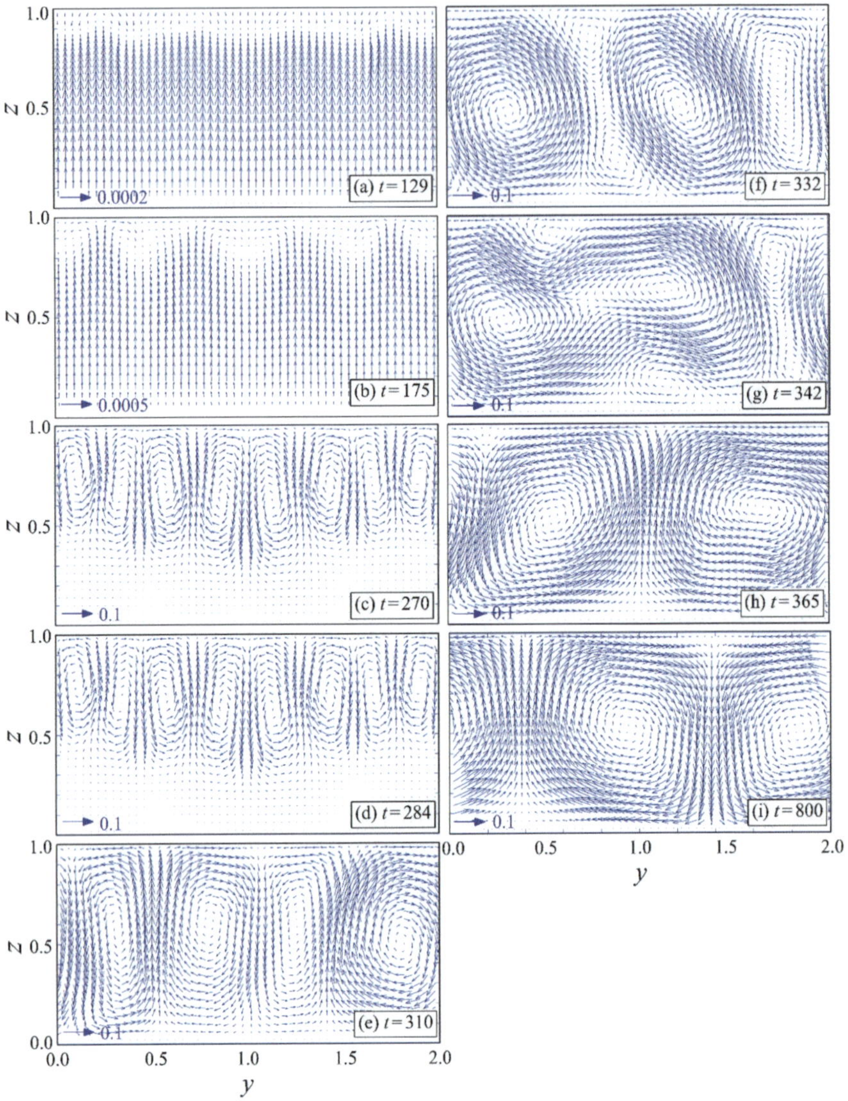

Fig. 9.12 A series **a–i** of snapshots of the velocity vector field for Fr = 50.0 computed by the FD method. The time and the reference velocity are shown in the bottom right and left corners, respectively (Reproduced from Ref. [8])

density and temperature profiles are nonmonotonic, with a negative slope in the inner area caused by higher gravity, which differs from the flattened profile seen for Fr = 50.0 (Fig. 9.13).

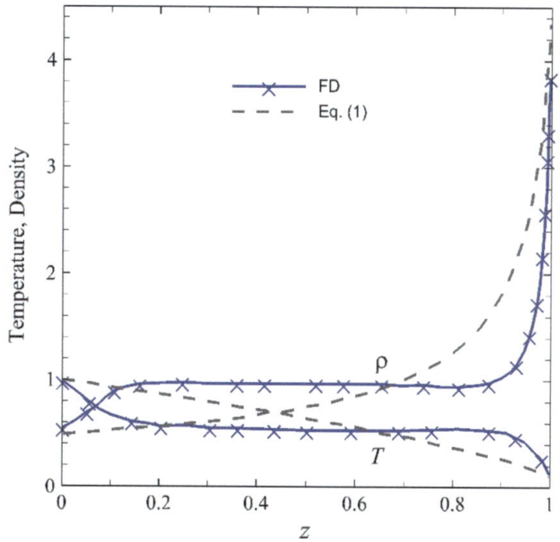

Fig. 9.13 The averaged density and temperature profiles of the final state for Fr = 50.0. The pure conduction solution [Eq. (9.1)] is shown by the dashed line (Reproduced from Ref. [8])

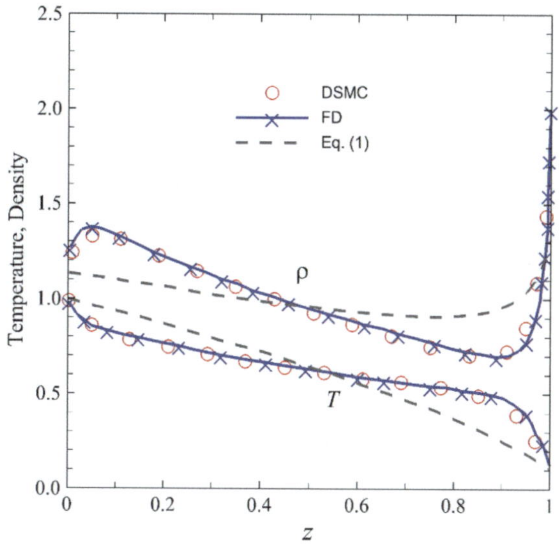

Fig. 9.14 The final state's averaged density and temperature profiles for (Fr = 2.0) compared with the pure conduction solution (Reproduced from Ref. [8])

9.3.3 Transition to Periodic and Quasiperiodic Regimes

A decrease in the Froude number leads to new convection regimes. For Fr = 1.5, the flow exhibits periodic oscillations in heat fluxes at the walls (Fig. 9.11b), the third stage of the transition process shown in Fig. 9.15. The attractor type is identified as a limit cycle, with the system approaching it for $t > 600$ (Fig. 9.16). This periodic regime

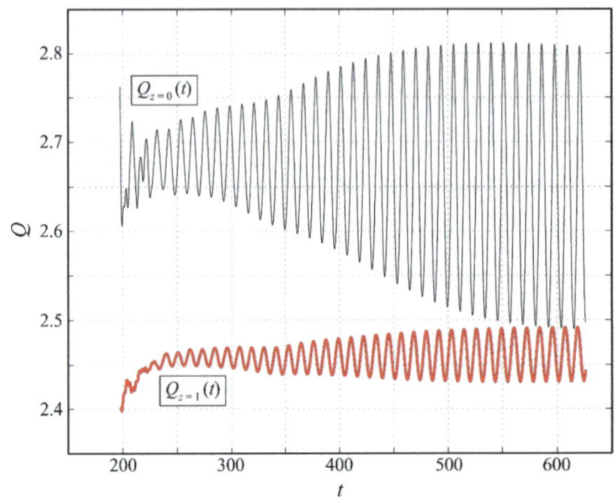

Fig. 9.15 The heat flux behavior during the transition to a periodic regime for Fr = 1.5 (Reproduced from Ref. [8])

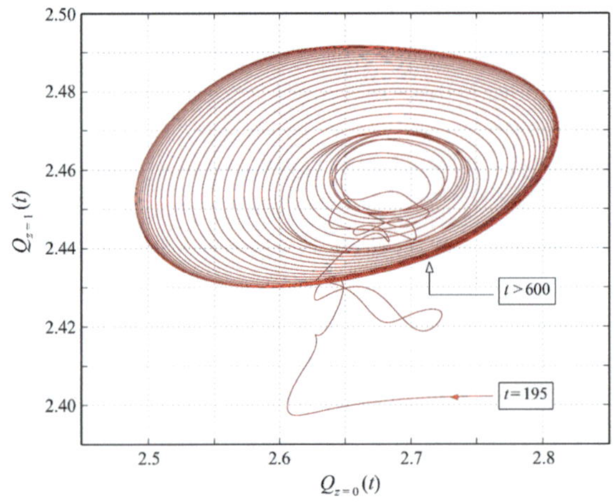

Fig. 9.16 The final orbit traced by the convection RB system in the plane for Fr = 1.5 (Reproduced from Ref. [8])

corresponds to small periodic swings of the two-roll flow configuration, resembling a pendulum.

For Fr = 1.2, the established regime shows a dominating mode of heat flux oscillations modulated by a low-frequency mode, with higher frequency modes also present. The flow oscillations are illustrated in three phases: left deviated, intermediate with a vertical flow core, and right deviated. The vortex configuration includes two main rolls and smaller vortex dipoles. Despite random terms in the flow oscillations (Fig. 9.17), the regime is considered quasiperiodic due to the dominant regularity in the attractor's orbit.

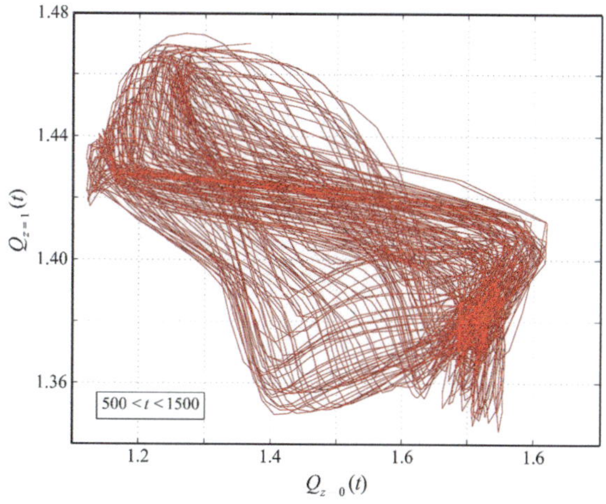

Fig. 9.17 The attractor's orbit in the phase plane for Fr = 1.2. [$Q_{Z=0}(t)$, the nondimensional averaged heat fluxes at the hot and cold walls: $Q_{Z=1}(t)$] (Reproduced from Ref. [8])

9.3.4 *Transition to a Chaotic Regime*

For Fr = 1.0, the flow degenerates into an irregular convection flow. To determine if the dynamics observed in the DSMC and FD calculations are permanent, the initial value problem needs to be solved over large time intervals. For Fr = 1.0, the heat transfer analysis (Fig. 9.11b) shows that the flow reaches a chaotic regime for $t > 300$. The FD solution was computed for a much larger time interval ($[0, t] = 2000.0$), while the DSMC code was run up to $t = 450.0$ due to lower efficiency than FD.

In Fig. 9.11b, the flow instability is triggered after a brief initial interval, and the heat flux curves gradually approach values slightly higher than those of pure heat conduction, establishing the heat transfer regime. Two sets of instantaneous vector fields (snapshots) of macroscopic velocity from DSMC and FD methods represent the transition period and the established chaotic state (Figs. 9.18a–e and 9.19a–e). The DSMC method excites the flow instability earlier than the FD method, as the perturbations in the DSMC simulation are larger due to random fluctuations from molecular motion. In contrast, the FD method's perturbations stem from round-off errors. After instability excitation near the cold wall, perturbations in both models grow and organize into small vortex dipoles. Initially, FD vortex configurations maintain symmetry with respect to the midline, but this symmetry is lost as vortices interact (Fig. 9.19a, b). DSMC vortex structures are asymmetric from the start due to the random initial perturbations. As vortex interactions intensify, large-scale vortices form, with sizes comparable to the distance between the hot and cold walls. Some transient states of the subsequent evolution of both models are presented in Figs. 9.18c, d and 9.19c, d. The final result is a permanent nonperiodic formation and destruction of structures dominated by large vortices. The last pictures (Figs. 9.18e and 9.19e) show states after the transition period (see Fig. 9.11b).

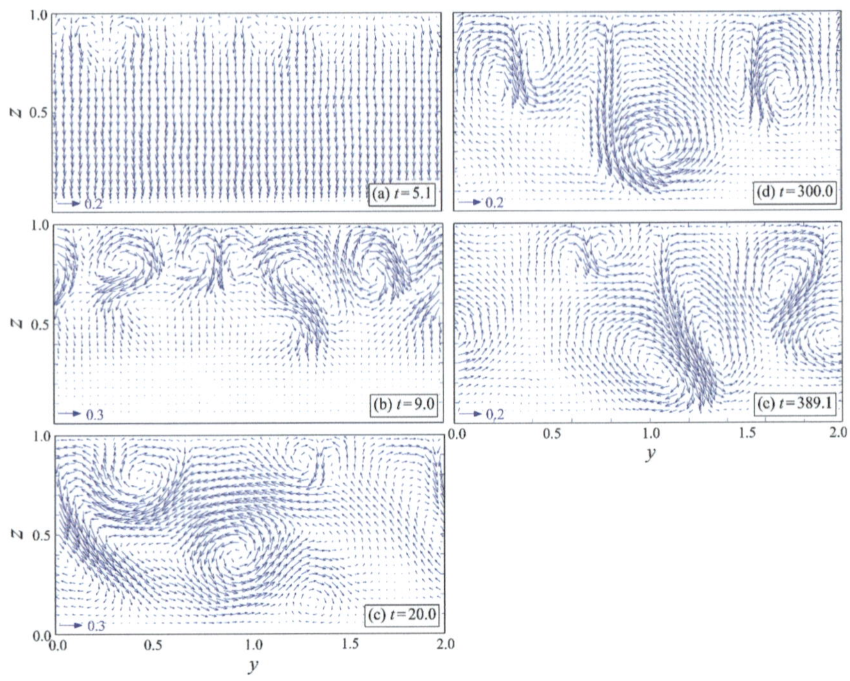

Fig. 9.18 A series of velocity vector field snapshots **a–e** representing the transient period and the permanently established chaotic state for Fr = 1.0 simulated by the DSMC method (Reproduced from Ref. [8])

Figure 9.20 shows the average density and temperature profiles at t = 320.0 after the transition period. The results closely match the exact solution of pure conduction, which contrasts with the profiles for larger Froude numbers (Fr = 50.0 (Fig. 9.13) and Fr = 2.0 (Fig. 9.14)). The negative slope of the density profile is characteristic of small Froude numbers.

Here, for Fr = 0.9 (Fig. 9.21) (after the transition period), the established flow is chaotic again, but the chaotic vortex structures do not reach the bottom hot wall. The largest vortex dipoles are less than 3L, compared to the size of the largest structures in chaotic convection for Fr = 1.0, comparable to L. Another key feature is that the chaotic oscillations of the heat flux at the hot wall have smaller amplitudes than those at the cold wall.

For the case Fr = 1.0 and Kn = 0.001, r = 0.2, the final state remains chaotic convection, with smaller vortex structures that do not reach the bottom wall (Fig. 9.22). The heat flux oscillations at the cold wall have larger amplitudes than at the hot wall (Fig. 9.23), contrasting with the case of r = 0.1. The reason is connected with the fact that the chaotic vortical structures exist near the cold wall and do not touch the hot one. The results suggest that as the temperature ratio r decreases, the chaotic convection becomes less intense; there is a bound below which (above which for r) the convection flow stabilizes in a stable vortex configuration.

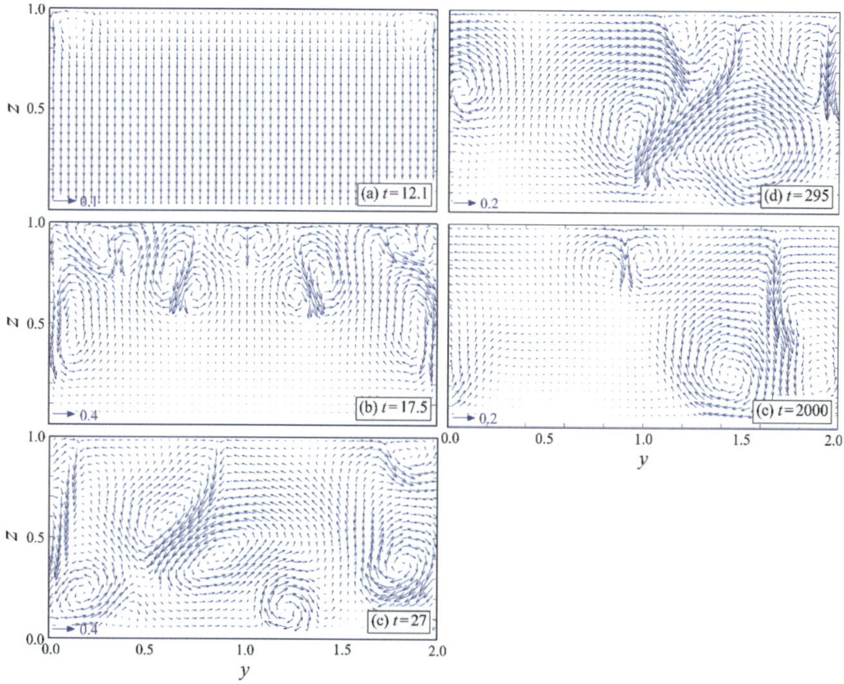

Fig. 9.19 A series of velocity vector field snapshots **a–e** representing the transient period and the permanently established chaotic state for Fr = 1.0 computed by the FD method (Reproduced from Ref. [8])

Fig. 9.20 The averaged density and temperature profiles of the instantaneous state at $t = 320.0$ for Fr = 1.0 (Reproduced from Ref. [8])

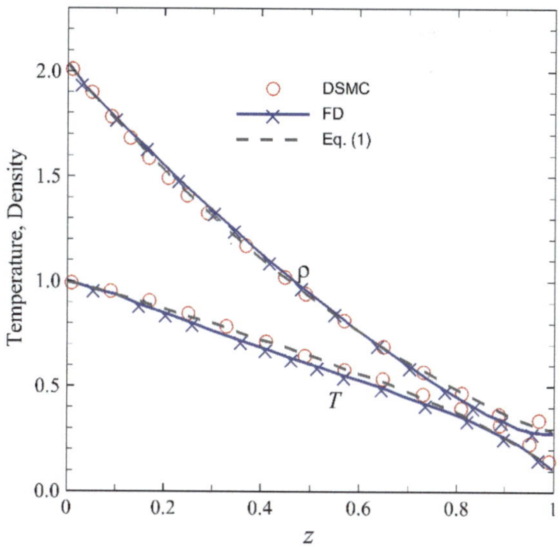

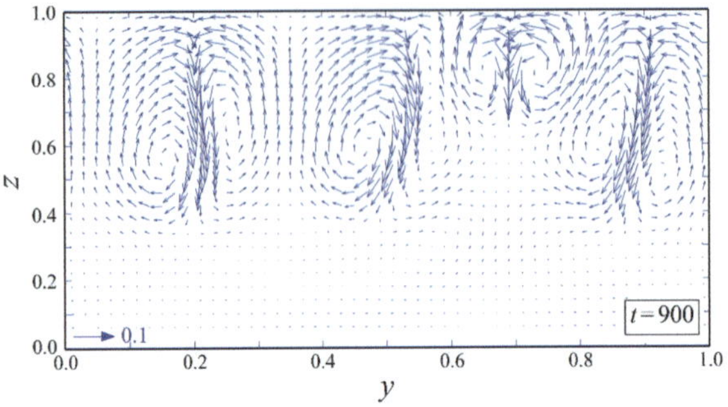

Fig. 9.21 A set of velocity vector fields for Fr = 0.9 at an instant after the transition period (Reproduced from Ref. [8])

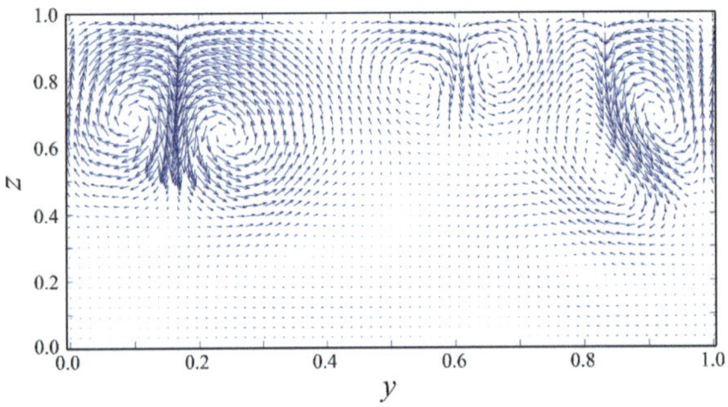

Fig. 9.22 Velocity vector field for the case (Kn = 0.001, Fr = 1.0, $r = 0.2$) at an instant after the transition period (Reproduced from Ref. [8])

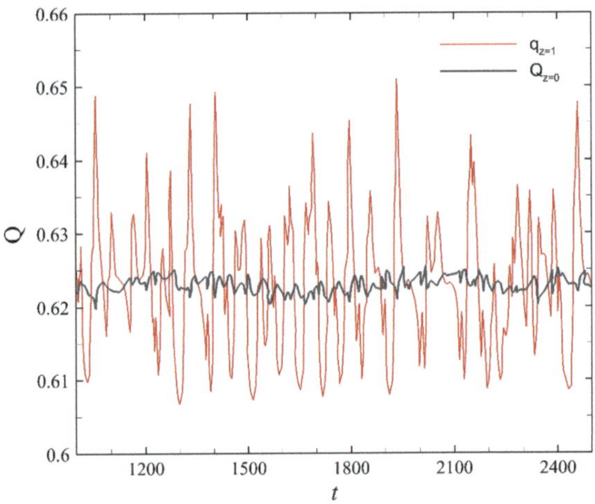

Fig. 9.23 The heat flux oscillations for the case (Kn = 0.001, Fr = 1.0, $r = 0.2$) (Reproduced from Ref. [8])

In the case with a larger aspect ratio A = 6.0, the flow instability for Kn = 0.001, Fr = 1.0, and r = 0.1 is excited later compared to the case with A = 2.0, as shown in Fig. 9.24a, b. In the basic case, this delay is due to the initial pure conduction state, which avoids the wave traveling between the walls caused by gravity. The larger aspect ratio maintains the initial symmetry longer (t > 150.0) than the smaller

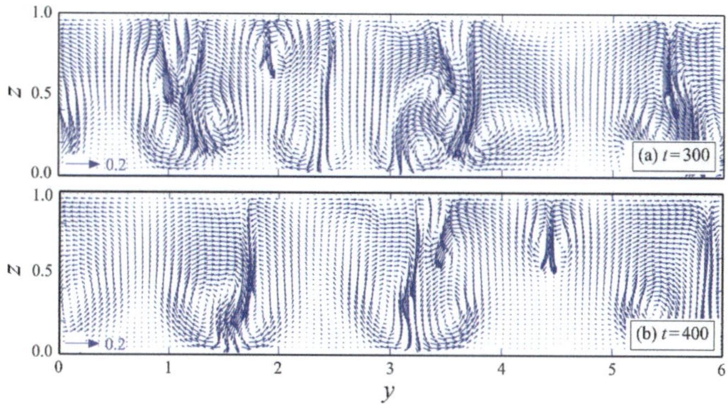

Fig. 9.24 Two snapshots **a** and **b** of velocity vector fields for Fr = 1.0, Kn = 0.001, r = 0.1 (Reproduced from Ref. [8])

aspect ratio case. Eventually, the flow loses symmetry and degenerates into chaotic spatiotemporal structures (Fig. 9.24), similar to the behavior observed in the A = 2.0 case.

9.4 Heat Transfer at Macroscopic and Microscopic Scales

Goshayeshi et al. [9] investigated convection in a rarefied gas using the Direct Simulation Monte Carlo (DSMC) method for a Rayleigh-Bénard (RB) problem at a temperature ratio r = 0.1and an aspect ratio of 2, under higher rarefaction degree Kn $\in$ [0.015–0.03]. The research focuses on heat transfer measurements both at macroscopic (bulk flow) and microscopic (molecular energy exchange) scales, finding good agreement between the two approaches for evaluating the Nusselt number (Nu). A modified Rayleigh number (Ra_m) is introduced to account for nonlinear temperature and density distributions in the pure conduction state.

Figure 9.25 shows the profiles of flow properties at a constant rarefaction degree (Kn = 0.02) across varying Froude numbers (Fr) from 1 to 9. The density profile changes as Fr increases, while the temperature profile remains constant. At Fr = 1, with large gravitational forces, the density distribution is negatively monotonic, preventing instability. At Fr = 2, a balance between thermal expansion and gravity creates a nonmonotonic density distribution, potentially allowing instability near the top plate for small Kn values. At Fr = 9, the density distribution becomes positively monotonic, where the denser gas is always above less dense gas layers, a condition necessary for instability, which can be quantified using a constant Rayleigh number under critical (Fr, Kn) conditions.

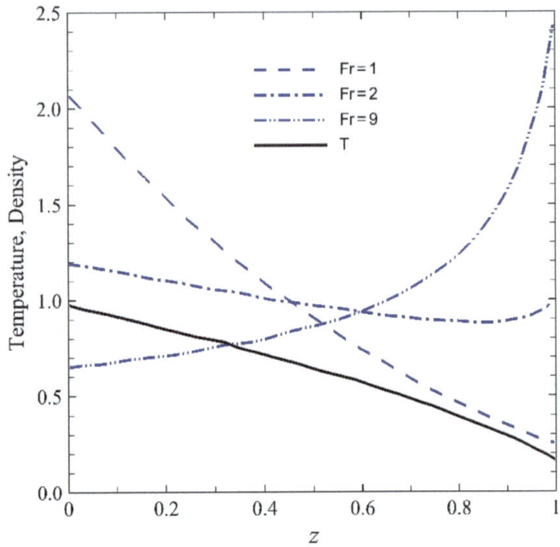

Fig. 9.25 The analytical solution in a rarefied RB cell for the pure conduction state: temperature (solid line) and density (dashed lines) profiles obtained at parametric values of $r = 0.1$, Kn = 0.02, and Fr = 1, 2, and 9 (Reproduced from Ref. [9])

In the Oberbeck-Boussinesq regime, Rayleigh-Bénard (RB) convection transitions from conduction to convection when the Rayleigh number (Ra) exceeds a critical value. In the no-slip regime, Rac = 1708 is determined by linear stability analysis. The average Rayleigh number for a hard-sphere gas has been modified by considering the pure conduction solution, which gives the temperature and density profiles as nonlinear functions based on the Kn and Fr parameters along the cell height. It gives lower Ra_m values compared to the traditional definition of Ra for cases with higher rarefaction and gravity acceleration.

$$Ra_m = \frac{512}{75\pi} \frac{\Delta T}{T_{avg}} \frac{1}{Fr_{avg} Kn^2} \tag{9.8}$$

Transition to convection at high Froude numbers occurs at a constant Rayleigh number, while at low Froude numbers, strong stratification suppresses convection. The latter limit corresponds to a condition where the density ρ(z) reaches a minimum at the cold wall. For small Fr, the minimum density does not occur within the domain and lies beyond the hot wall mathematically, but as Fr increases, it shifts towards z = 0. Convection becomes possible when the minimum density occurs for z < 1, as seen in Fig. 9.25 for Fr = 2, where the minimum density is close to the cold wall. The parametric region where convection is expected, constrained by low- and high-Froude asymptotes, is shown in Fig. 9.26 for various values of r.

Choosing a low-temperature ratio (r = 0.1) is crucial as DSMC simulations at low Kn are computationally expensive. (According to Fig 9.4) The convection zone extends only to a relatively high Kn. For example, Kn ≈ 0.03 makes simulations feasible for smaller temperature ratios. Thus, r = 0.1 allows practical DSMC studies of rarefied gas convection. The investigated flow regime (r = 0.1, Kn ∈ [0.015–0.03])

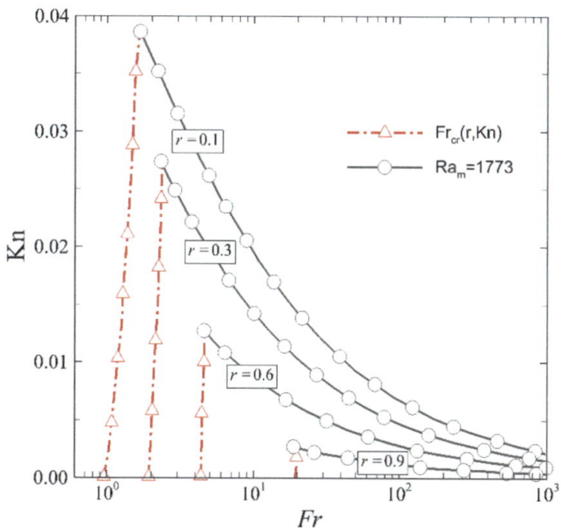

Fig. 9.26 Demonstration of the convection zone on the (Fr, Kn) plane at each temperature ratio and outlined via boundaries defined at low and high Froude numbers. These boundaries follow the Ra_m definition at high Froude numbers, and at low Froude numbers, they follow the critical Froude condition. The intersection points of these boundaries, denoted by (Fr_i, Kn_i), are located at (1.64, 0.039), (2.38, 0.027), (4.59, 0.013), and (19.51, 0.0026) for $r = 0.1$, 0.3, 0.6, and 0.9, respectively. No convection occurs for $Kn > Kn_i$ for the respective r values as stable stratification of the rarefied gas suppresses the onset of convection. Note the presence of a relatively small area with convection at $r = 0.9$, which is the typical temperature ratio in the conventional RB cases in the continuum regime for which the Oberbeck-Boussinesq approximation is valid. Convection at higher Knudsen values occurs only for low-temperature ratios. For instance at $r = 0.1$, the convection zone includes $Kn \in [0\text{–}0.03]$, limited by the rarefaction limit of $Kn = 0.03$ (Reproduced from Ref. [9])

is rarefied, with velocity-slip, temperature-jump effects, large density variations, and strong non-Oberbeck-Boussinesq effects.

The effect of rarefaction on RB convection is analyzed in Fig. 9.27. Three simulations with different Knudsen numbers, the (Fr, Kn) cases of (15.53, 0.015), (8.08, 0.02), and (2.87, 0.03) at $Ra_m \approx 2040$ are compared. As rarefaction increases, the system transitions from convection ($Kn = 0.015$) to conduction ($Kn = 0.03$) (panels (a)–(c) in Fig. 9.27).

Figure 9.27 shows the time-averaged Mach number for qualitative comparison. While the two-roll convection structure is mostly incompressible ($Ma \leq 0.2$), compressibility effects become noticeable near maximum convection ($Kn = 0.02$, $Ra_m \approx 6600$) with Mamax ≈ 0.3.

This study focuses on three stages of Rayleigh-Bénard convection: (i) onset, (ii) maximum convection, and (iii) cessation, which are discussed in the following sections.

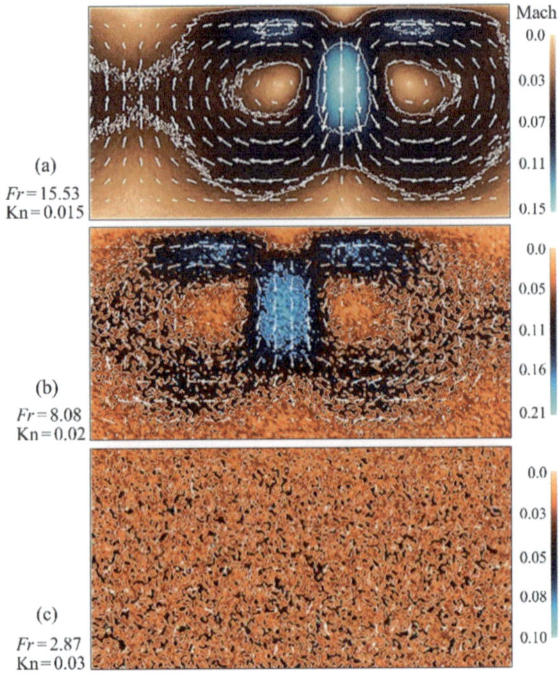

Fig. 9.27 Mach contour plots overlaid on velocity vectors at a constant $\text{Ra}_m \approx 2040$. Results are time averaged for 30 time instances once the steady-state solution, based on the temporal Nusselt Nu_f behavior, is reached (Reproduced from Ref. [9])

9.4.1 Onset of Convection

This section measures the critical modified Rayleigh number ($\text{Ra}_{\text{cr},I}$) for the onset of convection at the high-Froude limit. For each rarefaction level (Kn constant), conduction becomes unstable by increasing g (equivalent to decreasing Fr). As Fr increases and Ram decreases toward $\text{Ra}_{\text{cr},I}$, the transition time (t_{tr}) from conduction to convection, quantified by Nu, is delayed (Fig. 9.28). This transition is also observed in the total kinetic energy (E). Figure 9.28c, d show that E follows the same trend as heat transfer (Fig. 9.28a, b). Therefore, in this study, heat transfer measurements alone are used to characterize convection onset and suppression.

Figure 9.29 shows steady-state Nusselt values at the hot and cold walls for rarefied RB cases within Kn $\in$ [0.015–0.027]. Ram is incrementally increased, and the Nusselt number is measured. The critical modified Rayleigh number (Ra_{cr}) for convection onset is identified when it deviates from unity. The results show a small standard deviation in $\text{Ra}_{\text{cr,I}}$, determining its value as 1770 ± 15 for $0.015 \leq \text{Kn} < 0.030$.

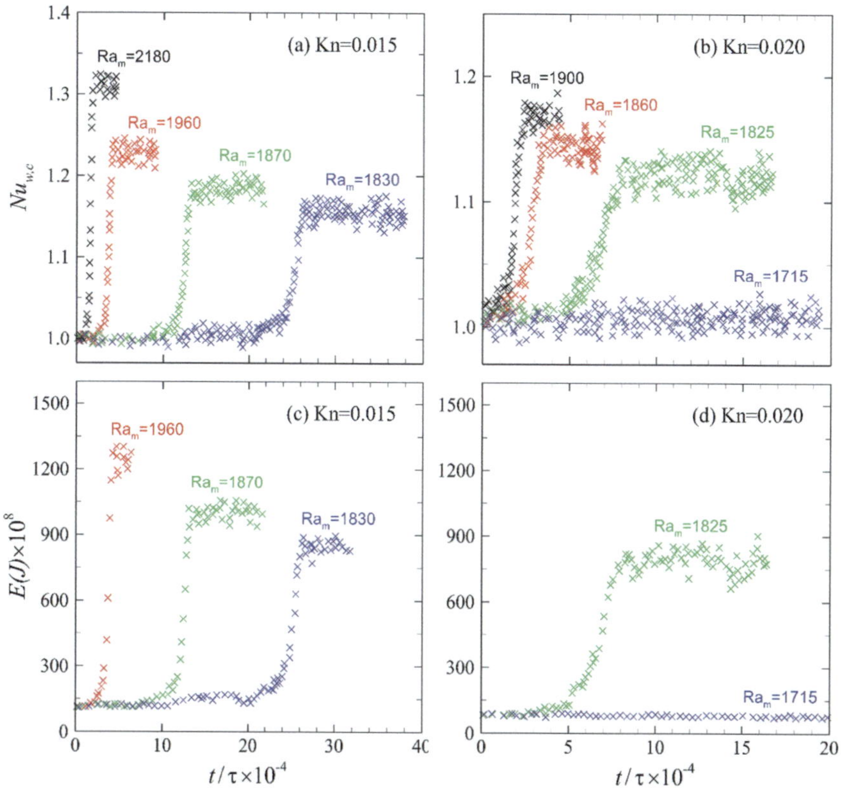

Fig. 9.28 Temporal behavior of the cold wall Nusselt number $Nu_{w,c}$ and the total kinetic energy E for a set of parameter combinations that are close to the instability limit at high Fr numbers. Note that the cold wall Nusselt number is opted for demonstration as it has lower fluctuations than the hot wall Nusselt number. It is observed that the transition time t_{tr}, normalized with the mean free flight τ, is larger for parameter regimes that are more continuum, thus lower Kn, and closer to the instability limit, e.g., $Ra_{cr,l} \approx 1770$ (Reproduced from Ref. [9])

9.4.2 Maximum Convection

This section examines maximum convection in a rarefied RB system under varying rarefaction conditions. Maximum Nusselt values are measured for different Knudsen numbers (Kn), with convection peaking at different Ram. Figure 9.30 shows a linear dependence of Nu_{max} on Kn $\in$ [0.02–0.03], but deviations occur for Kn < 0.02, where Nu_{max} is slightly higher. The linear trend predicts a conduction state ($Nu_{max} = 1$) near Kn = 0.03.

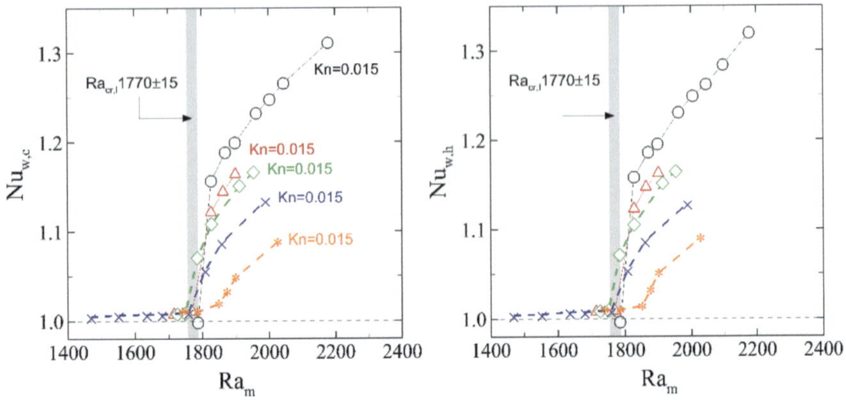

Fig. 9.29 Transition to convection in proximity of the high-Fr limit, identified via the Nusselt values at the cold and hot walls (Reproduced from Ref. [9])

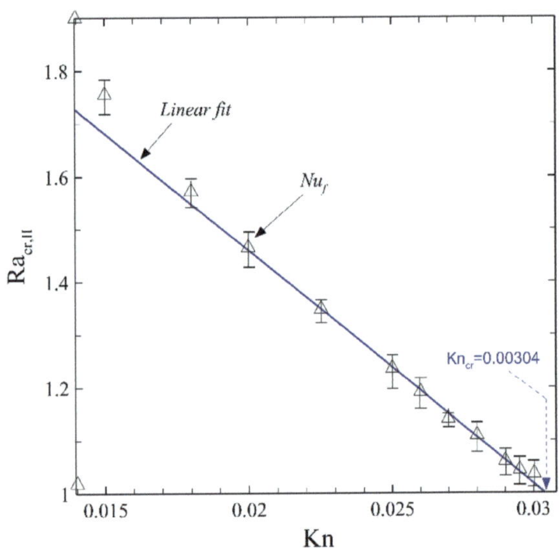

Fig. 9.30 Maximum Nusselt, Nu_{max}, obtained at each rarefaction degree Kn. Results indicate that the maximum convection of the rarefied RB system tends to decrease linearly for the range of $Kn \in [0.02–0.03]$. The intersection of this linear trend with the conduction limit $Nu_{max} = 1$ coincides with $Kn \approx 0.03$ (Reproduced from Ref. [9])

9.4.3 *Limit at Low Froude Numbers*

The suppression of convection at low Froude numbers (Fr) is analyzed through heat transfer at the walls. Figure 9.31 shows steady-state heat transfer magnitudes (Nu_{wall}) for rarefied RB cases. Convection ceases due to highly stratified flow, bounded by the critical Fr_{cr}. At each Kn, Ram increases with gravity (g), bringing the flow closer to the convection-to-conduction transition. Table 9.2 identifies and lists the critical modified Rayleigh number ($Ra_{cr,II}$) for this transition at low Fr.

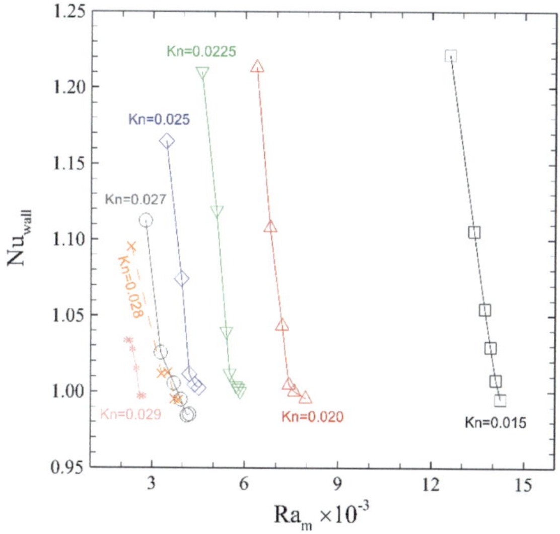

Fig. 9.31 The heat transfer magnitudes measured for parametric cases at the low-Fr limit and close to the critical Froude number (Fr_{cr}). Based on these points, the critical modified Rayleigh number $Ra_{cr,II}$ for which cessation of convection occurs is assessed for the rarefaction range of 0.015 Kn < 0.03. The results for $Ra_{cr,II}$ are summarized in Table 9.2 (Reproduced from Ref. [9])

Table 9.2 Assessment of the critical modified Rayleigh number $Ra_{cr,II}$ for the cessation of convection in the rarefaction range of 0.015 Kn < 0.03

Kn	0.015	0.02	0.0225	0.025	0.028	0.029
$Ra_{cr,II}$	14,091	7400	5671	4343	3285	2541

A set of DSMC simulations are performed at each rarefaction degree Kn, while gravity g is being increased incrementally. The critical modified Rayleigh number, for occurrence of the conduction state, i.e., the Nusselt number being close to unity, is then chosen from the simulated cases. In most of the measurements, the determination of $Ra_{cr,II}$ is done with precision about $Ra_{cr,II} \approx 200$ that is the parametric distance between two consecutive states in which the transition from the convective to the conductive state occurs [9]

Figure 9.32 shows the variation of $Ra_{cr,II}$ with rarefaction degree (Kn), revealing a linear decrease for Kn > 0.02, deviating from predictions based on Fr_{cr}. Assuming that at high rarefactions, the conduction limits for high ($Ra_{cr,I} = 1770$) and low ($Ra_{cr,II}$) Froude numbers asymptotically intersect, this trend is extrapolated. The two limits ($Ra_{cr,I}$, $Ra_{cr,II}$) meet near Kn $\approx$ 0.03, marking the conduction boundary due to rarefaction. Figure 9.15 also supports this, showing that extrapolated Nusselt values at maximum convection intersect the conduction limit near Kn $\approx$ 0.03.

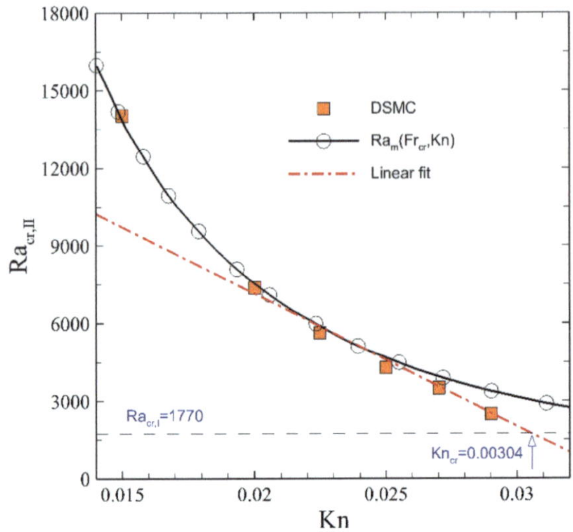

Fig. 9.32 Critical points in the transition from convective into conductive state for the low-Fr range. Red dots indicate DSMC data for $Ra_{cr,II}$ and the error bars are smaller than the size of the circular symbols. These symbols are compared with the profile of $Ra_m(Fr_{cr}, Kn)$ at the Fr_{cr} condition. Note that the linear trend existing at higher rarefactions intersects with the conduction condition of $Ra_m = 1770$ at approximately the hypothetical rarefaction limit of $Kn \approx 0.03$ (Reproduced from Ref. [9])

9.5 Conclusion

The transition from pure conduction to stable convection in the Rayleigh-Bénard flow of a rarefied gas is explored. The critical Knudsen numbers for this transition were similar: Kn = 0.029 (DSMC) and Kn = 0.028 (FD), with improved agreement as the Knudsen number decreases. The study identified coexisting attractors and hysteresis loops in the convection states as the Froude number was varied. The separation line, determined for fixed Knudsen numbers, is a lower bound for the convection zone, with an upper bound derived from the continuum model for pure conduction stability. For large Froude numbers, Ra = 1708 provides the upper bound, closely matching the numerical results. At small Froude numbers, the density profile's nonmonotonicity near the cold wall is key to the onset of instability, which is necessary for convection. This defines a boundary for the instability zone at small and large Froude numbers.

Two transient periods—towards a chaotic attractor at Fr = 1.0 and a stable two-roll configuration at Fr = 2.0—are computed using DSMC (molecular approach) and FD (continuum approach). The comparison shows that both methods capture the essential properties of the transition to chaotic fluid motion. FD computations show a transition from pure conduction to stable two-roll ($Fr \geq 2.0$), periodic (Fr = 1.5), quasiperiodic (Fr = 1.2), chaotic (Fr = 0.9–1.0), and six-roll configuration (Fr = 0. 8) as Fr decreases. It starts from $Fr = 1.5 \times 10^3$ and returns to pure conduction for Fr < 0.8. For r = 0.2 and Fr = 1. 0, the chaotic attractor remains, but with lower intensity and frequency of velocity pulsations. For attractors represented by stable two-roll configuration, the process of self-organization passes through several bifurcations. The corresponding stages of transition to chaotic attractor are not so well separated. The transition to chaotic motion seems to occur within a range of Knudsen numbers around 0.001–0.002.

Results show that maximum convective heat transfer decreases linearly for $0.02 \leq \mathrm{Kn} < 0.030$, intersecting the conduction state at $\mathrm{Kn} \approx 0.03$. A modified Rayleigh number ($\mathrm{Ra_m}$) was introduced to account for non-Boussinesq effects, considering nonlinear temperature distributions in the pure conduction state. The critical modified Rayleigh number for convection onset was Racr, $\mathrm{I} = 1770 \pm 15$for the rarefied range of $\mathrm{Kn} \geq 0.015$. For low Froude numbers, modified Rayleigh numbers $\mathrm{Ra}_{\mathrm{cr},II}$ for convection cessation were identified. Still, non-Oberbeck-Boussinesq effects and a highly stratified regime prevent defining a single parameter for this transition. Despite this, DSMC results show a linear dependency between $\mathrm{Ra_{cr,II}}$ and Kn for $\mathrm{Kn} > 0.02$. At higher rarefaction, this linear trend intersects with the onset of the conduction limit $\mathrm{Ra}_{\mathrm{cr},I} = 1770$ near $\mathrm{Kn} \approx 0.03$, marking the conduction limit due to rarefaction.

References

1. Chandrasekhar, S.: Hydrodynamic and Hydromagnetic Stability. Clarendon, Oxford (1961)
2. Garcia, A.: Hydrodynamic fluctuations and the direct simulation Monte Carlo method. In: Mareschal, M. (ed.) Microscopic Simulation of Complex Flows, p. 177. Plenum, New York (1990)
3. Garcia, A., Penland, C.: Fluctuating hydrodynamics and principal oscillation pattern analysis. J. Stat. Phys. **64**, 1121 (1991)
4. Stefanov, S., Cercignani, C.: Monte Carlo simulation of Bénard's instability in a rarefied gas. Eur. J. Mech. B/Fluids **11**, 543 (1992)
5. Stefanov, S., Roussinov, V., Cercignani, C.: Rayleigh–Bénard flow of a rarefied gas and its attractors. I. Convection regime. Phys. Fluids **14**, 2255 (2002)
6. Manela, A., Frankel, I.: On the Rayleigh-Bénard problem in the continuum limit. Phys. Fluids **17**(3), 036101 (2005)
7. Golshtein, E., Elperin, T.: Convective instabilities in rarefied gases by direct simulation Monte Carlo method. J. Thermophys. Heat Transf. **10**, 250 (1996)
8. Stefanov, S., Roussinov, V., Cercignani, C.: Rayleigh–Bénard flow of a rarefied gas and its attractors. II. Chaotic and periodic convective regimes. Phys. Fluids **14**, 2270 (2002)
9. Goshayeshi, B., Di Staso, G., Toschi, F., Clercx, H.J.H.: Numerical study of heat transfer in Rayleigh-Bénard convection under rarefied gas conditions. Phys. Rev. E **102**, 013102 (2020)

Chapter 10
Shock Wave Over a Cylinder in Rarefied Gas Flows

10.1 Introduction

This chapter presents the flow features for supersonic rarefied gas over a cylinder which is an interesting test cases in many investigated problems. Figure 10.1 shows a bow shock wave in front of a cylinder. The bow shock structure contains three parts, that is, normal, strong, and weak shocks. Normal and strong shocks are significantly affected by the body surface, while the weak shock part is far from the surface. At faraway distances from the body, the shock is converted to a Mach wave. The rarefied condition occurs where the flow length scale is comparable with respect to the upstream mean free path (λ_∞). The Knudsen number is defined as $\mathrm{Kn} = \lambda_\infty/\mathrm{D}$.

10.2 The Shock Wave Detection (SWD) Technique

Akhlaghi et al. [1] proposed a Schlieren imaging-based technique with clear advantages to overcome the limitations of previous shock detection methods. This method applies to all flow types (viscous, inviscid, continuum, rarefied) and requires no filtering. It is also independent of flow field generation methods (numerical or experimental) and accurately identifies the shock region and specific points (e.g., shock center, pre-/post-shock points). Unlike previous methods, the proposed technique can easily be implemented for existing flow fields containing shock waves and uniquely predicts the shock profile as a distinct line. These capabilities are particularly valuable for studying shock waves and examining rarefaction effects. This method determines shock wave coordinates using numerical Schlieren image processing. Schlieren visualization, developed by Toepler, relies on light deflection by a refractive index gradient, which is directly related to flow density gradients. This relationship follows a simplified Lorentz-Lorenz formula, i.e., $n - 1 = k\,\rho$, where

E. Roohi et al., *Advances in Direct Simulation Monte Carlo: From Micro-Scale to Rarefied Flow Phenomena*, https://doi.org/10.1007/978-981-96-8200-3_10

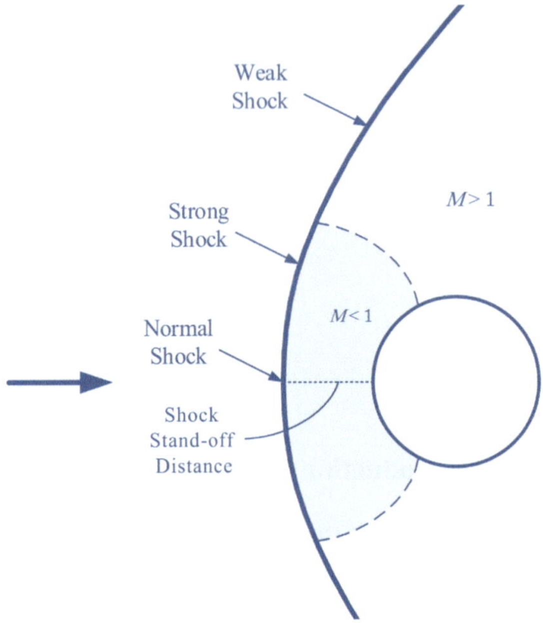

Fig. 10.1 Shock wave structure in front of a cylinder at supersonic flow (Reproduced from Ref. [2])

the Gladstone-Dale coefficient (k) is 0.23 for air under standard conditions. Consequently, Schlieren image intensity represents density gradients. In image processing, a linear or nonlinear spatial filter matrix is used to apply the edge detection filter. In this study, Roberts cross edge detection was used [1].

$$f = \left[\left(\nabla_x \rho_{i,j} \right)^2 + \left(\nabla_y \rho_{i,j} \right)^2 \right]^{\frac{1}{2}}$$
$$\nabla_x \rho_{i,j} = \frac{\rho_{i,j+1} - \rho_{i+1,j}}{\sqrt{2}h}, \quad \nabla_y \rho_{i,j} = \frac{\rho_{i+1,j+1} - \rho_{i,j}}{\sqrt{2}h} \tag{10.1}$$

A Gaussian distribution is introduced for a Schlieren function through the shock wave, making it possible to detect shock thickness and access any desired point within the shock wave [1].

$$G(x) = B_1 \exp\left[-\frac{(x - B_2)^2}{2B_3^2} \right] \tag{10.2}$$

The parameter $\sigma = G/B_1$ is a normalized form of G, where $\sigma = 0.1$ corresponds to aft shock locations and the beginning and end points correspond to $\sigma \approx 0$. The maximum Schlieren function value is identified in each row near the shock wave, defining the shock wave edge or center, and the location of the highest density gradient corresponds to the shock center edge. The center point or shock-center edge, which is characterized by $\sigma = 1$, indicates the point within the shock region where the

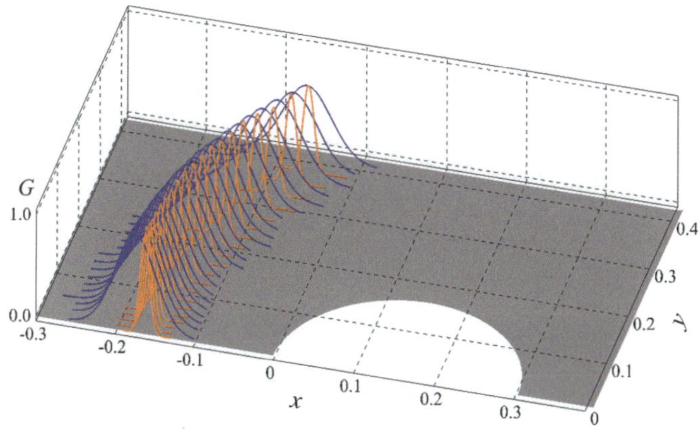

Fig. 10.2 Schlieren function distributions for Argon gas at free stream Mach numbers of 3 and Kn = 0.01 (orange) and Kn = 0.07 (blue) (Reproduced from Ref. [2])

maximum flow gradient occurs. The point with $\sigma = 0.5$ corresponds to a location between the shock-center edge and aft shock location, where the shock effects on the flow are still incomplete.

Figure 10.2 represents Schlieren function distributions along different streamlines. The flow field has been obtained through DSMC simulations for Argon gas. Distributions correspond to two different rarefied flow conditions. Rarefaction increases the shock width. The maximum values of Schlieren functions, which indicate shock edge, indicate that rarefaction also increases the shock standoff distance. At lower Mach number condition, the above effect is more significant.

10.3 Continuum and High-Speed Rarefied Gas Flow Over a Wedge

To evaluate the SWD technique for supersonic continuum flow, the flow fields are simulated using a CFD solver named INS code, employing the Roe upwind scheme for convective fluxes and a finite volume method for viscous fluxes. The method's ability to access any point within the shock wave is evaluated by comparing results with analytical gas dynamics relations, including θ-β (shock angle vs. deflection angle), θ-P^* (pressure ratio vs. deflection angle), and hodograph (normalized velocity components) diagrams.

Figure 10.3 compares SWD results with analytical relations for different locations σ. Solid lines correspond to the analytical relations for continuum flow at $M = 3$. While SWD results align well with analytical solutions near post-shock locations (lower values of σ), deviations appear within the shock region ($\sigma = 0.5$),

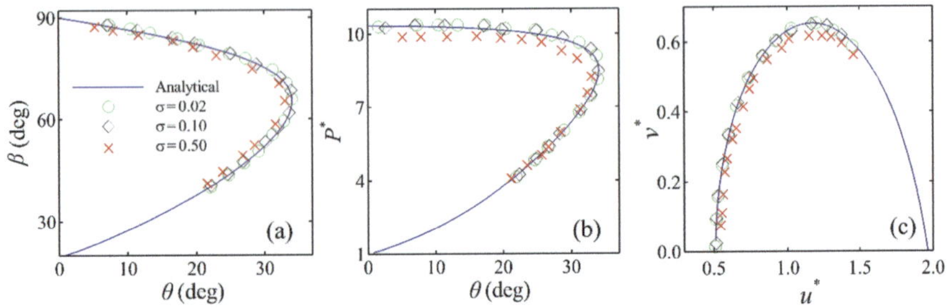

Fig. 10.3 The effect of different locations within the shock on **a** θ-β, **b** θ-P^*, and **c** hodograph diagrams (Reproduced from Ref. [1])

indicating continuous property changes across the shock. The SWD technique accurately predicts post-shock conditions by picking up the true aft shock location in the continuum flow regime, and it is reliable for both strong and weak shocks.

To evaluate the SWD technique's capability for analyzing the shock wave characteristics in supersonic rarefied gas flow, numerical simulations are performed using the DSMC-based RGS2D solver. The method discretizes time and space, dividing motion into free molecular movement and binary intermolecular collisions per time step, using the Generalized Hard Sphere (GHS) model and the No Time Counter (NTC) method for collision selection. Specular reflection is assumed at walls to prevent gas-surface interaction losses, and an unstructured grid with refined resolution in the shock region ensures accuracy.

Figure 10.4 analyzes the effects of different locations within the shock wave on θ-β, hodograph, δ-P^*, and θ-P^* diagrams for supersonic nitrogen flow (M = 3, T_∞ = 200 K, Kn = 0.12) over a wedge with different angles. The solid lines represent analytical solutions for continuum flow. The SWD technique closely matches analytical predictions at the post-shock edge, while the θ-P^* diagram (Fig. 4d) shows minimal sensitivity to location. Figure 4c highlights that the flow deflection angle varies along the shock despite a constant wedge angle. Overall, the SWD technique successfully captures oblique shock behavior, aligning well with analytical models.

10.4 Supersonic Rarefied Gas Flow Over a Cylinder

Supersonic nitrogen flow (M = 3, T_∞ = 200 K) over a cylinder (D = 0.3048 m) in rarefied conditions is considered to assess the SWD technique's performance in bow shock wave detection. Different rarefaction levels are analyzed by varying free stream pressure, corresponding to Knudsen numbers of 0.001 (near continuum), 0.01, and 0.1. Figure 10.5 presents pressure contours, showing that isobaric line accumulation is significant in the shock region but decreases as rarefaction effects become more pronounced.

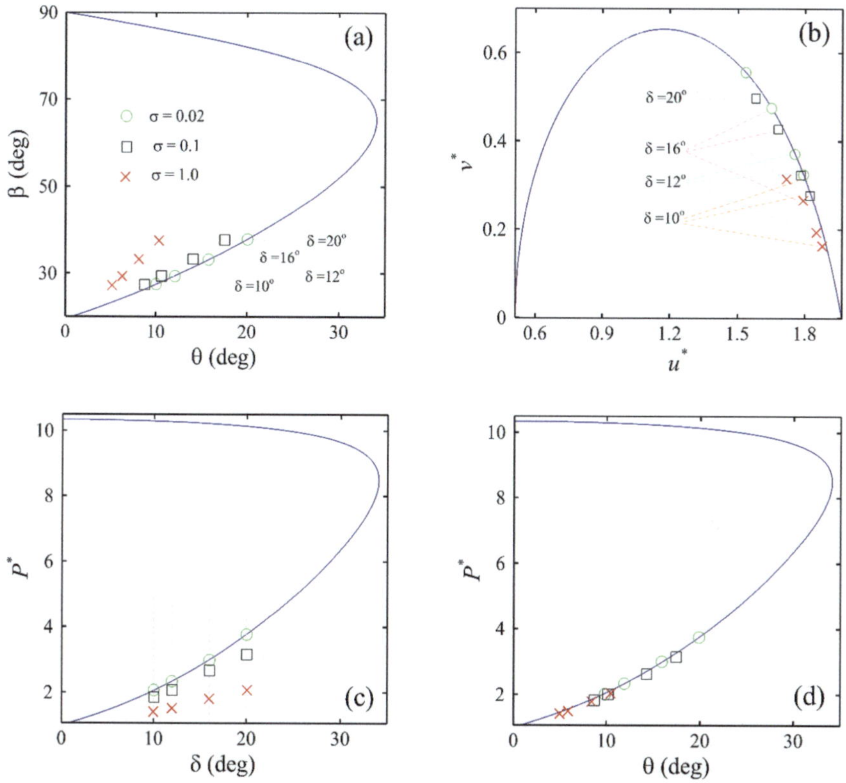

Fig. 10.4 The effect of within-shock location on **a** θ-β, **b** hodograph, **c** δ-P^*, and **d** θ-P^* diagrams for supersonic flow (M = 3, Kn = 0.12) over the wedge at different angles (Reproduced from Ref. [1])

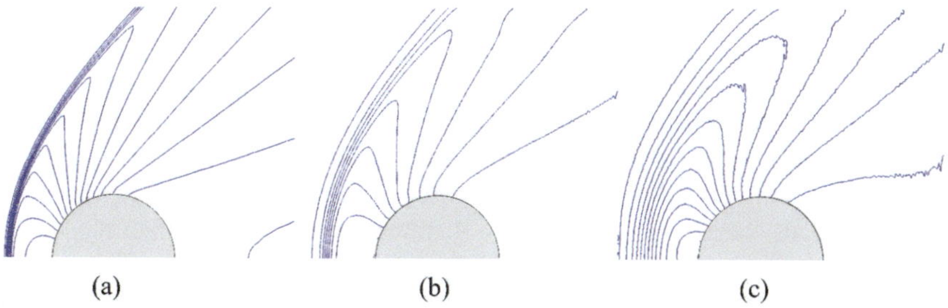

Fig. 10.5 Pressure fields for supersonic flow over the cylinder at different rarefaction levels; **a** Kn = 0.001, **b** Kn = 0.01, and **c** Kn = 0.1 (Reproduced from Ref. [1])

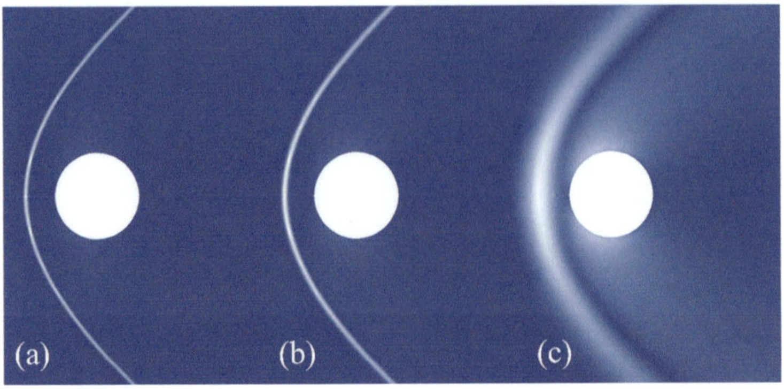

Fig. 10.6 Contours of Schlieren function for supersonic flow over the cylinder at different rarefaction levels; **a** Kn = 0.001, **b** Kn = 0.01, and **c** Kn = 0.1 (Reproduced from Ref. [1])

Fig. 10.7 Effect of rarefaction on the shock profiles obtained by the SWD technique (Reproduced from Ref. [1])

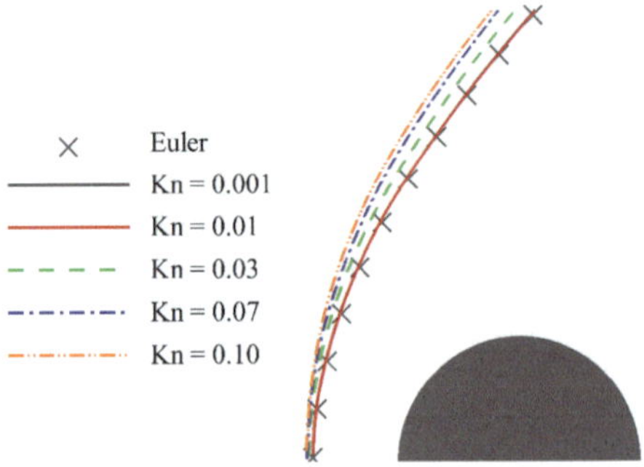

Figure 10.6 presents the Schlieren function fields corresponding to the pressure contours from Fig. 10.5, calculated using Eq. (10.1). The bow shock appears as highlighted curved regions. In the near-continuum case, the shock is visualized as a narrow band with constant width, whereas it expands under highly rarefied conditions.

Figure 10.7 compares shock wave profiles for continuum and rarefied flow conditions. The shock profile is defined as the locus of points with $\sigma = 1$. Star symbols represent the continuum flow regime from INS code results, while triangles indicate near-continuum conditions. For Kn ≤ 0.01, rarefaction does not affect the shock profile shape. However, for Kn > 0.01, rarefaction increases the shock stand-off distance, with a more pronounced effect under weak shock conditions.

Akhlaghi et al. [2] investigate the effect of rarefaction on shock polars using the proposed Shock Wave Detection (SWD) method to identify shock waves and their corresponding pre- and post-shock conditions from DSMC simulations. The study also explores how the Mach number affects shock polars and shock profiles.

10.5 Effect of Rarefaction

The effect of rarefaction (Knudsen number) on θ-β, pressure-deflection, and hodograph diagrams for nitrogen flow (M = 3) over a cylinder is presented in Figs. 10.8, 10.9 and 10.10. The analysis focuses on two shock locations: $\sigma = 0.5$ and $\sigma = 0.1$. The results align well with continuum theory at low rarefaction (Kn = 0.001, Kn = 0.01). However, rarefaction effects are noticeable at Kn = 0.03 and significant at Kn = 0.07, particularly at $\sigma = 0.1$ [2].

Rarefaction increases both weak and strong shock angles for a given flow deflection angle, a trend observed at both shock locations ($\sigma = 0.1$ and $\sigma = 0.5$). This shows that the observed physics corresponds to the nature of the shock wave at the rarefied condition and does not relate to incorrect locations within the shock. Notably, at high

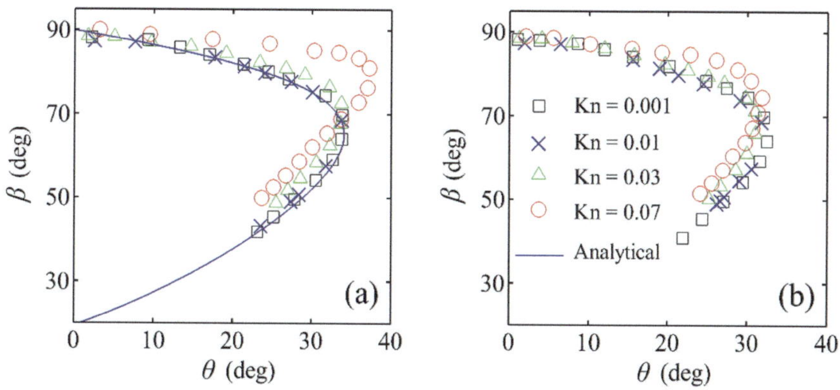

Fig. 10.8 Effect of rarefaction on θ-β diagrams; **a** σ = 0.1 and **b** σ = 0.5 (Reproduced from Ref. [2])

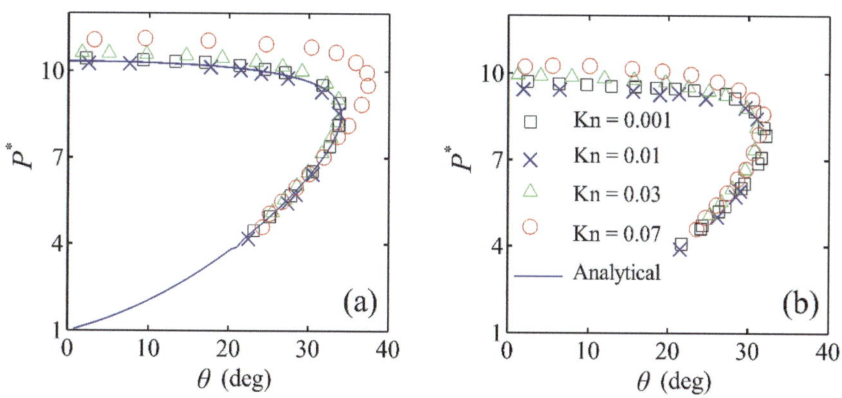

Fig. 10.9 Effect of rarefaction on pressure-deflection diagrams; **a** σ = 0.1 and **b** σ = 0.5 (Reproduced from Ref. [2])

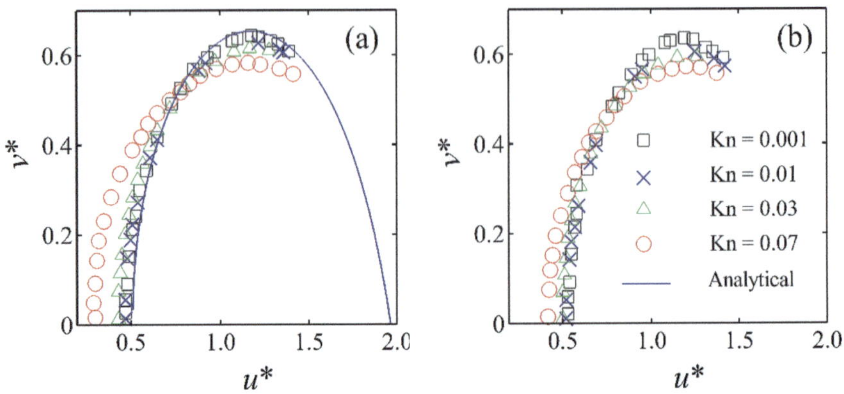

Fig. 10.10 Effect of rarefaction on hodograph diagrams; **a** σ = 0.1 and **b** σ = 0.5 (Reproduced from Ref. [2])

rarefaction, the maximum flow deflection angle also increases. The authors observed Similar trends for supersonic flow over a wedge, where flow rarefaction and shock angle increase near the wedge corner. Additionally, in highly rarefied conditions, the oblique shock over a wedge remains attached at higher wedge angles than in continuum flow.

Figure 10.9 illustrates the impact of flow rarefaction on the pressure-deflection (θ-P^*) diagram along the bow shock wave. Significant deviations from continuum theory appear only at high rarefaction levels (Kn = 0.07). Rarefaction has little effect on the weak shock branch of the θ-P^* polar but notably increases the pressure ratio in strong shock conditions at a given flow deflection angle. Similar behavior was observed in oblique shock waves over a wedge, where pressure increased behind the shock as the flow approached the wedge corner. The pressure-deflection diagrams confirm that rarefaction enhances shock wave effects in supersonic flow.

According to Fig. 10.10, At low rarefaction levels (Kn = 0.001 and Kn = 0.01), the results align well with the analytical hodograph diagram for continuum flow. However, rarefaction effects become more pronounced at Kn = 0.03, particularly at σ = 0.1. For a given flow deflection angle (by plotting the slopes of straight lines drawn from the origin to different curves), rarefaction decreases u^* and increases v^*. Similar to θ-β and θ-P^* diagrams, the hodograph polar confirms that rarefaction increases the maximum flow deflection angle, and this trend holds for both σ = 0.1 and σ = 0.5.

At higher rarefaction, reduced intermolecular collisions allow reflected molecules to move farther upstream, extending the subsonic region and increasing the shock stand-off distance. As a result, flow resistance rises, strengthening the shock wave and increasing shock angle and post-shock pressure at a constant flow deflection angle. According to Fig. 10.11, The subsonic flow region, enclosed by sonic lines

Fig. 10.11 Effect of rarefaction on the bow shock profiles and sonic lines in front of the cylinder at M = 3 (Reproduced from Ref. [2])

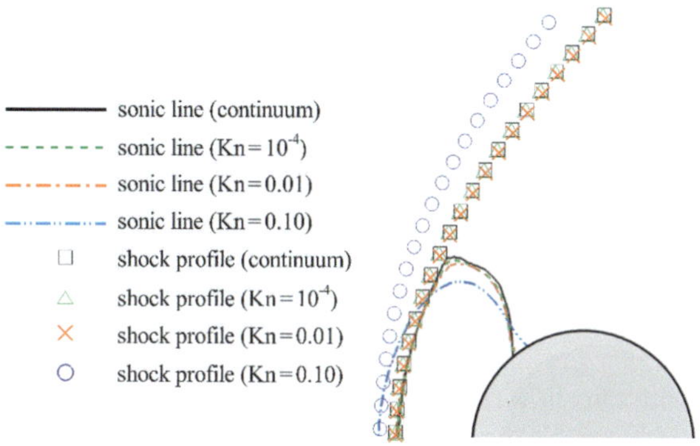

in front of the cylinder, is an obstacle to the upstream supersonic flow. However, rarefaction effects remain negligible up to Kn = 0.01.

Figure 10.12 compares flow streamlines for continuum flow and rarefied flow. In highly rarefied conditions, shock-induced variations are more gradual than continuum flow, with negligible differences near the stagnation point. Reduced molecular interactions weaken the body's influence on far-off streamlines for rarefied flow, resulting in a lower flow angle behind the shock. This contributes to a stronger weak shock in rarefied conditions, as shown in Fig. 10.9. Additionally, lower intermolecular collision rates allow molecules behind the strong shock to escape toward the weak shock region, transferring strong shock characteristics and intensifying the weak shock at higher rarefaction levels.

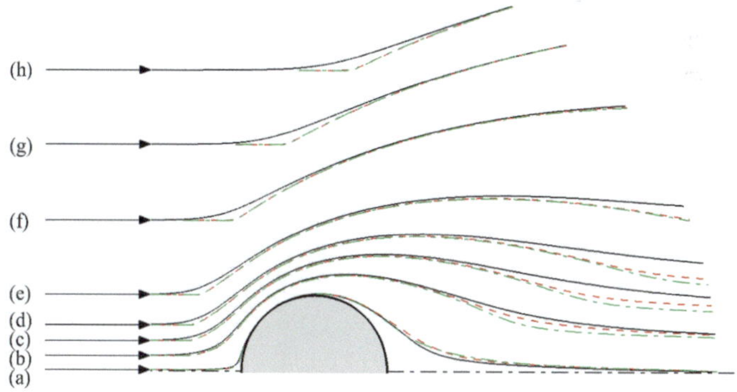

Fig. 10.12 Streamlines at vertical distances of $y/D = 0.03$ (*a*), 0.18 (*b*), 0.50 (*c*), 0.82 (*d*), 1.15 (*e*), 1.48 (*f*), 1.80 (*g*), 2.13 (*h*), for continuum flow (dashed-dotted), Kn = 0.01 (dashed), and Kn = 0.1 (solid) (Reproduced from Ref. [2])

10.6 Effect of Mach Number

Figures 10.13, 10.14 and 10.15 illustrate the θ-β, pressure-deflection, and hodograph diagrams for Argon gas flows at various Mach and Knudsen numbers at aft shock locations ($\sigma = 0.1$). According to these figures, rarefaction effects are stronger at higher Mach numbers, and the Rarefaction effects on the weak shock branch are negligible (Solid lines represent continuum flow solutions).

Figures 10.16 illustrate shock profiles at different Mach and Knudsen numbers for nitrogen gas flows. The symbols correspond to the Euler solution (squares) and near continuum flow condition with Kn = 0.001 (crosses). The results indicate that rarefaction increases the shock stand-off distance and the Shock profile slope. Rarefaction

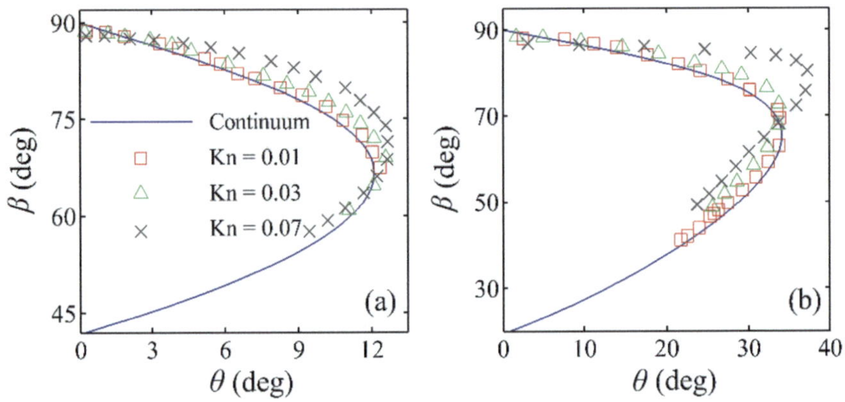

Fig. 10.13 *θ-β* diagrams for Nitrogen gas flows at various Mach numbers of **a** 1.5 and **b** 3.0 (Reproduced from Ref. [2])

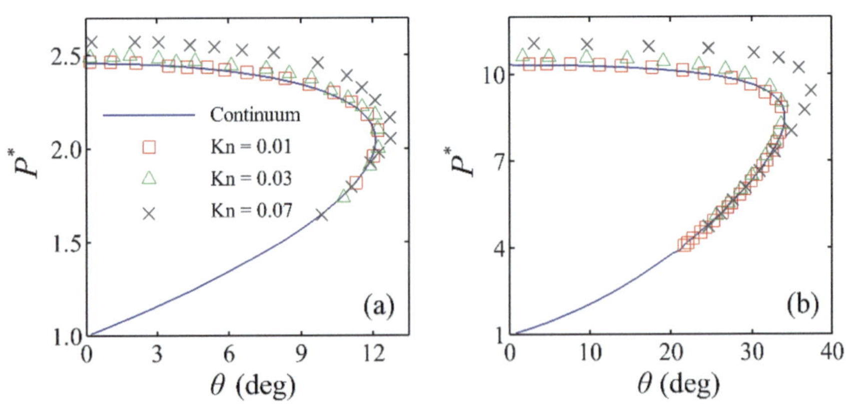

Fig. 10.14 Pressure-deflection diagrams for Nitrogen gas flows at various Mach numbers of **a** 1.5 and **b** 3.0 (Reproduced from Ref. [2])

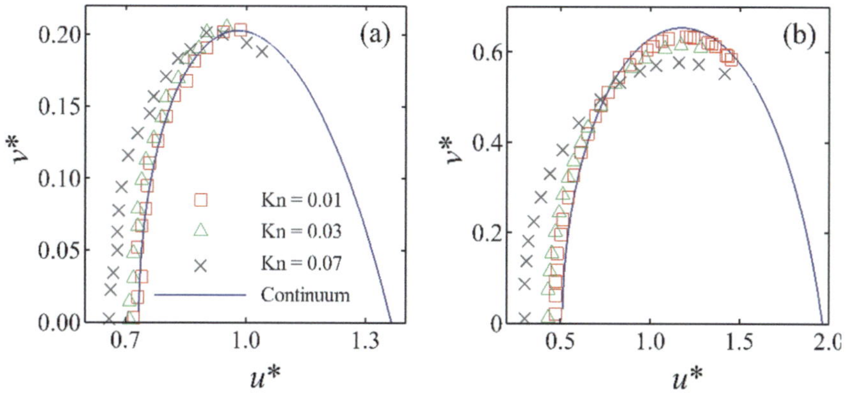

Fig. 10.15 Hodograph diagrams for Nitrogen gas flows at various Mach numbers of **a** 1.5 and **b** 3.0 (Reproduced from Ref. [2])

Fig. 10.16 Shock wave profiles for Nitrogen gas flows at different Knudsen numbers and Mach numbers (Reproduced from Ref. [2])

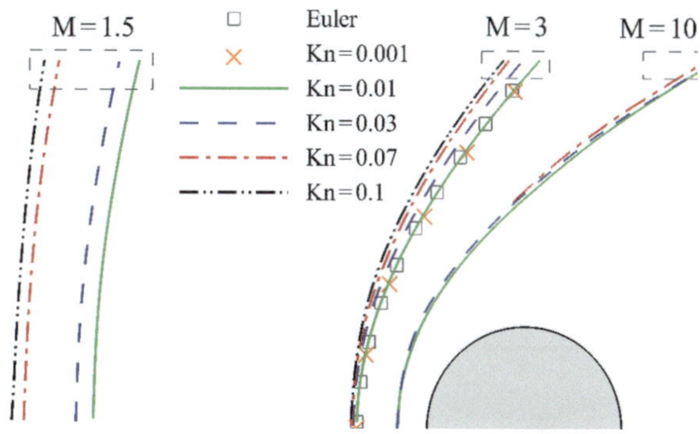

effects are stronger at lower Mach numbers and relatively weaker at higher Mach numbers.

10.7 Conclusion

The SWD technique is validated through comparisons with gas dynamics curves (θ-β and shock polar diagrams), showing strong agreement with analytical solutions in the continuum regime. Its effectiveness for rarefied gas flows is also demonstrated, confirming its reliability for detecting pre- and post-shock locations. The results showed that rarefaction strengthens shock characteristics, increasing shock angle, maximum flow deflection angle, and pressure ratio while reducing post-shock velocity components. Rarefaction effects are stronger in weak shock conditions in θ-β polar. Effects are more pronounced in strong shock conditions in Pressure-deflection and hodograph diagrams. Shock stand-off distance increases due to higher particle reflection effects in rarefied flows. Rarefaction has a greater impact at lower Mach numbers, increasing shock distance from the body for both weak and strong shocks.

References

1. Akhlaghi, H., Daliri, A., Soltani, M.R.: Shock-wave-detection technique for high-speed rarefied-gas flows. AIAA J. **55**(11), 3747–3756 (2017)
2. Akhlaghi, H., et al.: Shock polar investigation in supersonic rarefied gas flows over a circular cylinder. Phys. Fluids **33**(5) (2021)

MIX
Papier aus verantwortungsvollen Quellen
Paper from responsible sources
FSC® C105338

If you have any concerns about our products,
you can contact us on
ProductSafety@springernature.com

In case Publisher is established outside the EU,
the EU authorized representative is:
Springer Nature Customer Service Center GmbH
Europaplatz 3, 69115 Heidelberg, Germany

Printed by Libri Plureos GmbH
in Hamburg, Germany